Sustainability Solutions

An Interdisciplinary and Transdisciplinary Nexus

Sustainability Solutions addresses a wide range of global challenges with applied solutions in the interdisciplinary and transdisciplinary sustainability nexus. Titles in the series present a strong sustainable development focus, and topics which connect environment, economic and social applications are encouraged. Topics could include environmental impacts, vulnerability, risk assessment and reduction, human-environmental systems (e.g. food, water, health) and innovative solutions to climate change. Volumes in the series present novel solutions to societal challenges using innovative application, for example demonstrating a solutions-based approach using machine learning, artificial intelligence, decision support systems, advance geospatial technology. Case studies and shorter report volumes which present knowledge transfer for policymaking and governance are also encouraged.

Keywords:

Interdisciplinary environmental solutions, transdisciplinary knowledge transfer, geospatial technologies, natural resource management, decision support system, artificial intelligence, deep learning, smart technology, environmental science, economic, social and spatial, human-environment interactions, sustainable development

Uday Chatterjee · Anitabha Ghosh ·
Sujit Kumar Roy
Editors

Eco-Resilience: Climate Change, Land Degradation and Sustainable Solutions

Springer

Editors
Uday Chatterjee
Department of Geography
Bhatter College, Dantan (Autonomous)
Paschim Medinipur, West Bengal, India

Anitabha Ghosh
Department of Geography
Egra Sarada Shashi Bhusan College
Vidyasagar University
Egra, West Bengal, India

Sujit Kumar Roy
Institute of Water and Flood Management
Bangladesh University of Engineering
and Technology
Dhaka, Bangladesh

ISSN 3059-278X ISSN 3059-2798 (electronic)
Sustainability Solutions
ISBN 978-3-032-00707-0 ISBN 978-3-032-00708-7 (eBook)
https://doi.org/10.1007/978-3-032-00708-7

This Springer imprint is published by the registered company Springer Nature Switzerland AG
The registered company address is: Gewerbestrasse 11, 6330 Cham, Switzerland

Dedicated to
Young Scholars in the Field of Soil Science,
Atmospheric Science, Ecology,
Environmental Science, Hazard and Disaster
Management, Geospatial Science,
Geography and Policy Makers

Foreword

The climate change era confronts humanity with an urgent dual dilemma: the accelerating momentum of climate extremes and the quiet yet relentless advance of natural resource degradation. At the intersection of these formidable challenges lies the enduring resilience of land systems and resources—and our shared responsibility to protect and restore these in a holistic context of the ecosystem's ability to recover and sustain. *Eco-Resilience: Climate Change, Land Degradation and Sustainable Solutions* is a timely and remarkable contribution to this global imperative, presenting a rich confluence of perspectives from diverse scholars, seasoned practitioners, and researchers across disciplines.

What distinguishes this volume is its resolute focus on land degradation—not as a secondary symptom, but as a foundational concern intricately interlinked with climate instability, biodiversity loss, and food and water insecurity. Through evidence-based research and grounded case studies, the contributors offer a multidimensional understanding of eco-resilience: from regenerative land-use practices and adaptive agriculture to forward-looking policy frameworks and community-driven innovation. Each chapter speaks to both complexity and possibility, weaving scientific rigor with a resolute commitment to equity and sustainability.

Eco-resilience, as advanced in this book, transcends theory—it is a call to reimagine the architecture of development. It compels us to transition from extractive paradigms to regenerative pathways, from fragmented efforts to integrated systems thinking. This volume offers not only conceptual clarity but also actionable insights to inform policy discourse, guide programmatic interventions, and serve as a research baseline for advancing systemic resilience.

As a reference, this book will make key references to scholars, policy advocates, and institutions striving for transformative change. May its collective scholarship inspire deeper reflection, catalyze meaningful action, and fortify our resolve to restore ecological balance in a rapidly changing world.

New Delhi, India
July 2025

Prof. (Dr.) Anil Gupta
Senior ICIMOD Fellow—DRR
Professor—IIT Roorkee
Greater Noida Campus
PI—Integrated Centre for Adaptation
DRR and Sustainability (ICARS
CoE of DST-GOI)
Member—Sikkim Commission
of Glacial Hazards

Preface

Land degradation and climate change represent two of the most critical and interlinked environmental challenges confronting humanity in the twenty-first century. The accelerating pace of global environmental change threatens ecosystems, livelihoods, and the very foundation of sustainable development across diverse geographic and socio-economic contexts. It is within this complex nexus that the concept of eco-resilience (the capacity of ecosystems to absorb disturbances while maintaining core functions) emerges as a pivotal framework for understanding and addressing the dynamic interplay between climate variability, land degradation processes, and the imperative for sustainable restoration and management. This editorial volume, *Eco-Resilience: Climate Change, Land Degradation and Sustainable Solutions*, addresses the urgent intersection of climate change and land degradation, exploring pathways to enhance ecological resilience through sustainable practices. Brings together a comprehensive and multidisciplinary collection of research, case studies, and policy insights. The book is organized into three thematic parts designed to progressively deepen our understanding and response strategies to these urgent global challenges.

The *Introduction* lays the foundational concepts, framing the critical intersections between climate change and land degradation, and highlighting the urgency of advancing eco-resilience as a guiding principle. This part synthesizes current scientific knowledge and outlines the conceptual underpinnings that inform subsequent analyses. In the second part, *Climate Change Impacts on Landscapes Ecosystem and Land Degradation*, contributors present empirical research and reviews that elucidate the diverse manifestations of climate-induced land degradation across multiple spatial scales. From regional case studies to global assessments, this part underscores the variability and complexity of land system responses to changing climate drivers, while emphasizing vulnerable hotspots and emerging trends. The final part, *Addressing Climate Change, Eco-Resilience and Land Restoration Practices*, focuses on innovative strategies, sustainable solutions, and policy frameworks that aim to enhance ecosystem resilience and promote effective land restoration. It highlights interdisciplinary approaches integrating ecological, socio-economic, and governance dimensions, showcasing best practices, technological advancements, and participatory models that can inform resilient land management and climate adaptation.

In Part I, Chap. 1 is about understanding and addressing the interconnected impacts of climate change, land degradation, and eco-resilience to ensure sustainable ecosystem functionality. Chapter 2 addresses the interconnected crises of land degradation and climate change through sustainable practices, policy interventions, and global cooperation to ensure ecological and socio-economic resilience.

In Part II, Chap. 3 examines the intertwined impacts of land degradation and climate change on agriculture, and promotes context-specific, technology-driven, and community-based solutions for resilient and sustainable farming systems. Chapter 4 highlights how climate change threatens global ecosystems—particularly wetlands—and emphasizes the need for sustainable management to protect livelihoods and ecosystem services. Chapter 5 highlights that rapid urbanization in Guwahati has led to significant green space loss and increased urban heat, emphasizing the urgent need for ecologically balanced urban planning. Chapter 6 investigates how climate change and soil degradation jointly undermine crop production and rural livelihoods on the Mbum Plateau, of the Northwest Region, Cameroon, emphasizing the need for sustainable agricultural practices to build resilience. Chapter 7 analyzes climate-induced ecological degradation in the Barind Tract, the study highlights the link between climate stress and vegetation loss while advocating for targeted interventions to enhance eco-resilience and sustainability. Chapter 8 uses the RUSLE model and geospatial techniques to assess soil erosion in the Upper Bharathapuzha Basin, providing a framework for sustainable land management and eco-resilience through targeted conservation strategies. Chapter 9 examines how unchecked urbanization and development contribute to land degradation and climate change, advocating for sustainable practices, ecological restoration, and green consumption to mitigate these impacts. Chapter 10 utilizes MODIS NDVI time-series data with TIMESAT and BFAST tools to analyze vegetation phenology and long-term trends, Tawa River Basin, India, offering a framework for ecological monitoring and sustainable land management.

In Part III, Chap. 11 examines climate change impacts on land suitability using GIS, remote sensing, and participatory approaches to support resilient and sustainable land-use planning. Chapter 12 explains that implementing soil conservation and sustainable land management practices is essential for preventing soil degradation, ensuring agricultural productivity, and maintaining ecosystem health. Chapter 13 explores the role of mangrove ecosystems in Ecological Disaster Risk Reduction (Eco-DRR), emphasizing their ecological services, climate mitigation potential, and the importance of conservation policies and community participation. Chapter 14 discusses that community-based coastal land restoration initiatives, particularly involving mangrove ecosystems, are essential for enhancing resilience to climate change, protecting biodiversity, and supporting the livelihoods of vulnerable coastal communities. Chapter 15 discusses how advanced soil conservation techniques are crucial for combating land degradation and promoting sustainable land management amid escalating anthropogenic pressures and climate change. Chapter 16 highlights that community-based restoration initiatives are vital for addressing the intertwined challenges of climate change and land degradation in fragile hill ecosystems. Chapter 17 highlights that community-based adaptive land restoration strategies are essential for combating climate change and ensuring ecological and socio-economic resilience in fragile hilly regions like the Nilgiris. Chapter 18 discusses urbanization-driven land cover change in Bamenda III Municipality, Cameroon has intensified climate impacts, prompting ecosystem restoration and green infrastructure as Nature-based Solutions for climate change mitigation. Chapter 19 examines farmer-led adaptive strategies to combat soil degradation and sustain agricultural productivity in Sub-Saharan Africa, Bafou, Menoua Division, Cameroon, amid climate and environmental challenges. Chapter 20 presents a GIS-based, climate-sensitive land suitability framework for Thrissur, Kerala, to balance urban development with ecological sustainability through evidence-based, risk-informed planning. Chapter 21 examines the impact of climate change on wetland dynamics in the Ndop Plain, Cameroon, highlighting the importance of conservation for sustaining water resources and local livelihoods.

Collectively, these contributions provide an integrative understanding of the challenges and solutions critical for sustaining land resources under changing climatic conditions. This volume aims to inform researchers, policymakers, and practitioners engaged in environmental management, supporting evidence-based decision-making toward resilient landscapes and sustainable development. Together, these parts offer a holistic and nuanced exploration of how scientific understanding can be translated into actionable pathways for sustaining land productivity, biodiversity, and community well-being in the face of climate adversity. This volume seeks to serve as a valuable resource for researchers, policymakers, practitioners, and students engaged in environmental sciences, climate policy, land management, and sustainability studies.

This volume, *Eco-Resilience: Climate Change, Land Degradation and Sustainable Solutions*, is designed as a comprehensive and accessible resource for senior undergraduate and graduate students, academic researchers, and professionals working in the fields of sustainable development, environmental engineering, civil engineering, agricultural science, land management, and environmental policy. It provides an in-depth examination of the intricate linkages between climate change, land degradation, and the quest for sustainability, while presenting innovative and practical approaches to environmental restoration and resilience-building.

Medinipur, West Bengal, India — Uday Chatterjee
Medinipur, West Bengal, India — Anitabha Ghosh
Dhaka, Bangladesh — Sujit Kumar Roy

Acknowledgments

We, the editors of the volume *Eco-Resilience: Climate Change, Land Degradation and Sustainable Solutions*, extend our heartfelt gratitude to all the contributors, colleagues, researchers, and reviewers who played a pivotal role in the successful development and publication of this book. Your active engagement, scholarly insights, and unwavering commitment have enriched the quality and depth of this volume. This book is the outcome of collective efforts aimed at advancing the understanding of climate-induced challenges, particularly soil degradation and the urgent need for soil health restoration. It further explores the integration of geospatial tools, geostatistical analyses, and modeling approaches as sustainable pathways toward environmental resilience and land restoration. Your valuable contributions have helped create a reference that we believe will serve researchers, practitioners, and policymakers across disciplines. We are especially grateful to our families for their continuous support, patience, and encouragement throughout this demanding journey. Their understanding and motivation during difficult and busy phases were essential in bringing this project to completion.

Our sincere appreciation also goes to the publishing team at Springer Nature. Their expertise, professionalism, and dedication from meticulous editing to the final stages of design and production ensured that this volume meets the highest standards of academic publishing. This book stands as a testament to collaborative scholarship and our shared vision of building a sustainable and resilient future.

Medinipur, West Bengal, India — Uday Chatterjee
Medinipur, West Bengal, India — Anitabha Ghosh
Dhaka, Bangladesh — Sujit Kumar Roy

Contents

Editors and Contributors

About the Editors

Dr. Uday Chatterjee is an Assistant Professor at the Department of Geography, Bhatter College, Dantan, affiliated with Vidyasagar University, West Bengal, India. He earned his Doctoral Degree in Applied Geography from Ravenshaw University, Cuttack, Odisha. His research interests encompass Urban Planning, Social and Human Geography, Applied Geomorphology, Indigenous knowledge and Hazards and Disasters, Risk Assessment and Resilience, Community Response and Environmental Issues. Dr. Chatterjee has been actively involved in editing scholarly works, including the Special Issue on *Urbanism, Smart Cities, and Modelling for GeoJournal*, Springer, Special issue in *Global Social Welfare*, Springer and *Frontiers Environmental Science Journal*, *Geological Journal* Wiley, Editorial Board Member of *Scientific Report (Nature Portfolio)* journal and serving as a Book Series Editor for *Developments in Environmental Science*, Elsevier (https://www.elsevier.com/books-and-journals/book-series/developments-in-environmental-science) and *Sustainability Solutions-An Interdisciplinary and Transdisciplinary Nexus* (https://link.springer.com/bookseries/59999), Springer Nature. His extensive research and editorial contributions highlight his commitment to advancing geographical knowledge and its applications. His research work has been funded by the West Bengal Pollution Control Board (WBPCB), Government of West Bengal, India, Indian Council of Social Science

Research (ICSSR), Government of India. He has published research papers in reputed journals, edited books with reputed publishers like Springer, Elsevier, CRC Press, and authored books, book chapters, and conference proceedings. Currently, Dr. Chatterjee is doing an international project, in collaboration with Indonesia, Malaysia, and Japan, funded by APN (Asia Pacific-Global Change Research, Grant DOI https://doi.org/10.30852/p.22128), and visited Saudi Arabia for an International collaboration and conference session chair and Keio University, Tokyo, Japan as a Visiting Professor.

Anitabha Ghosh is a Research Scholar at the Coastal Environmental Studies Research Center, Department of Geography, Egra S. S. B. College, Egra, Purba Medinipur, West Bengal, under Vidyasagar University. He is a dedicated and accomplished academician with a rich background in Geography and Environmental Management has pursued an extensive educational journey, pursuing Ph.D. under the supervision of Dr. Dipak Bisai, from the Coastal Environmental Studies Research Center, Department of Geography, Egra S. S. B. College, under Vidyasagar University. His academic accolades include passing the UGC-NET and WBCSC SET exams multiple times, highlighting his proficiency in Geography and Environmental Science. His areas of research interest cover Applied Geomorphology, Remote Sensing and GIS, Hazard and Disaster Management, Environmental Issues, Watershed Management, Hydrological Modeling, LULC changes, Ecosystem Services, and Social and Human geography. Moreover, Anitabha has enhanced his expertise through numerous certifications in geospatial technologies, remote sensing, and programming languages like Python and R, obtained from renowned institutions including Presidency University, Jadavpur University, The Central University of Jharkhand, ISRO, and NIELIT. He is a Lifetime Member of the Indian Society of Geomatics (ISG). His active participation in national and international seminars and conferences underscores his ongoing engagement with contemporary issues in applied geography and environmental management.

Sujit Kumar Roy is a Researcher and Environmental Expert at the Climate Smart Agriculture Water and Management Project (CSAWMP), implemented by the Bangladesh Water Development Board with funding from the World Bank. He holds a Master of Science in Water Resource Development from the Institute of Water and Flood Management (IWFM), Bangladesh University of Engineering and Technology (BUET), and a Bachelor's degree in Environmental Science and Disaster Management from Noakhali Science and Technology University (NSTU). With a strong foundation in geospatial science and environmental management, Sujit has worked in various capacities, including GIS Analyst, Research Assistant, and Assistant Manager (Environment) in organizations such as Plan Plus Ltd., RAiN Forum, and IWFM, BUET. His research focuses on Remote Sensing and GIS, Climate Change Impact Assessment, Soil Moisture Downscaling, Hydrological and Hydro-climatic Modeling, Environmental Hazard Mapping, and Water Resource Management. He has published in high-impact journals like *Groundwater for Sustainable Development*, *Environmental Science and Pollution Research*, *Scientific Reports*, *Desalination, and Natural Hazards*. Sujit has also delivered specialized training on Google Earth Engine (GEE), Machine Learning for Landslide Prediction, and Statistical Analysis with R, and has presented his work at several national and international conferences. Technically proficient in ArcGIS, QGIS, RStudio, Python, GEE, and SPSS, he has significant experience in EIA/EMP preparation, water quality analysis, and integrated spatial modeling. Sujit is an executive member of the Coastal Environment Network. His academic and professional pursuits reflect a strong commitment to sustainable environmental solutions in South Asia.

Contributors

Mildred Endam Acha Department of Geography and Planning, The University of Bamenda, Bamenda, Cameroon

Anila Alex Department of Agronomy, ICAR- Indian Agricultural Research Institute, Hyderabad Academic Hub, Hyderabad, India

Neha Awasthi Khalsa College of Charitable Society, Amritsar, India

Anamika Barman Division of Agronomy, ICAR-Indian Agricultural Research Institute, New Delhi, India

Nisani Hussain Basha Department of Agronomy, Professor Jayshankar Telengana Agricultural University, Hyderabad, Telangana, India

Mahima Begum Department of Agronomy, Assam Agricultural University, Jorhat, India

Charmon Keri Bey Department of Agronomy, Assam Agricultural University, Jorhat, Assam, India

S. K. Bhardwaj Dr. Yashwant Singh Parmar University of Horticulture and Forestry, Nauni, Solan, India

Dipti Bisarya Department of Agronomy, School of Agriculture, Lovely Professional University, Phagwara, Punjab, India

Abhiroop Chowdhury Jindal School of Environment and Sustainability, O.P. Jindal Global University, Sonipat, Haryana, India

Kajal Das Division of Germplasm Evaluation, ICAR-National Bureau of Plant Genetic Resources, New Delhi, India

Aastha Dubey School of Agriculture, Galgotias University, Greater Noida, Uttar Pradesh, India

Dimpi Dutta Department of Agronomy, Assam Agricultural University, Jorhat, Assam, India

Pallabi Dutta Department of Education, Gauhati University, Guwahati, Assam, India

Suman Dutta Department of Genetics and Plant Breeding, School of Agriculture and Rural Development, Ramakrishna Mission Vivekananda Educational and Research Institute, Belur Math, Howrah, India

Tiomo Dongfack Emmanuel Department of Geography, Environment and Planning, University of Dschang, Dschang, Cameroon

Pooja Erapuram Department of Agronomy, Professor Jayshankar Telengana Agricultural University, Hyderabad, Telangana, India

Santosh L. Gaonkar Department of Chemistry, Manipal Institute of Technology, Manipal Academy of Higher Education, Manipal, Karnataka, India

S. Gayathri Department of Urban and Regional Planning, School of Planning and Architecture, Bhopal, India

M. S. Jangra Dr. Yashwant Singh Parmar University of Horticulture and Forestry, Nauni, Solan, India

J. Killivalavan Department of Geography, Bharathidasan University, Tiruchirappalli, Tamil Nadu, India

Moye Eric Kongnso Department of Geography, Environment and Planning, University of Dschang, Dschang, Cameroon

Suthari Vijay Kumar Department of Agronomy, Uttarbanga Krishi Viswavidyalaya, Coochbihar, West Bengal, India

Vinai Kumar Department of Crop Physiology, Chandra Shekar Azad University of Agriculture and Technology, Kanpur, Uttar Pradesh, India

Santanu Kundu Division of Agronomy, ICAR-Indian Agricultural Research Institute, New Delhi, India

Paulose N. Kuriakose Department of Urban and Regional Planning, School of Planning and Architecture, Bhopal, India

Tsafack Ngoufo Louis Serge Department of Geography, University of Yaounde I, Yaounde, Cameroon

Uzma Manzoor School of Agriculture, Galgotias University, Greater Noida, Uttar Pradesh, India

P. Masilamani Department of Geography, Bharathidasan University, Tiruchirappalli, Tamil Nadu, India

Nfor Delphine Mbongsi Department of Geography and Planning, University of Bamenda, Bamenda, Northwest Region, Cameroon

Ariane Nivelle Dongmo Mbougo Department of Geography, University of Dschang, Dschang, West Region, Cameroon

Risat Ara Meem Department of Environmental Sciences, Jahangirnagar University, Dhaka, Bangladesh

Sanjida Rahman Mishu Department of Geography and Environment, Eden Mohila College, Dhaka, Bangladesh

D. Monika Department of Natural Resources and Environmental Studies, National Dong Hwa University, Hualien, Taiwan

Rohan Nambiar Department of Agronomy, School of Agriculture, Lovely Professional University, Phagwara, Punjab, India

B. Namratha Department of Chemistry, The Yenepoya Institute of Arts, Science, Commerce and Management, Yenepoya (Deemed to be University), Mangaluru, Karnataka, India

Leinyuy Serge Nangsinyuy Department of Geography and Planning, The University of Bamenda, Bamenda, Cameroon

Debasmita Nath Department of Plant Breeding and Genetics, Assam Agricultural University, Jorhat, Assam, India

Loveline Kongla Nsahlai Department of Geography, HTTC, The University of Bertoua, Bertoua, East Region, Cameroon

Apurv Bhogibhai Patel Urban Planner—Jamnagar Area Development Authority (JADA), Jamnagar, Gujarat, India

Vijay Pooniya Division of Agronomy, ICAR-Indian Agricultural Research Institute, New Delhi, India

Raghav Prakash Jindal School of Environment and Sustainability, O.P. Jindal Global University, Sonipat, Haryana, India

Pradeep Kumar Rajput Department of Geography, School Basic Science, CSJMU Kanpur University, Kanpur, Uttar Pradesh, India

Bolagam Ravikumar Department of Horticulture, Lovely Professional University, Phagwara, Jalandhar, India

Boppidi Manasa Reddy Department of Agronomy, Professor Jayshankar Telengana Agricultural University, Hyderabad, Telangana, India

Chinmaya Sahoo Department of Agronomy, Dr. Rajendra Prasad Central Agricultural University, Pusa, Samastipur, Bihar, India

Ankit Saini Department of Agronomy, Dr. KSG Akal College of Agriculture, Eternal University, Baru Sahib, Sirmaur, Himachal Pradesh, India

M. Saranya Department of Agronomy, University of Agricultural Sciences, Dharwad, Karnataka, India

Abhijit Sarkar Department of Geography, Gauhati University, Guwahati, Assam, India

Pranjit Kumar Sarma Department of Geography, Mangaldai College, Darrang, Assam, India

Israt Jahan Sazuty Department of Geography and Environment, Eden Mohila College, Dhaka, Bangladesh

Maksud Hasan Shah Department of Agronomy, Bidhan Chandra Krishi Viswavidyalaya, Mohanpur, Nadia, West Bengal, India

Ankita Shinde Division of Agricultural Engineering, ICAR-Indian Agricultural Research Institute, New Delhi, India

Prasun Kumar Singh School of Agriculture, Galgotias University, Greater Noida, Uttar Pradesh, India

Shalini Singh School of Agriculture, Galgotias University, Greater Noida, Uttar Pradesh, India

Agnibha Sinha Department of Soil Science and Agricultural Chemistry, School of Agriculture, Lovely Professional University, Phagwara, Punjab, India

Dharmesh Shantilal Solanki Urban Planner—Junagadh Municipal Corporation (JuMC), Junagadh, Gujarat, India

J. Subhashini Department of Geography, Sri Vijay Vidyalaya College of Arts and Science, Dharmapuri, Tamil Nadu, India

Mbinkar Godlove Suika Department of Geography, Environment and Planning, University of Dschang, Dschang, Cameroon

Shivani Thakur Department of Soil Science and Agricultural Chemistry, School of Agriculture, Lovely Professional University, Phagwara, Punjab, India

R. Thanuja Krishnan Department of Geography, Bharathidasan University, Tiruchirappalli, Tamil Nadu, India

Md. Khayrul Islam Tuhin Department of Geography and Environment, Dhaka College, Dhaka, Bangladesh

Suiven John Paul Tume Department of Geography and Planning, University of Bamenda, Bamenda, Northwest Region, Cameroon

Part I
Introduction

Chapter 1
Eco-resilience, Climate Change, and Land Degradation-Global Environmental Change

Debasmita Nath, Dimpi Dutta, Charmon Keri Bey, Pallabi Dutta, Chinmaya Sahoo, and Ankit Saini

Abstract In an era marked by rapid environmental changes, the inter-relationship between eco-resilience, climate change, and land degradation has become increasingly critical to understand. The capability of ecosystems to endure and acclimatize to the disturbances, apart from maintaining its essential services and functions is known as Eco-resilience. Ecosystem faces challenges arising from climate change, such as adverse frequency of weather pattern, extreme heat, change in precipitation cycle, and last but not least change in climate or seasonal pattern, so this concept is indeed a necessity. Shift in climatic conditions not only threatened biodiversity and its surroundings, but also cause land degradation. It is often accelerated by factors like deforestation, overgrazing, and injudicious agricultural practices. Anthropogenic activities such as widespread deforestation, rapid growth of urban areas, accelerated growth of industrial sectors, and unsustainable use of agricultural practices are all significantly attributes in degrading the land quality or soil quality. In addition, green house gas emissions are also contributing to change in the climatic condition. This series and sequences of land degradation need immediate attention and effectively

D. Nath
Department of Plant Breeding and Genetics, Assam Agricultural University, Jorhat, Assam, India

D. Dutta · C. K. Bey
Department of Agronomy, Assam Agricultural University, Jorhat, Assam, India
e-mail: dimpi.dutta.adj23@aau.ac.in

P. Dutta
Department of Education, Gauhati University, Assam, India

C. Sahoo
Department of Agronomy, Dr. Rajendra Prasad Central Agricultural University, Pusa, Samastipur, Bihar, India

A. Saini (✉)
Dr. KSG Akal College of Agriculture, Eternal University, Baru Sahib, Sirmaur, Himachal Pradesh, India
e-mail: ankitsagr@eternaluniversity.edu.in

U. Chatterjee et al. (eds.), *Eco-Resilience: Climate Change, Land Degradation and Sustainable Solutions*, Sustainability Solutions,
https://doi.org/10.1007/978-3-032-00708-7_1

stopped at global level. So, collaborative approaches can be used to protect and amplify the resiliency of the ecosystem in the long term.

Keywords Eco-system · Anthropogenic activities · Climate change · Eco-resilience

1.1 Introduction

In the twenty-first century, climate change has threatened biodiversity both globally and on a regional scale (O'Neill et al., 2017; Oppenheimer et al., 2015). The fluctuation in the global temperature, precipitation patterns, and the increasing frequency of adverse weather conditions like heat waves, droughts, floods, etc., have impacted ecosystems and population worldwide (Dias, 2023). Ecosystems refer to the symbiotic relationship of flora, fauna, and different micro-organism groups with their surrounding environmental condition (Convention on Biological Diversity, 1992). It has two components: Biotic (living) and Abiotic (non-living). It is also predicted that global temperatures could increase upto 3.7 °C at the edge of twenty-first century (Pachauri et al., 2014). Over the past few decades, it has become evident that anthropogenic activities are also a potent contributor of change in climatic condition (Houghton et al., 2001). Table 1.1 shows the projected changes in temperature and rainfall intensity in India.

The important consequence of human activities is increase in the rate of emission of greenhouse gases especially carbon dioxide (CO_2), nitrous oxide (N_2O) and methane (CH_4) (IPCC, 2021). These emissions are mainly due to burning of fossil fuels such as coal, oil, and natural gas, which are used for generation of energy, transportation, and for different industrial activities (Houghton et al., 2001). These harmful gases capture heat in the environment, causing a rise in global temperature, which can lead to global warming and change in climatic condition over time (Díaz et al., 2015; IPCC, 2021). Notre Dame Global Adaptation Initiative (ND-GAIN) is

Table 1.1 Projected changes in temperature and rainfall pattern in India

		Change in temperature (°C)		Change in rainfall (%)	
Year	Season	Lowest	Highest	Lowest	Highest
2050	Annual	2.2	2.8	5.3	9.3
	Rabi	2.5	3.1	− 9.2	3.8
	Kharif	1.8	2.3	7.1	10.5
2080	Annual	3.5	5.5	7.4	9.9
	Rabi	4.1	6.3	− 24.8	− 4.5
	Kharif	2.9	4.6	10.1	15.1

Source Adopted from Lal et al. (2001)

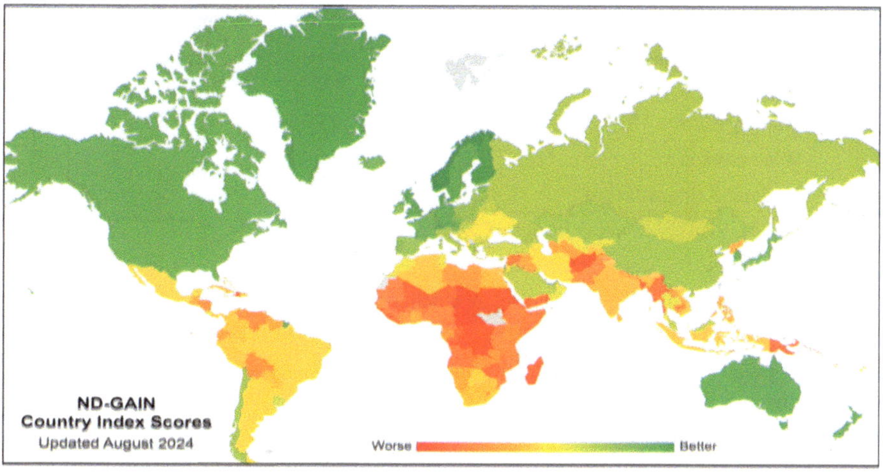

Fig. 1.1 Vulnerability and readiness of different countries towards climate change. *Source* Adapted from ND-GAIN Country Index Scores (2024)

a global source index that ranks countries based on its vulnerability to change in climate and its readiness to cope up with climatic changes. Developing countries are bearing a large chunk of these opposing impacts of changing climatic condition and are considered to be most vulnerable (Fig. 1.1) (Elasha et al., 2005) as they have less resources and adaptation capabilities in rebuilding itself from the aftermath shocks of adverse climate conditions, and India stands 105th rank in ND-GAIN index among the countries.

Land is the basic resource for human existence on earth and for the overall development of society (Xie et al., 2020). The frequency of land degradation has increased to a greater extent due to heavy use of natural resources to feed the growing population and for other developmental purposes like urbanization, industrialization, etc. (Hammad & Tumeizi, 2012). Continuous widespread land degradation causes havoc in the availability of healthy, productive land for agricultural purposes and production, ultimately impacting the food security in global communities. India is suffering from around 120.7 million hectare of land degradation annually out of 328.73 million hectares from different factors, including water erosion (NAAS, 2010), which is illustrated in Table 1.2. Deforestation is driven by the rapid need or demands of the agricultural land for crop production, log and timber for the embellishment and urbanization progress, which all lead to the removal of very vital vegetation that was helping in the protection of the land and keeping the soil structure intact and firm. The major contributing states for land degradation in India are Madhya Pradesh, Karnataka, Telangana, Odisha, Jammu and Kashmir, Jharkhand, Maharashtra, Rajasthan, and Gujarat (Desertification and land degradation atlas of India, 2016). Therefore, a global collaborative policy should be proposed to create awareness about the impacts of land degradation and its management to maintain the stability of environment.

Table 1.2 Factors responsible for degradation and the impacted areas in India

S.No	Source of degradation	Arable land (M ha)	Open forest (<40% canopy) (Mha)
1	Water erosion (> 10 Mg/ha/year)	73.27	9.30
2	Wind erosion	12.40	–
3	**Chemical degradation**		
	Soil exclusively affected from salinity	5.44	–
	Salt-affected + water-eroded soils	1.20	0.10
	Soil exclusively affected from acidic condition (pH < 5.5)	5.09	–
	Acidic (pH < 5.5) + water-eroded soil	5.72	7.13
4	**Physical degradation**		
	Mining and industrial waste	0.19	–
	Water-logging	0.88	–
Total		104.19	16.53
Grand total (Arable land + open forest)		120.72	

Source Soil Loss Map of India-IISWC, Dehradun; Wind Erosion Map of India CAZRI, Jodhpur; Salt-affected Soils of India, CSSRI, Karnal; NBSS&LUP, Nagpur; NRSA, Hyderabad, and others; Acid Soils of India, CSSRI, Karnal; NBSS&LUP, Nagpur; Wasteland Map of India-NRSA, Hyderabad

1.2 Climate Change and Agriculture

Other human activities that include farming, deforestation, unsustainable land management, and land-use changes cause an increase in greenhouse gas emissions (IPCC, 2021) and also lead to land degradation (Settele et al., 2015). Globally, the forest areas are decreasing at a rate of 31.6 million hectare annually, which ultimately affects the vegetation, nutrient balance, and wildlife habitat (Adawalle et al., 2018). Forest acts as a natural reservoir of carbon dioxide, and deforestation reduces their natural absorbance capacity of carbon (Díaz et al., 2015), causing carbon stored in forest to release back to the earth's atmospheric layers (Baccini et al., 2017). Moreover, the expansion of agriculture causes changes in land use pattern, which is also contributing to climate change and eliminated endemic species (Turner et al., 1993). In 2019, around 11 percent of CO_2 emissions were contributed by land use change (Le Quere et al., 2020). There is a transition in land use pattern over the years within an area by various human activities. Humans are engaging in deforestation, subsistence farming, intensive agriculture, and expansion of urban areas, which lead to extensive changes in the global landscapes (Fig. 1.2). On the basis of historical, environmental, and socio-economic condition, different regions of the globe are experiencing different stages of transition but their ultimate goal remains the same (De Fries et al., 2004).

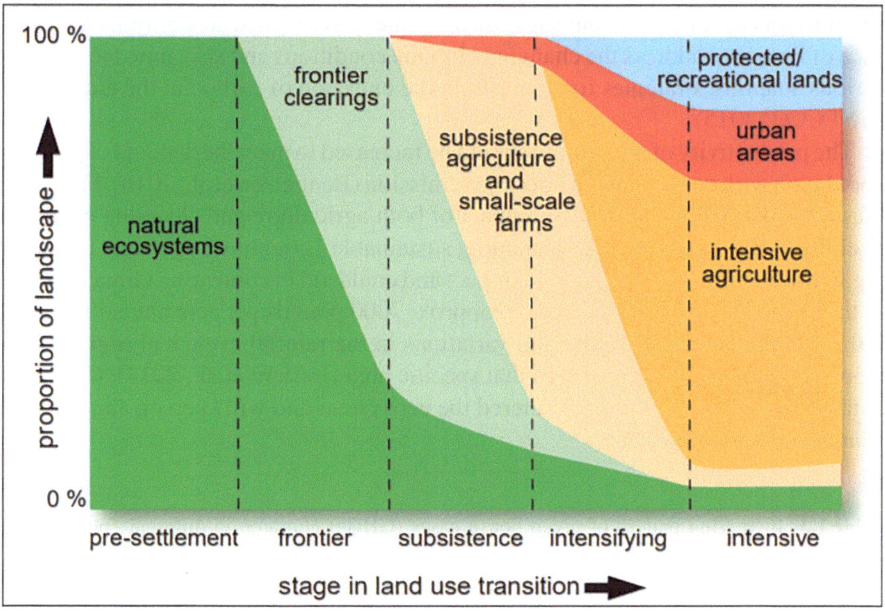

Fig. 1.2 Transformation in the land use pattern over the years. *Source* Adapted from De Fries et al. (2004)

The main goal of human being is to progress, evolve, and make their lives more comfortable and increase the standard of living, which ultimately leads to the extensive use of the naturally available resources in the vicinity for its immediate exploitation, whether it may be at the cost of environmental or natural degradation. There are several regions that persist in a one particular stage for a prolonged time period, while other regions are in a rapid transition phase (De Fries et al., 2004). Several agricultural practices cause erosion, nutrient, and fertilizer leaching to groundwater, which degrades the quality of groundwater. It is one of the major contributing factors of eutrophication in aquatic ecosystem (Bennett et al., 2001). Urbanization is also responsible for deterioration of water quality mainly where sewage treatment is not available. These factors ultimately result in land degradation and water pollution which may lead to health hazards (Bennett et al., 2001). A change in land use pattern and deforestation disrupt the normal surface water balance and thereby increases the rate of surface runoff (Costa et al., 2003).

Agriculture is the backbone of livelihood in rural areas of developing countries. The fluctuation in climatic condition has a significant negative impact on crop production. It is also responsible for contributing 30–40% of green house gas emission, thereby leading to global warming effect in earth's environment (Mishra et al., 2021; Thornton & Lipper, 2014). Although the food grain production has been increased to a remarkable level, important staple crops like rice and wheat production have declined due to climate change (Achanta, 1993). Table 1.3 illustrates the greenhouse gas emissions in different countries. An agreement was signed at twenty-first Conference of Parties to address the changing climatic conditions and also aimed to promote various adapting strategies for controlling the emission of carbon in the atmosphere (UNFCCC, 2015).

The productivity of agriculture should be increased to meet the demand of growing population with net zero green house gas emission (Bennetzen et al., 2016). Figure 1.3 illustrates the reciprocal adverse impact of both agriculture and changing condition of climate. But eventually implementing sustainable agricultural practices including organic agriculture have beneficial impact and could aid in combating climate change as they can able to sequester carbon (approx. 7000 pounds per acre annually) (NAS, 2015). Increasing temperature and variations in the rainfall cycle had impacted the crop, which might have grown in that specific area (Bellard et al., 2014). Change in climatic conditions has already altered the ecosystem and will keep on altering it on a large and uncertain scale.

Table 1.3 Increasing pattern of green house gases (GHGs) emission in different countries (Mt CO_2eq/yr)

Country			EU27	Russia	China	USA	India	Australia	South Africa	Saudi Arabia
Agriculture	Year	1970	14.75	5.07	4.63	3.21	2.46	0.31	0.15	–
		2023	9.63	5.81	15.06	6.31	30.58	3.42	1.55	0.27

Source Adopted from Crippa et al. (2023)

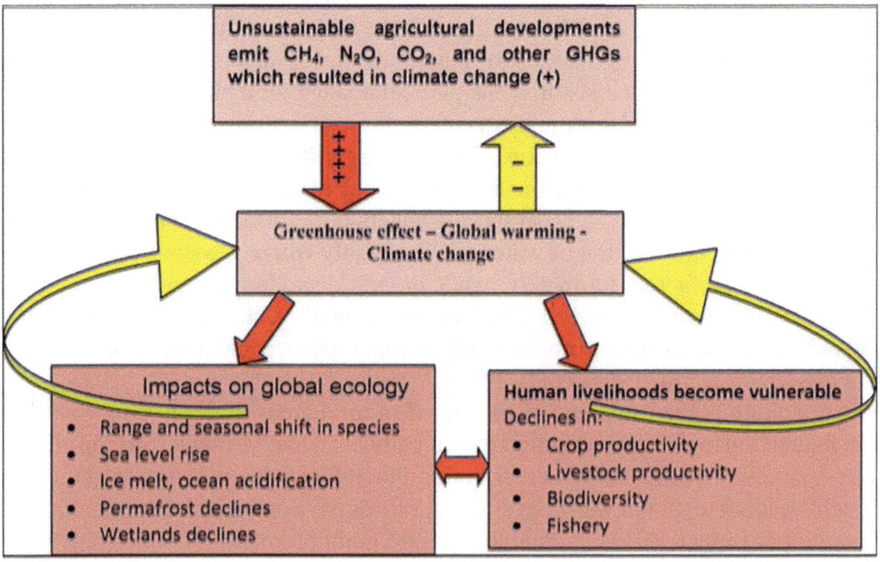

Fig. 1.3 Reciprocal affects of change in climate and agricultural emissions. *Source* Adopted from IPCC (2007), IPCC (2013) and Mekonnen (2019)

1.3 Climate Change and Land Degradation

Another undesirable driver of climate change is land degradation, which indicates the reduction in the land capacity to generate benefits for a particular use (Barrow, 1992; Blaikie & Brookfield, 1987). Land degradation alters soil properties, which in turn affects the soil's ability to sustain plant growth and development. Soil plays a pivotal role in securing food security, but climatic change has affected this by altering soil properties (Pimentel, 2006). Land that suffers from degradation due to poor management may become infertile. The various factors responsible for land degradation are wind and water erosion, soil carbon loss, disturbed nutrient balance, deterioration of soil structure, soil acidification etc. (Kumar & Das, 2014). The inter-relationship between eco-resilience, climate change, and land degradation has become complex to understand. Both land degradation and change in climatic condition have a serious impact on environment and those communities which are relied on the natural available resources for survival (IPCC, 2019). Figure 1.4 illustrates the interaction of change in climate, land degradation, and agricultural productivity.

Land is the most precious natural resource that plays an indispensible role in terrestrial ecosystem, i.e., it provides suitable environment to large group of fauna and flora. So, degradation of this important resource would affect the agricultural productivity as well as hampered other important services required in biodiversity (Lee, 2011). According to Food and Agriculture Organization (1999), land degradation is classified into four severity classes: light, moderate, severe, and very severe (Fig. 1.5). Severity of degradation is measured by considering both degree and extent of degradation (FAO, 1999). Approximately 3.5 billion hectare of global land area have been degraded at an annual rate of about 5–10 million hectare (Lal et al., 2012). Overall, around 1.5 billion people are facing challenges from land degradation (Gnacadja, 2012). Therefore, there is an urgent need for developing different policy for mitigating the adverse effects of degradation.

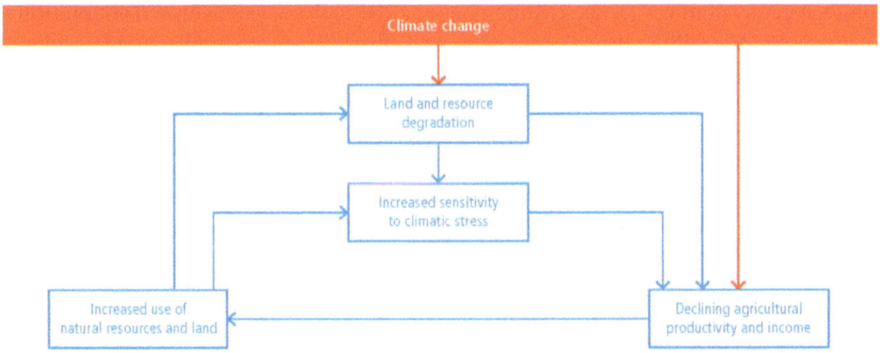

Fig. 1.4 Interaction of land degradation and climate change. *Source* IPCC reports

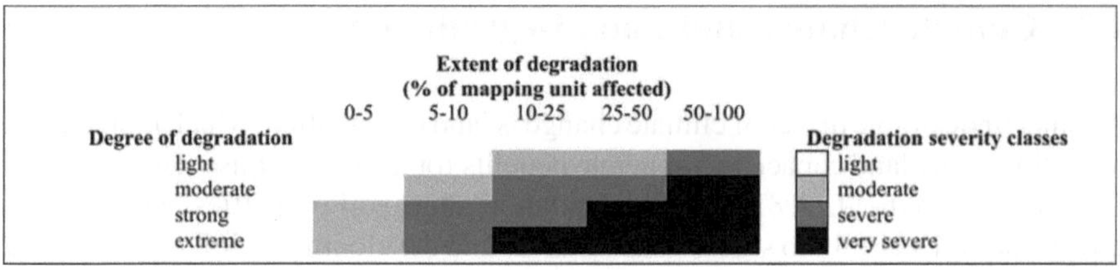

Fig. 1.5 Different severity classes of land degradation. *Source* Food and Agriculture Organisation (1999)

1.4 Climate Change and Biodiversity

The most adverse consequences of change in climate is the threat to biodiversity because it is an important component to balance the ecosystem and the basic resources they provide for livelihood (Dias, 2023). There is a positive and negative influence of climate variability on biodiversity. For instance, the heating problem of the atmosphere due to change in climate can have a direct impact on biodiversity (Meakin, 1992). Biodiversity, which means the presence of different forms of living things in an environment, is crucial for keeping healthy ecosystem and also supporting the well-being of human. It can be observed using three key factors: genetic, species, and ecosystem diversity (Dias, 2023). The term genetic diversity refers to presence of all genetic variations within the species which facilitates the evolution and adaptation to different conditions of environment. The diversity at species level, which means the wide range of species found in the ecosystem for maintaining its resilience and stability. Figure 1.6 illustrates the linkages among people, adaptability of ecosystem and climate variability. According to Darwin's (1859) theory, a region is considered to be more ecologically stable when it has many diverse species than too few. Ecosystems play an important role in protecting communities from the impacts of disasters caused due to climate change, and in turn, well-designed adaptation approaches can enhance ecosystem resilience (Munang et al., 2013). Ecosystem-based adaptation is local people-friendly approach as it integrates the indigenous traditional knowledge to cope up with adverse climatic condition (Reid et al., 2009a, 2009b).

Range of habitats, groups, and different environmental processes found in a specific area is known as Ecosystem diversity (Hooper et al., 2005). Biodiversity is noteworthy in ensuring the proper working and functioning of the ecosystem, and it has been proven and validated through a vast range of empirical research (Hooper et al., 2012). There is an adverse impact of biodiversity loss, caused by different climatic condition, on the functioning of the ecosystem as well as on the stability due to shrinkage in species abundance, which reduces the primary production, causes low rate of decomposition, and disrupts the soil nutrient cycle (Cardinale et al., 2012). Furthermore, biodiversity plays a pivotal role in enhancing the resilience of ecosystem, enabling them to recuperate from the instability and acclimatize to different changed environmental surroundings (Elmqvist et al., 2003). The existence of genetic diversity is a prerequisite criterion for breeding programme to

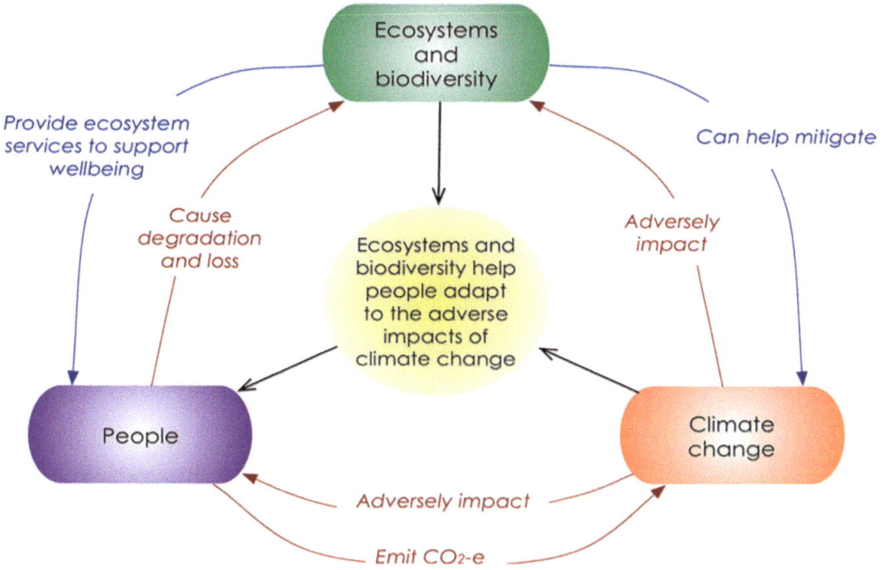

Fig. 1.6 Different linkages observed in the ecosystem adaptation and climate change. *Source* Adopted from Chong (2014)

develop new plant variety and animal breed that will show resistance to diverse biotic and abiotic conditions (Pimentel et al., 1997). Human activities have various negative consequences, for instance habitat loss, pollution, overexploitation, etc., which show adverse effects on biological diversity (Butchart et al., 2010) and lead to the annihilation of species (Pimm et al., 2014). Figure 1.7 illustrates the direct and indirect impacts of change in climate on biodiversity.

Biodiversity played a very significant role in providing important ecosystem services to human beings, i.e., from providing food and medicine to cultural and aesthetic benefits (Dias, 2023). Eco-resilience means the inherent capability of an ecosystem to resist the disruptive events such as natural disasters, human activities, or climate change and recover from disruptions while maintaining its essential functions, structures, and processes (Dias, 2023). In the domain of climatic changes, eco-resilience refers to the inherent ability that enables species to survive and provides continuous ecological services to biotic factors (Dias, 2023). Research has demonstrated that biodiversity at every level plays a crucial role in enhancing ecosystem resilience. One of the important approaches to support eco-resilience is the conservation of biodiversity rather than concentrating on selective species for conservation (Vasiliev, 2022). Numerous researches have displayed the optimistic interaction between biodiversity and ecological resilience. It signifies that the ecosystem having a vast range of diversity is more proficient in enduring and recovering from disturbances (Dias, 2023). This positive linkage between them has also been noticed in an aquatic ecosystem. Oliver et al. (2015) observed that the existence of broader

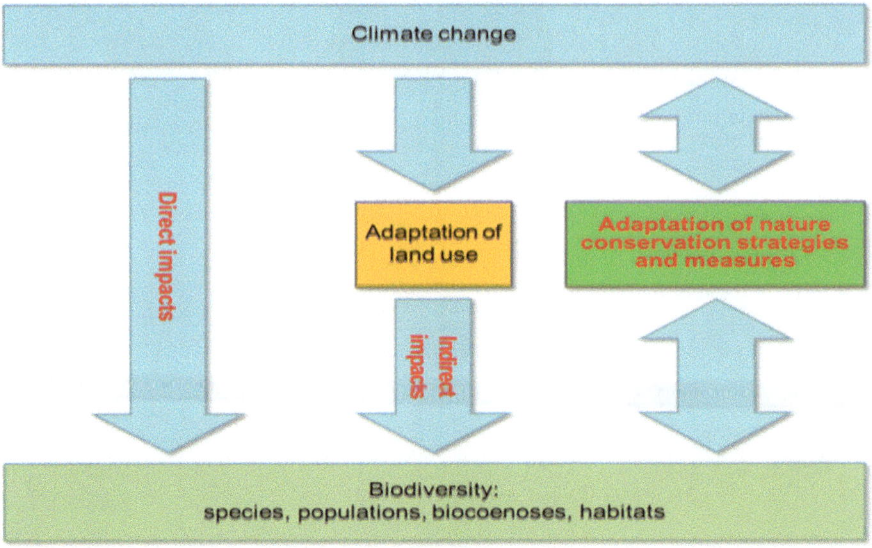

Fig. 1.7 Direct and indirect impacts of climate change on biodiversity. *Source* Adopted from Schliep et al. (2018)

species diversity could boost the drought resistance capability in experimental grassland ecosystems. The meta-analysis study in sixty-four experiments showed that ecosystem role could be amplified with species variability (Cardinale et al., 2012).

The association between biodiversity and eco-resilience can be described by two theories viz., "insurance hypothesis" and "redundancy hypothesis". "Insurance hypothesis" states that existence of higher biodiversity level acts as "buffer" against environmental fluctuations and interruptions. The above hypothesis highlights the significance of species diversity, as certain species may have superior adaptation capacity to various opposing environmental conditions or have better recovery rates, eventually improving ecosystem function under adverse conditions (Folke et al., 2004; Yachi & Loreau, 1999). Another hypothesis that underlines the relation between biodiversity and eco-resilience is the "redundancy hypothesis". This theory affirms that having numerous species with parallel functions (i.e., redundant species) in the ecosystem can assist in enhancing the recovery rate from challenging conditions (Walker, 1992). Redundant species can aid in upholding ecosystem balance by compensating for the species loss under unfavorable conditions. As changes in environmental conditions distress the ecosystem in numerous ways, therefore, it is a compelling need to restore biodiversity for maintaining ecological balance. Different conservation programs have been implemented to preserve the biodiversity for upholding and maintaining important natural services (Millennium Ecosystem Assessment, 2005).

So, the significance of both the theories implies that diverse ecosystem is more important as it is more stable and maintained under fluctuating environmental conditions.

1.5 Conclusion

The relation among eco-resilience, climate change, and land degradation is essential in comprehending and tackling the challenges arising in global environment. This chapter highlights the occurrence of climate change at an alarming rate and shows the adverse impact of climatic changes on the ecosystem and human communities. Apart from this, climate change also causes shrinkage of available land resources, becoming more and more apparent, contributes in degrading land condition and escalating the load on both natural and human system. This continuing degradation in turn severely affects the ability of the ecosystem to operate effectively while dismantling and disrupting their significant role in maintaining the ecological balance and also impairing their capacity to provide crucial services which are important for the well-being of humankind for example in ensuring food security and regulating water cycle system. Moreover, ecosystem facilitates in sequestration of carbon and helps in minimizing the impact of climate change.

However, eco-resilience indicates the built-in capability of the ecosystem to not only adapt to environmental fluctuations but also to cope up from vast disturbances while providing a pivotal pathway for minimizing the various impacts of climate change and land degradation. It is offering a way to acclimatize to these demanding challenges over a long period. The implementation of sustainable land management, conservation of biological diversity, and adopting a comprehensive strategy for adaptation to changing climatic condition, would help in enhancing the resilience of ecosystem. These approaches would also significantly reduce the vulnerability of both human communities and natural system by increasing their capacity to withstand and recuperate from the rising pressure imposed by environmental disturbances. Other strategies for increasing ecosystem resilience are the restoration of degraded lands, advancement of agro-ecological practices that combine sustainable techniques in farming and maintaining different ecological principles. The other approach is reducing the emission of greenhouse gases from various sources. All of these approaches inclusively playing impactful role in maintaining or even enhancing the necessary and essential functions of ecosystems.

Eventually, addressing the complex and interrelated challenges posed by land degradation and climate variability urges the implementation of integrated multi-dimensional and multi-disciplinary strategies that bring altogether a combination of social, ecological, and policy measures. By encouraging collaboration at global level and dedicated commitment for achieving sustainable development goals, it can be possible to build the required necessary condition for increasing eco-resilience as well as to mitigate the opposing consequences of global environmental changes. It also works toward conserving and securing the vitality and health of the ecosystem.

This collective efforts and work will help in affixing and preserving the natural system that could sustain life on earth for a long period of time and making sure that they continue to provide essential and crucial resources for future generations while enhancing the ability and capacity of societies to thrive in an increasingly adverse and uncertain climate condition.

Authors Contribution All authors have contributed equally to this work.

Funding The research was not supported by any funding.

Data Availability Statement The datasets used during the present study available from the corresponding author on reasonable request.

Potential Conflicts of Interest There is no conflict of interest between the authors.

Informed Consent Informed consent was obtained from all individual participants included in the study.

AI Disclosure Authors confirm that no AI has been used for writing, or editing of this chapter.

References

Achanta, A. N. (1993). An Assessment of the Potential Impact of Global Warming on Indian Rice Production, The Climate Change Agenda: An Indian Perspective, Tata Energy Research Institute, New Delhi.

Adewale, S., Godson, A., Sokan-Adeaga, A., Sokan-Adeaga, M., & Sokan-Adeaga, E. (2018). Resilience and sustainability of the ecosystem: An environmental health perspective. *London Journal Press, 18*(1).

Baccini, A., Walker, W., Carvalho, L., Farina, M., Sulla-Menashe, D., & Houghton, R. A. (2017). Tropical forests are a net source of carbon based on above-ground measurements of gain and loss. *Science, 358*(6360), 230–234.

Barrow, C. J. (1992). *World Atlas of Desertification (UNEP).*

Bellard, C., Leclerc, C., & Courchamp, F. (2014). Impact of sea level rise on the 10 insular biodiversity hotspots. *Global Ecology and Biogeography, 23*, 203–212.

Bennett, E. M., Carpenter, S. R., & Caraco, N. F. (2001). Human impact on erodable phosphorus and eutrophication: A global perspective: Increasing accumulation of phosphorus in soil threatens rivers, lakes, and coastal oceans with eutrophication. *BioScience, 51*(3), 227–234.

Bennetzen, E. H., Smith, P., & Porter, J. R. (2016). Agricultural production and greenhouse gas emissions from world regions—The major trends over 40 years. *Global Environmental Change, 37*, 43–55.

Blaikie, P., & Brookfield, H. (1987). *Land degradation and society*. Methuen.

Butchart, S., Walpole, M., Collen, B., Van Strien, A., Scharlemann, J., Almond, R., Baillie, J., Bertzky, B., Brown, C., Bruno, J., Carpenter, K., Carr, G., Chanson, J., Chenery, A., Csirke, J., Davidson, N., Dentener, F., Foster, M., Galli, A. and Watson, R. (2010). Global biodiversity: Indicators of recent declines. *Science (New York, N.Y.), 328*. 1164–1168. https://doi.org/10.1126/science.1187512

Cardinale, B. J., Duffy, J. E., Gonzalez, A., Hooper, D., Perrings, C., Venail, P., Narwani, A., Tilman, D., Wardle, D., Kinzig, A., Daily, G., Loreau, M., Grace, J., Larigauderie, A., Srivastava, D., & Naeem, S. (2012). Biodiversity loss and its impact on humanity. *Nature, 486*, 59–67.

Chong, J. (2014). Ecosystem-based approaches to climate change adaptation: Progress and challenges. *International Environmental Agreements, 14*, 391–405. https://doi.org/10.1007/s10784-014-9242-9

Convention on the Biological Diversity. (1992). The Earth Summit. Held in Rio de Janeiro on 5 June 1992 and entered into force on 29 December 1993. United Nations, General Assembly.

Costa, M. H., Botta, A., & Cardille, J. A. (2003). Effects of large-scale changes in land cover on the discharge of the Tocantins River, Southeastern Amazonia. *Journal of Hydrology, 283*, 206–217.

Crippa, M., Guizzardi, D., Pagani, F., Banja, M., Muntean, M., Schaaf, E., Becker, W., Monforti-Ferrario, F., Quadrelli, R., Risquez Martin, A., Taghavi-Moharamli, P., Köykkä, J., Grassi, G., Rossi, S., Brandao De Melo, J., Oom, D., Branco, A., San-Miguel, J., & Vignati, E. (2023). GHG emissions of all world countries. *Publications Office of the European Union, Luxembourg*. https://doi.org/10.2760/953332

Darwin, C. (1859). *On the origin of species by means of natural selection or the preservation of favoured races in the struggle for life [reprinted 1964]*. Harvard University.

DeFries, R., Asner, G., & Houghton, R. (2004). Trade-offs in land-use decisions: Towards a framework for assessing multiple ecosystem responses to land-use change. Ecosystems and Land Use Change (American Geophysical Union, Geophysical Monograph Series, Vol. 153, Washington, DC). https://doi.org/10.1029/153GM02

Desertification and land degradation atlas of India. (2016): Space Applications Centre. https://www.sac.gov.in/SACSITE/Desertification_Atlas_2016_SAC_ISRO.pdf

Dias, R. (2023). Ecological resilience in a changing world: Challenges and opportunities in biodiversity conservation in the face of climate change. *Seven Editora*. https://sevenpublicacoes.com.br/editora/article/view/1726

Díaz, S., Demissew, S., Carabias, J., Joly, C., Lonsdale, M., Ash, N., Larigauderie, A., Adhikari, J. R., Arico, S., Báldi, A., Bartuska, A., Baste, I. A., Bilgin, A., Brondizio, E., Chan, K. M. A., Figueroa, V. E., Duraiappah, A., Fischer, M., Hill, R., et al. (2015). The IPBES conceptual framework—Connecting nature and people. *Current Opinion in Environmental Sustainability, 14*, 1–16. https://doi.org/10.1016/j.cosust.2014.11.002

Elasha, B. O., Elhassan, N. G., Ahmed, H., & Zakieldin, S. (2005). Sustainable livelihood approach for assessing community resilience to climate change: Case studies from Sudan. AIACC Working Paper No. 17. Khartoum, Sudan: Higher Council for Environment and Natural Resources.

Elmqvist, T., Folke, C., Nyström, M., Peterson, G., Bengtsson, J., Walker, B., & Norberg, J. (2003). Response diversity, ecosystem change, and resilience. *Frontiers in Ecology and the Environment, 1*, 488–494.

FAO. (1999). *TERRASTAT CD-ROM (FAO Land and Water Digital Media)*. Food and Agriculture Organization of the United Nations.

Folke, C., Carpenter, S., Walker, B., Scheffer, M., Elmqvist, T., Gunderson, L., & Holling, C. S. (2004). Regime shifts, resilience, and biodiversity in ecosystem management. *Annual Review of Ecology, Evolution, and Systematics, 35*, 557–581.

Gnacadja, L. (2012). Moving to zero-net rate of land degradation. Statement by Executive Secretary. UN Convention to Combat Desertification, Rio de Janeiro. http://www.unccd.int/Lists/SiteDocumentLibrary/secretariat/2012/UNCCD%20ES%20Statement%20at%20PR%20in%20NY%20on%2026%20March%202012

Hammad, A., & Tumeizi, A. (2012). Land degradation: Socioeconomic and environmental causes and consequences in the eastern Mediterranean. *Land Degradation and Development, 23*, 216–226. https://doi.org/10.1002/ldr.1069

Hooper, D., Adair, C., Cardinale, B., Byrnes, J., Hungate, B., Matulich, K., Gonzalez, A., Duffy, J., Gamfeldt, L., & O'Connor, M. (2012). A global synthesis reveals biodiversity loss as a major driver of ecosystem change. *Nature, 486*, 105–108.

Hooper, D. U., Chapin, F. S., III., Ewel, J. J., Hector, A., Inchausti, P., Lavorel, S., Lawton, J. H., Lodge, D. M., Loreau, M., Naeem, S., Schmid, B., SetSlS, H., Symstad, A. J., Vandermeer, J., & Wardle, D. A. (2005). Effects of biodiversity on ecosystem functioning: A consensus of current knowledge. *Ecological Monographs, 75*(1), 3–35.

Houghton, J. T., Ding, Y., Griggs, D. J., Noguer, M., van der Linden, P. J., Dai, X., Maskell, K., & Johnson, C. A. (2001). *Climate change 2001: The scientific basis* (p. 881). Cambridge University Press.

IPCC. (2007). Climate change 2007: Impacts, adaptation and vulnerability. In M. L. Parry, O. F. Canziani, J. P. Palutikof, P. J. van der Linden & C. E. Hanson (Eds.), *Contribution of Working Group II to the Fourth Assessment Report of the Intergovernmental Panel on Climate Change* (976 pp.). Cambridge University Press.

IPCC. (2013). Climate change 2013: The physical science basis. In T. F. Stocker, D. Qin, G. K. Plattner, M. Tignor, S. K. Allen, J. Boschung, A. Nauels, Y. Xia, V. Bex & P. M. Midgley (Eds.), *Contribution of Working Group I to the Fifth Assessment Report of the Intergovernmental Panel on Climate Change* (1535 p.). Cambridge University Press.

IPCC. (2019). *Climate Change and Land. Summary for Policy Makers.* Retrieved from https://www.ipcc.ch/site/assets/uploads/2019/08/Fullreport.pdf

IPCC. (2021). Climate change 2021: The basis of physical science. In V. Masson-Delmotte, P. Zhai, A. Pirani, S. L. Connors, C. Péan, S. Berger, N. Caud, Y. Chen, L. Goldfarb, M. I. Gomis, M. Huang, K. Leitzell, E. Lonnoy, J. B. R. Matthews, T. K. Maycock, T. Waterfield, O. Yelekçi, R. Yu & B. Zhou (Eds.), *Contribution of Working Group I to the Sixth Assessment Report of the Intergovernmental Panel on Climate Change* (2391 p.). Cambridge University Press. Cambridge, United Kingdom and New York, NY, USA. https://doi.org/10.1017/9781009157896

Kumar, R., & Das, A. J. (2014). Climate change and its impact on land degradation: Imperative need to focus. *Journal of Climatology & Weather Forecasting, 2*, 108.

Lal, M., Nozawa, T., Emori, S., Harasawa, H., Takahashi, K., Kimoto, M., Abe-Ouchi, A., Nakajima, T., Takemura, T., & Numaguti, A. (2001). Future climate change: Implications for Indian summer monsoon and its variability. *Current Science*, 1196–1207.

Lal, R., Safriel, U., & Boer, B. (2012). Zero Net Land Degradation: A New Sustainable Development Goal for Rioþ 20. A Report Prepared for the Secretariat of the United Nations Convention to Combat Desertification. http://www.unccd.int/Lists/SiteDocumentLibrary/secretariat/2012/Zero%20Net%20Land%20Degradation%20Report%20UNCCD%20May%202012%20background.pdf

Le Quere, C., Jackson, R., Jones, M., Smith, A., Abernethy, S., Andrew, R., De-Gol, A., Willis, D., Shan, Y., Canadell, J., Friedlingstein, P., Creutzig, F., & Peters, G. (2020). Temporary reduction in daily global CO2 emissions during the COVID-19 forced confinement. *Nature Climate Change, 10*, 1–7.

Lee, D. K. (2011). Land and Soil in the Context of a Green Economy for Sustainable Development, Food Security and Poverty Eradication. The Submission of the UNCCD Secretariat to the Preparatory Process for Rioþ 20. UNCCD, Bonn. http://www.uncsd2012.org/content/documents/462unccd.pdf

Meakin, S. (1992). Rio Earth Summit 1992. *Summary of the United Nations Conference on Environment and Development.* https://sustainabledevelopment.un.org/milestones/unced

Mekonnen, Z. (2019). Observed and projected reciprocate effects of agriculture and climate change: Implications on ecosystems and human livelihoods. *IntechOpen.* https://doi.org/10.5772/intechopen.79118

Millennium Ecosystem Assessment (MA) (2005). Synthesis Island Press, Washington DC. Available on the Internet http://www.MAweb.org

Mishra, A., Bruno, E., & Zilberman, D. (2021). Compound natural and human disasters: Managing drought and COVID-19 to sustain global agriculture and food sectors. *Science of the Total Environment, 754*, Article 142210.

Munang, R., Thiaw, I., Alverson, K., Liu, J., & Han, Z. (2013). The role of ecosystem services in climate change adaptation and disaster risk reduction. *Current Opinion in Environmental Sustainability, 5*(1), 47–52.

NAAS. (2010). *Degraded and wastelands of India–status of spatial distribution.* National Academy of Agricultural Sciences.

NAS. (2015). *Climate intervention: Carbon dioxide removal and reliable sequestration*. National Research Council of the National Academies, The National Academies Press.

ND-GAIN. (2024). *Notre dame global adaptation initiative*. https://gain.nd.edu/our-work/country-index/

O'Neill, B., Oppenheimer, M., Warren, R., Hallegatte, S., Kopp, R., Portner, H., Scholes, R., Birkmann, J., Foden, W., Licker, R., Mach, K., Marbaix, P., Mastrandrea, M., Price, J., Takahashi, K., Van-persele, J., & Yohe, G. (2017). IPCC reasons for concern regarding climate change risk. *Natural Climate Change, 7*(1), 28–37.

Oliver, T. H., Heard, M. S., Isaac, N. J., Roy, D. B., Procter, D., Eigenbrod, F., Freckleton, R., Hector, A., Orme, C. D. L., Petchey, O. L., Proença, V., Raffaelli, D., Suttle, K. B., Mace, G. M., Martín-López, B., Woodcock, B. A., & Bullock, J. M. (2015). Biodiversity and resilience of ecosystem functions. *Trends in Ecology & Evolution, 30*(11), 673–684. https://doi.org/10.1016/j.tree.2015.08.009

Oppenheimer, M., Campos, M., Warren, R., Birkmann, J., Luber, G., O'Neill, B., Takahashi, K., Brklacich, M., Semenov, S., Licker, R., & Hsiang, S. (2015). *Emergent Risks and Key Vulnerabilities*. https://doi.org/10.1017/CBO9781107415379.024

Pachauri, R. K., Allen, M. R., Barros, V .R., Broome, J., Cramer, W., Christ, R., et al. (2014). Climate change 2014: Synthesis report. *Contribution of Working Groups I, II and III to the Fifth Assessment report of the Intergovernmental Panel on Climate Change: Ipcc*

Pimentel, D. (2006). Soil erosion: A food and environmental threat. *Environment, Development and Sustainability, 8*, 119–137.

Pimentel, D., Wilson, C., McCullum-Gomez, C., Huang, R., Dwen, P., Flack, J., Tran, Q., Saltman, T., & Cliff, B. (1997). Economic and environmental benefits of biodiversity. *BioScience*. https://doi.org/10.2307/1313097

Pimm, S. L., Jenkins, C., Abell, R., Brooks, T., Gittleman, J., Joppa, L., Raven, P., Roberts, C., & Sexton, J. (2014). The biodiversity of species and their rates of extinction, distribution, and protection. *Science (New York, N.Y.), 344*. 1246752. https://doi.org/10.1126/science.1246752

Reid, H., Alam, M., Berger, R., Cannon, T., Huq, S., & Milligan, A. (2009). *Community-based adaptation to climate change: An overview, in community-based adaptation to climate change*. IIED.

Reid, H., Phillips, J., & Heath, M. (2009). *Natural resilience: Healthy ecosystems as climate shock insurance. IIED briefing*. IIED.

Schliep, R., Walz, U., Sukopp, U., & Heiland, S. (2018). Indicators on the impacts of climate change on biodiversity in Germany—Data driven or meeting political needs? *Sustainability, 10*(11), 3959. https://doi.org/10.3390/su10113959

Settele, J., et al. (2015). Terrestrial and Inland Water Systems. In: Climate Change 2014: Impacts, Adaptation and Vulnerability. Part A: Global and Sectoral Aspects. In C. B. Field, V. R. Barros, D. J. Dokken, K. J. Mach, M. D. Mastrandrea, T. E. Bilir, M. Chatterjee, K. L. Ebi, Y. O. Estrada, R. C. Genova, B. Girma, E. S. Kissel, A. N. Levy, S. MacCracken, P. R. Mastrandrea & L. L.White (Eds.), *Contribution of Working Group II to the Fifth Assessment Report of the Intergovernmental Panel on Climate Change* (pp. 271–360). Cambridge University Press, Cambridge, United Kingdom and New York, NY, USA.

Thornton, P. K., & Lipper, L. (2014). How does climate change alter agricultural strategies to support food security? (Vol. 1340): International Food Policy Research Institute.

Turner, B. L., Clark, W. C., Kates, R. W., Richards, J. F., Mathews, J. T., & Meyer, W. B. (1993). *The earth as transformed by human action*. Cambridge University.

UNFCCC. (2015). https://www.unwomen.org/en/how-we-work/intergovernmentalsupport/climate-change-and-the-environment/united-nations-framework-convention-on-climate-change/cop-21

Vasiliev, D. (2022). *The role of biodiversity in ecosystem resilience IOP conference series: Earth environmental science* (1072). https://doi.org/10.1088/1755-1315/1072/1/012012

Walker, B. (1992). Biological diversity and ecological redundancy. *Conservation Biology, 6*, 18–23.

Xie, H., Zhang, Y., Wu, Z., & Lv, T. (2020). A bibliometric analysis on land degradation: Current status, development, and future directions. *Land, 9*, 28. https://doi.org/10.3390/land9010028
Yachi, S., & Loreau, M. (1999). Biodiversity and ecosystem productivity in a fluctuating environment. *The Insurance Hypothesis, 96*(4), 1463–1468. https://doi.org/10.1073/pnas.96.4.1463

Chapter 2
The Nexus Between Land Degradation and Climate Change

Vinai Kumar, Dipti Bisarya, Shivani Thakur, Agnibha Sinha, Rohan Nambiar, and Chinmaya Sahoo

Abstract The intricate interplay between land degradation and climate change poses a profound challenge to global sustainability, threatening ecosystems, biodiversity, and human livelihoods on an unprecedented scale. Land degradation, driven by unsustainable practices, deforestation, and overexploitation of resources, erodes soil fertility, disrupts ecosystems, and intensifies greenhouse gas emissions. Simultaneously, climate change exacerbates these impacts through rising temperatures, extreme weather events, and altered precipitation patterns, perpetuating a relentless cycle of mutual reinforcement. This chapter explores the dynamic nexus between these crises, dissecting their extensive impacts on food security, migration, and socio-economic stability. By analyzing diverse global case studies, it showcases innovative solutions such as reforestation, climate-smart agriculture, and ecosystem restoration as transformative pathways to disrupt this destructive cycle. Additionally, it underscores the critical role of robust policy frameworks, international cooperation, and community-driven initiatives in fostering resilience and achieving land-degradation neutrality. Through a compelling synthesis of scientific insights and strategic interventions, this chapter envisions a sustainable future where humanity and nature coexist in balance and harmony.

V. Kumar
Department of Crop Physiology, Chandra Shekar Azad University of Agriculture and Technology, Kanpur, Uttar Pradesh, India

D. Bisarya · R. Nambiar
Department of Agronomy, School of Agriculture, Lovely Professional University, Phagwara, Punjab, India

S. Thakur · A. Sinha (✉)
Department of Soil Science and Agricultural Chemistry, School of Agriculture, Lovely Professional University, Phagwara, Punjab, India
e-mail: agnibhasinha.official@gmail.com

C. Sahoo
Department of Agronomy, Dr. Rajendra Prasad Central Agricultural University, Pusa, Samastipur, Bihar, India

U. Chatterjee et al. (eds.), *Eco-Resilience: Climate Change, Land Degradation and Sustainable Solutions*, Sustainability Solutions,
https://doi.org/10.1007/978-3-032-00708-7_2

Keywords Land degradation · Climate change · Sustainability · Ecosystem restoration · Resilience

2.1 Introduction

Land degradation and climate change are two of the most critical challenges facing our planet today. Together, they threaten ecosystems, biodiversity, food security, and human livelihoods, making it imperative to address their complex interactions (IPCC, 2019; Nguyen et al., 2023). Land degradation is defined as "persistent decline in the condition of land, often caused by direct or indirect human activities, including anthropogenic climate change". It is characterized by the long-term reduction or loss of biological productivity, ecological integrity, and value to humans (Olsson et al., 2019). This degradation may manifest through various interconnected processes such as soil erosion, nutrient depletion, loss of biodiversity, deforestation, and declining vegetative health (Grinand et al., 2019). Climate change, on the other hand, refers to a long-term alteration in the state of the climate, including changes in temperature, precipitation, and other atmospheric conditions, identifiable through shifts in the mean or variability of its characteristics, as determined by statistical analysis. These changes persist over extended periods, typically decades or longer, and can result from natural internal processes, external forcings such as variations in solar cycles, volcanic eruptions, and persistent anthropogenic influences, including changes in atmospheric composition and land use (IPCC, 2022).

The impacts of land degradation and climate change are profound and far-reaching. Globally, land degradation adversely affects approximately 3.2 billion people, with biodiversity loss and a decline in ecosystem services accounting for over 10% of the annual global gross product (Mani et al., 2021; Scholes et al., 2018). According to the *Global Land Outlook* (UNCCD, 2022), 20–40% of the world's land is degraded. Soil degradation is particularly severe in arid and semi-arid regions (Wang et al., 2023). Each year, around 24 billion tons of fertile soil are lost, significantly undermining agricultural productivity and food security (FAO, 2020). Simultaneously, climate change is a universal crisis, with global temperatures having already risen by 1.1 °C above pre-industrial levels. This warming trend has intensified extreme weather events, disrupted ecosystems, and threatened biodiversity. Climate change disproportionately affects vulnerable populations, particularly those in developing countries who lack the resources to adapt. Land degradation and climate change are intricately linked. Degraded land releases stored carbon into the atmosphere, contributing to climate change. For instance, deforestation not only reduces carbon sequestration but also emits significant amounts of carbon dioxide. Conversely, climate change exacerbates land degradation through increased droughts, floods, and other extreme weather events, which accelerate soil erosion, salinity, and desertification. This interplay creates a vicious cycle: land degradation worsens climate change, which in turn accelerates further degradation. This cycle poses a severe threat to sustainable development and the achievement of global goals such as food security and poverty reduction (Olsson et al., 2019). This chapter aims to explore the intricate relationship between climate change and land degradation, analysing the processes involved, systematically assessing scientific literature, and evaluating mitigation options and case studies. It addresses the bidirectional impacts between climate change and land degradation while emphasizing sustainable solutions and identifying future research directions.

2.2 Status and Current Trend of Land Degradation

Anthropogenic activities have altered over 70% of the Earth's land area from its natural state, leading to unprecedented environmental degradation and making a substantial contribution to global warming. Land degradation is the result of human activities that exploited land, leading to a decline in its utility, biodiversity, soil fertility, and overall health. This issue impacts billions who rely directly on agriculture and has widespread consequences that affect everyone globally. These effects encompass food shortages, rising food price volatility due to reduced cropland productivity, amplified global climate change driven by carbon and nitrous oxide emissions from degrading land, and the risk of social instability caused by forced migration. As per the reports released by "United Nations Convention to Combat Desertification (UNCCD)" nearly 100 million hectares of land were degraded each year between 2015 and 2019, totalling about 403 million hectares in just these 4 years.

The accelerated rate of degradation has affected the lives of approximately 1.3 billion individuals. To achieve a land-degradation-neutral world, it is crucial to restore approximately 1.5 billion hectares of land by 2030. As per IPCC (2019), each year, 24 billion tons of fertile soil are lost to erosion, and 12 million hectares of land are degraded—equivalent to 23 hectares per minute (Masson-Delmotte et al., 2019). During 2014–15, India experienced an economic loss equivalent to 2.54% of its GDP, amounting to approximately 3177.39 billion rupees, due to land degradation and changes in land use patterns (TERI, 2016). Between 2015 and 2019, 30.51 million hectares of India's total reported land experienced degradation, as per data from the United Nations Convention to Combat Desertification (UNCCD, 2019). This means that 9.45% of the country's landmass was degraded as of 2019 compared to 4.42% in 2015. Table 2.1 given below consists various case studies on land degradation worldwide.

Land degradation has been a persistent challenge for humanity since the widespread adoption of agriculture during the Neolithic era, approximately 10,000–7500 years ago, accompanied by a rise in population. Reports suggested that atmospheric greenhouse gas (GHG) levels—particularly carbon dioxide (CO_2) and methane (CH_4)—began increasing over 3000 years ago due to the expansion of agriculture, deforestation, and the domestication of wild animals (Olsson et al., 2019). This shows that as the population rises and climate changes, the process of land degradation becomes more severe.

Table 2.1 Different case studies on land degradation

S.No	Site/location	Case study of land degradation	Reference
1	India	Between 2015 and 2019, 30.51 million hectares of land were degraded, rising from 4.42% in 2015 to 9.45% in 2019. Causes include deforestation, soil erosion, and water scarcity in drylands	UNCCD, (2019)
2	Romania	A GIS-based analysis using remote-sensing data identified 7.76% of land as degraded. Most degraded areas overlap with drylands, while high-altitude forests showed vegetation recovery due to effective conservation	Ontel et al. (2023)
3	China (Loess Plateau)	Intensive agricultural practices led to severe soil erosion and loss of fertility. Restoration efforts, such as reforestation and terracing, significantly improved soil health and vegetation cover	Qiang-guo (2001)
4	United States (Great Plains)	Dust Bowl phenomena in the 1930s covered entire Plains for almost a decade. It led to severe soil erosion due to poor agricultural practices Major crops were affected due to rainfall deficit, high temperature, wind intensity, and dust storms, causing agricultural depression. Conservation tillage and crop rotation have since been adopted to restore soil quality	National Drought Mitigation Centre https://drought.unl.edu/dustbowl/

(continued)

Table 2.1 (continued)

S.No	Site/location	Case study of land degradation	Reference
5	Calabar South, Nigeria	Soil erosion significantly impacted land, with approximately one meter of soil lost annually over 250,000 hectares. Around 194 million hectares were affected by water erosion, where 6.1 million hectares were found degraded in the study area. Deforestation further exacerbated the issue, with about 6 million hectares of forest cleared annually, primarily for agricultural purposes	Eni (2012)

Therefore, attention must be given to manage the relationship between human beings and land to achieve the goal of land degradation neutrality. Despite the fact that land degradation is a severe problem, there are no reliable global maps representing the extent and severity of land degradation. As land degradation is a complex process that differs in type, scale, and across spatial and temporal dimensions, there is no defined procedure for quantifying land degradation objectively and consistently over larger areas. Although various approaches exist for the assessment of varied aspects of land degradation, yet these models face challenges in prediction accuracy because of the intricate interactions and the significant influence of land management on land degradation outcomes. Table 2.2 given below shows various methods used by different researchers for quantification of extent of degradation worldwide.

2.3 Climate Change: A Catalyst for Land Degradation

Climate change serves as a catalyst for land degradation by increasing greenhouse gas (GHG) emissions and reducing the Earth's ability to absorb carbon. It accelerates existing degradation processes and introduces new patterns. Studies highlight the role of human-induced global warming in increasing the frequency and intensity of heavy precipitation events and rising heat stress, both critical drivers of land degradation (IPCC, 2022). These factors have led to soil erosion, desertification, and loss of vegetation in affected regions (UNCCD, 2022). Sea level rise further worsens coastal erosion, particularly in vulnerable low-lying areas (IPCC, 2022). Projections suggest that ongoing global warming will intensify floods, droughts, cyclones, and rising sea levels, all of which contribute to degradation. Thawing permafrost in polar regions and shifting storm patterns have introduced degradation in previously unaffected

Table 2.2 Quantification of global extent of degradation using various methods

S.No.	Methods/indicators	Findings	References
1	Vegetation index-related method and indicator	For tropical regions, it's a better method to capture land degradation	Jiang et al. (2024)
2	Overlay multiple methods and indicators	Suitable for studying extent of land degradation, and areas with high amount of overlay values can be considered hotspot for land degradation	Jiang et al. (2024)
3	Multi-metric indices	These indices often appear scientifically robust but can include subjective elements, lose crucial information when complex data are condensed into a single value, introduce bias through data categorization, and unevenly emphasize certain properties due to weighting	Prince et al. (2018)
4	Mechanistic models for soil erosion	Mechanistic models provide viable results when biophysical mechanism is known and data is available. But not precise enough to pick up on small-scale or subtle changes in degradation	Tamene and Le (2015)
5	Remote sensing-derived indicators	This approach is scalable for global applications, providing a robust framework for monitoring and mitigating land degradation using objective, data-driven methods	Ontel et al. (2023)
6	Expert opinion	Measures actual and potential degradation, but not globally consistent	Oldeman et al. (1990), Gibbs and Salmon (2015)

areas (Schuur et al., 2015). In polar regions, thawing permafrost releases significant amounts of greenhouse gases while destabilizing landscapes and threatening cold-adapted species (Schuur et al., 2015). For cyclone-prone regions, the combination of rising sea levels and stronger storms is expected to cause widespread land degradation, threatening both ecosystems and livelihoods (IPCC, 2022). Coastal erosion due to sea level rise also poses global risks, with severe consequences for human settlements and ecosystems (Magnan et al., 2022). Climate change has profound impacts on ecosystems worldwide, disrupting their structure, function, and biodiversity. These impacts vary across ecosystems, with terrestrial ecosystems being among the most affected due to their direct interaction with climate variables such as temperature, precipitation, and extreme weather events. Climate change significantly impacts terrestrial ecosystems, altering their composition, structure, and functioning. Rising temperatures, shifting precipitation patterns, and increased frequency of extreme weather events have led to habitat degradation, species migration, and biodiversity loss. Forest ecosystems, for instance, face intensified threats from wildfires, pest outbreaks, and droughts, particularly in boreal and tropical regions (IPCC, 2022). Grasslands and savannas experience reduced productivity and species shifts due to prolonged droughts and invasive species favored by elevated carbon dioxide levels. These changes compromise critical ecosystem services such as carbon sequestration, water regulation, and soil fertility, underscoring the urgent need for adaptive and mitigative strategies to protect terrestrial ecosystems and their vital functions.

2.3.1 Linkages Between Climate Variables and Land Degradation Processes

Climate variables like temperature, precipitation, and extreme weather events significantly impact soil quality, vegetation cover, and ecosystem health (Fig. 2.1).

- **Temperature Increases**: Rising temperatures accelerate soil moisture loss, intensify evapotranspiration, and weaken vegetation cover, increasing vulnerability to erosion. In arid and semi-arid regions, prolonged heatwaves contribute to desertification by exacerbating water scarcity and reducing productivity (IPCC, 2022). Increased temperatures also favor the proliferation of pests and pathogens, further stressing ecosystems.
- **Precipitation Patterns**: Changes in rainfall intensity and distribution are central to land degradation. Heavy precipitation causes soil erosion and runoff, stripping topsoil and reducing agricultural productivity (Pimentel and Burgess 2013). Prolonged droughts reduce vegetation cover, particularly in drylands where water scarcity is severe.

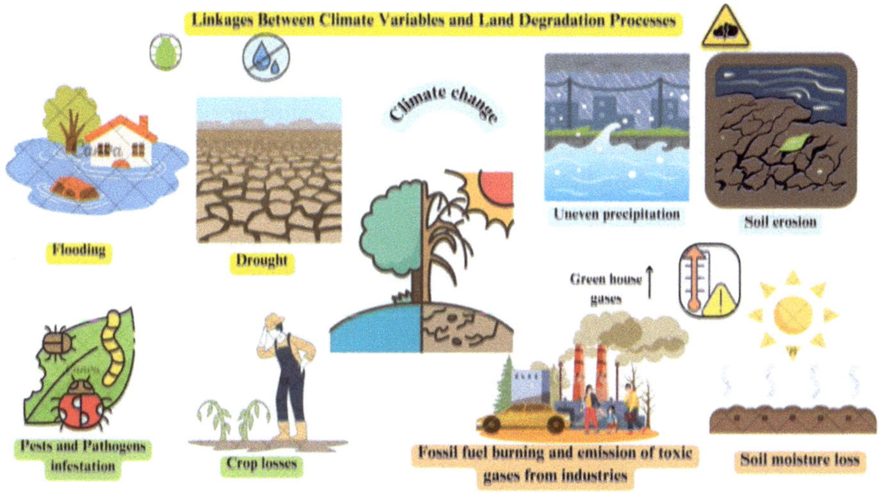

Fig. 2.1 Linkages between climate variables and land degradation processes

- **Extreme Weather Events**: Flooding erodes soil and damages vegetation, while droughts deplete groundwater and weaken ecosystems, increasing their susceptibility to further degradation (UNCCD, 2022).

2.3.2 Examples of Climate-Induced Land Degradation

- **Droughts and Desertification**: Prolonged droughts, driven by increasing global temperatures and reduced precipitation, exacerbate desertification in arid and semi-arid regions. For instance, the Sahel region in Africa has experienced severe droughts, leading to soil degradation, loss of vegetation cover, and declining agricultural productivity (IPCC, 2022). The lack of water reduces plant growth, leaving soils vulnerable to wind erosion, and accelerates the expansion of deserts into previously arable lands (Burezq, 2020).
- **Rising Sea Levels and Salinization**: Sea level rise, a consequence of melting ice sheets and thermal expansion, causes saltwater intrusion into coastal agricultural lands. This results in soil salinization, rendering the land unsuitable for crop production. In Bangladesh, rising sea levels have increased soil salinity in coastal regions, threatening food security for millions (Magnan et al., 2022). Coastal ecosystems such as mangroves are also at risk, reducing their ability to protect against further erosion and flooding.
- **Thawing Permafrost**: In Arctic regions, rising temperatures have led to permafrost thawing, destabilizing landscapes, and releasing stored greenhouse gases like methane and carbon dioxide. Thawed permafrost disrupts soil stability, contributing to erosion and altering hydrological systems, which impacts both natural ecosystems and infrastructure (Schuur et al., 2015).
- **Extreme Weather and Erosion**: Increasing frequency and intensity of extreme weather events, such as heavy rainfall and cyclones, have accelerated soil erosion. In tropical regions, intense rainfall events strip away nutrient-rich topsoil, reducing land productivity and causing sedimentation in rivers (Pimentel & Burgess, 2013). Cyclones in South Asia have caused significant coastal erosion, with long-term consequences for agriculture and settlements (IPCC, 2022).
- **Wildfires**: Rising temperatures and prolonged dry periods increase the frequency and intensity of wildfires, leading to loss of vegetation and soil fertility. In Australia, the 2019–2020 bushfires destroyed millions of hectares of forest, leaving soils exposed to erosion and reducing the carbon sequestration potential of these ecosystems (Kala, 2023).

2.4 Land Degradation: A Driver of Climate Change

Land degradation directly drives climate change by reducing the land's capacity to act as a carbon sink. When vegetation cover is lost due to deforestation, overgrazing, or unsustainable farming practices, the carbon stored in plant biomass and soils is released into the atmosphere. Hence, degraded soils lose their ability to retain organic carbon and further increase greenhouse gas emissions. Desertification, a severe form of land degradation, is particularly problematic due to its far-reaching impacts on ecosystems, livelihoods, and climate systems. This process is characterized by the persistent degradation of dryland ecosystems, which often results from unsustainable agricultural practices like overgrazing, deforestation, and climate variability. It diminishes soil fertility, reduces water retention capacity, and leads to the loss of vegetation cover, making affected regions more vulnerable to extreme weather events and less capable of supporting biodiversity. Furthermore, desertification exacerbates poverty and food insecurity for millions of people, particularly in arid and semi-arid regions, while also contributing to significant carbon emissions as degraded soils release stored CO_2 into the atmosphere. In arid and semi-arid regions, the loss of vegetation accelerates soil erosion and diminishes water availability, creating a cycle of degradation that amplifies climate impacts. For example, as soil dries out, it releases stored CO_2, contributing to the warming of the local climate. This warming exacerbates land degradation, creating a feedback loop that is difficult to break. The impact of land degradation on climate change is not limited to emissions alone. Degraded lands often experience altered albedo (reflectivity), which can affect local and regional climate systems. For instance, the loss of vegetation in tropical regions can lead to higher surface temperatures and changes in rainfall patterns, further intensifying the effects of climate change (Fig. 2.2).

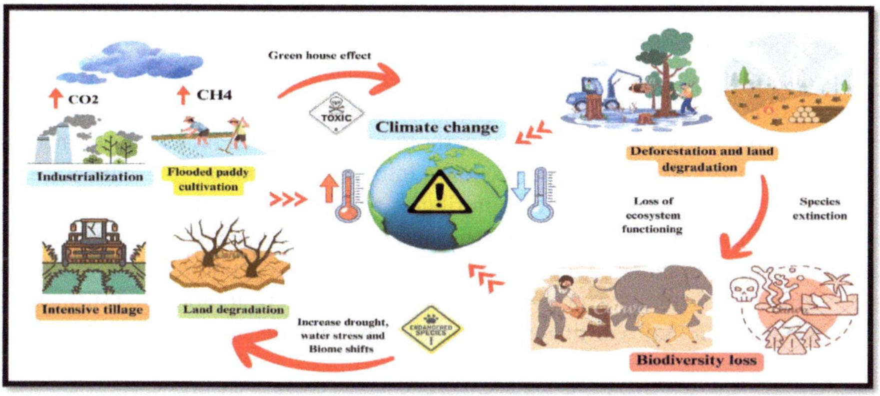

Fig. 2.2 Interactions between biodiversity, climate change, and land use

2.4.1 Land-Use Changes and Loss of Vegetation Cover Contributing to Carbon Emissions

Land-use changes, such as deforestation, agricultural expansion, and urbanization, significantly contribute to global carbon emissions. These changes often lead to a reduction in vegetation cover, which serves as a critical carbon sink. Forests and grasslands store vast amounts of carbon in their biomass and soil. When these ecosystems are cleared or degraded, the stored carbon is released into the atmosphere as carbon dioxide (CO_2), exacerbating climate change. For example, deforestation in tropical regions accounts for approximately 10% of global CO_2 emissions annually. Large-scale land clearing for agriculture, such as palm oil plantations or cattle grazing, not only reduces the carbon sequestration capacity of forests but also accelerates soil degradation. Furthermore, urban expansion often replaces vegetative land with impervious surfaces, which not only reduces carbon uptake but also contributes to the urban heat island effect. Efforts to mitigate these emissions include reforestation, afforestation, and the adoption of sustainable agricultural practices. These strategies aim to restore vegetation cover, enhance carbon storage, and reduce the overall impact of land-use changes on the climate.

2.4.2 Degraded Lands as Sources of Greenhouse Gases (GHGs)

Degraded lands, characterized by soil erosion, nutrient depletion, and loss of biodiversity, are significant sources of greenhouse gas emissions. When land becomes degraded, its ability to store carbon diminishes, leading to increased emissions of CO_2, methane (CH_4), and nitrous oxide (N_2O). For instance, agricultural practices on degraded soils often require more fertilizers to maintain productivity, which can result in higher N_2O emissions-a potent greenhouse gas. Peatlands are a prime example of ecosystems that, when degraded, become major sources of GHGs. These wetlands store immense quantities of carbon; however, when drained or burned for agriculture, they release this stored carbon into the atmosphere. In Southeast Asia, for instance, peatland degradation contributes significantly to global CH_4 and CO_2 emissions. Restoring degraded lands through techniques such as soil conservation, organic matter enrichment, and re-establishing native vegetation can help reverse these emissions. Moreover, policies that incentivize land restoration and sustainable land management play a critical role in mitigating climate change (Table 2.3).

Table 2.3 Key findings of some studies on land degradation: a driver of climate change

Study	Key findings	Reference
Greenhouse gas emissions from tropical forest degradation	Gross emissions from activities associated with tropical forest degradation are significant contributors to global GHG emissions	Pearson et al. (2017)
Peat land degradation and GHG emissions	Damaged peat lands contribute 4–5% of global GHG emissions. Rewetting 60% of degraded peatlands could transform them into net carbon sinks by 2100	Leifeld et al. (2019)
Global land use change and its impact on GHG emissions	Land use changes, including urbanization and deforestation, drive GHG emissions. Sustainable practices and forest restoration can mitigate these effects	Smith et al. (2017)
Deforestation and its role in climate change	Land use change, particularly deforestation, contributes 12–20% of global GHG emissions. Forest degradation, such as logging and burning, is significant	Harris et al. (2015)
Carbon emissions from degraded lands	Annual GHG emissions from degraded lands reached up to 4.4 billion tonnes of CO_2 from 2000 to 2009	IPCC (2019)

2.4.3 Case Studies Illustrating Feedback Loops Between Land Degradation and Climate Change

- **The Amazon Rainforest**: Once a major carbon sink, parts of the Amazon are now emitting more CO_2 than they absorb due to deforestation and wildfires. These activities degrade the forest ecosystem, reducing its capacity to sequester carbon. As the climate warms, the frequency and intensity of droughts and fires increase, further exacerbating deforestation and creating a feedback loop between land degradation and climate change.
- **The Sahel Region, Africa**: Land degradation in the Sahel, driven by overgrazing and unsustainable farming, has resulted in significant soil erosion and loss of vegetation. These changes reduce the land's Art ability to store carbon and retain water, contributing to desertification. As temperatures rise and rainfall patterns shift due to climate change, the region faces further land degradation, perpetuating the cycle.
- **Melting Permafrost in the Arctic**: Permafrost stores vast amounts of organic carbon. As global temperatures rise, permafrost thaws, releasing CO_2 and CH_4 into the atmosphere. This release accelerates climate change, which, in turn, leads to further permafrost thawing, illustrating a dangerous feedback loop.

2.5 Socio-economic Impacts of the Nexus

The interplay between land degradation and climate change represents one of the most pressing global challenges of our time. This nexus profoundly influences ecosystems, economies, and societies, exacerbating vulnerabilities in already fragile regions. This chapter explores the socio-economic impacts of this intricate relationship, offering insights into its far-reaching consequences and potential pathways for mitigation and adaptation. Land degradation, characterized by the loss of productivity and ecosystem services, often results from unsustainable land-use practices, deforestation, overgrazing, and urbanization. Climate change, marked by rising temperatures, altered precipitation patterns, and an increased frequency of extreme weather events, intensifies these processes. Conversely, degraded land releases stored carbon, contributing to greenhouse gas emissions and further accelerating climate change. The key dimensions of this nexus include:

- **Agricultural Productivity**: Climate change and land degradation directly reduce crop yields and livestock productivity. Rising temperatures can shorten growing seasons and increase evapotranspiration, thereby stressing water availability for crops. Soil erosion strips away nutrient-rich topsoil, diminishing fertility and requiring more intensive input use to maintain productivity. Salinization, a by-product of poor irrigation and climate variability, further degrades arable land. Livestock are also affected by decreased pasture quality and water scarcity, reducing milk and meat production. Smallholder farmers, especially in regions like sub-Saharan Africa and South Asia, are particularly vulnerable due to limited access to climate-resilient seeds, irrigation, and market infrastructure. These reductions in productivity contribute to increased food prices and heightened risk of hunger and poverty.
- **Economic Losses**: Declining agricultural output has cascading effects on broader economies, especially in nations where agriculture constitutes a large share of GDP. For instance, a 1% reduction in rainfall can result in up to a 5% decrease in GDP in highly agriculture-dependent countries. Beyond crop and livestock production, land degradation leads to the loss of critical ecosystem services such as carbon storage, pollination, and water purification, which are difficult and costly to replace. This degradation can undermine tourism, forestry, and fisheries, further straining national budgets. Infrastructure damage from climate-exacerbated disasters like floods and landslides adds to these costs, requiring expensive recovery and adaptation measures.
- **Migration and Displacement**: As productive land becomes uninhabitable due to degradation and climate impact, populations are forced to migrate. These climate migrants are often driven from rural areas to cities, increasing urban sprawl, unemployment, and strain on housing, healthcare, and public services. According to the World Bank, by 2050, over 216 million people could become internal climate migrants. In regions like the Sahel and parts of South Asia, competition over dwindling natural resources has already led to conflicts and social unrest.

Migration also disrupts traditional livelihoods, contributing to a loss of cultural heritage and intergenerational knowledge related to land management.
- **Public Health**: Land degradation and climate change intersect to create multiple public health crises. Dust storms arising from desertified regions lead to respiratory diseases such as asthma and bronchitis. Poor water quality resulting from soil erosion and chemical runoff contributes to waterborne diseases. Malnutrition becomes widespread as agricultural yields decline and food prices rise, particularly affecting children and pregnant women. Additionally, warmer temperatures and stagnant water from irrigation and floods promote the spread of vector-borne diseases like malaria and dengue. In regions lacking adequate healthcare infrastructure, these challenges severely overwhelm existing systems, increasing mortality and reducing overall well-being.

2.5.1 Socio-economic Inequalities

The nexus disproportionately impacts marginalized communities, deepening existing socio-economic inequalities. Indigenous populations, women, and low-income households often bear the brunt of these challenges due to their dependence on natural resources and limited adaptive capacities.

Gendered Impacts

- Women, often primary caregivers and agricultural workers in rural areas, experience heightened vulnerabilities as water and fuelwood become scarcer.
- Empowering women through education, access to resources, and participation in decision-making processes is crucial for building resilience.

Regional Disparities

- Sub-Saharan Africa, South Asia, and parts of Latin America are hotspots for land degradation and climate change impacts. These regions face compounded challenges, including poverty, weak governance, and limited infrastructure.

2.6 Strategies to Combat Land Degradation and Climate Change

Addressing land degradation and climate change is an urgent global mandate requiring integrated, multidimensional, and transformative strategies to safeguard the planet and its ecosystems. Land degradation undermines agricultural productivity, disrupts critical ecosystem services, and threatens biodiversity, while climate change intensifies extreme weather events, further stressing fragile land systems. These interconnected crises demand immediate, decisive, and comprehensive action. To tackle these challenges, a multidimensional approach is essential, incorporating cutting-edge technological advancements, large-scale sustainable land management, and resilient, forward-looking policy frameworks. Only through coordinated global efforts and unwavering commitment can we protect the environment and ensure the long-term survival and well-being of human societies. Below are categorized strategies—mitigation, adaptation, and integrated approaches—designed to address these pressing challenges.

2.6.1 Mitigation Strategies: Reducing Causes

Mitigation involves actions to address the root causes of climate change and land degradation, focusing on reducing greenhouse gas (GHG) emissions and restoring ecosystems. This encompasses strategies such as ecosystem restoration and nature-based solutions (NBS), with key approaches including:

Reforestation and Afforestation: Reforestation, the process of planting trees in deforested areas, improves soil fertility, prevents erosion, and enhances water retention, while afforestation, which involves establishing forests in previously unforested areas, creates carbon sinks that reduce atmospheric GHG levels. Both methods contribute to biodiversity conservation by providing habitats for wildlife, regulate hydrological cycles by minimizing runoff and improving groundwater recharge, and help stabilize microclimates through evapotranspiration and temperature regulation (Bastin et al., 2019).

Reforestation rejuvenates degraded land by increasing organic matter and supporting soil microbial activity, while afforestation combats desertification in arid regions. Notable initiatives, such as India's *Green India Mission* and China's *Great Green Wall*, have significantly reduced land degradation and increased carbon sequestration. Bastin et al. (2019) estimated that global reforestation efforts could sequester up to 205 gigatons of carbon, while Griscom et al. (2017) suggested that large-scale projects could offset about 30% of human-made emissions. Brazil's commitment to restoring 12 million hectares of degraded land under the Bonn Challenge highlights the global dedication to ecosystem restoration.

Wetland Restoration: Wetlands are vital ecosystems that regulate biodiversity, carbon storage, and flood control. However, global wetland areas have decreased by 35% since 1970, primarily due to land-use changes (Convention on Wetlands, 2021; Nath et al., 2024). Studies highlight that wetlands are widely recognized for their unparalleled ability to sequester carbon, with rates significantly exceeding those of any other terrestrial ecosystem globally. This remarkable capacity is attributed to several factors, including their high levels of above- and belowground biomass productivity, the anoxic (oxygen-depleted) conditions of wetland soils that slow down the decomposition of organic matter, and the frequent deposition of sediments that bury and preserve organic carbon. These unique characteristics not only enhance carbon storage but also ensure its long-term retention, making wetlands indispensable in mitigating climate change (Chmura et al., 2003; Lolu et al., 2020; Mitsch & Gosselink, 2015).

Grassland and Savanna Restoration: Grasslands, which support global livestock production and biodiversity, have suffered significant degradation, with approximately 50% affected by overgrazing and land-use changes. Restoration strategies such as rotational grazing and reintroducing native vegetation enhance soil health, increase water retention, and promote biodiversity (Conant et al., 2017). Research highlights that rotational grazing can increase soil organic matter, storing 10–30 Mg C ha^{-1}, while also preventing further degradation. Restoration efforts in Brazil's Cerrado savanna have transformed degraded farms into native ecosystems, boosting carbon sequestration and ecosystem resilience. Such initiatives demonstrate the potential of grassland restoration in mitigating climate change and enhancing land productivity.

Mangrove Conservation: Mangroves are highly effective carbon sinks, protecting coastlines from erosion, enhancing biodiversity, and storing significant amounts of carbon. Donato et al. (2011) emphasized that intact mangroves store four times more carbon per hectare than tropical forests. However, mangrove habitats face severe threats, with projections suggesting that over half could disappear by 2050 due to deforestation and degradation. Efforts to conserve and restore mangroves could prevent approximately 32 teragrams of CO_2 emissions annually between 2020 and 2030. Mangrove reforestation projects, which deliver higher blue carbon benefits than natural regeneration, underscore the importance of proactive conservation measures. These efforts not only address climate change but also support local livelihoods and biodiversity conservation.

2.6.2 *Adaptation Strategies: Building Resilience*

Adaptation focuses on managing and reducing the impacts of climate change and land degradation by enhancing the resilience of ecosystems and communities. The key adaptation strategies include: adopting Agroecological Practices, Climate-Resilient Agriculture, Efficient water management, Soil Restoration and Management, Diversification of Livelihoods, Strengthening Local Knowledge and Community Involvement, Policy Support and Capacity Building.

Agroecological Practices: Agroecology integrates ecological principles into agricultural practices, enhancing the resilience of farming systems. Techniques like intercropping, agroforestry, and crop diversification promote soil health, water retention, and pest control, reducing dependency on chemical inputs. These practices improve ecosystem services while building resilience to extreme weather events. For instance, agroforestry systems, which combine trees with crops or livestock, sequester carbon, reduce erosion, and buffer against climate variability (Glover et al., 2012).

Climate-Resilient Agriculture: Climate-resilient agriculture (CRA) encompasses strategies to strengthen the resilience of agricultural systems, ecosystems, and communities against the impacts of climate change and land degradation. Key practices include precision farming, conservation tillage, and the development of drought- and heat-resistant crop varieties. For example, no-till farming reduces soil disturbance, minimizing erosion and improving water retention. CRA also mitigates greenhouse gas emissions, conserves natural resources, and enhances adaptive capacity, ensuring food security, improving livelihoods, and sustaining productivity under increasingly harsh conditions. Notable innovations include the development of drought-tolerant maize (DTM) cultivars in Sub-Saharan Africa, such as WE 2101, WE 2103, and WE 2104, which have increased yields by 30% in drought-prone regions. Similarly, Southeast Asian farmers have adopted heat-resilient rice varieties developed by the International Rice Research Institute (IRRI), such as IR18C1004 and IR18C1005, which thrive under rising temperatures while reducing irrigation needs. These advances not only secure food supplies but also conserve water and enhance resilience to climate stressors.

Integrated Water Resource Management (IWRM): Efficient water management is essential for agriculture in regions vulnerable to drought and water scarcity. Practices such as rainwater harvesting, drip irrigation, and mulching reduce water loss and improve water-use efficiency. In India, rainwater harvesting systems have been widely adopted in arid areas, enhancing groundwater recharge and ensuring water availability for crops (Kerr, 2007).

Soil Restoration and Management: Healthy soils are critical for climate resilience and agricultural productivity. Restoring degraded soils through the application of organic matter, cover crops, and rotational grazing enhances soil fertility, water infiltration, and carbon storage. Soil management practices not only mitigate land degradation but also increase crop yields and reduce vulnerability to climate-related shocks (Lal, 2004).

Diversification of Livelihoods: Encouraging diversified agricultural livelihoods reduces the risks associated with climate change. Farmers can adopt integrated farming systems that combine crops, livestock, and aquaculture to create multiple income streams. In Vietnam, rice-fish farming systems have improved food security and reduced vulnerability to flooding (World Bank, 2018).

Strengthening Local Knowledge and Community Involvement: Indigenous knowledge and community-driven approaches are vital in building climate resilience. Local practices, such as traditional water conservation methods or pest management systems, often align with sustainable resource use. Empowering communities to participate in decision-making ensures that adaptation strategies are context-specific and sustainable (Leal Filho et al., 2022; Nyong et al., 2007).

Policy Support and Capacity Building: Supportive policies and capacity-building initiatives enable the widespread adoption of climate-resilient agricultural practices. Governments and international organizations must prioritize access to climate finance, extension services, and training programs for farmers. For instance, the Global Environment Facility (GEF) has funded projects that integrate climate adaptation into agricultural policies in developing countries.

2.6.3 *Integrated Mitigation and Adaptation Approaches*

To effectively combat land degradation and climate change, integrated strategies that combine both mitigation and adaptation are essential. These approaches work synergistically to reduce greenhouse gas emissions, restore ecosystems, and build resilience to climate impacts. Below are key integrated approaches:

Sustainable Land Management: Sustainable Land Management (SLM) incorporates practices that enhance land productivity while maintaining the ecological balance of the environment. SLM promotes soil conservation, water management, and biodiversity protection, addressing both climate change and land degradation. Through methods such as agroforestry, crop rotation, and conservation tillage, SLM reduces soil erosion, enhances carbon sequestration, and improves water retention. These practices not only mitigate climate change by capturing carbon in soils but also adapt agricultural systems to be more resilient to climate variability. For example, agroecological practices improve soil health and crop diversity, reducing vulnerability to extreme weather events like droughts or floods. By adopting SLM, farmers can restore degraded land and ensure long-term agricultural productivity, making it a key strategy for sustainable land use. In Ethiopia, the Tigray region adopted sustainable land management practices to combat desertification and restore degraded lands. Techniques such as terracing and water harvesting resulted in improved soil fertility, increased crop yields, and recharged groundwater resources (UNCCD, 2017).

Landscape Restoration: Landscape restoration focuses on rehabilitating degraded ecosystems at the landscape level, improving ecological functions, and reversing the effects of deforestation, desertification, and soil erosion. This approach involves large-scale interventions, such as reforestation, afforestation, wetland restoration, and the restoration of grasslands. Landscape restoration efforts enhance biodiversity, restore water cycles, and increase carbon sequestration. A prominent example of landscape restoration is the Great Green Wall in the Sahel region of Africa, which aims to combat desertification by planting trees and restoring degraded land. This ambitious project aims to restore 100 million hectares of degraded land by planting trees across 11 countries. By 2020, more than 20 million hectares had been restored, improving food security for millions and sequestering significant amounts of carbon (UNCCD, 2021). By increasing the capacity of landscapes to store carbon and regulate hydrological processes, landscape restoration serves as a dual strategy for climate change mitigation and adaptation. Restoring these ecosystems helps protect communities from climate impacts like flooding, while simultaneously capturing and storing carbon, reducing the overall carbon footprint.

Urban Greening Initiatives: Urban greening initiatives, such as green roofs, urban forests, and the creation of parks and green spaces, contribute to both climate change mitigation and adaptation in urban areas. These initiatives reduce the urban heat island effect, increase carbon sequestration, and improve air quality. Urban green spaces help manage stormwater, reduce flooding risks, and enhance biodiversity within cities. For instance, green roofs not only lower the temperature in cities but also provide insulation, improving energy efficiency and reducing reliance on air conditioning. Additionally, urban greening can enhance community resilience to climate change by providing safe spaces for recreation, supporting mental health, and strengthening social cohesion. Examples like Singapore's "Garden City" initiative and New York City's green roof program demonstrate how urban greening can create sustainable, climate-resilient cities. By integrating green infrastructure, cities can mitigate climate change impacts while adapting to rising temperatures and increased storm events (National Parks Board, Singapore, 2022; NYC Environmental Protection, 2023) (Fig. 2.3).

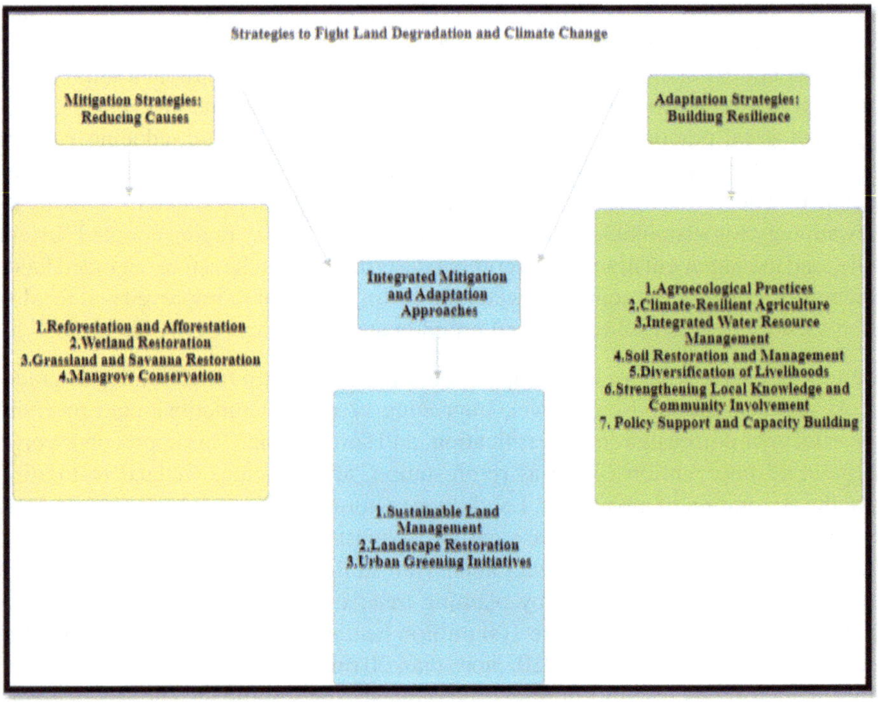

Fig. 2.3 Strategies to fight land degradation and climate change

2.7 Policy Interventions and Global Frameworks

Global frameworks and policy actions are essential for tackling the interconnected problems of climate change and land degradation. Governments, corporations, and communities can coordinate efforts to successfully alleviate these issues by putting comprehensive strategies into place. The United Nations' Land Degradation Neutrality (LDN) framework, the incorporation of the land-climate relationship into Nationally Determined Contributions (NDCs), and regional policies for sustainable land use and adaptation are important endeavours.

- **United Nations initiatives: the land degradation neutrality framework**

Adopted under the "United Nations Convention to Combat Desertification (UNCCD)", "Land Degradation Neutrality (LDN)" framework seeks to strike a balance between restoration, sustainable management techniques, and land degradation. More than 127 nations had pledged by 2022 to meet the LDN targets by 2030 (UNCCD, 2022). In order to sequester roughly 2.5–3 billion tons of CO_2 equivalent, India, for example, has committed to restoring 26 million hectares of degraded land by 2030. The LDN framework encourages nations to pinpoint areas of degradation, rank restoration efforts, and use scientific methods to track advancements. For instance, Niger's "Regreening the Sahel" program has improved agricultural yields by 15% and benefited 2.5 million people by restoring over 5 million hectares of land through farmer-managed natural regeneration (Reij et al., 2009).

- **Nationally Determined Contributions (NDCs) and the land-climate nexus**

A key element of the Paris Agreement, the "Nationally Determined Contributions (NDCs)", gives nations a way to include land management techniques in their plans for combating climate change. About 80% of NDCs that have been submitted as of 2023 make reference to land-based solutions, such as sustainable agriculture, afforestation, and reforestation (UNFCCC, 2023).

As part of their NDCs, nations like Brazil and Indonesia have placed a strong emphasis on halting deforestation and rehabilitating degraded forests. Brazil has committed to restoring 12 million hectares of forests by 2030 and ending illegal deforestation by 2028 in its updated NDC. Reforestation and renewable energy projects are also incorporated into Ethiopia's Climate-Resilient Green Economy strategy, which aims to reduce CO_2 emissions by 255 million tonnes per year by 2030.

2.7.1 Local and Regional Policies for Sustainable Land Use

Addressing the particular problems of climate change and land degradation in certain areas requires localized policy. Significant effects have been shown by policies that support community-based restoration, sustainable agriculture, and participatory land use planning. The Grain-for-Green Program (GGP) in China, for example, provides incentives to farmers who transform damaged croplands into grasslands or woods. The initiative prevented soil erosion by 25%, repaired 32 million hectares of land, and stored 1.2 billion tonnes of carbon by 2020 (Zhao et al., 2021). By 2030, the Great Green Wall project in 11 African nations hopes to repair 100 million hectares of degraded land. Over 20 million hectares have been restored so far, boosting biodiversity and generating 350,000 jobs (FAO, 2020). In order to counteract desertification, regional water management strategies like India's Jal Shakti Abhiyan also emphasize watershed development and rainwater collection. The program revitalized 100,000 water bodies by 2023, improving agricultural output and groundwater recharge.

2.7.2 Synergies with Global Goals

The Sustainable Development Goals (SDGs), especially SDG 13 (Climate Action) and SDG 15 (Life on Land), are greatly aided by global frameworks and policy actions. These policies tackle interrelated issues by combining land restoration with climate mitigation techniques, guaranteeing a sustainable future for everybody. Addressing the twin issues of climate change and land degradation requires effective policy interventions and international frameworks. Governments and stakeholders can develop sustainable solutions by utilizing programs such as the LDN framework, coordinating NDCs with land management plans, and putting localized policies into effect. These initiatives support global climate goals, improve livelihoods, and increase ecological resilience.

2.7.3 Recent Data and Global Commitments

- **Global soil degradation**: The FAO (2020) reported that 33% of global soil is degraded, affecting nearly 1.5 billion people and resulting in a $400 billion annual economic loss.
- **Carbon storage potential**: Ecosystem restoration under the UN Decade on Ecosystem Restoration aims to sequester 26 gigatons of CO_2 by restoring 350 million hectares of degraded land.
- **Renewable energy growth**: According to IRENA (2024), renewable energy capacity grew by 12% globally in 2023, with solar energy alone contributing 60% of the new capacity.

In order to combat land degradation and climate change, the climate induced 4 emphasizes the crucial interactions between socioeconomic activities, technological breakthroughs, and effective land restoration programs. Along with community-driven initiatives and historic projects like the Great Green Wall, it highlights important advancements like GIS, AI, and bioengineering (Table 2.4).

2.8 Conclusion

Land degradation and climate change are intricately linked and interdependent, posing one of the major concerns of the twenty-first century by negatively affecting ecosystems, biodiversity, food security, and human livelihoods worldwide. Land-use changes, degraded lands, and their interconnected feedback loops significantly contribute to global greenhouse gas emissions and climate change. Conversely, climate change exacerbates land degradation by increasing temperatures, altering rainfall patterns, and triggering extreme weather events like droughts and floods, which lead to soil erosion, desertification, and salinization. To address these issues, a multifaceted approach is required. Restoring vegetation, implementing sustainable land management practices, and promoting climate-smart agriculture can help reduce emissions and restore ecosystems. These efforts improve soil health, enhance carbon sequestration, and reduce the vulnerability of land to climate impacts. Policy interventions play a crucial role in mitigating land degradation and climate change. Governments must adopt strong legislative frameworks that encourage land restoration, the use of renewable energy, and the adoption of nature-based solutions. Global cooperation between governments, scientists, and local communities is essential to create effective, equitable strategies that protect both the environment and human well-being. By prioritizing resilience and equity, societies can reduce the adverse effects of land degradation and climate change. A holistic approach that combines sustainable land use practices, innovative solutions, and international collaboration can help to fight against land degradation and climate change, ensuring a sustainable future for all.

Table 2.4 Integrating technology, community engagement, and success stories in combating land degradation and climate change

Key focus areas	Way of implementation	Explanation and examples	References
Role of technology and innovation	Utilization of remote sensing and GIS for monitoring land degradation	Targeted restoration efforts are made possible by the use of GIS technologies to locate hotspots for deforestation and soil erosion. For instance, keeping an eye on the Amazon rainforest's destruction	Ahmad et al. (2021)
	Application of artificial intelligence (AI) in predictive modelling and risk assessment	Artificial intelligence plays a pivotal role in enhancing predictive modelling and risk assessment to combat land degradation and climate change. By leveraging big data and machine learning, AI enables early detection of environmental threats and supports proactive land management strategies. AI-based models help with proactive farming practices by forecasting drought events. AI-powered precision farming in the Midwest of the United States is one example	Dhal and Kar (2024)
	Bioengineering approaches for soil restoration and carbon capture	Bioengineering approaches offer sustainable and innovative solutions for soil restoration and carbon capture in the fight against land degradation and climate change. Techniques such as vegetative barriers, biochar application, and microbial soil enhancers improve soil structure, fertility, and water retention. Mycorrhizal fungi and biochar are utilized to capture carbon and improve soil fertility. For instance, the use of biochar in Indian agricultural areas. It promotes long-term ecosystem resilience and supports climate mitigation while restoring degraded landscapes	Liu et al. (2024)

(continued)

Table 2.4 (continued)

Key focus areas	Way of implementation	Explanation and examples	References
Socioeconomic dimensions and community engagement	Empowering indigenous communities in sustainable land management	Agroforestry is one example of an indigenous practice that supports sustainability objectives. For instance, Kenyan Maasai pastoralists are protecting their grazing grounds	Zhao et al. (2024)
	Education and capacity-building programs for farmers and stakeholders	Training courses and workshops raise awareness of sustainable practices. Sub-Saharan African farmer field schools are one example	Garima et al. (2024)
	Addressing climate migration and its link to land degradation	Policies lessen the effects on host regions and assist climate refugees. For instance, Bangladesh's climate refugee resettlement initiatives	Tejerina et al. (2024)
Success stories and case studies	Examples of successful land restoration projects (e.g., Great Green Wall, China's Loess Plateau)	In Africa, the Great Green Wall has improved livelihoods by restoring millions of hectares. For instance, China's agricultural output rose as a result of the restoration of the Loess Plateau	Cropper (2024); Zhao et al. (2021)
	Innovative local interventions combating climate-induced degradation	Community-driven solutions such as check dams reduce erosion and restore water tables. Example: Check dam projects in Rajasthan, India	Rackelmann et al. (2024)
Future directions and recommendations	Prioritizing investments in land restoration as a climate mitigation strategy	For long-term advantages, governments and organizations are investing in restoration projects. For instance, the focus on reforestation in the European Green Deal	Chaturvedi et al. (2024)
	Strengthening global and regional collaborations	Initiatives across borders address common issues like desertification. Sahara and Sahel Observatory collaborations are one example	Mukherji (2023)
	Integrating traditional ecological knowledge with modern practices	Efficiency is increased when modern technology is combined with traditional irrigation techniques. For instance, in Iran, Qanat systems are used with drip irrigation	Chen et al. (2024)

References

Ahmad, A., Ahmad, S. R., Gilani, H., Tariq, A., Zhao, N., Aslam, R. W., & Mumtaz, F. (2021). A synthesis of spatial forest assessment studies using remote sensing data and techniques in Pakistan. *Forests, 12*(9), 1211.

Bastin, J. F., Finegold, Y., Garcia, C., Mollicone, D., Rezende, M., Routh, D., Zohner, C. M., & Crowther, T. W. (2019). The global tree restoration potential. *Science, 365*(6448), 76–79.

Burezq, H. (2020). Combating wind erosion through soil stabilization under simulated wind flow condition-Case of Kuwait. *International Soil and Water Conservation Research, 8*(2), 154–163.

Chaturvedi, V., Ghosh, A., Garg, A., Avashia, V., Vishwanathan, S. S., Gupta, D., Sinha, N. K., Bhushan, C., Banerjee, S., Datt, D., & Bansal, J. (2024). India's pathway to net zero by 2070: Status, challenges, and way forward. *Environmental Research Letters, 19*(11), Article 112501.

Chen, Y., Yu, P., Wang, L., Chen, Y., & Chan, E. H. W. (2024). The impact of rice-crayfish field on socio-ecological system in traditional farming areas: Implications for sustainable agricultural landscape transformation. *Journal of Cleaner Production, 434*, Article 139625.

Chmura, G. L., Anisfeld, S. C., Cahoon, D. R., & Lynch, J. C. (2003). Global carbon sequestration in tidal, saline wetland soils. *Global Biogeochemical Cycles, 17*(4).

Conant, R. T., Cerri, C. E., Osborne, B. B., & Paustian, K. (2017). Grassland management impacts on soil carbon stocks: A new synthesis. *Ecological Applications, 27*(2), 662–668.

Convention on Wetlands. (2021). Global Wetland Outlook: Special Edition 2021. Gland, Switzerland: Secretariat of the Convention on Wetland. https://www.global-wetland-outlook.ramsar.org/report-1

Cropper, J. (2024). 'Growing a World Wonder': The Great Green Wall and the history of environmental decline in the Sahel, 1450–2022. *Environment and History, 30*(2), 291–313.

Dhal, S. B., & Kar, D. (2024). Transforming agricultural productivity with AI-Driven forecasting: Innovations in food security and supply chain optimization. *MDPI Forecasting, 6*(INL/JOU-24-81560-Rev000).

Donato, D. C., Kauffman, J. B., Murdiyarso, D., Kurnianto, S., Stidham, M., & Kanninen, M. (2011). Mangroves among the most carbon-rich forests in the tropics. *Nature Geoscience, 4*(5), 293–297.

Eni, I. (2012). Effects of land degradation on soil fertility: A case study of Calabar South, Nigeria. *Environmental Land Use Planning, 18*(1), 166–171.

FAO. (2020). Action against desertification: Desertification and land degradation. Food and Agricultural Organization of the United Nations (FAO).

Garima, Olaru, D., Smith, B., & Siddique, K. H. (2024). Decision-making in grain supply chains for sustainable food system. *Sustainability, 16*(24), 10816.

Gibbs, H. K., & Salmon, J. M. (2015). Mapping the world's degraded lands. *Applied Geography, 57*, 12–21. https://www.researchgate.net/deref/. https://doi.org/10.1016/j.apgeog.2014.11.024

Glover, J. D., Reganold, J. P., & Cox, C. M. (2012). Plant perennials to save Africa's soils. *Nature, 489*(7416), 359–361.

Grinand, C., Bégué, A., Montfort, F., Leroux, L., & Randrianarivo, T. (2019). An analysis of land use changes and land degradation in Mozambique. *NITIDAE & CIRAD, Maputo.*

Griscom, B. W., Adams, J., Ellis, P. W., Houghton, R. A., Lomax, G., Miteva, D. A., Schlesinger, W. H., Shoch, D., Siikamäki, J. V., Smith, P., & Woodbury, P. (2017). Natural climate solutions. *Proceedings of the National Academy of Sciences, 114*(44), 11645–11650.

Harris, Z. M., Spake, R., & Taylor, G. (2015). Land use change to bioenergy: A meta-analysis of soil carbon and GHG emissions. *Biomass and Bioenergy, 82*, 27–39.

IPCC. (2019). *Final government distribution: Land degradation (IPCC SRCCL).* https://www.ipcc.ch/site/assets/uploads/2019/08/2e.-Chapter-4_FINAL.pdf

IRENA. (2024). Renewable capacity statistics 2024. In *International renewable energy agency, Abu Dhabi*, 2024. [Online]. Available: https://www.irena.org/-/media/Files/IRENA/Agency/Publication/2024/Mar/IRENA_RE_Capacity_Statistics_2024.pdf?rev=a587503ac9a2435c8d13e40081d2ec34

Intergovernmental Panel on Climate Change (IPCC). (2022). Climate Change 2022 – Impacts, adaptation and vulnerability: *Working group II contribution to the sixth assessment report of the intergovernmental panel on climate change* (1st ed.). Cambridge University Press. https://doi.org/10.1017/9781009325844

International Water Management Institute (IWMI). (2010). *Water-Smart Agriculture in Ethiopia.* https://www.iwmi.org/About_IWMI/Strategic_Documents/Annual_Reports/2010/Annual_Report_2010.pdf

Jiang, K., Teuling, A. J., Chen, X., Huang, N., Wang, J., Zhang, Z., Gao, R., Men, J., Zhang, Z., Wu, Y., & Cai, L. (2024). Global land degradation hotspots based on multiple methods and indicators. *Ecological Indicators, 158*, Article 111462. https://doi.org/10.1016/j.ecolind.2023.111462

Kala, C. P. (2023). Environmental and socioeconomic impacts of forest fires: A call for multilateral cooperation and management interventions. *Natural Hazards Research, 3*(2), 286–294.

Kerr, J. (2007). Watershed management: Lessons from common property theory. *International Journal of the Commons, 1*(1), 89–109.

Lal, R. (2004). Soil carbon sequestration impacts on global climate change and food security. *Science, 304*(5677), 1623–1627.

Leal Filho, W., Barbir, J., Gwenzi, J., Ayal, D., Simpson, N. P., Adeleke, L., Tilahun, B., Chirisa, I., Gbedemah, S. F., Nzengya, D. M., & Sharifi, A. (2022). The role of indigenous knowledge in climate change adaptation in Africa. *Environmental Science & Policy, 136*, 250–260.

Leifeld, J., Wüst-Galley, C., & Page, S. (2019). Intact and managed peatland soils as a source and sink of GHGs from 1850 to 2100. *Nature Climate Change, 9*(12), 945–947. https://doi.org/10.1038/s41558-019-0615-5

Liu, Z., Chen, Z., Wang, Z., Zhang, D., & Ni, B. J. (2024). Recycling and repurposing of waste carbon nanofiber polymers: A critical review. *Environmental Science: Nano.*

Lolu, A. J., Ahluwalia, A. S., Sidhu, M. C., Reshi, Z. A. & Mandotra, S. K. (2020). Carbon sequestration and storage by wetlands: Implications in the climate change scenario. *Restoration of wetland ecosystem: A trajectory towards a sustainable environment*, pp.45–58.

Magnan, A. K., Oppenheimer, M., Garschagen, M., Buchanan, M. K., Duvat, V. K., Forbes, D. L., Ford, J. D., Lambert, E., Petzold, J., Renaud, F. G., & Sebesvari, Z. (2022). Sea level rise risks and societal adaptation benefits in low-lying coastal areas. *Scientific Reports, 12*(1), 10677.

Mani, S., Osborne, C. P., & Cleaver, F. (2021). Land degradation in South Africa: Justice and climate change in tension. *People and Nature, 3*(5), 978–989.

Masson-Delmotte, V., Pörtner, H. O., Skea, J., Buendía, E. C., Zhai, P., & Roberts, D. (2019). Climate change and land. *IPCC Report.* https://www.ipcc.ch/srccl/chapter/chapter-4/

Mitsch, W. J., & Gosselink, J. G. (2015). *Wetlands.* Wiley.

Mukherji, A. (2023). *Climate Change 2023 synthesis report.*

Nath, A. J., Sileshi, G. W., Bania, J. K., & Nath, A. (2024). Threats to inland wetlands and uncertainty around global soil carbon stocks and sequestration rates. *Science of the Total Environment, 955*, Article 177190.

Nguyen, T. T., Grote, U., Neubacher, F., Do, M. H., & Paudel, G. P. (2023). Security risks from climate change and environmental degradation: Implications for sustainable land use transformation in the Global South. *Current Opinion in Environmental Sustainability, 63*, Article 101322.

NYC Environmental Protection. (2023). *Green Roof Tax Abatement Program.* https://nyc-business.nyc.gov/nycbusiness/description/green-roof-property-tax-abatement-program/about

Nyong, A., Adesina, F., & Osman Elasha, B. (2007). The value of indigenous knowledge in climate change mitigation and adaptation strategies in the African Sahel. *Mitigation and Adaptation Strategies for Global Change, 12*, 787–797.

Oldeman, L. R., Hakkeling, R. T. A., & Sombroek, W. G. (1990). *World map of the status of human-induced soil degradation: an explanatory note* (27-pp.).

Olsson, L., Barbosa, H., Bhadwal, S., Cowie, A., Delusca, K., Flores-Renteria, D., Hermans, K., Jobbagy, E., Kurz, W., Li, D., & Sonwa, D.J., 2019. Land degradation: IPCC special report on climate change, desertification, land 5 degradation, sustainable land management, food security, and 6 greenhouse gas fluxes in terrestrial ecosystems. In *IPCC special report on climate change, desertification, land 5 degradation, sustainable land management, food security, and 6 greenhouse gas fluxes in terrestrial ecosystems* (p. 1). Intergovernmental Panel on Climate Change (IPCC). https://www.ipcc.ch/site/assets/uploads/sites/4/2019/11/07_Chapter-4.pdf

Ontel, I., Cheval, S., Irimescu, A., Boldeanu, G., Amihaesei, V. A., Mihailescu, D., Nertan, A., Angearu, C. V., & Craciunescu, V. (2023). Assessing the recent trends of land degradation and desertification in Romania using remote sensing indicators. *Remote Sensing, 15*(19), 4842. https://doi.org/10.3390/rs15194842

Pearson, T. R., Brown, S., Murray, L., & Sidman, G. (2017). Greenhouse gas emissions from tropical forest degradation: An underestimated source. *Carbon Balance and Management, 12*, 1–11.

Pimentel, D., & Burgess, M. (2013). Soil erosion threatens food production. *Agriculture, 3*(3), 443–463.

Prince, S., Von Maltitz, G., Zhang, F., Byrne, K., Driscoll, C., Eshel, G., Kust, G., Martínez-Garza, C., Metzger, J. P., Midgley, G., & Moreno-Mateos, D. (2018). Status and trends of land degradation and restoration and associated changes in biodiversity and ecosystem functions. *Intergovernmental Science-Policy Platform on Biodiversity and Ecosystem Services (IPBES).*

Qiang-guo, C. A. I. (2001). Soil erosion and management on the Loess Plateau. *Journal of Geographical Sciences, 11*, 53–70.

Rackelmann, F., Sparkes, E., Sabino Siemons, A. S., Hashweh, D., Pineda Fernandez, D. M., Werners, S.E., Orr, B. J., Andreeva, O., & Walz, Y. (2024). Promoting synergies between land degradation neutrality and climate change adaptation: A supplement to the National Adaptation Plan technical guidelines. United Nations Convention to Combat Desertification (UNCCD).

Reij, C., Tappan, G., & Smale, M. (2009). Agroenvironmental transformation in the Sahel: Another kind of" Green Revolution". Intl Food Policy Res Inst.

Scholes, R., Montanarella, L., Brainich, A., Barger, N., ten Brink, B., Cantele, M., Erasmus, B., Fisher, J., Gardner, T., Holland, T. G., & Kohler, F. (2018). Summary for policymakers of the assessment report on land degradation and restoration of the Intergovernmental Science-Policy Platform on Biodiversity and Ecosystem Services.

Schuur, E. A., McGuire, A. D., Schädel, C., Grosse, G., Harden, J. W., Hayes, D. J., Hugelius, G., Koven, C. D., Kuhry, P., Lawrence, D. M., & Natali, S. M. (2015). Climate change and the permafrost carbon feedback. *Nature, 520*(7546), 171–179.

Singapore National Parks Board 2022. *Garden City Movement*. https://www.nparks.gov.sg/portals/annualreport/garden-city-fund.html

Smith, R. M., Kaushal, S. S., Beaulieu, J. J., Pennino, M. J., & Welty, C. (2017). Influence of infrastructure on water quality and greenhouse gas dynamics in urban streams. *Biogeosciences, 14*(11), 2831–2849.

Tamene, L., & Le, Q. B. (2015). Estimating soil erosion in sub-Saharan Africa based on landscape similarity mapping and using the revised universal soil loss equation (RUSLE). *Nutrient Cycling in Agroecosystems, 102*, 17–31. https://doi.org/10.1007/s10705-10015-19674-10709

The Energy and Resources Institute (TERI). (2016). Vol I: Macroeconomic assessment of the costs of land degradation in India. Economics of desertification, land degradation and drought in India. Prepared for Ministry of Environment, Forest and Climate Change New Delhi

Tejerina, B., de Almeida, C. M., & Acuña, C. (2024). *Climate change, sustainability and socio-ecological practices.*

UNCCD. (2017). *The Global Land Outlook*. https://www.unccd.int/resources/global-land-outlook/glo1

UNCCD. (2019). Land climate: Key message from the IPCC special report on climate change and land by the United Nations Convention to Combat Desertification

UNCCD. (2021). *The Great Green Wall Implementation Status and Way Ahead to 2030.* https://www.unccd.int/resources/publications/great-green-wall-implementation-status-and-way-ahead-2030.

UNCCD. (Ed.). (2022). *United Nations convention to combat desertification–Global land outlook* (2nd ed.). UNCCD secretariat.

UNFCCC. (2023). *Global stocktake information portal* [online]. UNFCCC. Accessed April 2023. Available from: https://unfccc.int/topics/global-stocktake/information-portal

Wang, J., Zhen, J., Hu, W., Chen, S., Lizaga, I., Zeraatpisheh, M., & Yang, X. (2023). Remote sensing of soil degradation: Progress and perspective. *International Soil and Water Conservation Research, 11*(3), 429–454.

World Bank. (2018). *Promoting climate-resilient agriculture in Vietnam.*

Zhao, Q., Yu, L., & Chen, X. (2024). Land system science and its contributions to sustainable development goals: A systematic review. *Land Use Policy, 143*, Article 107221.

Zhao, X., Taheripour, F., Malina, R., Staples, M. D., & Tyner, W. E. (2021). Estimating induced land use change emissions for sustainable aviation biofuel pathways. *Science of the Total Environment, 779*, Article 146238.

Part II
Climate Change Impacts on Landscapes Ecosystem and Land Degradation

Chapter 3
Impact of Climate Change and Land Degradation on Agriculture

Aastha Dubey, Prasun Kumar Singh, Shalini Singh, Uzma Manzoor, Chinmaya Sahoo, and Ankit Saini

Abstract Two interconnected global issues that have a significant impact on agricultural output, food security, and rural livelihoods are land degradation and climate change. This chapter provides a thorough analysis of how land degradation processes like deforestation, soil erosion, nutrient depletion, and salinisation interact with climatic variability, which includes rising temperatures, changed rainfall patterns, extended droughts, and extreme weather events that make agroecosystems more vulnerable. The chapter demonstrates how various climatic stressors and unsustainable land management practices combine to exacerbate soil fertility loss, water scarcity, and food insecurity across a range of agro-ecological settings. It does this by drawing on regional and global data, as well as case studies from the Jaggar Watershed in India and the Mount Elgon watershed in Uganda. This evaluation emphasizes the interrelated effects of environmental and socioeconomic factors on agricultural sustainability by combining global trends with local experiences. In vulnerable farming systems, it identifies scalable, context-sensitive land and water management techniques that have proven successful in reducing degradation and boosting resilience. The chapter emphasizes the increasing importance of precision agriculture, remote sensing technologies, and involving multiple stakeholders in building long-lasting resilience in farming systems, while also reiterating the key roles of climate-smart agriculture, sustainable land management, and community-based efforts. By providing research-backed comparisons that help shape both global policies and local plans for climate-friendly and environmentally sustainable food systems, the insights contribute to the worldwide discussion on sustainable agriculture and land management.

A. Dubey · P. K. Singh · S. Singh · U. Manzoor
School of Agriculture, Galgotias University, Greater Noida, Uttar Pradesh, India

C. Sahoo (✉)
Department of Agronomy, Dr. Rajendra Prasad Central Agricultural University, Pusa, Samastipur, Bihar, India

A. Saini
Department of Agronomy, Dr. KSG Akal College of Agriculture, Eternal University, Baru Sahib, Sirmaur, Himachal Pradesh, India

U. Chatterjee et al. (eds.), *Eco-Resilience: Climate Change, Land Degradation and Sustainable Solutions*, Sustainability Solutions,
https://doi.org/10.1007/978-3-032-00708-7_3

Keywords Climate change · Land degradation · Food security · Sustainable agriculture · Eco-resilience

3.1 Introduction

Agriculture has long been the foundation of human civilization, providing life-sustaining foods, raw materials, and livelihoods for billions worldwide. Therefore, the sustainability of agriculture especially relies on consistent climatic patterns and productive land resources. Recently, climate change—reflected in higher temperatures, erratic rainfall patterns, extended droughts, and severe natural disasters—has put unprecedented strain on agricultural systems. At the same time, land degradation from soil erosion, nutrient depletion, salinization, deforestation, and unsustainable land use has accelerated, together compromising food security, harming ecosystem services, and endangering rural livelihoods. The IPCC (2014) claims that especially in ecologically sensitive and socio-economically marginal regions, these synergistic environmental factors have significantly increased the vulnerability of agricultural output. With rainfed and smallholder systems in arid and semi-arid areas particularly vulnerable, agriculture stays one of the most climate-sensitive sectors. Rising agricultural failures in Sub-Saharan Africa, South Asia, and Southeast Asia have been mostly connected to erratic rainfall patterns, deteriorating soil quality, and water shortages (Carroll et al., 2018). With significant socio-economic consequences, recent projections indicate that more than 33% of the world's arable land has already been degraded by unsustainable land management, overgrazing, and deforestation (WMO, 2022). The combined effects of climate change and land degradation have disturbing consequences for world food systems, natural resource sustainability, and rural livelihoods, given that the world population is expected to exceed 9.7 billion by 2050 (UN, 2019).

These connected issues show that we urgently need combined strategies that include climate-resilient agriculture, sustainable land management, and supportive policies. Restoring degraded landscapes has been highlighted internationally as a key action for climate change mitigation and adaptation, with growing focus under world environmental governance systems, including the UNFCCC and UNCCD. Acknowledging this, it is vital to assess how the interrelated processes of land degradation and climate change jointly endanger the sustainability of agroecosystems and the well-being of rural people depending on them.

The current chapter addresses this need by means of a globally comparative, regionally contextualized review backed by chosen case study data, providing a thorough synthesis of the combined impacts of these stressors. It investigates how land degradation and climate change interact to influence soil health, water resources, agricultural production, and rural livelihoods by means of systematic study of literature, global trends, and regionally particular experiences. The chapter emphasises scalable, sustainable land and water management techniques appropriate for various agro-ecological settings by identifying both universal patterns and context-dependent differences. It also describes adaptive, evidence-based suggestions for encouraging agricultural resilience and sustainable land management under shifting climate conditions.

This objectives of the chapter are: (i) to evaluate the combined effects of climate change and land degradation on agricultural output, soil health, water resources, and rural livelihoods; (ii) to identify and analyze the environmental, climatic, and socio-economic drivers contributing to land degradation and agricultural vulnerability globally and regionally; (iii) to assess the consequences of these combined challenges for food security, ecosystem sustainability, and economic stability in vulnerable agroecosystems; (iv) to investigate and recommend sustainable, eco-resilient land and water management practices and agricultural strategies to address these challenges; and (v) to provide evidence-based, comparative insights to guide international policy discussions and sustainable land management interventions.

This chapter provides insightful analysis for improving agricultural sustainability and resilience by placing regional experiences in a worldwide context and stressing both common vulnerabilities and different adaptive responses. Its inclusive, comparative viewpoint—blending global trends with regionally nuanced analysis—complements continuous international initiatives to tackle the converging problems of land degradation and climate change while pointing out practical solutions for food and livelihood security in vulnerable areas all around.

3.2 Materials and Methods

This study adopted a multi-scale, integrative methodology involving an extensive literature review of global and regional research, supported by illustrative case studies. Presented in Fig. 3.1, the methodological framework emphasizes the sequential steps of evidence collection, comparative analysis, thematic synthesis, and identification of scalable adaptive solutions pertinent for various agro-ecological settings.

3.3 Understanding Climate, Climate Change, and Their Role in Agriculture

Climate forms the foundation of agricultural systems by influencing temperature, precipitation, solar radiation, and wind patterns, all of which directly impact crop growth and productivity. While often used interchangeably, climate and weather are distinct phenomena. Weather encompasses specific atmospheric conditions such as temperature, pressure, humidity, wind speed and direction, precipitation, heat waves, snowfall, cloud cover, and sunshine, observed at a particular point (Kumar, 2019). On

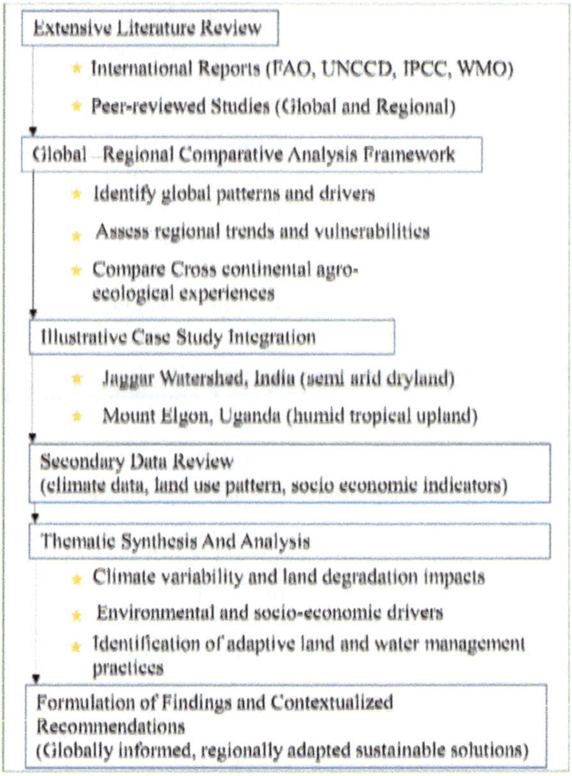

Fig. 3.1 This methodological framework shows the integrative, global-to-regional approach used to synthesize data on the combined effects of land degradation and climate change on agriculture

the other hand, climate refers to the average weather patterns over a prolonged period, typically spanning about 30 years. Persistent changes in climate over several decades or longer, driven by natural factors or human activities, are referred to as climate change (Solomon et al., 2007). Climate change encompasses rising temperatures, shifts in rainfall cycles, sea level rise, saltwater intrusion, and the occurrence of floods and droughts. Globally, these climatic shifts adversely affect agriculture, ultimately disrupting the global food supply (Pathak et al., 2012). Variability and changes in climate pose significant challenges to agricultural production, including both annual and perennial horticultural crops.

Climate change, as defined by UNFCCC, is "a change that is attributed directly or indirectly to human activity which alters the composition of the global atmosphere and that is in addition to natural climate variability observed over comparable times". Climate change can also be described as long-term alterations in temperature and weather patterns, primarily driven by human activities like greenhouse gas emissions and deforestation, which have progressively disrupted the stability of these climatic factors. Agriculture is a significant contributor to climate change, accounting for approx. 20% of the annual rise in anthropogenic greenhouse gas emissions (IPCC, 1995). The agricultural system has grown increasingly complex

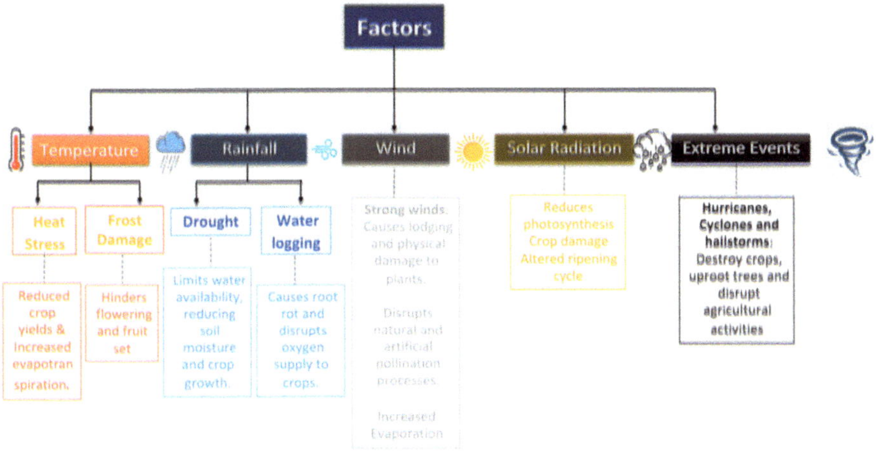

Fig. 3.2 Climate factors impacting agriculture

due to escalating temperatures, shifting precipitation patterns, and the heightened occurrence of extreme weather events.

For instance, changes in rainfall patterns induced by climate change may cause either drought or flood, which directly affects crop growth, though periodic rainfall is responsible for irrigation and preservation of soil moisture. High temperatures and heat waves can also interfere with the stages of crop growth, decreasing yields. The interplay between traditional climatic roles and emerging challenges due to climate change highlights the evolving dynamics that agriculture must adapt to. Understanding these changes is crucial to mitigate risks and ensure food security in a changing climate. Figure 3.2 depicts the factors impacting climate change.

3.3.1 Temperature

Temperature is one of the most critical climatic factors for agriculture, as it plays a direct role in the growth, development, and productivity of all crops. It influences essential physiological processes such as photosynthesis, respiration, transpiration, germination, and flowering. Optimal temperature ranges are therefore crucial for maximizing the growth and yield of staple crops like wheat and rice, which sustain a significant portion of the global population. However, deviations from these optimal ranges, particularly prolonged heat stress, can substantially reduce crop yield and quality.

3.3.2 Rainfall

Rainfall is also a significant climatic parameter affecting agriculture (Goswami et al., 2006). It is the only source of water in rainfed agriculture; thus, its distribution, intensity, and time of occurrence have significant importance for determining planting and harvesting schedules. When there is enough rainfall, it facilitates crop growth; excess rain leads to waterlogging and soil erosion and may leach nutrients from the soil, while there may be drought stress or complete crop failure due to insufficient rainfall.

Rainfall is a primary climatic parameter affecting agriculture by determining soil moisture, groundwater recharge, and crop scheduling (Goswami et al., 2006). In rainfed systems, the distribution, intensity, and timing of rainfall govern sowing, growth, and harvest. Adequate rainfall supports healthy crop growth, while excess leads to waterlogging, and insufficient rainfall results in drought stress. Humidity, being directly related to rainfall, affects evapotranspiration rates, pest and disease incidence, and post-harvest storage conditions. Particularly in rainfed systems, the variability and intensity of rainfall affect soil erosion, plant nutrient availability, pest infestation, and crop yields. Geneti (2019) claims that water stress during important growth phases lowers light interception and photosynthetic activity, therefore requiring consistent and well-distributed rain to preserve crop production.

Equally important is the number of rainy days during the growing period, which can be as decisive as the total seasonal rainfall in determining crop production (Geneti, 2019). Rainfall and humidity, by their effect on these interrelated processes, make their contributions indispensable for sustaining the agricultural system and optimizing crop performance.

3.3.3 Wind

Wind plays an important role in agriculture, affecting various processes like pollination and seed dispersal, as well as evaporation. Even though soft breezes may bring many positive results, stronger winds often lead to some unwanted outcomes: crop lodging, greater soil erosion, and more harm to infrastructure like greenhouses and irrigation systems. The patterns of seasonal winds—often known as trade winds and monsoons—also have their impacts on the agricultural calendar of different areas. While wind offers significant benefits to farming, its strength poses serious challenges. Strong winds damage crops, worsen soil erosion, and cause desiccation, which can lower agricultural productivity or even force farmers out of their land in extreme cases (Stigter et al., 2005). Such challenges can be very effectively overcome through strategies like planting trees for the creation of windbreaks and shelterbelts.

3.3.4 Solar Radiation

Solar radiation has a crucial role in agriculture and crop physiology because it is the major energy input for photosynthesis, plant growth, and development (Sivakumar, 2023; Villa et al., 2022) and hence directly determines crop growth rates and biomass production. Changes in radiation levels, caused by factors like cloud cover and atmospheric conditions, alter the duration of growing seasons and, therefore, the potential yield of crops. The radiation climate is an important element in agricultural meteorology that affects plant transpiration rates. Crop canopies impose certain interactions with solar radiation via mechanisms like reflection and transmission that depend on factors such as leaf area index and leaf distribution (Monteith, 1965). For maximum yields, it is important to increase and prolong the period of foliar interception of solar radiation (Villa et al., 2022). These interactions must be understood to enhance crop yields and develop efficient methods in agriculture.

3.4 Land Degradation

Land degradation, a complex global issue that impacts the environment, agriculture, and society, significantly impacts sustainability and human welfare (Prokop, 2020). Broadly speaking, it is a drop in the biological or economic output of land caused by processes including soil erosion, desertification, deforestation, and pollution (UNCCD, 2014). Various landscapes, including croplands, pastures, rangelands, and forested and urbanized areas, are impacted by this degradation, which causes soil fertility loss, water cycle disturbance, lower carbon sequestration, biodiversity loss, and negative consequences on ecological balance and livelihoods.

The decline of essential natural resources like soil, water, plants, and air results from both human-induced and natural causes (Prokop, 2020). Key processes are desertification, deforestation, organic matter loss, and soil erosion, which together diminish food security, livelihoods, biodiversity, and ecosystem services (Abdel-Rahman, 2023; Mahata & Sharma, 2021). Affecting more than 2.6 billion people in more than 100 nations, this common problem has the most serious consequences in Asia and Africa (Barman et al., 2013). Figure 3.3 is depicting the process of land degradation.

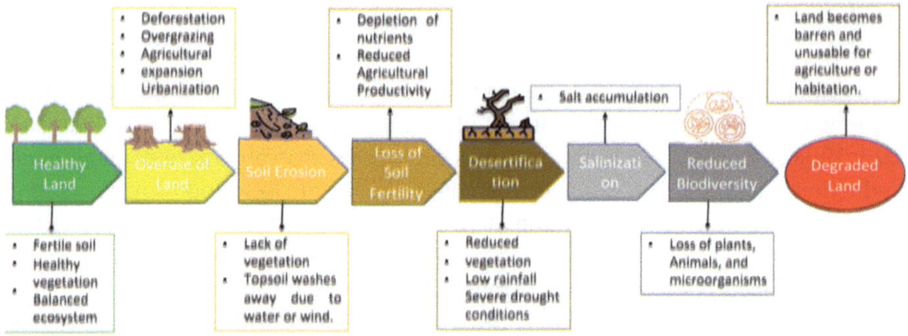

Fig. 3.3 Process of land degradation

3.5 Causes of Land Degradation

3.5.1 Soil Erosion

Water as well as wind drives erosion of soil, which is one of the most widespread and vital factors for land degradation that causes loss of soil, organic matter, and nutrients. High rainfall and runoff primarily cause water erosion by removing nutrient-enriched topsoil. This is particularly critical in sloping areas and areas not covered with adequate vegetation because of the absence of protective plant cover, which accelerates the loss of soil. Wind erosion prevails in the arid and semi-arid regions, with strong winds removing and transporting loose dry soil particles. This results in decreased soil fertility, diminished agricultural output, and sediment deposition in waterways, which adversely impacts ecosystems and farming activities.

The impacts of soil erosion worldwide vary but remain profound. Therefore, it's time to apply direct conservation strategies and sustainable land management solutions geared toward ameliorating the effects of soil erosion and restoring land productivity. Table 3.1 shows the region-wise causes of soil erosion.

3.5.2 Overgrazing and Deforestation

Overgrazing and deforestation are primary drivers of land degradation, contributing to soil erosion, desertification, and biodiversity loss. Overgrazing by livestock removes vegetation cover, leaving the soil vulnerable to erosion and compaction. This degradation extends beyond pasturelands, exacerbating desertification in arid and semi-arid regions, where vegetation plays a vital role in retaining water and stabilizing the soil. Studies in Greece, China, and Mongolia showed the consequences of overgrazing, which include increased soil erosion, loss of plant cover, and lower biomass productivity (Cao et al., 2013; Hilker et al., 2014; Kairis et al.,

Table 3.1 Region-wise causes of soil erosion

S.No.	Region	Type of erosion	Causes	References
1	Nepal	Water erosion	Intense rainfall, conventional tillage, poor soil structure, and steep slopes	Chalise et al. (2019)
2	Sub Saharan Africa	Wind erosion	Harsh climate, fragile soils, and steep terrains	Kiage (2013)
3	Ethiopia	Water erosion	Rapid population growth, deforestation, overgrazing, and inadequate conservation practices	Tadesse and Hailu (2024)
4	China	Water and wind erosion	Complex landforms, erodible soils, intense rainfall, and human activity	Zheng and Wang (2014)
5	Northeastern India	Water erosion	Steep terrain, heavy rainfall, and unsustainable land use practices like Jhum cultivation	Choudhury et al. (2022)

2015). For example, in Mongolia, it is estimated that the increased livestock population accounts for an increase in vegetation loss of 80% (Hilker et al., 2014). The dynamics of overgrazing are complex and often closely linked with climatic variables such as drought events and changes in precipitation levels (Middleton, 2018). Some researchers believe that policy measures like the privatization of grazing rights might accelerate land degradation more significantly than overgrazing itself does (Cao et al., 2013).

Furthermore, clearing forests for timber, fuel, and agriculture reduces water retention and nutrient-holding capabilities since it removes essential tree cover. Without vegetation, soil erosion becomes more severe, local water cycles are disrupted, and biodiversity suffers. Deforestation stands out as a leading cause of land degradation around the world. As noted by Hosonuma et al. in 2012, the main drivers include both commercial and subsistence agriculture, while other contributing factors are logging, fuelwood collection, timber harvesting, and uncontrolled fires, all of which escalate the rate of destruction. Every year, 10 million hectares of forest cover are destroyed in the world, with the trend expected to remain constant until at least 2025 (Worku, 2023).

3.5.3 Unsustainable Practices

Land degradation involves several events like desertification, erosion, and loss of organic matter, which severely impact soil productivity, natural resources, and biodiversity (Mahata & Sharma, 2021; Worku, 2023). The drivers' deep understanding of these challenges and the implementation of sustainable land management practices will help mitigate adverse effects.

Intensive and unsustainable farming practices are one of the major factors causing soil degradation. These practices are implemented to meet the needs of a rapidly expanding population. Overcultivation, monocropping, improper use of chemical fertilizers, poor irrigation, and mechanical tilling have deleterious effects on the soil's structure, degrade the essential nutrient content, and reduce the level of organic matter (Alori et al., 2020; Kopittke et al., 2019). These methods often favour short-term productivity over long-term sustainability, which has potentially severe environmental impacts.

For example, tilling without applying conservation measures can lead to soil erosion, while the injudicious use of fertilizers can lead to pollution, acidification, and salinization (Kopittke et al., 2019). Mechanical tilling and irrigation mismanagement also increase soil density along with nutrient loss, which lowers the fertility and biodiversity of soil (Sigar, 2023). Kopittke et al. (2019) highlighted that agricultural intensification has greatly reduced the proportion of organic matter, essential for sustaining soil health and ecosystem equilibrium.

Traditional agriculture attains high crop yields but relies on unsustainable agriculture, causing the degradation of agroecosystems and thus threatening future productivity (Komatsuzaki & Ohta, 2013). Issues such as this one need to be fought for by transitioning into sustainable agriculture where short-term achievement of maximum yield is balanced with maintaining soil health and long-term ecological sustainability.

3.5.4 Water Logging and Salinization

While irrigation is an essential factor in agricultural production, poor irrigation management causes severe land degradation in terms of salinization and waterlogging. Salinization is the most common form of degradation caused by the accumulation of salts in the soil due to irrigation water application. Salinization and waterlogging collectively degrade soil health, decrease agricultural productivity, and create significant challenges for sustainable land use. Overcultivation, deforestation, overgrazing, and excessive reclamation of grasslands have led the area of degraded land due to waterlogging, salinization, and desertification to increase from 2400 km^2 in 1992 in Tongyu County, Northeast China, to 4214 km^2 in 2002, according to Gao and Liu (2010). On a global scale, approximately 1 billion hectares are impacted by these challenges, jeopardizing food security in 75 nations (Dagar & Minhas, 2016).

Salinization has a huge impact on the quality of soil, the growth of crops, and consequently their productivity in arid and semi-arid regions (Hailu & Mehari, 2021). The causes of salinization are irrigation with low-quality water, groundwater tables, and natural processes (Harper et al., 2021). Soils high in salinity will impact plant growth by causing osmotic stress, lowering the absorption of water, and creating an imbalance in nutrients (Hailu & Mehari, 2021). Yearly, globally, abandonment of 0.3–1.5 M ha of arable land occurs due to salinization (Harper et al., 2021).

Waterlogging is one of the principal causes of land degradation when irrigation exceeds the capacity or improper drainage mechanisms increase the groundwater table, thus lowering the oxygen supply to the roots of plants and elevating the salinity level in the soil (Bowonder et al., 1987). It is highly prevalent in irrigated areas of arid and semi-arid regions due to interference with the natural drainage system and poor management of soil, water, and crops (Jayan & Sathyanathan, 2010). In India, water stagnation and salinity in agricultural areas are common due to such practices. Bowonder et al. (1987) highlighted that waterlogging affects approximately 22 million hectares globally, threatening food security and ecosystem sustainability.

These impacts have been mitigated by the use of remote sensing techniques, which include the Normalised Differential Salinity Index and Salinity Index in monitoring and mapping salinization and waterlogging risks (Jabbar & Chen, 2008; Modaihsh et al., 2014). These methods assist in evaluating temporal changes in affected areas, which can then facilitate the use of preventive measures and sustainable drainage systems (Dwivedi et al., 1999). More integrated approaches, such as mechanical, biological, ecological, and engineering methods are needed in rehabilitating degraded lands (Gao & Liu, 2010). For example, Kerala, India, has incorporated integrated rice-shrimp farming in waterlogged areas, which has become sustainable, but these ecosystems remain under threat of pollution, eutrophication, and loss of biodiversity (Jayan & Sathyanathan, 2010). Soil salinization and waterlogging have to be combated to achieve long-term soil health and agricultural sustainability.

3.5.5 Urban and Industrial Expansion

It contributes significantly to land degradation with urban and industrial expansion due to construction, mining, and industrial waste discharge. Urban expansion causes soil sealing, in which natural soil surfaces get covered by impermeable materials, resulting in the loss of high-quality arable land and deterioration of the quality of soil (Dai et al., 2022; Oliveira et al., 2018). This transformation not only makes agricultural productivity lower but even disrupts soil functions pertinent to ecological balance.

Industrial development worsens the degradation of land due to soil, water, and air pollution. Industrial wastes release heavy metals and polycyclic aromatic hydrocarbons into the soil. Moreover, poor waste disposal and mining lead to soil erosion and compaction, among other factors that degrade the quality of land. Industrial and other human activities increase the susceptibility of urban soils to compaction, erosion, contamination, and atmospheric deposition. Together, these processes reduce the capacity of the land to support vegetation and deliver ecosystem services (Ferreira et al., 2019). Urban expansion often results in the permanent loss of productive agricultural lands. The cumulative effects of urban and industrial development have caused considerable degradation of land resources, highlighting the need for sustainable urban planning and industrial practices to mitigate these effects and preserve soil health for future generations. Significant deterioration in land resources through the cumulative impacts of urban and industrial developments is now necessitating proper strategies of urban development as well as sound industrial practices aimed at protecting these for later uses.

3.5.6 Desertification

Desertification refers to land degradation in arid, semi-arid, and dry subhumid regions caused by both natural and human factors. It affects two-thirds of countries, covers one-third of the Earth's surface, and impacts the lives of billions of people, accounting for one-sixth of the global population. In recent years, the processes of desertification and land degradation have accelerated. The United Nations Convention to Combat Desertification defines desertification as "land degradation in arid, semi-arid and dry sub-humid areas resulting from various factors, including climatic variations and human activities" (UNCCD, 2014). Here, "land" denotes the terrestrial bio-productive system, while land degradation refers to the loss of biological or economic productivity and complexity of rainfed cropland, irrigated cropland or range, pasture, forest, and woodlands. This degradation arises from diverse land uses or a combination of processes linked to human activities and settlement patterns, like soil erosion by wind and water, the deterioration of the soil's physical, chemical, and biological properties, and the long-term depletion of natural vegetation.

3.6 Global Land Degradation Hotspot

Global land degradation has been estimated to affect around 29% of the Earth's land area and, hence, about 3.2 billion people, thus making it an environmental issue of great importance (Le et al., 2014). Regions with extreme degradation have been identified using different approaches like patterns in biomass productivity, indicators related to vegetation and soil, and integrated methods (Le et al., 2014; Wuepper et al., 2021). It also includes a variety of agroecologies and cover types. These involve key areas, such as the margins of the Sahara, the coast of the Mediterranean, Eastern Europe, South Asia, and Northwest China, as well as parts of America (Zhang, 2021).

About 3.9% of the degraded areas experienced worsened trends since the year 2000. It is interesting that about 22.8% show positive trends, indicating that targeted interventions can reverse degradation trends (Zhang, 2021). Intensity is still difficult to determine using degradation trends, especially for tropical and high-latitude regions, and hence needs intensified monitoring with varied methods and indicators (Jiang et al., 2024).

In the case of India, geospatial techniques have been used to examine the degradation pattern across the semi-arid zone of Rajasthan, and it was found that close to 44% of this area exhibited moderate to high vulnerability to degradation (Malav et al., 2022). This particular case study highlights a trend at the broader global level about the hotspots of degradation and underscores the significance of regional assessments aimed at finding degraded areas requiring sustainable management.

3.7 Impact of Climate Change on Agriculture

Climate change hugely impacts agricultural systems by influencing critical climatic factors such as temperature, precipitation, wind patterns, and the occurrence of El Niño events and extreme weather. It interferes with the already established growing conditions, promotes natural disasters like floods and droughts, and poses major challenges to crop yields, soil quality, and availability of water resources. It shows how various climate change-related effects, such as inconsistent temperatures, unusual rainfalls, and harsh climatic occurrences, change agricultural settings and challenge world food security.

3.7.1 Temperature Variations

3.7.1.1 Heat Stress on Crops and Livestock

Rising temperatures due to climate change severely affect agriculture and livestock production and pose a great threat to the world's food security. With high temperatures, the risk of heat stress to crops is increased. Heatwaves result in premature crop ripening and lower yields, as well as decreased quality. It also causes morphological and physiological changes in crops. Table 3.3 shows the disruptions of physiological stress through heat stress. Low-temperature increases of 1–2 °C can have a significant effect on the yields of cereals and oilseeds and show how sensitive the agricultural systems are to changes in the climate (Sajid et al., 2021). Temperature fluctuations have been a critical factor for crop growth and development in terms of the rising impact on Growing Degree Days, which is a critical metric associated with the estimation of rising heat summation needed for various development and maturity periods of crops (Alagesan et al., 2020). As temperatures increase, the total of Growing Degree Days reduces the growing season and associated yield potentials of fruits and vegetables. It may also be argued that excessive temperature leads to physiological disorders in plants. Insufficient GDD will delay crop maturity. However, high GDD would advance the growing season of a crop. Frequently, it causes poor grain filling, thereby decreasing harvest quality.

On a global level, it is expected that increasing temperatures will lead to a reduction in crop yields by about 1.5% each decade, with possible losses ranging from 0 to 4% (Lobell & Gourdji, 2012). While higher CO_2 levels might boost productivity by roughly 1.8% per decade, the rise in temperatures is likely to worsen pest problems, change weed populations, and shorten the duration of crop growth (Malhi et al., 2021). Tropical regions are especially vulnerable, facing significant yield reductions and increased poverty levels (Kurukulasuriya & Rosenthal, 2003). While moderate temperature increases may marginally benefit some crops, warming beyond 1 °C is generally associated with negative impacts on productivity (Jarvis et al., 2010).

Yield losses: Cereals and oilseed crops are particularly vulnerable, with yield losses reaching up to 30% due to decreased grain filling and physiological stress (Sajid et al., 2021). For instance, wheat productivity is reduced when high temperatures coincide with grain-filling stages, as exemplified during the 2022 Indian heatwave, which caused a 15–25% yield reduction in wheat (Singh et al., 2023; Kajal et al., 2015). Many studies show the strong impact of temperature on agricultural productivity. For instance, Shah and Srivastava (2017) reported that temperature contributes to 67% of variability in crop performance whereas only 33% is attributed to soil and nutrient management during the growing season, and they also highlighted that it is expected that for every 1 °F increase in temperature, crop yields may decline by 3–5%. Seasonal variation and temperature changes affect the impact of temperature on agricultural yield, making it funnel-shaped.

Impact on livestock: The effects of heat stress extend beyond crops to livestock. High temperatures reduce productivity, reproductive performance, and overall health in livestock, causing significant economic losses (Mittal et al., 2019). Heat stress diminishes estrus behavior in dairy animals, milk production, and egg-laying capacity in poultry (Mittal et al., 2019). These challenges underscore the multifaceted threats rising temperatures pose to both agricultural sustainability and livestock systems globally.

Impact on the economy: Economically, high temperatures have profound implications for agriculture and worker productivity. Extreme heat events can cause substantial losses in agricultural profits, with potential increases in economic losses by up to 26.65% without adaptation strategies (Wang & He, 2024). However, adopting heat-tolerant rice varieties can mitigate losses, as demonstrated in Japan, where such measures reduced welfare losses from 264 to 118 billion yen (Akune et al., 2015). In Europe, heatwaves have led to direct economic losses in agriculture ranging from $59–90 per worker (Orlov et al., 2019).

3.7.1.2 Effects of Frost and Extreme Cold

Sudden frost events and extreme cold significantly impact agricultural production, posing substantial risks to crop yield, livestock health, and food security (Sivakumar, 2020). Frost-sensitive crops, such as citrus and tropical fruits, often suffer severe

Table 3.2 Regional cold stress impacts on agriculture

S.No.	Region	Challenges to crops	References
1	Mediterranean region	Freezes to − 5 to − 10 °C severely damage citrus and olive plantations, while spring frosts threaten horticultural crops in temperate zones	Sakai and Larcher (1987)
2	Nepal's Terai region	Cold waves frequently coincide with foggy days, negatively affecting winter crops and livestock	Shrestha et al. (2022)
3	Gangetic Plains of India	Frost between December and January with a decreasing duration trend of 0.5–1.5 days per decade disrupting local food systems, impacting both crop output and livestock management	Attri et al. (2020)

yield losses during flowering and fruiting stages, while extreme cold weakens livestock health and reduces fodder availability. These events directly damage crops and animals and indirectly limit agricultural production capacities (Sivakumar, 2020). Table 3.3 lists the physiological challenges crops face under cold stress. For instance, tropical and subtropical plants are particularly vulnerable, as temperatures below 8–10 °C restrict their altitudinal range.

Plants have evolved cold-acclimation mechanisms, such as protein accumulation and membrane modifications, to mitigate these adverse effects, but these adaptations are often insufficient under prolonged or severe frost conditions (Soualihou et al., 2022). Frost events also influence regional agriculture patterns. Low temperatures also play a critical role in shaping plant distribution and productivity in tropical mountains like Mediterranean regions and temperate zones (Sakai & Larcher, 1987). For example, the combination of low temperatures and frost in Mediterranean climates can severely disrupt agricultural productivity, while cold waves in tropical and subtropical regions constrain crop choices and production levels. In regions such as the Gangetic Plains, frost can disrupt local food systems, impacting both crop output and livestock management. Table 3.2 illustrates the impacts of regional cold stress on plants.

3.7.2 Impacts of Wind

Severe outcomes occur in agricultural systems due to strong winds, particularly in arid and semi-arid regions. Wind erosion removes topsoil, leading to a severe loss of soil fertility as the topsoil carries away the nutrient-rich material from it

Table 3.3 Key physiological disruptions under climate stress

S.No.	Physiological impacts	Factors	References
1	Reduces photosynthesis, impairs growth, and disrupts plant physiology and biochemistry	High temperature	Sajid et al. (2021), Khan et al. (2013)
2	Reduced seed germination, increased transpiration, and diminished photosynthesis	High temperature	Sajid et al. (2021)
3	Deline in vegetative and reproductive growth	Cold stress	Soualihou et al. (2022)
4	Anthesis and ripening were adversely affected, with deviations from optimal GDD	High temperature	Sanchez et al. (2014)
5	Altered phenophases like early budburst, flowering, and fruiting	High temperature	Pawar and Sakhale (2022)
6	A spongy tissue in mango, fruit cracking in litchi, impacting quality and yield	High temperature	Datta (2013)
7	Reduces water potential, disrupts electron transport chains, and generates reactive oxygen species	Water stress	Shumilina et al. (2018)
8	Disrupt chlorophyll biosynthesis, photosynthesis, and CO_2 assimilation	Drought	Hussain et al., 2019
9	Decreases stomatal conductance, cell membrane stability, and water use efficiency, leading to disruption in normal functioning	Drought	Hussain et al. (2019)
10	Altered metabolism, growth and development, leading to plant death	Drought	Kaur and Kumar (2020)
11	Reduces growth and altered carbohydrate metabolism	Flood	Patel et al. (2014)
12	Reduced photosynthesis, stomatal closure, disruption in carbohydrate metabolism, and reduced stomatal conductance	Flood	Kozlowski (2002)

(Colazo & Buschiazzo, 2015). Maize and sugarcane crops, having shallow roots or tall structures, are susceptible to wind damage and have lower productivity with increased vulnerability to drought conditions (Lyles, 1977). Moreover, the erosion caused by the wind alters the texture of the soil. Fine-grained particles such as silt and clay disappear, which disorganize the soil and its water-holding capacity (Pierre et al., 2018). In addition, wind patterns affect the efficiency of pollination and seed dispersal. Changes in wind speed can decrease the efficiency of pollination for crops that rely on wind for pollination, which may lead to reduced crop yield and quality. In contrast, strong winds can also enhance the spread of pest populations and pathogen spores, which may enhance the risk of crop diseases (Chakraborty et al., 2015).

Land management also comes under the influence of wind erosion; however, vulnerability differs with field type, especially among fields like fallow or groundnut. Therefore, changes in agricultural practice are important; these include management of crop residue and tillage. In stilling (decrease in wind speeds), for example, though it may be beneficial in the reduction of soil erosion and crop growth conditions, it may at the same time cause pest movement and disrupt seed dispersal mechanisms, which is critical for certain crop propagation (Chakraborty et al., 2015). In general, the alteration of wind patterns due to climate change has significant implications for crop yield, productivity, and food security. While some changes, such as reduced wind speeds, may offer short-term benefits in specific areas, long-term disruptions can lead to decreased agricultural productivity and increased vulnerability to pests and disease outbreaks (Mahato, 2014; Shil, 2018).

3.7.3 Pests and Diseases

Variations in temperature, rainfall, and extreme weather conditions due to climate change are the most influential factors for agricultural pests and diseases. Thus, changes will affect the distribution, reproduction, and abundance of the insects, mostly positive but sometimes negative, according to the weather factors involved.

High temperatures are likely to enhance pest reproduction and population increase, hence widening the range of tropical species. In temperate regions, high temperatures tend to enhance the population growth of pests, but in the tropics, too much heat can result in the extinction of some pests (Katsaruware-Chapoto et al., 2017). Researchers in Australia have identified rainfall variability as the primary factor influencing insect populations, and Cooler winter temperatures and wind-driven migration are also known to influence the distribution of pests (Drake, 1994). High precipitation may enhance the severity of some diseases and facilitate the movement of pests, which leads to increased pest populations (Huo ZhiGuo et al., 2012). Conversely, during drought periods, reduced rainfall may limit the development of pests; however, it may simultaneously cause stress to crops and increase vulnerability to other pest attacks. The interaction of climatic factors, host vegetation, and pathogens determines the severity of pest and disease occurrences. Climatic variables like temperature, humidity, rainfall, and wind speed significantly influence the prevalence of pests and diseases in rice farming. These factors have led to yield losses of up to 35% for farmers (Mangala & Padmapriya, 2019). Erratic trends in precipitation, such as flooding events, can lead to significant financial losses, as in the case of shallot farming in Indonesia, where flooding caused an economic loss of more than 1.6 billion IDR (Marseva et al., 2022).

Overall, variations in temperature, rainfall, and extreme weather events could be both positive and negative in terms of the prevalence of pests and diseases; therefore, agricultural practices must adapt to these changing conditions.

3.7.4 Unpredictable Rainfall

Erratic rainfall patterns disturb the planting schedules, leading to crop failures. The late-season rains during the harvest time reduce grain quality and increase post-harvest losses. Climate change results in unpredictable rainfall patterns that have massive impacts on agriculture and the livelihoods of farmers. Studies in Bihar, India, indicate that the trend of changing rainfall patterns—unseasonal rain and increased rain intensity—has resulted in yield losses and decreased agricultural productivity and has brought economic loss to the farmers (Soz & Raza, 2024). Equally, studies conducted in Wiyumiririe in Laikipia County in Kenya present declining total rainfalls, increased frequency and intensity of droughts, and reduced wet periods (Muchiri et al., 2020). Both extremes of high and low rainfall affect crop choice and yields in Malaysia; more so, over-rainfall during the ending phases of crop cycles is highly destructive (Alam et al., 2011). Research from the semi-arid region of Nigeria also presented a 10-day shift in planting periods with negative correlations between crop yield and dry spells (Yamusa et al., 2015).

3.7.5 Drought

Variation in precipitation patterns, including delayed and poor rainfall, is one factor that leads to drought with adverse effects on agricultural production. Long dry spells drain soil moisture, delay the germination of crops, and lower yield outputs. Besides, the lack of water limits irrigation capacity in most rain-fed agriculture areas, thus stressing crops and hindering their growth. Drought exerts highly significant influences on plant physiology and agricultural productivity. It affects many morpho-physiological and biochemical processes. Table 3.3 provides physiological disruptions caused by water stress. Along with biological, drought also has high economic implications. Crop yield anomalies, reduction in agricultural GDP, and price variations have been the effects of these droughts. In South Asia, the severity of drought strongly correlates with economic losses (Mohammed et al., 2020). Vulnerable to drought due to its Mediterranean climate, agriculture production, prices, and trade in Turkey suffer adversely (Dellal & McCarl, 2010). South Korea lost $4.438 million and recovery costs of around $5.180 million in the 2018 drought cycle but lighter impact in 2019 (Song et al., 2024). The Sistan region in Iran likewise lost about 80% of its agricultural activities due to drought experienced in 1999–2005, amounting to 1,561 billion USD in direct losses (Ebrahimzadeh & Esmaelnejad, 2013).

Alteration of precipitation patterns because of climate change is now being considered with great importance. Higher temperatures and unbalanced rainfall result in higher instances and intensities of droughts, which intensify the agricultural conditions and compound the challenges faced by farming communities all over the world.

3.7.6 Flood

Heavy rainfall, often an after-effect of the shifting pattern of rainfall in regions because of climate change, causes floods. Water gathers excessively and deprives the crops of oxygen, and eventually, it leads to waterlogging and root rot. Flooding damages the crops but also erodes topsoil and sweeps away essential nutrients from land, making it less fertile. Flooding is one of the major environmental stresses that affects agricultural production globally (Patel et al., 2014; Striker, 2012). It triggers changes in the morphological, biochemical, and anatomical traits of plant root systems. Long-term flooding leads to soil anaerobiosis that hampers crucial physiological activities listed in Table 3.3. Flooding also impacts the properties of soil through the leaching of organic matter and other essential nutrients that significantly alter crop yield (Eni et al., 2011). The riparian forest ecosystems are significant in controlling flooding and biodiversity, thus showing a high vulnerability to changes in flooding regimes (Kozlowski, 2002).

Floods cause heavy economic loss in agriculture and also badly affect crop production. In Pakistan, consecutive floods from 2010 to 2014 caused damage to crops, livestock, and infrastructures (Rehman, 2015). In Assam, India, the major losses due to floods were suffered like paddy and jute, and the fishery sector suffered the most in the Lower Brahmaputra Valley (Ahmed & Rahman, 2021). The flood-induced sand deposition in Dhemaji, India reduces soil fertility and affects the yield of paddy crops with an annual loss of US$13–35 per hectare (Das, 2012). Economic assessments of flood damage to agriculture are very important to improve flood management strategies. However, the prevalent methods only concentrate on crop damage, ignoring aspects such as farm buildings and loss of activity. A large damage function and a farm-level approach are necessary to represent an accurate scope of agricultural loss due to flooding (Brémond et al., 2013).

Alteration of rainfall patterns due to climate change is growing in importance, thus enhancing the frequent occurrence and intensity of floods that can aggravate the agricultural crisis in flood-prone areas. With the growing prevalence and intensity of flooding, establishing flood-tolerant crops and improved management techniques would be important in ensuring food security in affected regions (Striker, 2012).

3.7.7 Extreme Weather Events

Other important climate-related extreme weather events—cyclones, hurricanes, heatwaves, and hailstorms—can greatly devastate agriculture and food security, particularly in developing countries (Sivakumar, 2020). They may cause damage to crops, animals, and trees or indirectly disrupt the production capacity. Changes in the patterns of frequency and intensity of these events have been observed during the last fifty years, affecting the yields of crops in general, and in India specifically

(Ray et al., 2019). Their impacts are felt acutely in rainfed agriculture due to the interannual variability in precipitation (Swaminathan & Rengalakshmi, 2016).

Extreme hydrological events, such as floods and droughts, combined with the effects of climate change, significantly impact agriculture and economies, particularly in Asia. The region experiences the highest frequency of such events, with countries like China and India incurring the greatest economic losses (Gao et al., 2019). These extreme weather events are closely linked to substantial crop yield reductions in developing nations, although the extent varies by region and crop type (Marmai et al., 2022). In Central Vietnam, for instance, agriculture and fisheries have been severely affected by extreme buy extreme weather and climate change, with estimated losses ranging from 3597.72 to 18,891.2 billion VND under different climate scenarios (Do et al., 2021). Similarly, in southern Africa, severe weather events, including the passage of cyclone Freddy in February 2023, have impacted regions such as Madagascar, southern Malawi, Mozambique, and Zimbabwe (WMO, 2024).

These are increasing in frequency due to changes in atmospheric and oceanic conditions, driven by shifts in climate patterns. This, together with increased global temperatures, altering wind patterns, and changed ocean currents, results in greater intensification and unpredictability of such events, continuing to challenge agricultural systems globally.

3.7.8 Other Climate Variabilities

El Niño Southern Oscillation: The two prominent phases of the ENSO (El Niño Southern Oscillation) are the El Niño and La Niña phenomena, which greatly impact global agricultural output due to impacts on climatic conditions. Generally, El Niño is characterized by higher dryness, whereas La Niña produces cooler and wetter conditions. Such climatic fluctuations affect agricultural productivity in different ways depending on the region and crop. As the transition from El Niño and La Niña led to rising average global temperature anomalies, numerous regions, including parts of China, Europe, Mexico, North Africa, the Middle East, Siberia, South America, and the United States, experienced heatwaves (WMO, 2024). In Malaysia, climate variability, due to climate change, associated with El Niño has negative effects on the agricultural sector, bringing reduced crop yields and negative impacts on the regional economy (Tawang et al., 2002). Similarly, Bhatla et al. (2021) highlighted about Indian scenario, in western India, there are lesser rainfalls during El Niño years, which reduces the production of rice, maize, and sugarcane but are generally beneficial for rainfed crops during La Niña events, thus leading to improved yields of crop production and the Indian Ocean Dipole (IOD) also plays a part here, with positive IOD years often coinciding with El Niño and being generally associated with poor yields.

In Java, Indonesia, and La Niña events resulted in a rise in maize production, although the supply of paddy remains more or less unaffected (Utami et al., 2011). In contrast, El Niño Southern Oscillation phases modulate winter wheat and maize yields in the central USA through changes in precipitation and temperature regimes associated with both El Niño and La Niña events (Mauget & Upchurch, 1999). Such region-specific impacts depict a rather complex relationship between El Niño Southern Oscillation events and agricultural productivity, where often El Niño causes quite unfavorable conditions—like drought and reduced yields—while La Niña usually brings better conditions for crop growth, mostly in regions that depend on rain. However, even in neutral years, some crops like pulses only show very marginal improvements.

Carbon Dioxide: According to the WMO report (2024), atmospheric CO_2 levels reached 417.9 ± 0.2 ppm in 2022, representing 150% of pre-industrial levels. This rising concentration has significant implications for crop productivity, as carbon dioxide plays a vital role in photosynthesis. Elevated atmospheric CO_2 influences crop growth by reducing stomatal conductance, stomatal aperture, and transpiration, which enhances photosynthesis and growth through improved water use efficiency (Stockle et al., 2009). However, these benefits also come with increased demands for water and fertilizers to sustain growth, with higher nitrogen inputs required to achieve optimal productivity under changing climatic conditions (Ramirez & Kallarackal, 2015).

3.8 Impact of Land Degradation on Agriculture

3.8.1 Decline in Soil Health and Fertility

Land degradation caused by continuous use of land without restoring soil nutrients leads to soil infertility over time. This reduction in the health of the soil directly affects the production of agriculture worldwide and food security. Land degradation influences almost 33% of the Earth's surface, and it affects 2.6 billion people across more than 100 countries (Abebaw, 2019). According to Eswaran et al. (2019), the global agriculture sector incurs a loss of 75 billion tons of soil per year, which costs $400 billion across the world. Furthermore, in Africa, soil erosion has led to a yield reduction of 2–40% with an average loss of 8.2%. Land degradation is multifaceted and is caused by a variety of complex interactions between physical, chemical, biological, and socio-economic factors (Abebaw, 2019), which eventually result in loss of nutrients and erosion in the soil and low crop yields. Despite it being not that threatening to food security on a global scale it presents significant problems in regions with insecure property rights, poor soils, and limited information and market access (Wiebe, 2003). In Asia, which is the most impacted region, India alone has 147 million hectares affected by various forms of degradation, including water erosion, acidification, and salinity (Bhattacharyya et al., 2015). Both natural and anthropogenic factors contribute to this issue, with inappropriate agricultural practices and socio-economic pressures playing major roles (Bhattacharyya et al., 2015). Black soil regions of central and peninsular India are particularly vulnerable because of inherent soil properties and unsustainable management practices (Sharma, 2020).

To resolve these challenges, scientists highlight the need for integrated watershed management, conservation measures, and sustainable agricultural practices (Roy et al., 2022). Long-term experimentation is crucial to establish quantitative criteria for assessing soil quality, degradation, and resilience (Lal, 1997).

3.8.2 Impact on Yield and Quality

The physicochemical attributes of soil and its quality serve to determine the production potential of farming. Oftentimes, poor farming practices lead to depletion of arable land making it barren, which in turn results in food shortages. Farmers often experience lower crop yields, which in turn affects food security and agricultural sustainability. Land degradation not only reduces the number of crops produced but also the quality of the crops. Vegetation such as plants do not grow well therefore, they produce low-quality crops bound to be stressed due to poor water holding capacity. Moreover, the lost nutrients from soil degradation degrades the quality of the crops, making them more susceptible to pests, diseases, and different weather conditions.

Crop productivity loss and food insecurity in developing countries are the main impacts of land degradation. Effects include crop yield reduction, increased cost of production, and expansion of land considered not cultivable (Tun et al., 2015). According to Taddese (2018), Land degradation negatively affects crops, water availability, and biodiversity in Ethiopia. Studies in Senegal and Myanmar present that crop yields in areas that have been highly degraded are 3–12 times lower than yields recorded from less degraded areas (Sonneveld et al., 2015; Tun et al., 2015). The major causes of land degradation include population pressure, deforestation, inappropriate use of land, and climatic factors (Taddese, 2018).

In Central Asia, land degradation reduced agricultural profits by 27% in 2009–2010 (Mirzabaev et al., 2018). Crop yields in Myanmar's Dry Zone were 3–12 times lower in highly degraded areas than in less degraded areas (Tun et al., 2015). Land degradation (chemical and physical deterioration of soil), the rates of soil erosion increased from 14.2 to 54.6 t ha^{-1} yr^{-1} between 2000 and 2012 in Myanmar (Tun et al., 2015). The poor quality of the soil brings a decrease in crop yields, production volume, and farm incomes (Calzadilla & Carr, 2020). Despite the fertilizers, the increased levels of land degradation overlap with reduced crop yields (Sonneveld et al., 2015). In the African context, soil erosion has resulted in yield reductions that fall between 2 and 40% (Eswaran et al., 2019). Measures to combat land degradation are regulation of overgrazing, reduction of population pressure on resources, contour farming, afforestation, soil fertility through organic matter inclusion, and soil and water conservation infrastructures.

3.8.3 *Impact on Water Productivity*

Land degradation is very strongly affecting the water resources especially by decreasing water productivity and disrupting the water cycle. Land degradation leads to a reduction in the capability of soil to hold water and increases surface runoff. This runoff washes away essential nutrients from the soil and contributes to the depletion of groundwater, which is crucial for irrigation. Over time, as groundwater tables drop, access to water for both agricultural and domestic purposes becomes increasingly limited. Land degradation is one of the leading causes of a decrease in water productivity, with an estimated 50% of the global arable land being affected (Bossio et al., 2008). But it doesn't only affect water quantity—it degrades water quality. The expansion of natural ecosystems to cropland and pastureland has improved water recharge and increased streamflow in some areas (Scanlon et al., 2007), but has also caused poor water quality associated with salt mobilization and leaching of fertilizers (Scanlon et al., 2007). It further reduces the availability of clean water for irrigation, complicating water management in agricultural systems.

The impact of land degradation directly affects water resources, which results in lower crop yields and pressures farmers to seek alternative sources of water. To maintain water productivity and ensure affected regions have continued food security, restoration of degraded lands with sustainable practices combined with better water management will be needed.

3.8.4 *Economic and Livelihood Implications*

Land degradation has vital economic implications for farmers and rural people since it brings about a rise in the cost of production in agriculture while reducing profit-making in agriculture. Declining land productivity pushes farmers to spend more money on fertilizers, pesticides, and other inputs for controlling land degradation. This, in turn brings high input costs that reduce profits, and agricultural practice often becomes unsustainable. It ultimately makes the farmers give up completely and perpetuates poverty and unemployment in the country. Recent studies have indicated that land degradation has reduced the net farm revenues by 4.8-fold as compared to the non-degraded areas (Mirzabaev et al., 2018).

The cost of land degradation is very high in terms of the economy. Its restoration cost globally, in terms of soil erosion, is estimated to be US $400 billion a year (Eswaran et al., 2019). For instance, the annual loss due to degradation in India was INR 3654/ha, while increasing degradation by 1% loss was INR 104 per hectare per year (Gorain et al., 2023). Similarly, in Central Asia, land degradation was associated with a loss of 27% in agricultural income during 2009–2010 (Mirzabaev et al., 2018). A decrease of 1% point in soil organic matter causes a reduction of 0.07 units in net farming income, in Ivory Coast, West Africa (Angaman et al., 2024).

The economic impacts of land degradation differ among socioeconomic groups. For instance, the losses at the middle and higher ends of the income range in Central Asia are much greater because the usage of inputs and investments in agriculture is greater there (Mirzabaev et al., 2018). Moreover, with the deterioration of soils, dependence on fertilizers increases and pushes farmers under increased costs. For example, one-third of annual nitrogen fertilizer consumption for maize agriculture in the United States has been used in 'compensating' soil fertility loss over the last century, with an extra cost of half a billion dollars each year (Jang et al., 2020).

Threats to the livelihoods of the rural population are increased input costs and reduced agricultural returns, which are reasons enough to necessitate the addressing of land degradation through sustainable farming practices, conservation of resources, and soil restoration technologies.

3.9 The Combined Impact of Climate Change and Land Degradation

Climate has an important role in shaping land and its ability to support agricultural productivity. Land's health and fertility can be maintained in a stable and predictable climate with adequate rainfall and favorable temperatures, supporting crop production. Once land degradation occurs, the land's water-holding capacity, nutrient-retaining capacity, and plant-supporting capacity diminish, thereby creating a cycle where land becomes more vulnerable to climatic stresses. For instance, poor soil health areas are more susceptible to erosion during heavy rainfalls or droughts—both components are influenced by changing climate patterns. It is this interaction of climate and land degradation that amplifies the challenges faced by farmers to restore soil quality and maintain agricultural productivity. For example, land surface degradation in northern China and southern Mongolia caused changes in regional climate, including changes in precipitation and temperature (Zhang et al., 2005a, 2005b). The same was found in the Sahel, East Asia, and the central United States, where land degradation contributed to reduced rainfall, evapotranspiration, soil moisture, and increased surface temperatures (Xue & Fennessy, 2002). Figure 3.4 shows the combined impacts of climate change and land degradation.

3.9.1 Impacts of Extreme Weather Events

Climate change-induced extreme weather events (droughts, floods, and heatwaves) can worsen land degradation. In many degraded soils, there is lower water holding capacity and consequently higher runoff with heavy rainfall and rapid evaporation during dry spells. This amplifies the effects of natural disasters since the soil is unable to retain rainfall or hold moisture for crops during dry periods. Hence, farmers

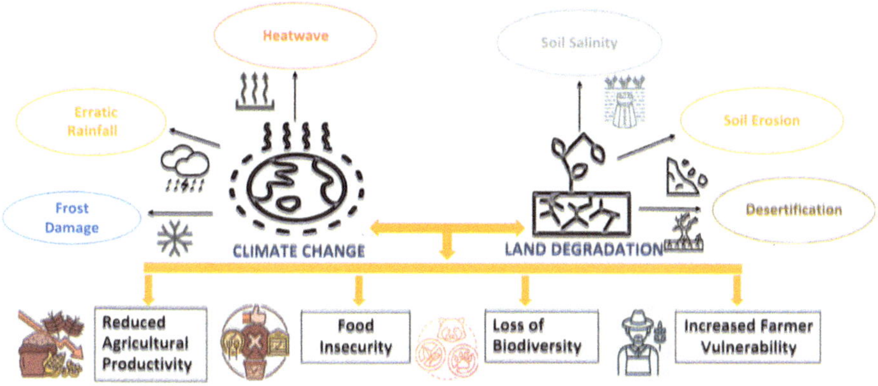

Fig. 3.4 Combined impacts of climate change and land degradation

are adversely affected directly by extreme weather events and subsequently hit even harder indirectly through land degradation (more crop failures/yields resulting in financial costs). Land degradation exacerbates vulnerability to climate change impacts and reduces adaptability (Webb et al., 2017). According to McDonald (1996), Desertification (the process of large-scale dryland degradation) affects more than 250 million people spread across 100 countries worldwide, and in regions with land degraded by desertification, droughts are of great importance for the deterioration of land, with climate change being among major reasons responsible for this. In addition, Land degradation is also a driver of climate change, since its rapid release of carbon dioxide due to cleared vegetation lowers the ability to carbon sequester (Ioraş et al., 2014).

3.9.2 Impact on Crop Productivity and Food Security

The collective impacts of changing climate and land degradation represent substantial challenges to crop production. Degraded soils with less fertility and more erodibility face reduced water and nutrient retention. These conditions are further exacerbated by climate change, leading to more droughts and floods that are more extreme. This, in turn, reduces crop yields and hence food availability while increasing the price of food. All these extra stresses decrease the value chain output of both farmers and consumers while lowering the yields and, besides, the nutritional quality of the produce. This finally threatens food security at both local and global levels. Degraded land is, according to several studies, leading to the threat of soil erosion, salinization, and organic carbon loss in the tropics, which degrades agricultural productivity (Roy et al., 2022).

3.9.3 Increased Vulnerability of Agricultural Systems

The cumulative effects of climate and land degradation form a feedback loop whereby each factor feeds the other, exacerbating each other. Climate change reduces the fertility of the soil, nutrients drain from the earth, and cultivable land becomes desert. At the same time, the land makes the agricultural system more vulnerable to erratic climatic changes. Therefore, due to this kind of vulnerability, rural economies and livelihoods bear the major cost, where more money goes towards fertilizers, irrigation, and other inputs. This compensation for reduced soil health and uncertain weather patterns needs to be taken into account by the farmers themselves. For example, the United States uses one-third of nitrogen fertilizer in corn to make up for the loss of long-term fertility of arable lands, which amounts to over half a billion dollars per year (Jang et al., 2020). The financial problems and reduced productivity may push farmers to poverty, which can lead to increased rural unemployment and exacerbate food shortages. Land degradation in Central Asia has resulted in a 27% loss in the profits of farmers during 2009–2010 (Mirzabaev et al., 2018), thus indicating the economic vulnerability brought by the combined impacts of climate change and land degradation.

3.9.4 Socio-economic Impacts on Rural Livelihoods, Migration, and Employment

Rural communities worldwide are becoming more vulnerable to the socio-economic consequences of land degradation and climate change. Declining soil fertility and erratic weather patterns have caused lower agricultural output, therefore aggravating poverty and food insecurity among smallholder farmers. Rising vulnerability has, for example, been connected to climate change in Ethiopia's Upper Blue Nile area, which has driven rural people to move in quest of other livelihoods (Berhanu et al., 2024). Around 3 billion people worldwide suffer from land degradation, which causes more movement, instability, and insecurity in many societies (UN, 2024). According to the Intergovernmental Science-Policy Platform on Biodiversity and Ecosystem Services (IPBES), land degradation and climate change act as threat multipliers for already precarious livelihoods, leaving them very sensitive to extreme climatic events with outcomes including poverty and food insecurity (IPCC, 2019a, 2019b).

Limited finances greatly impede the use of climate-resilient practices. A study in Southwest Nigeria found that 80% of smallholder farmers are financially vulnerable to climate change risks, needing an average of $106.55 to increase their adaptive capacity (Ojo et al., 2024). These financial constraints hinder the widespread adoption of sustainable practices; therefore, they strengthen the weaknesses of poor farming communities. Migration caused by climate change is becoming more and more worrying. Without quick action to lower global greenhouse gas emissions, the World Bank estimates that climate change might drive up to 216 million people across six world areas to move within their countries by 2050 (World Bank, 2021). Climate anomalies in Uganda, especially heat stress, have been noted as main causes of environmental migration; short hot spells raise temporary migration and long-term heat stress causes permanent migration via agricultural livelihood routes (Call & Gray, 2020a, 2020b). Land degradation is a major cause of migration in Central Asia; drought and desertification are responsible for yearly losses of around USD 6 billion. The annual number of people moving in search of employment totals 2.5–4.3 million, representing 10–15% of the economically active population (UNCCD, 2023a, 2023b).

3.10 Case Studies

3.10.1 Case Study 1: Impact of Climate Change and Land Degradation on Agriculture in Jaggar Watershed, Rajasthan, India (Javed et al., 2012).

Background

Climatic changes and the effects of land degradation on agricultural sustainability are quite vividly represented by the case study of Jaggar Watershed, spread over an area of approximately 352.82 km^2 in Karauli district. It is rainfed and hence susceptible to climatic variability and environmental degradation. The present study indicates changes using remote sensing, GIS, and socio-economic surveys.

Geographical and Climatic Context

The watershed, which is a part of the Gambhir River Basin, has uncertain rainfall and high temperatures. Between 1977 and 2007, rainfall decreased by 269 mm, with an annual average of only 689.2 mm. Along with this, maximum and minimum temperatures increased, enhancing vulnerability in the region. Undulating terrain and a fragile ecosystem enhance the risk of soil erosion and loss of arable land.

Land Use and Land Cover Changes

Satellite data showed notable land use changes from 1989 to 2009. Largely because of water shortage and soil degradation, agricultural land fell from 34.2 to 29.65%, losing around 1558 ha. Reflecting lower rainfall and falling groundwater, water bodies shrank dramatically from 411 hectares in 1989 to only 95 ha in 2009. Driven by soil erosion and unsustainable land use, wasteland grew from 1.69 to 10.5% in the interim. Under pressure from grazing, fuelwood collection, and ongoing droughts, open forest areas declined by 5.5%. The data shown in Table 3.4 shows notable deterioration of agricultural and water resources over the two-decade period, along with a notable rise in wasteland.

Table 3.4 Land use and land cover changes in Jaggar Watershed, Rajasthan

Land use type	Area (1989) (ha)	Area (2009) (ha)	% change
Agricultural land	12,068	10,510	– 12
Water bodies	411	95	– 76.9
Wasteland	598	3710	+ 520

Source Adapted from Javed et al. (2012)

Socio-economic Impacts

Studies on the socio-economics of ten villages showed significant consequences of environmental decline. In some cases, more than 240 ft, groundwater depletion pushed people to draw water from increasingly deep wells. In areas with limited water, women and children spend four to six hours each day gathering it. Declining agricultural output by 35% pushed farmers to abandon water-intensive crops like sugarcane and pigeon pea in favour of lower-value replacements such sesame and mustard. Although NREGA's execution shortcomings diminish its effectiveness, more than half of the households also earned less than INR 45,000 year and many relied on government initiatives.

Drivers of Land Degradation

Climatic variability, human activity, and geomorphological elements are the main causes of land degradation in the Jaggar Watershed. Soil erosion risks have increased as a result of erratic rainfall patterns, rising temperatures, and repeated drought events. At the same time, overgrazing and deforestation for fuel contribute to land degradation. The undulating sandy topography of the area hastens soil depletion and wasteland growth even more.

Recommendations

Integrated watershed management is essential to overcome these obstacles. Priority actions should be the establishment of fair irrigation systems, groundwater recharge initiatives, and rainwater collecting systems. Resilience can be improved by agricultural adaptation policies, including encouraging drought-resistant crop types, soil conservation techniques, and agroforestry systems. Maintaining long-term eco-resilience will depend on strengthening policy frameworks and supporting community-led natural resource management.

Conclusion

The Jaggar Watershed is a perfect example of the cascading impacts of climate change and land degradation on agriculture and rural livelihoods. The combined effects of declining rainfalls, high temperatures, and inappropriate use of land and water resources have so far led to decreased agricultural productivity and increased socio-economic vulnerabilities. Such a reversal can be achieved only through the integrated management of land and water resources. These include promoting efficient use of water, crop diversification to increase resilience, and enabling local communities to undertake sustainable practices. The region seeks to achieve improved agricultural sustainability and socio-economic stability by eliminating the root causes of degradation.

3.10.2 Case Study 2: Impact of Climate Change and Land Degradation on Agriculture in the Mount Elgon Region, Uganda (Jiang et al., 2014).

Background

This area of Uganda demonstrates the enormous impacts of climate change and land degradation on agriculture in Africa. Soils and landscapes in this region are highly degraded through erosion and unsustainable land use due to mountainous topography and diversities of ecosystems. This paper will explore soil erosion trends, climate variability, and changes in land use based on advanced modeling techniques.

Geographical and Climatic Context

Mount Elgon covers an area of 365 km^2 in the catchment Manafwa, whose terrain range is from 1084 to 2455 m elevation, with slopes steep enough that the risk for erosion is great. The humid subtropical climate of the region experiences annual precipitation of 1500 mm. Volcanic soils, fertile enough to support intensive agriculture, are highly prone to erosion due to steep slopes and population pressures.

Land Use and Soil Erosion Trends

Rapid population growth in the Mount Elgon region resulted in significant deforestation, particularly below 2000 m, where forests were converted into croplands and grazing areas. More than 56% of the area was designated as high to very high erosion risk; thus, this shift exposed the sharp volcanic slopes to notable erosion hazards. The average soil loss was 101 t/ha/year. Though improved practices caused erosion rates to fall momentarily between 2000 and 2006, they climbed back by 2012 due to population pressure and uneven rainfall distribution.

While average annual soil loss rates also stayed at 101 t/ha/year for these years, soil erosion risk stayed consistently high at 56% in both 2000 and 2012, as Table 3.5 indicates. Though Jiang et al. (2014) acknowledged the trend, the table does not show the temporary drop in erosion rates during the mid-2000s since exact numbers for 2006 were not reported.

Table 3.5 Land use and soil erosion risk in Mount Elgon

Category	2000	2012
High to very high erosion risk (%)	56	56
Average annual soil loss (t/ha/year)	101	101

Adapted from Jiang et al. (2014)

Methodology

The study employed the Revised Universal Soil Loss Equation (RUSLE) model with data from:

1. DEM for the slope analysis.
2. Climate Data for rainfall erosivity estimation.
3. Satellite Imagery for the NDVI of vegetation cover.
4. Soil Maps for the erodibility factors.

Socio-economic Impacts

Diminished soil fertility, lower agricultural production, and growing food insecurity showed the effects of land degradation. Declining arable land and deteriorating water quality caused by sediment pollution in downstream rivers troubled subsistence farmers. These combined pressures highlighted local living conditions and agricultural sustainability.

Recommendations

Priority actions to reduce these effects are the encouragement of agroforestry systems, terracing, and soil and water conservation structures, including check dams. Essential is reforestation in erosion-prone areas together with community education initiatives and enhanced policy enforcement on land use. It is also advised that financial incentives for local conservation projects and sustainable practices be included as well.

Conclusion

The Mount Elgon region is the representation of all issues facing agriculture through climate change and land degradation in mountainous ecosystems. Even with enhanced agricultural production technologies, erosion due to high population pressure and misuse of land still affects it. Sustainable management, reforestation, and community approaches must be integrated to overcome these adverse impacts and provide environmental resilience.

3.10.3 Comparative Analysis of Case Studies: Jaggar Watershed, India, and Mount Elgon Region, Uganda

Climate change and land degradation are of prime importance to the agricultural systems throughout the world, but more significantly, in places where agriculture strongly depends on natural resources. It can be argued that the case studies of Jaggar Watershed in India and the Mount Elgon area in Uganda epitomize a model for commonality and distinction both in their geographical settings, the socio-economic implications, and approach towards mitigation. The semi-arid zone of Rajasthan faces erratic rainfall averaging 689.2 mm per year, rising temperatures, and worsening water scarcity, leading to soil degradation and reduced agricultural productivity. On the other hand, the humid subtropical climate of the Mount Elgon region offers abundant rainfall averaging 1500 mm per year, but steep mountainous terrain and volcanic soils make it highly susceptible to severe soil erosion. Anthropogenic activities, including deforestation, overgrazing, and unsustainable agriculture increase the rates of land degradation in both areas. The population dynamics of Mount Elgon have resulted in the high deforestation rates intended to gain extra agricultural lands and pastures. In Jaggar Watershed, there is a 5.5% decline in forest cover and a remarkable rise in wasteland areas.

Socio-economic impacts of land degradation in both regions are evident. Water scarcity and low fertility of soil threaten the very viability of agriculture. For example, the agricultural viability of Jaggar Watershed has reduced by 35%, with Mount Elgon experiencing an average annual loss of 101 t/ha/year where the erosion risk is categorized as high to very high in 56% of the area. Both regions, poverty and livelihood vulnerability were increased, along with food insecurity, communities are either dependent on relief from the government or must move. In contrast, both regions offer an opportunity for mitigating measures. It is important to consider an integrated approach towards sustainable land and water management. Rainwater harvesting and groundwater recharge have been emphasized as the key activity for Jaggar Watershed, but in Mount Elgon, terracing and check dams should be employed for soil conservation. Some common practices applied include agroforestry, reforestation, and drought-resistant crops in stabilizing degraded lands with improved resilience.

The case studies of the Jaggar Watershed and Mount Elgon well reflect the sequential impacts of climate change and land degradation on agricultural systems, making them highlighting the need for innovative policies, active participation from the community, and sustainable practices. Such measures can contribute to sustainable environmental and agricultural resilience and offer lessons applicable in similar contexts across the globe.

3.11 Results and Discussion

3.11.1 Key Findings

Research on the combined effects of climate change and land degradation on agricultural systems shows globally synthesised results that consistently harm food production, natural resource sustainability, and rural livelihoods. Across various continents and agro-ecological zones, higher temperatures, changes in rainfall patterns, and a growing incidence of severe weather events have greatly changed the circumstances under which farming is practiced. Many studies from Sub-Saharan Africa, South Asia, Central Asia, and portions of South America show significant drops in crop yields, more frequent crop failures, shorter growing seasons, and more pest and disease outbreaks as main results of these climate changes (Mohammed et al., 2020; Soz & Raza, 2024).

At the same time, land degradation keeps depleting the natural resource foundation on which farming relies. Nearly 29% of the Earth's land surface is thought to be already degraded, which negatively impacts the lives of more than 3.2 billion people (Le et al., 2014). Worldwide, soil erosion is still the most common kind of land degradation. Water erosion has greatly depleted topsoil and soil nutrients in humid and sub-humid tropical regions, including the Ethiopian highlands, Nepal, and sections of Western China. By contrast, in dry and semi-arid areas like the Sahel, portions of Central Asia, and Northern Africa, wind erosion predominates and causes significant soil loss, sand encroachment, and desertification (Kiage, 2013; Tadesse & Hailu, 2024).

Climate and degradation stresses have caused measurable drops in agricultural production and productivity. Average soil loss rates from cultivated lands in Ethiopia's highlands surpass 80 tonnes per hectare per year (Taddese, 2018). In degraded and drought-prone regions, Central Asia has recorded agricultural profitability declines of as much as 27% (Mirzabaev et al., 2018). Similar trends of falling soil fertility and productivity losses have been noted in the drylands of West Africa, floodplains of South Asia, and mountainous areas of Latin America, showing the universal, transcontinental character of this issue. Socio-economic factors aggravate these issues even more. High population densities, ongoing rural poverty, unstable land tenure systems, and poor natural resource management increase the strain on land and water resources. Many areas show clear effects from fast land conversion, overgrazing, deforestation, and poor land management practices aggravating soil degradation, deforestation, and biodiversity loss (Alam et al., 2011; Herrmann & Hutchinson, 2005).

Various adaptive and restorative land management techniques have shown success in enhancing agricultural resilience and land productivity in reaction to these ongoing issues. Rainwater harvesting systems in arid India and Ethiopia, integrated watershed management projects in Nepal, China, and parts of East Africa, and agroforestry interventions in the Sahel and Central America have all generated observable benefits, including reductions in soil erosion, improved water retention, and restored soil fertility (Olsson et al., 2005; Roy et al., 2022; Zhang et al., 2021).

3.11.2 Discussion

The information gathered from global, regional, and local studies clearly shows that climate change and land degradation are affecting farming systems everywhere. Rising temperatures, more evapotranspiration, erratic rainfall patterns, and more frequent droughts and floods have upset food production systems, so contributing to food insecurity and economic vulnerability on several continents. Documented from the semi-arid drylands of Africa to the humid highlands of Asia and the irrigated plains of South America, these patterns show the consistent biophysical effects of climatic stressors on agroecosystems (Soz & Raza, 2024; Tadesse & Hailu, 2024). At the same time, land degradation processes—especially soil erosion, salinization, and nutrient depletion—have greatly reduced soil productivity and compromised the long-term viability of agricultural land. Although the rates and severity of degradation vary by area, the basic processes and their cumulative impacts are surprisingly consistent all over. For example, while wind erosion-driven land degradation affects large areas in the Sahel, Northern China, and Central Asia (Kiage, 2013; Zhang et al., 2021), the Ethiopian highlands and portions of Rwanda suffer soil loss rates over 80 tonnes per hectare annually (Taddese, 2018).

These trends reveal that while the biophysical problems brought on by climate change and land degradation are shared all over, the particular manifestations and intensities of these pressures are shaped by local ecological, socio-economic, and institutional circumstances. In arid and semi-arid areas like Niger, Chad, and Rajasthan, recurrent droughts and overgrazing drive desertification and land degradation; steep slopes combined with heavy rain in mountainous regions like Nepal and Ethiopia make landscapes very susceptible to soil erosion and landslides. Intensive irrigation, excessive use of chemical fertilizers, and groundwater depletion have aggravated soil salinity and lowered land productivity in densely populated agricultural areas of South and Southeast Asia (Alam et al., 2011; Herrmann & Hutchinson, 2005).

Though some regional differences exist, several adaptive techniques have shown success in several settings. In dry India, the Ethiopian highlands, and the hills of Nepal, rainwater collecting and integrated watershed management initiatives have effectively reduced soil and water deterioration (Roy et al., 2022). In the drylands of the Sahel, portions of Southeast Asia, and Central America, agroforestry systems have restored soil organic content, improved moisture retention, and moderated microclimates (Olsson et al., 2005). These interactions across different regions highlight the potential to adapt existing land and water management methods to other vulnerable farming systems. Importantly, this synthesis reveals how socio-economic elements aggravate biophysical vulnerabilities. High rural population densities, poverty, insecure land tenure, and poor enforcement of land use rules compound land degradation processes and restrict the ability of farming communities to invest in long-term conservation initiatives. This socio-environmental interaction is clear in Sub-Saharan Africa, where overgrazed communal rangelands and encroachment on marginal agricultural lands have accelerated soil erosion, as well as in South Asia and parts of Latin America, where the expansion of settlements, deforestation, and over-irrigation have caused land and water degradation (Alam et al., 2011; Herrmann & Hutchinson, 2005).

By examining these processes comparatively, the chapter contributes to a broader global understanding of how interconnected environmental and socio-economic drivers shape the sustainability and resilience of agricultural systems. The study shows that although difficulties vary in size and particular expression, fundamental patterns are remarkably similar across continents. It strengthens the need for knowledge sharing as well since effective interventions in one area sometimes provide insightful analysis for other regions under similar environmental and institutional constraints. This integrated knowledge emphasizes the need for comprehensive strategies combining sustainable land and water management with socio-economic development and governance reforms. The converging experiences recorded across Africa, Asia, and Latin America imply that efficient land rehabilitation, community-led restoration projects, and climate-smart agricultural practices can be adapted successfully to different local settings, therefore improving resilience and sustainable food security all around.

3.12 Future Prospects and Research Needs

3.12.1 Innovative Technologies for Soil and Climate Management

Agroecological technologies that use organic and inorganic nutrient sources, including zaï, half-moons, and stone bunds have been described to sustain food production systems and livelihoods in semiarid West Africa (Zougmoré et al., 2014). Conservation tillage, mulching, water harvesting, and nutrient management practices have been reported to maintain better soil health and productivity (Agus et al., 2015). Such practices enhance water use, conservation of soil, and biodiversity in the ecosystem (Clements et al., 2011). Effective Soil management, especially integrating organic matter, increases water-holding capacity, improves soil fertility, and conserves microbial activity and is important in climate adaptation (Agus et al., 2015). These soil management techniques, combined with advanced climate modeling and forecasting technologies, have the potential to address the integrated problems of soil degradation, climate change, and food security (Lal, 2014).

3.12.2 Precision Farming and Remote Sensing

As a backbone of precision farming, remote sensing has been an increasingly crucial insight for measuring, analyzing, and improving agricultural practices (Meng et al., 2011). Remote sensing supports the early detection of crop issues and disease mapping, and also assessment of nutrient demand by monitoring field structure, crop growth, and environmental conditions such as soil nutrition and moisture (Robert, 2003). Polarimetric SAR systems are a class of sensors that have provided promising results in the determination of crop biophysical parameters in the microwave remote sensing domain (Kühbauch & Hawlitschka, 2003). Nevertheless, there are challenges that need to be overcome in order to take full advantage of satellite imagery, including spatial resolution, timeliness, and imagery management (Robert, 2003). Future developments, such as the fusion of multi-source remote sensing data with agro-crop models, will enable more precise farming practices (Meng et al., 2011).

It holds the potential for more sustainable and profitable agriculture, with the added benefit of protecting environmental quality.

3.12.3 Integrating AI, Machine Learning, and Carbon Sequestration Technologies

Advancements in digital technologies, particularly artificial intelligence (AI) and machine learning (ML), offer promising avenues to enhance climate-resilient agricultural practices and mitigate land degradation. Machine learning and artificial intelligence algorithms can optimize resource use, predict crop output, and monitor soil health. AI-driven models, for instance, have been used to assess soil organic carbon levels, so facilitating targeted soil management strategies (Sambrook et al., 2023). Including artificial intelligence into carbon capture and storage (CCS) systems helps to increase the efficiency of carbon sequestration processes by means of monitoring and operational optimisation as well (Smith et al., 2020). Future research should focus on developing affordable and scalable applications of these technologies, therefore ensuring access for smallholder farmers in resource-limited settings. Cooperative efforts among governments, private enterprises, and local communities will help to support the adoption of such innovations and build the necessary capacity for their effective execution.

3.12.4 Strengthening Resilient Agricultural Systems

Current agroecological practices, which include crop diversification, intercropping, and conservation agriculture are important in developing sustainable agricultural systems. The utilization of drought-tolerant seeds, erosion-sensitive soil plantations, as well as water conservation irrigation systems has been effective in managing risks (Amiraslani & Caiserman, 2018). It is apparent that better management of organic matter combined with new technologies such as biochar and nanotechnology can improve soil health, increase its resilience, and positively contribute to its ability to cope with climate change. Hence, better coordination among the various stakeholders and complete databases of actual site data recording land management details will facilitate scaling up successful interventions. As opined by Berry et al. (2018), the adaptation of measures like groundcover management and total grazing pressure management will be encouraged while strengthening the rural agricultural economy, promoting farmers' training.

3.12.5 Collaboration Between Farmers, Scientists, and Policymakers

This chapter recognizes the need for farmer, scientist, and policymaker to collaborate in response to the problems of climate change in agriculture. Some best practices are incorporating the land degradation processes into adaptation planning, conducting vulnerability assessment, and encouraging knowledge exchange (Webb et al., 2017). For better and more effective solutions, there is a need to integrate both the global and the local perspectives on land management (Berry et al., 2018). Therefore, policymakers should encourage the use of sustainable practices, incentives should be provided to environmentally friendly technologies, and make them available resources to farmers. On the other hand, the scientists should develop their research in soil rehabilitation, carbon storage, and climate change. The stakeholders will be able to develop sound frameworks to deal with the future conditions through the application of local knowledge and adaptive management practices (Berry et al., 2018).

Therefore, with these measures being integrated, agricultural systems can be able to respond to land degradation and the impacts of climate change in a sustainable way to improve food security and resilience.

3.12.6 Feasibility, Limitations, and Trade-Offs of Adaptation Strategies

Although several land management and climate adaptation plans have been suggested and put into practice across agro-ecological systems, their more general use usually runs into major feasibility issues, practical constraints, and socio-economic trade-offs. Though their adoption is quite context-dependent, these strategies show potential for reducing the effects of land degradation and climate change.

Financial and Institutional Constraints

Financial and institutional feasibility remains a key constraint, particularly in low-income, resource-limited settings. Often, smallholder farmers lack the financial resources needed to buy agroforestry, soil and water conservation technologies, or enhanced irrigation systems. A study in Southwest Nigeria found that 80% of smallholder farmers are financially vulnerable to climate change risks, with 37% under severe financial constraints. The study projected that an average of $106.55 could improve farmers' adaptive capacity, therefore lowering their susceptibility to climate-related shocks (Fatoki et al., 2020). Apart from financial constraints, inadequate institutional support, poor land tenure policies, and restricted access to extension services hinder the broad implementation of these plans (IPCC, 2019a, 2019b).

Technical Limitations

Technical limitations also affect the suitability and scalability of many adaptation measures. While effective in hilly and erosion-prone areas, like Mount Elgon in Uganda, terracing and other labour-intensive practices are less relevant in dry or flat regions because of their maintenance requirements. Likewise, although rainwater collecting has been shown to be good in semi-arid regions of India, its use in humid highlands might result in too much runoff capture, therefore harming downstream environments (FAO, 2019).

Trade-offs and Unintended Consequences

Trade-offs must also be considered when implementing adaptation strategies. For instance, agroforestry could lower erosion, increase biodiversity, and improve soil fertility but might momentarily limit land available for yearly food crops, therefore impacting short-term household food security (Rahman et al., 2023). Large-scale projects like reforestation and check dam building often call for land acquisition, which could displace rural families or change current land-use patterns and cause social strife (UNCCD, 2023a, 2023b).

Furthermore, some activities could have unintentional environmental effects. If not properly designed at the watershed level, rainwater-collecting systems could change natural hydrological cycles and impact downstream water availability (FAO, 2019). Though beneficial for carbon sequestration and erosion control, afforestation with non-native or fast-growing species could harm local biodiversity and water supplies (IPCC, 2019a, 2019b).

Need for Context-Specific Approaches

Policymakers, development organizations, and community leaders must acknowledge these constraints and compromises if they are to create socially acceptable, context-appropriate, and environmentally responsible land management and climate adaptation initiatives. To guarantee the viability and sustainability of climate-resilient agricultural systems, future initiatives should stress participatory strategies, long-term capacity building, safe land tenure arrangements, and flexible, locally adapted solutions.

3.13 Conclusions

The combined effects of climate change and land degradation have been shown in this chapter to pose a common, growing difficulty for agricultural systems all around. The synthesis of global, regional, and local data verifies that increasing temperatures, changed precipitation patterns, extended droughts, and severe weather events are directly compromising the stability and output of agroecosystems. When combined with ongoing soil erosion, nutrient depletion, salinization, and other types of land degradation, these climatic stressors endanger not only food security and rural livelihoods but also the long-term viability of the natural resources necessary for agricultural resilience. One important realization from this study is the awareness that although the intensity and local expressions of these issues differ by area and agro-ecological context, the basic environmental and socio-economic forces stay surprisingly constant across continents. Consistently aggravating land degradation trends are rapid population growth, poverty, insecure land tenure, and poor governance systems, which therefore hinder the ability of vulnerable rural communities to respond to environmental pressures. The comparative study offered here supports the requirement for integrated, locally relevant, and scalable land and water management strategies able to handle several interacting stresses.

The results of this chapter have significant consequences for scientific research and policy at both regional and global levels. The chapter provides a useful comparison of the worldwide aspects of land degradation and climate change effects on agriculture by including data from various agro-ecological systems—including drylands, humid highlands, and irrigated lowlands. This wider perspective helps to improve the scientific knowledge of how biophysical and socio-economic systems interact to shape agricultural vulnerability under changing environmental conditions. Moreover, the synthesis emphasizes a portfolio of adaptive techniques that have proven effective in restoring damaged lands and improving agroecosystem resilience. The important lessons learnt from community-led land restoration projects, integrated watershed management, agroforestry, and rainwater collecting are valuable worldwide because they offer adaptable methods that can improve farming sustainability and food security in at-risk regions.

By emphasizing the need for cross-regional knowledge exchange and evidence-based policy interventions, this chapter, in a broader sense, adds to the increasing global conversation on sustainable land management, climate adaptation, and rural development. The chapter highlights the common problems facing agricultural landscapes all over the world by stressing the need for cooperative, inclusive, and scientifically informed land and resource management in the framework of fast-changing climate and environmental change by placing regional case evidence inside a worldwide comparative framework.

Funding The research was not supported by any funding.

Potential Conflicts of Interest There is no conflict of interest.

References

AbdelRahman, M. A. (2023). An overview of land degradation, desertification, and sustainable land management using GIS and remote sensing applications. *Rendiconti Lincei. Scienze Fisiche e Naturali, 1–42*. https://doi.org/10.1007/s12210-023-01155-3

Abebaw, W. A. (2019). Review on impacts of land degradation on agricultural production in Ethiopia. *Journal of Resources Development and Management, 57*. https://doi.org/10.7176/JRDM

Agus, F., Husnain, H., & Yustika, R. D. (2015). Improving agricultural resilience to climate change through soil management. *Jurnal Penelitian dan Pengembangan Pertanian, 34*(4), 30945. https://doi.org/10.21082/jp3.v34n4.2015.p147-158

Ahmed, A. R., & Rahman, A. (2021). Measuring agriculture loss of flood-affected farmers in Assam with special reference to Lower Brahmaputra Valley. *RRJAAS, 10*(2). https://www.rroij.com/open-access/measuring-agriculture-loss-of-floodaffected-farmers-in-assam-with-special-reference-to-lower-brahmaputra-valley.pdf

Akune, Y., Okiyama, M., & Tokunaga, S. (2015). Economic evaluation of dissemination of high temperature-tolerant rice in Japan using a dynamic computable general equilibrium model. *Jarq-Japan Agricultural Research Quarterly, 49*, 127–133. https://doi.org/10.6090/jarq.49.127

Alagesan, A., Vaishnavi, P., & Karthikeyan, R. (2020). Development of computational decision making tool for predicting the growth and development of rice crop using location specific diurnal air temperature data. *Reliability and Risk Assessment in Engineering: Proceedings of INCRS, 2018*, 439–444. https://doi.org/10.1007/978-981-15-3746-2_41

Alam, M. M., Toriman, M., Siwar, C., & Talib, B. (2011). Rainfall variation and changing pattern of agricultural cycle, 82–89. https://doi.org/10.3844/ajessp.2011.82.89

Alori, E. T., Adekiya, A. O., & Adegbite, K. A. (2020). Impact of agricultural practices on soil health. *Soil Health*. https://doi.org/10.1007/978-3-030-44364-1_5

Amiraslani, F., & Caiserman, A. (2018). Multi-stakeholder and multi-level interventions to tackle climate change and land degradation: The case of Iran. *Sustainability, 10*(6), 2000. https://doi.org/10.3390/su10062000

Angaman, K. V., Mirzabaev, A., & Niang, B. B. (2024). Economic impacts of land degradation: Evidence from Côte d'Ivoire. *Land Degradation & Development, 35*, 1541–1552. https://doi.org/10.1002/ldr.5004

Attri, S. D., Sudheer, K., Chattopadhyay, N., Tiwari, S., & Anita, K. (2020). The study of frost occurrence and risk analysis in Indo-Gangetic Plains of India during recent decades. *Mausam, 71*(1), 95–102. https://doi.org/10.54302/mausam.v71i1.9

Barman, D., Mandal, S. C., Bhattacharjee, P., & Ray, N. (2013). Land degradation: Its control, management and environmental benefits of management about agriculture and aquaculture. *Environment and Ecology, 31*, 1095–1103.

Berhanu, A. A., Ayele, Z. B., Dagnew, D. C., Melese, T., Fenta, A. B., & Kassie, K. E. (2024). Smallholder farmers' vulnerability to climate change and variability: Evidence from three agroecologies in the Upper Blue Nile, Ethiopia. *Heliyon, 10*(7). https://doi.org/10.1016/j.heliyon.2024.e28277

Berry, E., Metternicht, G., & Baumber, A. (2018). 'This country just hangs tight': Perspectives on managing land degradation and climate change in far west NSW. *The Rangeland Journal, 41*(3), 197–210. https://doi.org/10.1071/RJ18030

Bhatla, R., Bhattacharyya, S., Verma, S., Mall, R. K., & Singh, R. S. (2021). El Nino/La Nina and IOD impact on Kharif season crops over western agro-climatic zones of India. *Theoretical and Applied Climatology, 151*, 1355–1368. https://doi.org/10.1007/s00704-023-04361-z

Bhattacharyya, R., Ghosh, B., Mishra, P. K., Mandal, B., Rao, C. S., Sarkar, D., Das, K., Anil, K. S., Lalitha, M., Hati, K. M., & Franzluebbers, A. J. (2015). Soil degradation in India: Challenges and potential solutions. *Sustainability, 7*, 3528–3570. https://doi.org/10.3390/su7043528

Bossio, D., Noble, A., Molden, D., & Nangia, V. (2008). Land degradation and water productivity in agricultural landscapes, 20–32. https://doi.org/10.1079/9781845933876.0020

Bowonder, B., Ramana, K. V., Ravi, C. S., & Srinivas, C. V. (1987). Land use, waterlogging and irrigation management. *Land Use Policy, 4*, 331–341. https://www.ui.adsabs.harvard.edu/link_gateway/1987LUPol...4..331B/10.1016/0264-8377(87)90032-9

Brémond, P., Grelot, F., & Agenais, A. (2013). Review article: Economic evaluation of flood damage to agriculture—Review and analysis of existing methods. *Natural Hazards and Earth System Sciences, 13*, 2493–2512. https://doi.org/10.5194/nhess-13-2493-2013

Call, M., & Gray, C. (2020). Climate anomalies, land degradation, and rural out-migration in Uganda. *Population and Environment, 41*(4), 507–528. https://doi.org/10.1007/s11111-020-00349-3

Calzadilla, A., & Carr, T. W. (2020). Land degradation and food security: Water erosion impacts and adaptation options. https://www.gtap.agecon.purdue.edu/resources/res_display.asp?RecordID=6088

Cao, J., Yeh, E. T., Holden, N. M., Qin, Y., & Ren, Z. (2013). The Roles of overgrazing, climate change and policy as drivers of degradation of China's grasslands. *Nomadic Peoples, 17*, 82–101. https://doi.org/10.3167/np.2013.170207

Carroll, E., Chang, J., Lodi, L., Rapsomanikis, G., Zimmermann, A., & Blandford, D. (2018). The state of agricultural commodity markets 2018: agricultural trade, climate change and food security. https://openknowledge.fao.org/server/api/core/bitstreams/2263cf55-1ad8-4121-929d-bd9d34f786e0/content

Chakraborty, D., Saha, S., & Singh, R. K. (2015). Change in near-surface wind velocity: What implications will it have on ecosystem and agriculture? *Current Science, 108*(5), 761.

Chalise, D., Kumar, L., & Kristiansen, P. (2019). Land degradation by soil erosion in Nepal: A review. *Soil Systems, 3*(1), 12. https://doi.org/10.3390/soilsystems3010012

Choudhury, B. U., Nengzouzam, G., Ansari, M. A., & Islam, A. (2022). Causes and consequences of soil erosion in northeastern Himalaya, India. *Current Science*. https://doi.org/10.18520/cs/v122/i7/772-789

Clements, R., Haggar, J., Quezada, A., & Torres, J. (2011). Technologies for climate change adaptation: agricultural sector. https://backend.orbit.dtu.dk/ws/portalfiles/portal/5706575/Technologies_for_Climate_Change_Adaptation_Agriculture_sector.pdf

Colazo, J. C., & Buschiazzo, D. E. (2015). The impact of agriculture on soil texture due to wind erosion. *Land Degradation & Development, 26*, 62–70. https://doi.org/10.1002/ldr.2297

Dagar, J. C., & Minhas, P. (Eds.). (2016). *Agroforestry for the management of waterlogged saline soils and poor-quality waters*, 201. https://doi.org/10.1007/978-81-322-2659-8

Dai, S., Ma, Y., & Zhang, K. (2022). Land degradation caused by construction activity: Investigation, cause and control measures. *International Journal of Environmental Research and Public Health, 19*. https://doi.org/10.3390/ijerph192316046

Das, K. (2012). *Farm productivity loss due to flood-induced sand deposition: a study in Dhemaji, India*. South Asian Network for Development and Environmental Economics. https://econpapers.repec.org/RePEc:snd:wpaper:73

Datta, S. (2013). Impact of climate change in Indian horticulture-A review. *International Journal of Science, Environment and Technology*, 2(4), 661–671. https://doi.org/10.11648/j.bio.20210906.12

Dellal, I., & McCarl, B. A. (2010). The economic impacts of drought on agriculture: The case of Turkey. *Options Méditerranéennes, 95*, 169–174. http://ressources.ciheam.org/om/pdf/a95/00801342.pdf

Do, V. Q., Phung, M. L., Truong, D. T., Pham, T. T. T., Dang, V. T., & Nguyen, T. K. (2021). The impact of extreme events and climate change on agricultural and fishery enterprises in Central Vietnam. *Sustainability, 13*(13), 7121. https://doi.org/10.3390/su13137121

Drake, V. A. (1994). The influence of weather and climate on agriculturally important insects: An Australian perspective. *Crop & Pasture Science, 45*, 487–509. https://doi.org/10.1071/AR9940487

Dwivedi, R. S., Sreenivas, K., & Ramana, K. V. (1999). Inventory of salt-affected soils and waterlogged areas: A remote sensing approach. *International Journal of Remote Sensing, 20*, 1589–1599. https://doi.org/10.1080/014311699212623

Ebrahimzadeh, I., & Esmaelnejad, M. (2013). Climate changes and the role of recent droughts on agricultural economy of Sistan. *Romanian Review of Regional Studies, 9*(1). https://www.researchgate.net/publication/330764865_CLIMATE_CHANGES_AND_THE_ROLE_OF_RECENT_DROUGHTS_ON_AGRICULTURAL_ECONOMY_OF_SISTAN

Eni, D. I., Atu, J. E., Oko, C. O., & Ekwok, I. (2011). Flood and its impact on farmlands in itigidi, abi local government area, cross river state, nigeria. *International Journal of Humanities and Social Sciences, 1*(9), 98–104. https://www.researchgate.net/publication/326683694_Flood_and_its_impact_on_farmlands_in_Itigidi_Abi_Local_Government_Area

Eswaran, H., Lal, R., & Reich, P. F. (2019). Land degradation: An overview. *Response to Land Degradation*. https://doi.org/10.1201/9780429187957-4

FAO. (2019). *Rainwater Harvesting for Agriculture in the Near East and North Africa: Status, Challenges and Outlook*. Food and Agriculture Organization of the United Nations. https://www.fao.org/3/ca6145en/CA6145EN.pdf

Fatoki, O. A., Adesope, A. A., Awe, F., & Arowolo, O. V. (2020). Assessment of smallholder farmers' vulnerability to climate change in Ogun State, Nigeria. *Russian Journal of Agricultural and Socio-Economic Sciences, 6*(102), 136–143. https://doi.org/10.18551/rjoas.2020-06.16

Ferreira, C. S., Kalantari, Z., Salvati, L., Canfora, L., Zambon, I., & Walsh, R. P. (2019). Urban areas. *Soil Degradation, Restoration and Management in a Global Change Context*. https://www.researchgate.net/publication/337012746_Soil_Degradation_Restoration_and_Management_in_a_Global_Change_Context

Gao, J., & Liu, Y. (2010). Determination of land degradation causes in Tongyu County, Northeast China via land cover change detection. *International Journal of Applied Earth Observation and Geoinformation, 12*, 9–16. https://doi.org/10.1016/j.jag.2009.08.003

Gao, L., Tao, B., Miao, Y., Zhang, L., Song, X., Ren, W., He, L., & Xu, X. (2019). A global data set for economic losses of extreme hydrological events during 1960–2014. *Water Resources Research, 55*, 5165–5175. https://doi.org/10.1029/2019WR025135

Geneti, T. Z. (2019). Review on the effect of moisture or rain fall on crop production. *Civil and Environmental Research, 11*(2). https://doi.org/10.7176/CER

Gorain, S., Kuriachen, P., Kumar, C. V., & Suresh, A. (2023). Land degradation and its impact on agricultural productivity: The case of India. *Land Degradation & Development, 35*, 196–212. https://doi.org/10.1002/ldr.4909

Goswami, B., Mahi, G., & Saikia, U. S. (2006). Effect of few important climatic factors on phenology, growth and yield of rice and wheat—A review. *Agricultural Reviews, 27*, 223–228.

Hailu, B., & Mehari, H. (2021). Impacts of soil salinity/sodicity on soil-water relations and plant growth in dry land areas: A review. *Journal of Natural Science Research, 12*, 1–10. https://doi.org/10.7176/JNSR/12-3-01

Harper, R. J., Dell, B., Ruprecht, J. K., Sochacki, S. J., & Smettem, K. R. J. (2021). Salinity and the reclamation of salinized lands. *Soils and Landscape Restoration*. https://doi.org/10.1016/B978-0-12-813193-0.00007-2

Herrmann, S. M., & Hutchinson, C. F. (2005). The changing context of the desertification debate. *Journal of Arid Environments, 63*(3), 538–555. https://doi.org/10.1016/j.jaridenv.2005.03.003

Hilker, T., Natsagdorj, E., Waring, R. H., Lyapustin, A. I., & Wang, Y. (2014). Satellite observed widespread decline in Mongolian grasslands largely due to overgrazing. *Global Change Biology*. https://doi.org/10.1111/gcb.12365

Hosonuma, N., Herold, M., de Sy, V., De Fries, R. S., Brockhaus, M., Verchot, L. V., Angelsen, A., & Romijn, E. (2012). An assessment of deforestation and forest degradation drivers in developing countries. *Environmental Research Letters*. https://doi.org/10.1088/1748-9326/7/4/044009

Huo ZhiGuo, H. Z., Li MaoSong, L. M., Wang Li, W. L., Xiao JingJing, X. J., Huang DaPeng, H. D., & Wang ChunYan, W. C. (2012). *Impacts of Precipitation Variations on Crop Diseases and Pests in China*. https://doi.org/10.3864/j.issn.0578-1752.2012.10.006

Hussain, S., Rao, M. J., Anjum, M. A., Ejaz, S., Zakir, I., Ali, M. A., Ahmad, N., & Ahmad, S. (2019). Oxidative stress and antioxidant defense in plants under drought conditions. *Plant Abiotic Stress Tolerance: Agronomic, Molecular and Biotechnological Approaches*. https://doi.org/10.1007/978-3-030-06118-0_9

Ioras, F., Bandara, I., & Kemp, C. (2014). Introduction to climate change and land degradation. In *Climate change and restoration of degraded land* (pp. 15–52).

IPCC. (1995). Intergovernmental panel on climate change (IPCC). *CC BY, 4*. https://www.ipcc.ch/site/assets/uploads/2018/02/ipcc_sar_wg_I_full_report.pdf

IPCC. (2014). Climate change 2014 impacts, adaptation, and vulnerability. In *Contribution of working group II to the fifth assessment report of the intergovernmental panel on climate change*. https://www.researchgate.net/publication/307002491_Climate_Change_2014_Impacts_Adaptation_and_Vulnerability_Working_Group_II_Contribution_to_the_Fifth_Assessment_Report_of_the_IPCC_Australasia_Intergovernmental_Panel_on_Climate_Change

IPCC. (2019). *Climate Change and Land: An IPCC Special Report on Climate Change, Desertification, Land Degradation, Sustainable Land Management, Food Security, and Greenhouse Gas Fluxes in Terrestrial Ecosystems*. https://www.ipcc.ch/srccl/

IPCC. (2019). *Special Report on Climate Change and Land*. Retrieved from https://www.ipcc.ch/srccl/

Jabbar, M. T., & Chen, X. (2008). Land degradation due to salinization in arid and semi-arid regions with the aid of geo-information techniques. *Geo-Spatial Information Science, 11*, 112–120. https://doi.org/10.1007/s11806-008-0013-z

Jang, W. S., Neff, J. C., Im, Y., Doro, L., & Herrick, J. E. (2020). The hidden costs of land degradation in US Maize agriculture. *Earth's Future*. https://doi.org/10.1029/2020EF001641

Jarvis, A., Ramirez, J., Anderson, B., Leibing, C., & Aggarwal, P. (2010). Scenarios of climate change within the context of agriculture. *Climate Change and Crop Production*. https://doi.org/10.1079/9781845936334.0009

Javed, A., Jamal, S., & Khandey, M. Y. (2012). Climate change induced land degradation and socio-economic deterioration: A remote sensing and GIS based case study from Rajasthan, India. *Journal of Geographic Information System, 4*(3), 219–228. https://doi.org/10.4236/jgis.2012.43026

Jayan, P. R., & Sathyanathan, N. (2010). Overview of farming practices in the water-logged areas of Kerala, India. *International Journal of Agricultural and Biological Engineering, 3*, 28–43. https://doi.org/10.3965/j.issn.1934-6344.2010.04.018-033

Jiang, B., Bamutaze, Y., & Pilesjö, P. (2014). Climate change and land degradation in Africa: A case study in the Mount Elgon region, Uganda. *Geo-Spatial Information Science, 17*(1), 39–53. https://doi.org/10.1080/10095020.2014.889271

Jiang, K., Teuling, A. J., Chen, X., Huang, N., Wang, J., Zhang, Z., & Gao, R. (2024). Global land degradation hotspots based on multiple methods and indicators. *Ecological Indicators, 158*, Article 111462. https://doi.org/10.1016/j.ecolind.2023.111462

Kairis, O., Karavitis, C. A., Salvati, L., Kounalaki, A., & Kosmas, K. (2015). Exploring the impact of overgrazing on soil erosion and land degradation in a dry Mediterranean Agro-Forest Landscape (Crete, Greece). *Arid Land Research and Management, 29*, 360–374. https://doi.org/10.1080/15324982.2014.968691

Kajal, M., Yadav, V. K., Chhokar, R. S., & Sharma, R. K. (2015). Management practices to mitigate the impact of high temperature on wheat. *Journal of Wheat Research, 7*. https://ebook.icar.gov.in/index.php/JWR/article/view/46814

Katsaruware-Chapoto, R. D., Mafongoya, P. L., & Gubba, A. (2017). Responses of insect pests and plant diseases to changing and variable climate: A review. *The Journal of Agricultural Science, 9*, 160. https://doi.org/10.5539/jas.v9n12p160

Kaur, S., & Kumar, P. (2020). Morpho-physiological and biochemical response of plants under drought stress. *Journal of Pharmacognosy and Phytochemistry, 9*(4), 352–357.

Khan, M. I. R., Asgher, M., & Khan, N. A. (2013). Rising temperature in the changing environment: A serious threat to plants. *Climate Change and Environmental Sustainability, 1*(1), 25–36. https://doi.org/10.5958/j.2320-6411.1.1.004

Kiage, L. M. (2013). Perspectives on the assumed causes of land degradation in the rangelands of Sub-Saharan Africa. *Progress in Physical Geography, 37*, 664–684. https://doi.org/10.1177/0309133313492543

Komatsuzaki, M., & Ohta, H. (2013). Sustainable agriculture practices. *Designing our Future, United Nations*. https://doi.org/10.18356/4f898e93-en

Kopittke, P. M., Menzies, N. W., Wang, P., McKenna, B. A., & Lombi, E. (2019). Soil and the intensification of agriculture for global food security. *Environment International, 132*, Article 105078. https://doi.org/10.1016/j.envint.2019.105078

Kozlowski, T. T. (2002). Physiological-ecological impacts of flooding on riparian forest ecosystems. *Wetlands, 22*, 550–561. https://doi.org/10.1672/0277-5212(2002)022[0550:PEIOFO]2.0.CO;2

Kühbauch, W., & Hawlitschka, S. (2003). Remote sensing-a future technology in precision farming. In *Applications of SAR Polarimetry and Polarimetric Interferometry, 529*.

Kumar, M. (2019). Effect of climate change on indian economy and agriculture. *International Journal of Scientific Research and Review*. ISSN NO: 2279-543X,07(01).

Kurukulasuriya, P., & Rosenthal, S. (2003). Climate change and agriculture. A review of impacts and adaptations. *World Bank Environment Department Paper, 91*. https://hdl.handle.net/10986/16616

Lal, R. (1997). Degradation and resilience of soils. *Philosophical Transactions of the Royal Society B, 352*, 997–1010. https://doi.org/10.1098/rstb.1997.0078

Lal, R. (2014). Climate strategic soil management. *Challenges, 5*(1), 43–74. https://doi.org/10.3390/challe5010043

Le, Q., Nkonya, E. M., & Mirzabaev, A. (2014). Biomass productivity-based mapping of global land degradation hotspots. *EcoRN: Land Use Land Cover Change*. https://doi.org/10.1007/978-3-319-19168-3_4

Lobell, D., & Gourdji, S. M. (2012). The influence of climate change on global crop productivity1. *Plant Physiology, 160*, 1686–1697. https://doi.org/10.1104/pp.112.208298

Lyles, L. (1977). Wind erosion: Processes and effect on soil productivity. *Transactions of the ASABE, 20*, 880–0884.

Mahata, S., & Sharma, V. N. (2021). The global problem of land degradation: A review. *National Geographical Journal of India, 67*, 216–231. https://doi.org/10.48008/ngji.1773

Mahato, A. (2014). Climate change and its impact on agriculture. *International Journal of Scientific and Research Publications, 4*(4), 1–6.

Malav, L. C., Yadav, B., Tailor, B. L., Pattanayak, S., Singh, S. V., Kumar, N., et al. (2022). Mapping of land degradation vulnerability in the semi-arid watershed of Rajasthan, India. *Sustainability, 14*(16), 10198. https://doi.org/10.3390/su141610198

Malhi, G. S., Kaur, M., & Kaushik, P. (2021). Impact of climate change on agriculture and its mitigation strategies: A review. *Sustainability, 13*(3), 1318. https://doi.org/10.3390/su13031318

Mangala, R. R., & Padmapriya, A. (2019). Visualizing the impact of climatic changes on pest and disease infestation in rice. *International Journal of Recent Technology and Engineering, 8*(3), 8413–8421. https://doi.org/10.35940/ijrte.C6522.098319

Marmai, N., Franco Villoria, M., & Guerzoni, M. (2022). How the Black Swan damages the harvest: Extreme weather events and the fragility of agriculture in development countries. *PLoS ONE, 17*(2), Article e0261839. https://doi.org/10.1371/journal.pone.0261839

Marseva, A. D., Putri, E. I. K., & Ismail, A. (2022). Estimasi Kerugian Ekonomi Petani Bawang Merah (Studi Kasus Kabupaten Brebes). *Jurnal Ilmiah Inovasi, 22*(3), 291–296. https://doi.org/10.25047/jii.v22i3.3598

Mauget, S. A., & Upchurch, D. R. (1999). El Niño and La Niña Related Climate and Agricultural Impacts over the Great Plains and Midwest. *Journal of Production Agriculture, 12*, 203–215. https://doi.org/10.2134/jpa1999.0203

McDonald, E. (1996). Interactions of desertification and climate. *Eos, Transactions American Geophysical Union, 77*, 379–379. https://www.ui.adsabs.harvard.edu/link_gateway/1996EOSTr..77..379M/10.1029/96EO00261

Meng, J. H., Wu, B. F., Du, X., Zhang, F., Zhang, M., & Dong, T. (2011). A review and outlook of applying remote sensing to precision agriculture. *Remote Sensing for Land & Resources, 23*(3), 1–7. https://doi.org/10.6046/gtzyyg.2011.03.01

Middleton, N. (2018). Rangeland management and climate hazards in drylands: dust storms, desertification and the overgrazing debate. *Natural Hazards, 92*, 57–70. https://www.link.springer.com/article/10.1007/s11069-016-2592-6

Mirzabaev, A., Strokov, A., & Krasilnikov, P. (2018). *The Impact of Land Degradation on Agricultural Profits and Poverty in Central Asia.* https://doi.org/10.1016/j.landusepol.2022.106530

Mittal, P. K., Gottam, G. S., Gupta, B., & Bilochi, D. R. (2019). The effect of climate change on productivity and reproductive and health performance of livestock: A review. *Journal of Entomology and Zoology Studies, 7*, 04–09.

Modaihsh, A. S., Sallam, A. S., Ghoneim, A. M., & Mahjoub, M. O. (2014). Assessing salt-affected degraded soils using remote sensing. Case study: Al-Qassim region, Saudi Arabia. *Journal of Food Agric. Environ, 12*, 383–388.

Mohammed, A., Deng, X., Kabenge, M., Li, Z., & Juma, D. (2020). Spatiotemporal patterns of drought, its impact on crop production and the economy in South Asia. *Authorea Preprints.* https://doi.org/10.22541/au.158989996.65197123

Monteith, J. L. (1965). Radiation and crops. *Experimental Agriculture, 1*(4), 241–251. https://doi.org/10.1017/S0014479700021529

Muchiri, P. M., Karanja, F., Ogara, W. O., & Maweu, J. M. (2020). Analysis of historical rainfall variability, droughts and wet Seasons. A case of Wiyumiririe Laikipia County. *Journal of Sustainability, Environment and Peace, 3*(1) 15–20.

Ojo, M. P., Ayanwale, A. B., Adelegan, O. J., Ojogho, O., Awoyelu, D. E. F., & Famodimu, J. (2024). Climate change vulnerability and adaptive capacity of smallholder farmers: A financing gap perspective. *Environmental and Sustainability Indicators, 24*, 100476. https://doi.org/10.1016/j.indic.2024.100476

Oliveira, E., Tobias, S., & Hersperger, A. M. (2018). Can strategic spatial planning contribute to land degradation reduction in urban regions? State of the art and future research. *Sustainability, 10*(4), 949. https://doi.org/10.3390/su10040949

Olsson, L., Eklundh, L., & Ardö, J. (2005). A recent greening of the Sahel—trends, patterns and potential causes. *Journal of Arid Environments, 63*(3), 556–566. https://doi.org/10.1016/j.jaridenv.2005.03.008

Orlov, A., Sillmann, J., Aaheim, A., Aunan, K., & de Bruin, K. (2019). Economic losses of heat-induced reductions in outdoor worker productivity: A case study of Europe. *Economics of Disasters and Climate Change, 3*, 191–211. https://doi.org/10.1007/s41885-019-00044-0

Patel, P. K., Singh, A. K., Tripathi, N., Yadav, D., & Hemantaranjan, A. (2014). Flooding: abiotic constraint limiting vegetable productivity. *Advances in Plants & Agriculture Research, 1*(3), 00016. https://doi.org/10.15406/apar.2014.01.00016

Pathak, H., Aggarwal, P. K., & Singh, S. D. (2012). Climate change impact, adaptation and mitigation in agriculture: Methodology for assessment and applications, India.

Pawar, P., & Sakhale, B. (2022). *Cascading Impacts of Change in Carbon Dioxide Concentration, Temperature and Precipitation on Horticulture Crops: A Review.* https://doi.org/10.18805/ag.R-2579

Pierre, C., Kergoat, L., Hiernaux, P., Baron, C., Bergametti, G., Rajot, J. L., Abdourhamane Touré, A., Okin, G. S., & Marticorena, B. (2018). Impact of agropastoral management on wind erosion in Sahelian croplands. *Land Degradation & Development, 29*, 800–811. https://doi.org/10.1002/ldr.2783

Prokop, P. (2020). Remote sensing of severely degraded land: Detection of long-term land-use changes using high-resolution satellite images on the Meghalaya Plateau, northeast India. *Remote Sensing Applications: Society and Environment, 20*, Article 100432. https://doi.org/10.1016/j.rsase.2020.100432

Rahman, S. A., Samsudin, Y. B., & Bhatta, K. P. (2023). The Role of agroforestry systems for enhancing biodiversity and provision of ecosystem services in agricultural landscapes in Southeast Asia. In J. C. Dagar et al. (Eds.), *Agroforestry for Sustainable Intensification of Agriculture in Asia and Africa* (pp. 303–320). Springer Nature Singapore.https://doi.org/10.1007/978-981-19-4602-8_10

Ramirez, F., & Kallarackal, J. (2015). Responses of fruit trees to global climate change. *Springer.* https://doi.org/10.1007/978-3-319-14200-5

Ray, K., Arora, K., & Srivastav, A. K. (2019). Weather extremes and agriculture. *The International Archives of the Photogrammetry, Remote Sensing and Spatial Information Sciences, 42*, 493–497. https://doi.org/10.5194/isprs-archives-XLII-3-W6-493-2019

Rehman, A. (2015). Flood disaster in Pakistan and its impact on agriculture growth (a review). *Journal of Economics and Sustainable Development, 6*, 39–42.

Robert, P. C. (2003). Precision agriculture: New developments and needs in remote sensing and technologies. *Ecosystems' Dynamics, Agricultural Remote Sensing and Modeling, and Site-Specific Agriculture, 5153*, 85–94. https://doi.org/10.1117/12.514608

Roy, P., Pal, S. C., Chakrabortty, R., Islam, A. R., Chowdhuri, I., & Saha, A. (2022). Impact of ineffective measures on the increasing land degradation in a monsoon-dominated region of India: Issues and policy implications. *Land Degradation & Development, 33*, 3174–3185. https://www.ui.adsabs.harvard.edu/link_gateway/2022LDeDe..33.3174R/10.1002/ldr.4380

Sajid, M., Kifayat, M., Salman, M., Usman, M., Rahman, S. U., & Saddique, M. (2021). Effects of high temperature on morphological and physiological stages of different cultivated crops. *Mediterranean Journal of Basic and Applied Sciences (MJBAS), 5*(4), 44–50. https://doi.org/10.46382/MJBAS.2021.5405

Sakai, A., & Larcher, W. (1987). Low temperature and frost as environmental factors. In *Frost survival of plants. Ecological studies* (Vol. 62). Springer, Berlin, Heidelberg. https://doi.org/10.1007/978-3-642-71745-1_1

Sambrook, R. A., Pigozzi, B. W., & Thomas, R. N. (2023). Smallholder farmer resilience to extreme weather events in a global food value chain. *Climatic Change, 169*, 45. https://doi.org/10.1007/s10584-023-03586-1

Sanchez, B., Rasmussen, A. N., & Porter, J. R. (2014). Temperatures and the growth and development of maize and rice: A review. *Global Change Biology, 20*. https://doi.org/10.1111/gcb.12389

Scanlon, B. R., Jolly, I., Sophocleous, M., & Zhang, L. (2007). Global impacts of conversions from natural to agricultural ecosystems on water resources: Quantity versus quality. *Water Resources Research, 43*(3). https://doi.org/10.1029/2006WR005486

Shah, R., & Srivastava, R. (2017). Effect of global warming on Indian agriculture. *Sustainable Environment, 2*(4), 366. https://doi.org/10.22158/se.v2n4p366

Sharma, R. (2020). Soil health and land degradation related problems and remedial measures with special reference to central and peninsular India. *International Journal of Chemical Studies, 8*, 771–785. https://doi.org/10.22271/chemi.2020.v8.i2l.8861

Shil, S. (2018). Weather parameters and it's impact on agricultural production-A review. *Innovative Farming, 3*(4), 141–149.

Shrestha, S., Peel, M. C., & Moore, G. A. (2022). Cold waves in Terai region of Nepal and farmer's perception of the effect of fog events and cold waves on agriculture. *Theoretical and Applied Climatology, 151*, 29–45. https://doi.org/10.1007/s00704-022-04262-7

Shumilina, J. S., Kuznetsova, A. V., Frolov, A. A., & Grishina, T. V. (2018). Drought as a form of abiotic stress and physiological markers of drought stress. *Journal of Stress Physiology & Biochemistry, 14*(4), 5–15.

Sigar, C. P. (2023). Impact of Tillage on Soil Sustainability. *IJSDR, 8*(5). https://ijsdr.org/papers/IJSDR2305148.pdf

Singh, D., Yadav, A., Singh, S., Yadav, R. K., Sen, M., Yadav, A. K., Gehlot, T., Mishra, A., Pandey, V., & Singh, A. K. (2023). Heat waves and its impact on crop production and mitigation techniques: A review. *International Journal of Environment and Climate Change, 13*(9), 377–382. https://doi.org/10.9734/ijecc/2023/v13i92243

Sivakumar, M. (2023). Importance of solar radiation and the need for improved respect to sun by agrometeorologists. *Journal of Agrometeorology, 25*(1), 51–60. https://doi.org/10.54386/jam.v25i1.1971

Sivakumar, M. V. (2020). Climate extremes and impacts on agriculture. *Agroclimatology: Linking Agriculture to Climate, 60*, 621–647.

Smith, P., Calvin, K., & Nkem, J. (2020). Which practices co-deliver food security, climate change mitigation and adaptation, and combat land degradation and desertification? *Global Change Biology, 26*(3), 1532–1575. https://doi.org/10.1111/gcb.14878

Solomon, S., Manning, M., Marquis, M., & Qin, D. (2007). Climate change 2007-the physical science basis: Working group I contribution to the fourth assessment report of the IPCC, 4.

Song, Y., Joo, J., Kim, H., & Park, M. (2024). Development and applicability evaluation of damage scale analysis techniques for agricultural drought. *Water, 16*(10), 1342. https://doi.org/10.3390/w16101342

Sonneveld, B. G., Keyzer, M. A., & Ndiaye, D. S. (2015). Quantifying the impact of land degradation on crop production: The case of Senegal. *Solid Earth, 7*, 93–103. https://doi.org/10.5194/se-7-93-2016

Soualihou, S., Duan, F., Li, X., & Zhou, W. (2022). CROP PRODUCTION UNDER COLD STRESS: An understanding of plant responses, acclimation processes, and management strategies. *Plant Physiology and Biochemistry, 190*, 47–61. https://doi.org/10.1016/j.plaphy.2022.08.024

Soz, S., & Raza, M. S. (2024). Disruption in agricultural pattern due to unpredictable weather conditions and its effect on farmer's family of Kishanganj District of Bihar. *Journal of Climate Change, 10*(1), 35–42. https://doi.org/10.3233/JCC240005

Stigter, C. J., Oteng, S. B. B., Ohuwasemire, K. O., Al-amin, N. A., Kinama, J. M., & Onyewotu, L. O. Z. (2005). Recent answers to farmland degradation illustrated by case studies from African farming systems. *Annals of Arid Zone, 44*(3 & 4). https://epubs.icar.org.in/index.php/AAZ/article/view/66127

Stockle, C. O., Marsal, J., & Villar, J. M. (2009). Impact of climate change on irrigated tree fruit production. In *VI International Symposium on Irrigation of Horticultural Crops, 889*, 41–52. https://doi.org/10.17660/ActaHortic.2011.889.2

Striker, G. G. (2012). Flooding stress on plants: Anatomical, morphological and physiological responses. *Botany, 1*(1), 3–28. https://doi.org/10.5772/32922

Swaminathan, M. S., & Rengalakshmi, R. (2016). Impact of extreme weather events in Indian agriculture: Enhancing the coping capacity of farm families. *Mausam, 67*(1), 1–4. https://doi.org/10.54302/mausam.v67i1.1135

Taddese, S. (2018). The impacts of land degradation on crop productivity in Ethiopia. A review. *Journal of Environment and Earth Science, 8*, 102–106.

Tadesse, A., & Hailu, W. (2024). Causes and consequences of land degradation in Ethiopia: A review. *International Journal of Science and Qualitative Analysis, 10*(1), 10–21. https://doi.org/10.11648/j.ijsqa.20241001.12

Tawang, A. B., Ahmad, T. A. B. T., & Abdullah, M. (2002). *Stabilization of upland agriculture under El Nino induced climatic risk: Impact assessment and mitigation measures in Malaysia*. CGPRT Centre. https://hdl.handle.net/20.500.12870/5921

Tun, K. K., Shrestha, R. P., & Datta, A. (2015). Assessment of land degradation and its impact on crop production in the Dry Zone of Myanmar. *International Journal of Sustainable Development & World Ecology, 22*, 533–544. https://doi.org/10.1080/13504509.2015.1091046

UN. (2024). *Three billion people globally impacted by land degradation*. Retrieved from https://www.un.org/en/three-billion-people-globally-impacted-land-degradation

UNCCD. (2014). United Nations Convention to Combat Desertification (UNCCD): Issues and challenges. *E-Int. Rel, 30*. https://www.e-ir.info/2014/04/30/united-nations-convention-to-combat-desertification-issues-and-challenges/

UNCCD. (2023). *New study links climate change, land degradation and migration in Central Asia*. United Nations Convention to Combat Desertification. Retrieved from https://www.unccd.int/news-stories/stories/new-study-links-climate-change-land-degradation-and-migration-central-asia

United Nations. (2019). *World Population Prospects 2019*. United Nations Department of Economic and Social Affairs. Desa. https://population.un.org/wpp/Publications/Files/WPP2019_10KeyFindings.pdf

Utami, A. W., Jamhari, J., & Hardyastuti, S. (2011). El Nino, La Nina, Dan Penawaran Pangan Di Jawa, Indonesia. *Jurnal Ekonomi Pembangunan: Kajian Masalah Ekonomi Dan Pembangunan, 12*, 257–271.

Villa, B. D., Petry, M. T., Santos, M. S., Martins, J. D., Lago, I., Moura, M. B., Eggers, H. S., Melo, G. L., Tonetto, F., Ferrazza, C. M., Oliveira, J. A., Werle, R., & Castro, R. P. (2022). Effects of minimum and maximum limits of solar radiation and its temporal and geographic interactions. *Journal of Agricultural Science, 1*, 241–251.

Wang, Y., & He, Z. (2024). Online or offline: High temperature, sales channel adjustment, and agricultural profit. *International Journal of Production Economics, 269*, Article 109153. https://doi.org/10.1016/j.ijpe.2024.109153

Webb, N. P., Marshall, N. A., Stringer, L. C., Reed, M. S., Chappell, A., & Herrick, J. E. (2017). Land degradation and climate change: Building climate resilience in agriculture. *Frontiers in Ecology and the Environment, 15*(8), 450–459. https://doi.org/10.1002/fee.1530

Wiebe, K. D. (2003). Linking land quality, agricultural productivity, and food security. *USDA-ERS Agricultural Economic Report*, (823).

Worku, A. (2023). Review on drivers of deforestation and associated socio-economic and ecological impacts. *Vegetable crops of Russia. Advances in Agriculture, Food Science and Forestry, 11*(1). https://doi.org/10.51268/2736-1799.23.11.098

World Bank. (2021). Groundswell report. Retrieved from https://www.worldbank.org/en/news/feature/2021/09/13/ground-swell-report

World Meteorological Organization (WMO). (2022). State of the global climate 2021.

World Meteorological Organization (WMO). (2024). State of the global climate 2023.

Wuepper, D., Borrelli, P., Panagos, P., Lauber, T., Crowther, T. W., Thomas, A. R., & Robinson, D. A. (2021). A 'debt' based approach to land degradation as an indicator of global change. *Global Change Biology, 27*, 5407–5410. https://doi.org/10.1111/gcb.15830

Xue, Y., & Fennessy, M. (2002). Under what conditions does land cover change impact regional climate.

Yamusa, A. M., Abubakar, I. U., & Falaki, A. M. (2015). Rainfall variability and crop production in the North-western semi-arid zone of Nigeria. *Journal of Soil Science and Environmental Management, 6*, 125–131. https://doi.org/10.5897/JSSEM14.0444

Zhang, J., Dong, W., & Fu, C. (2005). Impact of land surface degradation in northern China and southern Mongolia on regional climate. *Chinese Science Bulletin, 50*, 75–81. https://doi.org/10.1007/BF02918748

Zhang, Y. Z., Hu, Y. F., Han, Y. Q., & Zhan, S. (2021). Spatial distributions and evolutions of global major ecological degradation regions and research hot spot regions. *Acta Ecologica Sinica, 41*, 7599–7613. https://doi.org/10.5846/stxb202002150260

Zheng, F., & Wang, B. (2014). *Soil erosion in the Loess Plateau Region of China*. https://doi.org/10.1007/978-4-431-54481-4_6

Zougmoré, R., Jalloh, A., & Tioro, A. (2014). Climate-smart soil water and nutrient management options in semiarid West Africa: A review of evidence and analysis of stone bunds and zaï techniques. *Agriculture & Food Security, 3*, 1–8. https://doi.org/10.1186/2048-7010-3-16

Chapter 4
Climate Change and Case Studies of Vulnerable Ecosystems and Wetlands

Anila Alex, M. Saranya, Chinmaya Sahoo, and Ankit Saini

Abstract Climate change, the new normal, induced by human, is threatening various ecosystems, causing loss of biodiversity, land degradation, disruption in the timing of natural events and cycles, disease spread, reduced ecosystem services, and harming food webs. Changes in the mean conditions, which result into wildfires, ocean acidification, flood and landslides, typhoons, drought are costing the economy. Climate change is predicted to cost the world economy, $23 trillion by 2050 to $178 trillion over the next five decades, and this cost is disproportionate, where the major brunt is borne by the poor. Piarre grasslands of North America, Boreal Forest of Canada, Freshwater basins of the US, Mediterranean marine ecosystems, China's mountain ecosystems, and Island ecosystem of The Bahamas are some of the highly vulnerable ecosystems that have been documented. Wetlands, the largest carbon sink and regulator of water cycle, whose biogeochemistry and operation can be significantly impacted by changes in hydrology and rising temperatures. Wetlands' transformation from a carbon sink to a carbon source demonstrates how this might change advantageous services into detrimental ones. Large numbers of disadvantaged people who depend on ecosystem services for a portion of their livelihoods are supported by wetlands' high productivity. To build resilience to the current situation, sustainable management is essential. This is achieved by comprehending the natural dynamics of climate impacts and addressing them at their source. Those who rely on these

A. Alex
Department of Agronomy, ICAR- Indian Agricultural Research Institute, Hyderabad Academic Hub, Hyderabad, India

M. Saranya
Department of Agronomy, University of Agricultural Sciences, Dharwad, Karnataka, India

C. Sahoo
Department of Agronomy, Dr. Rajendra Prasad Central Agricultural University, Pusa, Samastipur, Bihar, India

A. Saini (✉)
Department of Agronomy, Dr. KSG Akal College of Agriculture, Eternal University, Baru Sahib, Sirmaur, Himachal Pradesh, India
e-mail: ankitsagr@eternaluniversity.edu.in

U. Chatterjee et al. (eds.), *Eco-Resilience: Climate Change, Land Degradation and Sustainable Solutions*, Sustainability Solutions,
https://doi.org/10.1007/978-3-032-00708-7_4

services for their daily lives prioritize their immediate survival over the long-term goal of sustainable ecosystem management.

Keywords Wetlands · Climate change · Ecosystem services · Vulnerable ecosystems · Sustainable management

4.1 Introduction

The rapid man-led climate change caused basically because of fossil fuel burning and their emissions have become the new normal of the twenty-first century, which profoundly is changing the biosphere, where life and living beings exist. Such emissions create problems for the life and sustainability of natural ecosystems and the people who rely on them. In the recent times, the effects of these threats have become the daily news headlines as these changes in natural ecosystems threaten biodiversity worldwide, affecting global food production. "Compared to the industrial revolution, the average temperature of the earth's surface has increased by 1.2 °C", as studied by the Intergovernmental Panel on Climate Change (IPCC) in 2021. The International Panel on Climate Change (IPCC) Special Report on 1.5 °C put forward in 2018, warns that "allowing the planet to warm beyond 1.5 °C will result in climate change impacts including intense droughts, water scarcity, severe fires, rising sea levels, flooding, melting polar ice, catastrophic storms and declining biodiversity". Since Earth is a system and changes in one place can have an impact on changes in all other places, global warming is only the beginning of several other hazards that follows.

Major land and water organisms are responding to temperature changes, precipitation changes, changes in atmospheric and oceanic concentration of gases, and water balance. Different ecosystems differ in their sensitivity and response to climate change by changing individual characteristics, timing of certain biological events, ecosystem services, and changes in geographic forages. Ecosystem services and man who depend on them are facing the risk due to climate change, which is also tweaking ecosystem productivity, quickening the spread of invasive species, and wild forms, altering species interaction with themselves and their surroundings. When the rate of climate change exceeds the time it takes for species to adapt, extinctions also happen locally and globally as a consequence. Ecological vulnerability implies the extent to which a species or ecosystem is prone to the effects of stressors, taking into account factors such as exposure, sensitivity, and recovery capacity at the population level. It encompasses the assessment of organisms, their relationships, and the abiotic environment, making it a complex concept to evaluate (AI-generated definition based on: Science of the Total Environment, 2010). Vulnerability to climate change, as defined by the IPCC, "is the degree to which a system is susceptible to or unable to withstand the adverse effects of climate change, including climate extremes and variability. It is determined by a system's sensitivity, adaptive ability, and the type, amount, and rate of climatic change to which it is exposed".

"Many terrestrial and freshwater species have shifted their geographic ranges, seasonal activities, migration patterns, abundances, and species interactions in response to ongoing climate change," as per the 5th Assessment Report of Intergovernmental panel on climate change (IPCC, 2014). "Approximately 20–30% of plant and animal species are at risk of extinction when global mean temperatures climb 2–3 °C over pre-industrial levels", according to estimates based on long-term documented variations across the regions (Fischlin et al., 2007). Marine and coastal ecosystems (MCE) provide direct and indirect benefits, and spreads across 4% of land area of Earth and 11% of the oceans with regard to coastal and 70.8% of the surface by Marine systems. Ocean acidification, sea level, and sea temperature rise are all signs of climate change, which presents a serious risk, along with anthropocentric pressures like pollution from multiple sources (Mostofa et al., 2013), overfishing, illegal and underregulated fishing (Agnew et al., 2009), and habitat destruction (Crain et al., 2009). According to Ruiz and colleagues (1999), the effects of climate change include the loss of biodiversity and the invasion of non-native species. Mountains are a treasure house of a multitude of ecosystem services like carbon sequestration, air purifier, water provider, food, recreator (Grêt-Regamey et al., 2012; Liu et al., 2019). Nevertheless, they are one of the most fragile ecosystems, and as per Iglesias and coworkers (2018), exceedingly vulnerable to climate change. Human pressures through mining, as per Bokar and colleagues (2020), massive tourism, intensive agriculture expansion, as per Peters and colleagues (2019), and significant socio-economic shifts are impacting the balance of the ecosystem. An island ecosystem refers to a unique environment found on tropical islands, characterized by high biodiversity, endemic species, and distinct evolutionary processes influenced by isolation (AI-generated definition based on: Biodiversity and Climate Change Adaptation in Tropical Islands, 2008). Many marine animals, including seals, seabirds, and turtles, which are susceptible to predators, use them as breeding sites. Surprisingly, island species have experienced the majority of historical extinctions of plants and animals. Therefore, their sustainable management is of grave importance.

Wetlands are transitional zones between terrestrial and aquatic ecosystems that function as nature's kidneys, regulating water and infiltrating waste from the landscape. These ubiquitous ecosystems are so inextricably interned with our culture and identity. The Ramsar Convention defines wetlands as "areas of marsh, fen, peat land or water whether natural or artificial, permanent or temporary, static or flowing, fresh, brackish or saline, where at low tides the depth of water does not exceed six meters". "Over the past forty years, India has lost nearly one-third of its natural wetlands due to rapid urbanisation, agricultural expansion and pollution, underscoring the need to protect these picturesque landscapes", a study as per NGO Wetlands International South Asia (WISA). Therefore, we cannot continue to deny climate change because it is real and already occurring, and one can ignore it at their own expense.

Our book chapter aligns rightly with the theme of the book, "Eco-resilience: Climate Change, Land Degradation and Sustainable solutions" as Land degradation and climate change each contribute to the other's acceleration, and sensitive ecosystems, such as above mentioned and wetlands, due to extreme weather events can further hasten loss of productive land and disrupt harmonious living. This gives rise to the importance of resilience and sustainable solutions-interventions that are long-term and timely, with a joint participation of Government, NGO's and local communities. Its not too late to make a difference provided we have ambitious Government policies and engaging communities. For sure, climate change comes with life-changing consequences, and it will take both prevention and cure for the world to take this challenge.

4.2 Materials and Methods

This chapter explores the different vulnerable ecosystems on Earth and some of the case studies to establish the fact that climate change is not in the near future, but an ongoing phenomenon. The chapter is structured as follows. We have consolidated several vulnerable ecosystems, namely the terrestrial, freshwater, marine, mountain, and island ecosystems, with their most relevant case studies in Sect. 4.3. We have covered case studies from different part of the globe to confirm that this brunt called climate change is borne by all mankind. In Sect. 4.4, we have focussed totally on wetlands, the most important and endangered ecosystem on earth, and discussed with reference to their importance, threats, and three case studies that prove the relevance of this particular book chapter. We have used Figures for better understanding. The analysis provides insights into the susceptibility of various ecosystems and how anthropogenic pressures are aggravating climate change into a catastrophe. Finally, Sect. 4.5 concludes our chapter.

4.3 Vulnerable Ecosystems

4.3.1 Terrestrial- Grasslands and Forest Ecosystems

4.3.1.1 Importance and Threats

Grasslands are "land covered with herbaceous plants with less than 10% tree and shrub cover" as per UNESCO. Also, White in 1983 defines wooded grassland as land covered with "10–40% tree and shrub cover". They serve as a vital source of feed for animals, a home for wildlife, a reservoir of water and carbon, and a key player in the conservation of plant genetic resources at place. With global carbon stocks estimated to be between 525 and 634 Pg C, grassland soils are also an extremely important repository of carbon. Although "grassland biomass C is a small fraction of the whole system C storage, unlike forests, the soil here is a sink that retains 80 to 94% of the total grassland C pool as soil organic carbon (SOC) and plant C (roots, other belowground biomass)" as per studies of Carvalhais and colleagues (2014) and Yang and colleagues (2022). The World Resources Institute estimates that, "excluding Greenland and Antarctica, the world's grassland area is 52.5 million km^2, or 40.5% of the total terrestrial area" (based on IGBP statistics, 2000). Figure 4.1 illustrates how different geographic features and grassland types (such as steppe, savanna, and prairie) affect the causes of global grassland degradation and the extent of degradation (Ji et al., 2022; Macdougall, 2008). According to estimates, "the area of the declining trend in grassland's net primary productivity accounts for 25.3% of the global grassland area" as studied by Sun and colleagues. A critical necessity in "fighting climate change with grasslands" has been emphasized by the Food and Agricultural Organization (FAO, 2010).

In order to ensure food security, forests are crucial. Forest foods have provided protection against starvation and malnutrition during periods of seasonal food scarcity or emergencies brought on, for instance, by floods, droughts, or forest conflicts. They are a treasure house of many under-utilized crops and plants of medical value. About 25% of prescribed medications contain plant-based active

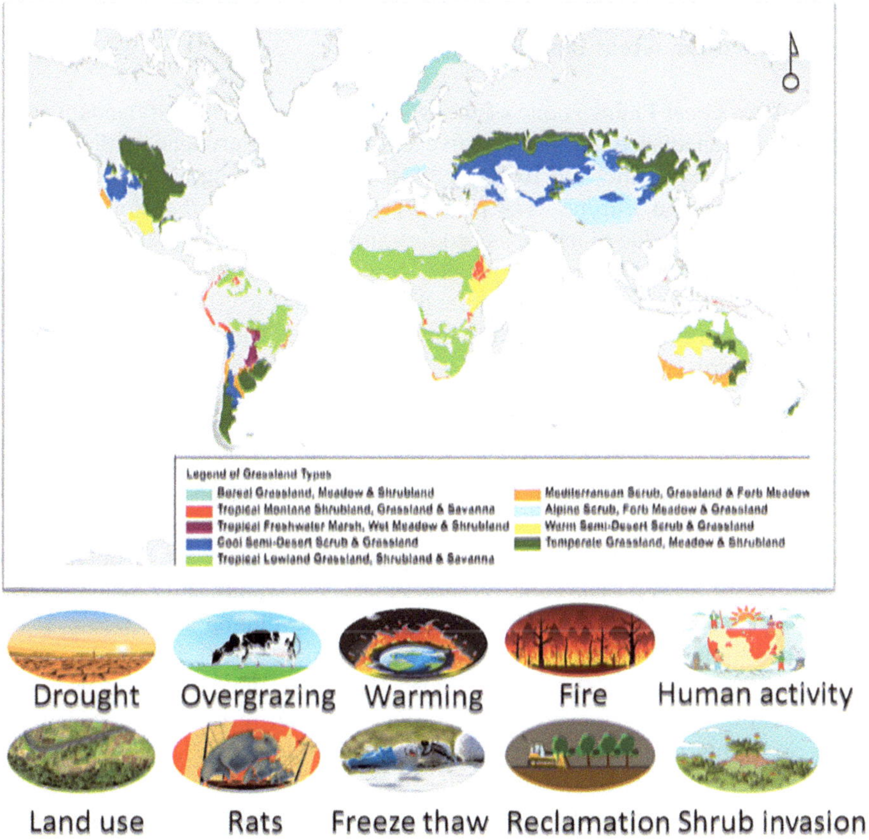

Fig. 4.1 Types of grasslands worldwide and factors causing their degradation. *Source* Adapted from Ji et al. (2022), Macdougall (2008)

ingredients, and the annual projected value of plant-based medications is close to US $45,000 million. "An estimated 30–40 million pastoralists around the world, who herd about 4000 million cattle, goats, and sheep" (Speedy & Pugliese, 1992), depend on forests for rangeland and fodder. It's interesting to note that non-wood forest resources frequently produce higher, more stable earnings than the same area could yield via logging or agriculture. On a global scale, FAO estimates that NWFPs (non-wood forest products) generates US $88 billion. Thirty million people are employed in forest-based industries in India alone. Trees recycle essential nutrients like nitrogen back into the soil, assist bring rain, and have deep roots that prevent erosion from wind and water. Trees may flourish widely, enabling production on marginal soils even in areas where agricultural products could fail. Additionally, trees absorb and store carbon dioxide (CO_2) as they develop. Sadly but, "the forest sector accounts for 17.3% of total carbon emissions, of which a substantial part is due to deforestation driven by agricultural conversion, including pasture and rangelands" (FAO, 2009).

4.3.1.2 Case Study of Kentucky Bluegrass (Poa Pratensis) in the NGP-Northern Great Plains

The Northern Great Plains is one of the world's largest prairie grasslands, encompassing two Canadian provinces and five US states. As European settlers arrived in North America in the mid to late 1800s, Kentucky bluegrass, a popular and well-known grass in Europe, was probably first brought to the Northern Great Plains. Surprisingly, due to its use as bedding and packaging material, the grass was frequently inadvertently spread by people. As per the data collected and combined from different places for a time span extending from 2011 to 2015, proves that "Kentucky bluegrass cover has increased in the Great Plains, with 86% of foliar canopy covering in North Dakota, 63% in South Dakota, 40% in Kansas, 38% in Nebraska, and 32% in Montana" (USDA-NRCS, 2018). The potential contributor to this aggravation was climate change. "North Dakota's growing season has extended by 12 days for the past 120 years" (Badh et al., 2009). Kentucky bluegrass, being photosynthetically active in the early spring and late fall, favoured its growth and spread (DeKeyser et al., 2015). Additionally, its higher vegetative reproductive capacity during the fall helped it to gain an advantage during long, cool fall seasons by producing more rhizomes (Etter, 1951). "The concentration of CO_2 has significantly increased from 280 parts per million in 1750, when the Industrial Revolution started, to over 400 parts per million" (Dusenge et al., 2019; Thirkell et al., 2020). Hence, In contrast to Big bluestems (C_4 grass), whose photosynthetic rate stayed constant, Kentucky bluegrass, a C_3 grass, increased its photosynthetic rate by 141% at elevated CO_2 levels, according to a prior study by He and colleagues (1992). Furthermore, the precipitation in the NGP has increased due to climate change, as per an earlier study that confirmed a highly positive correlation between rising precipitation in the great plains and increase in Kentucky bluegrass cover (Wand et al., 1999).

Between 1984 and 2007, Shannon's diversity index, which is an indicator of population diversity in a certain habitat, decreased at NGP from 2.5 to 1.6. According

Fig. 4.2 Variation in the frequency of Kentucky bluegrass in 28 North Dakota native prairie locations. The overall increase in Kentucky bluegrass invasion has been calculated by subtracting data from the grass samples collected in 2007 and 1984. *Source* Adapted from DeKeyser et al. (2015)

to Fig. 4.2, Kentucky blue grass was more common at 22 of the 28 sites studied in central North Dakota, and the rise was more noticeable than the decline at the remaining six sites. Across all sites, a 35% rise in the total frequency of the grass was noted throughout the 23-year period. Forage availability during warmer and drier seasons may be restricted, and long-term ecosystem resilience may be diminished, if plant communities go from being very diversified to being dominated by a single cool-season grass (Ohrtman et al., 2015; Smart et al., 2013). These factors make Kentucky bluegrass domination in the NGP is now a concern to livestock production and an ecological issue.

Ellsworth (2012) reported a spread of invasive grasses such as guinea grass (*Megathyrsus maximus*), fountain grass (*Pennisetum setaceum*), and buffelgrass (*Cenchrus ciliaris*), which not only outcompete native species but also exacerbate wildfire risks in Hawai‘i. Multifaceted approaches like managed grazing by livestock to reduce the fine fuels provided by invasive grasses, efforts to restore native ecosystems through for instance, in the Waianae Kai Forest Reserve on O‘ahu, a combination of mowing, herbicide application, and outplanting of native species has been employed to suppress guinea grass, innovative weed control techniques that minimizes collateral damage to native plants like that of Herbicide Ballistic Technology (HBT) developed by University of Hawai‘i's College of Tropical Agriculture and Human Resources (CTAHR) which allows precise application of herbicides to target invasive species and collaborative efforts among nonprofit organizations like Hawai‘i Wildfire Management Organization (HWMO), O‘ahu Invasive Species Committee (OISC), Big Island Invasive Species Committee (BIISC), government agencies, and local communities enhance resilience against invasive grasses and associated wildfire risks in Hawai‘i.

4.3.1.3 Case Study of Spruce Budworm Outbreak in the Boreal Forests of Canada

The temperature of insect microenvironment has a major influence on how quickly they progress through their life cycle. Case in point is that of spruce budworm (SBW), which feeds exclusively on young balsam fir, *Abies balsamea* (L.) Mill., and spruces, *Picea* spp., tree species found in the Canadian boreal forests. "The land area of Canada's boreal forests (including other wooded land types) is 309 Mha" (Brandt et al., 2013), which is almost "30% of the world's total boreal forested area" (Brandt, 2009). The country's distinct edaphic and climate scenarios, and stand structure, make these forests more susceptible to drought than other types of forests, as studied by Dial and coworkers (2022) and Lenton and colleagues (2008). By the end of this century, increasingly severe droughts are expected to affect Canada's boreal forests, according to meteorological analyses (Wang et al., 2014).

Generally speaking, the spruce budworm is highly fecund, quickly develops, and has a higher early instar survival as the temperature of its microhabitat rises towards its ideal value (Lucuik, 1984). Another survival skill is their ability to grow and develop fast in comparison to their natural enemies during severe conditions like drought. *Apanteles fumiferanae*, *Nosema fumiferanae*, *Erynia* (Zoophthora) *radicans* (Perry & Fleming, 1989) are the few major natural enemies to spruce budworms. Increased defoliation can trigger another threat, a common site of this decade-forest fires. According to Wein (1990), a drier, warmer environment combined with greater stress from prolonged defoliation may hasten death of host trees. Increased tree mortality increases the fuel load, paving the way to intense forest fires. Stocks (1987) describes "how spruce budworm defoliation opens up the canopy and allows understory vegetation to grow by creating a moist green layer that can effectively prevent fire spread by initially separating the dry surface fuels from the crown fuels. However, so much dead and down material has accumulated on the ground, after roughly five years of crown breakage and windthrow, that the surface fuel can overcome the dampening effect of the understory vegetation and cause fire". Navarro and colleagues (2018) conducted a study at Eastern Canadian Boreal forests and reported three main spruce budworm outbreak periods over the last century. Figure 4.3 shows that the first outbreak lasted from 1905 to 1930 affecting almost 40% of trees under study area. The second lasting from 1935 to 1965 was the longest infestation, and the third outbreak and most powerful outbreak from 1968 to 1988, affecting nearly 50% of the studied trees. The authors have confirmed that understanding natural disturbance cycles at multiple levels is crucial, particularly in light of potential future climate change to better understand such outbreaks.

Similar studies were done was Haynes et al. (2014) investigating how climate change over the past 212 years (1800–2011) has affected the outbreak patterns of five major pine-defoliating insect species in Bavaria, Germany wherein they observed a clear strong warming trend in all seasons after 1975 which resulted in increased voltinism causing faster reproduction, higher population growth and more frequent outbreaks of forest defoliators. Hlásny et al. (2021) in another study on *Ips typographus,* the European spruce bark beetle has proved their ecological and

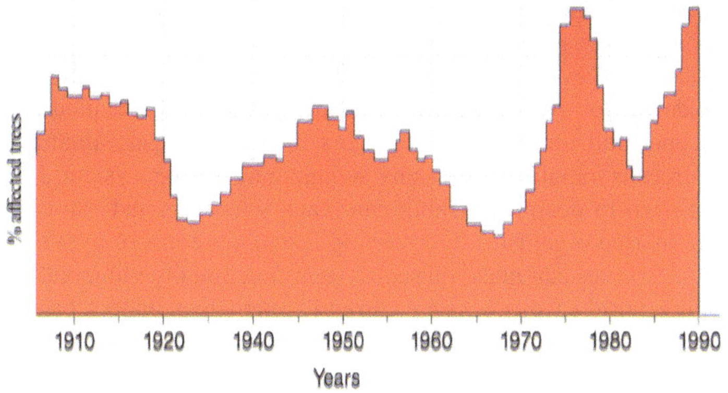

Fig. 4.3 The percentage of trees in the study area that are impacted by SBW outbreaks. *Source* Adapted from Navarro et al. (2018)

economic impacts across European forests mostly due to climate change increasing their population and activity.

4.3.2 Freshwater and Marine Ecosystems

4.3.2.1 Importance and Threats

Despite making up only one-hundredth of the Earth's surface, freshwater ecosystems such as lakes, rivers, wetlands, streams, and subterranean aquifers provide billions of people with vital ecosystem services like food, water, and energy. They also serve as "habitat for one-third of all vertebrate species, including approximately 40% of fish species" (UNEP, February 2022). "With a Freshwater Vulnerability Index (FVI) of 0.36–0.38 based on resource availability of 1091–1184 m^3/capita, rainfall variation coefficient of 0.206–0.220, water exploitation rate of 72–76%, and ecological degradation of 30–39% from 2000 to 2015, the freshwater sector is currently categorized as a moderately vulnerable sector" (Boojhawon & Surroop, 2021).

We have always learned and knew that 70% of the Earth's surface is made up of oceans. A story covered by UNEP reports that "from deep oceans to coastal reefs, from mudflats to sea grass beds, ocean and marine systems offer us essential necessities including storm surge protection, renewable energy, and carbon capture for tackling climate change". As per the Food and Agricultural Organization (FAO), "seafood makes up almost 20% of the animal protein consumed by over 3 billion people. Seafood offers a nutritious alternative that may supplement diets based on cereals since it contains micronutrients such as long-chain polyunsaturated omega-3 fatty acids that are rare in other dietary sources". Intake of seafood has increased twofold throughout the past half-century, and almost six crore people are employed by the fisheries and aquaculture industries. About three hundred crores of people who live along the coast depend on the ocean and its resources for their livelihood. Additionally, they are essential to preserving culture and traditions, especially in tiny seaside towns. As the global population grows with exploding demands, exploring deeper and further into the waters for fish, oil, gas, minerals, and new genetic resources is inevitable, but sadly is damaging this ecosystem. Fishing gears that are employed in deep sea fishing can reach sea floors and can cause a negative effect on marine organisms. Overfishing is another concern by which selective removal of one species can make this ecosystem functionally vulnerable. More than one-third of the world's fish stocks are currently overfished, underscoring the significance of long-term, sustainable management grounded in scientific understanding of the marine environment.

4.3.2.2 Case Study of Harmful Algal Blooms (HAB) in the US Freshwaters

Chapra et al. conducted a study in the year (2017) in the US freshwater basins and projected the effect of global temperature rise on the cyanobacterial harmful algal blooms. Presence of cyanobacterial hazardous algal blooms (CyanoHABs) makes the water unfit for consumption. This raises the necessity for water treatment, which is a costly business impacting human welfare. Considering majority of cyanobacteria are indigestible to planktivorous fishes and zooplankton (Arnold, 1971; DeMott, 1989; Fulton & Paerl, 1987), they form a "dead end" in the aquatic food chain, adversely affecting both commercial and recreational fishing. According to the projections, lake surface-layer temperatures will rise by 2–5 °C, and there will probably be an increase in nutrient levels due to shifting demographics that will drive nutrient transport. As a result, CyanoHAB concentrations will rise, and the average number of days per year that CyanoHAB occurs will increase from roughly 7 days under current conditions to over 16–23 days in 2050 and 18–39 days in 2090.

The data regarding the results of human health and water sampling on the HAB-related illness for the years 2009–2010 were compiled by Hilborn and colleagues (2014). There were 11 documented outbreaks of waterborne diseases linked to algal blooms, all of which happened in freshwater lakes. At least 61 people were impacted by the epidemics, which happened in three states of New York, Ohio, and Washington. The report recorded that out of the total individuals affected, 59% were female, as shown in Table 4.1. Based on age group, Fig. 4.4 shows the number and percent of people impacted by waterborne disease outbreaks linked to HAB in the United States from 2009 to 2010. Dermatological symptoms like rash, irritation, swelling, sores, gastrointestinal symptoms like vomiting, diarrhoea, respiratory symptoms like coughing and sneezing, and neurologic symptoms like tingling and confusion were among the health impacts. Three main modes of exposure were observed as through contact, ingestion, and aerosol inhalation, with contact being the most common route to the infection. Water was checked for microbes like cyanobacteria, *Escherichia coli*, and toxins. In every epidemic investigation where cyanotoxin testing was conducted,

Table 4.1 Number and percentage of individuals impacted by waterborne disease outbreaks linked to HAB, by sex, in the United States, 2009–2010

	Number of persons	Proportion of persons
Sex		
Male	24	41
Female	34	59

Source Adapted from Hilborn et al. (2014)

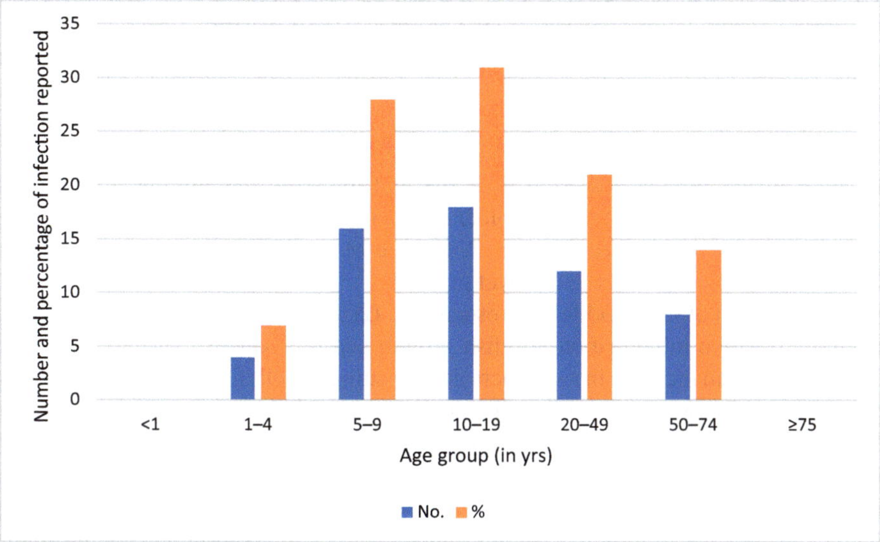

Fig. 4.4 By age group, number and percentage of individuals that were infected with HAB-associated waterborne disease outbreak in the United States, 2009–2010. *Source* Adapted from Hilborn et al. (2014)

Microcystins were found. In four outbreaks, out of the 11 disease outbreaks that were recorded, microcystin values of $\geq$ 20 μg/mL were observed, which is higher than the WHO recommendation for moderate health concerns. Anatoxin-a, cylindrospermopsin, and saxitoxin were also found during these epidemic studies. Remarkably, in 75% of outbreaks where recognized neurotoxins (saxitoxin and anatoxin-a) were detected, neurologic symptoms were documented.

4.3.2.3 Case Study of Mediterranean Marine Ecosystems

"Due to its semi-enclosed location and dense population, the Mediterranean Sea is a hotspot for anthropogenic stresses, and its delicate ecosystems are vulnerable to the effects of climate change" (Coll et al., 2012; Giorgi, 2006). According to Vargas-Yáñez and coworkers (2008), the basin is warming two to three times more quickly than the oceans across the world, and "by the end of the century, this trend is predicted to pick up speed" (Soto-Navarro et al., 2020). The evidence of changes to the geophysical environment at Western Mediterranean L'Estartit is portrayed in Table 4.2. Catalan sea, which is a representative section of the western Mediterranean basin and lies between the eastern Iberian coast and the Balearic Islands. Regional effects of global warming are many in Catalan sea. There is a mean change of 3.4 mm $year^{-1}$ in the sea level rise of Western Mediterranean over the past century (1999–2009) and an increase in surface 50 m and 80 m sea water temperature of about 1.1 °C and 0.7 °C, respectively, over the last 35 years. Along with that, there is a progressive salinization from surface to bottom at 0.0008 psu (practical salinity units) mean rate of change per year. According to the statistics, "the water's stratification has been strengthened during the past 30 years by an increase in the temperature differential between the sea surface and depth". Rise in atmospheric CO_2 brought on by human activity is another aftereffect of global warming causing gradual acidification in the oceans, lowering their pH and lowering the amount of dissolved calcium carbonates, the building blocks of calcifying organisms like corals and mussels causing habitat destruction. A change of − 0.14 to − 0.1 pH units have been recorded by Touratier and Goyet in 2011.

The research of Gilli and fellow researchers (2010) reports temperature increases having a positive influence on jellyfish growth and reproductive success, which in turn leading to an increase in swimmers getting stung by jellyfish. Another interesting phenomenon known as "meridionalization" which is the migration of fish from the warm waters to the cold ones as observed in case of ornate wrasse scientifically named as *Thalassoma pavo* which is a common warmwater fish that lodge in along the coast of the Catalan Sea that bears cold waters, was a few decades ago, exclusively seen in the warm southernmost Mediterranean waters. Mass coral reef fish mortality

Table 4.2 Studies done by various researchers to prove that changes are occurring to the geophysical environment of Mediterranean Sea

Geophysical variable	Time span	Depth	Change	Mean rate of change (yr^{-1})	Reference
Seawater temperature	1974–2008	Surface 50 m	+ 1.1 °C	+ 0.032 °C	Present study
	1974–2008	Surface 80 m	+ 0.7 °C	+ 0.021 °C	Present study
Salinity	1950–2000	Surface-Bottom	+ 0.035 to 0.04 psu	0.0008 psu	Rixen et al. (2005)
Sea level	1990–2009	Surface	+ 6.5 cm	+ 3.4 mm	Vargas-Yáñez et al. (2008); present study
Stability of water column	1974–2006	Surface 80 m	From 75 to 164 d	+ 1.25 d	Coma et al. (2009)
Acidification	Preindustrial to 2001	Surface-Bottom	− 0.14 to − 0.1 pH units	–	Touratier and Goyet (2011)

(–) Data not available; psu = Practical salinity units; d = days
Source Adapted from Calvo et al. (2011)

Table 4.3 Common pathogens in fresh and marine waters with possible human illnesses

Pathogen type	Example	Human illness
Bacteria	*Vibrio cholerae, Leptospira, E. coli*	Gastroenteritis, cholera, skin infections, leptospirosis, sepsis
Viruses	Norovirus, Hepatitis A, Adenovirus	Diarrhea, liver damage, respiratory illness
Protozoa	*Giardia, Cryptosporidium*	Persistent diarrhea, dehydration, chronic intestinal damage
Amoebae	*Naegleria fowleri*	Rare but often fatal brain infection (PAM)

Source Adapted from Seymour and McLellan (2025)

in the Red Sea has been caused by the warming events' quick onset and a common infection, *Streptococcus iniae*, fatally infected thermally stressed fish from dozens of species after a sharp increase in water temperature (4.2 °C increase in sea surface temperature [SST] in 2.5 days).

Seymour and McLellan (2025) review how climate change is intensifying the abundance, spread, and health impacts of harmful microorganisms (like pathogenic bacteria, viruses, toxic algae, and amoebae) in marine and freshwater ecosystems. Climate change mainly through surface warming is transforming aquatic ecosystems into hotspots for dangerous microorganisms, as mentioned in Table 4.3. This, along with extreme weather events like floods, rainfall, and hurricanes flushes sewage, manure, and pathogens into water bodies, elevate nutrient levels causing harmful algal blooms (HABs) posing rising risks to public health, food security, and local economies demanding urgent, integrated responses in water management and disease surveillance.

4.3.3 Mountain and Island Ecosystems

4.3.3.1 Importance and Threats

Mountains, which make up 27% of the planet's surface, support about 15% of the world's population and serve as essential water towers, biodiversity hotspots, and climate regulators. Mountains are extraordinarily different, even beyond their shared traits of steep slopes and high relative relief, or very significant topographic diversity (Ives et al., 1997). From near sea level to the highest point on the planet, they can be found on every continent and at every elevation. On a worldwide basis, mountains might be most valuable because they are the sources of all the main rivers in the world, some of which are also used to generate hydroelectric power. The ecosystem is home to several unique species worldwide and plays a crucial part in the water cycle. Around 60 crores people, or 10% of the total world population make their living on islands. Many islanders have distinctive traditions, and the abundant natural resources in their immediate surroundings directly or indirectly contribute to their financial, ecological, and social well-being. Small islands compose of a good deal of marine and coastal areas, which are significant revenue sources. Convention on Biological Diversity estimates "US$375 billion worth of goods and services produced by coral reefs annually in the world including aid for marine fisheries, which provide the principal protein source for many island populations, especially amongst SIDS (Small Island Developing States)". Table 4.3 shows the list of 39 SIDS recognised by UNCED.

Mountainous areas are more vulnerable to deterioration due to both climate change and human stresses. When forests are cleared for infrastructure, farming, or communities on steep slopes, significant soil erosion and habitat loss may result. Water quality downstream is harmed by pollution and erosion. The amount and timing of water supplies to cities, farmers, industries, and power plants are all under risk due to climate change. Mountain species, both man and animals, are being forced to migrate or adapt as a result of rapidly rising temperatures. "Subsistence farming in high-elevation areas in Nepal and India is being impacted by decreased snow cover and shifting precipitation patterns" (Ogra & Badola, 2015; Smadja et al., 2015). The frequency and location of natural hazards are impacted by the reduced land-surface stability caused by glacial retreat. Due to the limited, specialized, and localized nature of island species populations, island ecosystems are particularly susceptible to climate change. Although they offer many advantages to islanders, coral reefs, the biggest biologically formed structures on Earth, are extremely affected by changes in seawater's temperature and composition. 90% of bird species that have gone extinct since the seventeenth century, and 75% of animal species are thought to be insular. Additionally, 23% of island species are currently listed as endangered, compared to 11% for the rest of the globe. This demonstrates how susceptible this ecosystem is to being extinct (Table 4.4).

4.3.3.2 Case Study of Alpine Mountain Hazards in China

In Tianmo Valley, China, a compound disaster event caused by a combination of rainfall and glacial melting occurred on July 25 and September 4, 2010. A 400-m stretch of roadway from Bomi County to Lhasa was destroyed as a result of the debris flow, which had a volume of 450,000 m^3 and was carried along a gully before being finally flushed into the Parlung Tsangpo River. In 2019, Wang and colleagues carried out a thorough terrain analysis to identify the different contributing elements to this disaster. "The mean temperature over the last 20 years (from 1991 to 2010) has increased by about 1 °C at a rising rate of 0.05 °C $year^{-1}$", according to the data from Bomi meteorological station, which is located 45 km southeast of Tianmo Gully. Analysis of Thematic mapper, which compares images, verified that temperature rise has resulted in a 13.6% reduction in the glacier area in the Tianmo ravine's source area from 2000 to 2006 (Deng et al., 2017). A field study employing UAV surveys demonstrated that the glaciation region was the debris flow's source area. Temperature rise due to global warming caused snowmelt and glacial

Table 4.4 List of Small Island Developing States (SIDS) as identified by UNCED

1. Antigua and Barbuda	15. Haiti	29. St. Kitts and Nevis
2. Bahamas	16. Jamaica	30. St. Lucia
3. Barbados	17. Kiribati	31. St. Vincent and the Grenadines
4. Belize	18. Maldives	32. Seychelles
5. Cabo Verde	19. Marshall Islands	33. Solomon Islands
6. Comoros	20. Micronesia (Federated States of)	34. Suriname
7. Cook Islands	21. Mauritius	35. Timor-Leste
8. Cuba	22. Nauru	36. Tonga
9. Dominica	23. Niue	37. Trinidad and Tobago
10. Dominican Republic	24. Palau	38. Tuvalu
11. Fiji	25. Papua New Guinea	39. Vanuatu
12. Grenada	26. Samoa	
13. Guinea-Bissau	27. São Tomé and Príncipe	
14. Guyana	28. Singapore	

Source Adapted from https://www.un.org/ohrlls/content/list-sids
List of Small Island Developing States (SIDS) as identified by UNCED. https://www.un.org/ohrlls/content/list-sids

recession, generating runoff in the watershed. Additionally, "the melting of internal ice particles that were present in moraine, caused a number of changes in the glacial till, decreasing the moraine's shear strength and making it the starting point for debris flows" (Wang et al., 2019). Moraine is a substance left behind by a moving glacier when it melts. Furthermore, the Indian monsoon season, which lasted from June to September, produced a lot of precipitation, which was the main cause of the start of the debris flow in this area. In conclusion, a number of causes, including the melting of the glacier, the abundance of antecedent precipitation, and the instability of rock and soil due to alternate wetting and drying cycles, all contributed to the disaster as represented in Fig. 4.5, eventually justifying that Climate change was the culprit of these events.

Similar study was done by Thiney et al. (2019) to examine the projected changes in the distribution of six globally notorious invasive plant species *Ageratina adenophora, Ageratum conyzoides, Chromolaena odorata, Lantana camara, Mikania micrantha,* and *Parthenium hysterophorus* within Bhutan. Bhutan has experienced noticeable warming trends leading to altitudinal upward shifts of invasive weeds into higher elevations that were once too cold for them. Native species, especially those adapted to cold, nutrient-poor, or stable environments, often struggle to compete with fast-growing, aggressive invasives entering habitats previously dominated by them. For instance, *Ageratina adenophora* and *C. odorata* crowding out native grasses essential for livestock, and *M. micrantha* increasing fire risk. It finally

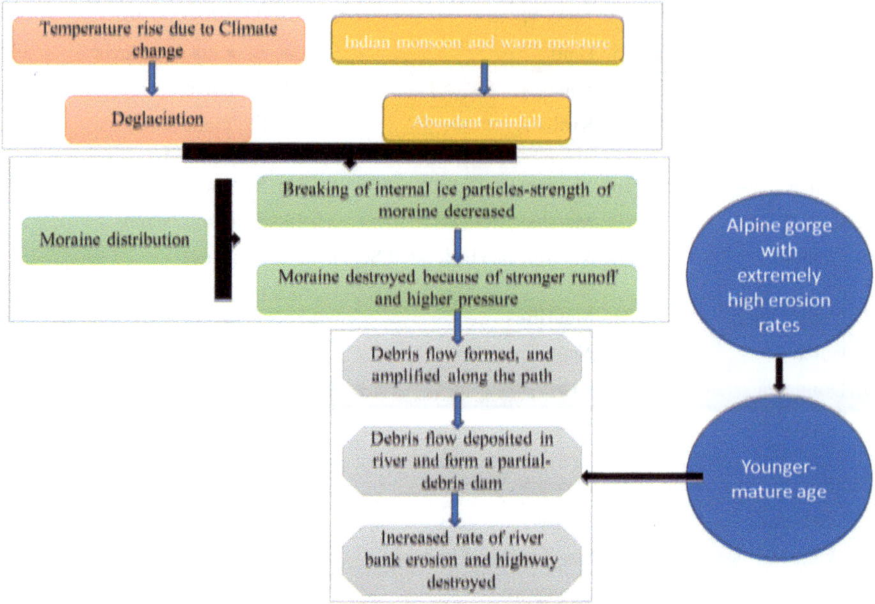

Fig. 4.5 An illustration of compound mountain hazards taking place in the Parlung Tsangpo River. *Source* Adapted from Wang et al. (2019)

reaches a stage of "summit trap" wherein the native species are pushed higher and higher until they have nowhere to go, pushing them to extinction. Loss of livestock productivity, loss of income for rural farmers, and increased labour for weed control, reduced availability of native fodder, potential need to buy external fodder, and increased vulnerability of food security in mountain farming communities are the impacts of such invasions. Bhutan has strong forest conservation laws (maintaining 60% forest cover), but with regard to specific invasive species management policies, they are reactive rather than proactive.

4.3.3.3 Case Study of Bahamas Island

The Bahamas is one of the SIDS-Small Island Developing States, which are "a group of countries that face unique social, economic and environmental vulnerabilities", as defined by Agenda 21 at the Earth Summit held in Rio de Janeiro, Brazil, in June 1992. With 700 islands and a land area of 10,010 km^2, the Bahamas is an enormous archipelago (CIA, 2018). "In 2019, the Bahamas' tourism sector contributed 43.3% of the nation's GDP, supported 52.2% of all jobs, and produced 81.6% of all revenue connected to tourist exports" (WTTC, 2020). Five major storms have struck the Bahamas in the past five years, including Hurricane Matthew of Category 4 in the year 2016 and three years later in 2019, Hurricane Dorian of Category 5 that brought

USD 3.4 billion in losses. Regardless of their magnitude, these storms have altered visitor attitudes, upset the national government, and decreased revenue from tourism. Pathak and coworkers (2021) studied on islands of New Providence (NP) and Paradise Islands (PI), 2 islands with the highest tourist intake, and found that storm surge and SLR—Sea level rise posed a threat to the islands' tourism industry. All of the beaches on the two islands now have beach widths of less than 50 m, making them vulnerable to erosion brought on by sea level rise. Unfortunately, much of the resources in this area are vulnerable to erosion-related problems. "Some studies estimate a total loss of beaches by 2100 that would result in a 56.17% drop in the number of visitors to the Bahamas and a USD 981.75 million drop in visitor spending" (Pathak et al., 2021). Sadly, 86% of tourists cite the beaches of NP and PI as the primary reason they travel to the islands who will be disappointed if right actions are not taken.

According to studies, about 537 square miles of coral reefs may disappear from the Bahamas as a result of warming waters. Human activities like fishing pressure and invasive lionfish increase the likelihood of coastal development and dredging, which puts the coral reefs in NP under serious risk (Arkema & Ruckelshaus, 2017). "Many ecosystem services, including storm protection, local finfish fisheries, habitat for spiny lobster (the Bahamas' principal export), and visitor spending, would be impacted by the loss of the surrounding coral reefs, which are primarily found on the island's northern side" (Arkema & Ruckelshaus, 2017). Debris hill, contaminated waters, and drops in fishery production are examples of post-disaster difficulties that worsen these effects and increase the losses to the nation's tourism sector and social-economic status. Figure 4.6 represents the different climate proxies, hazards caused, and possible aftereffects of the same. Therefore, the Bahamas, like many other SIDS, clearly need to diversify from a single economic basis, like tourism, which is highly sensitive to climate change. Vulnerable communities heavily affected by such extreme climate events are low-income tourist workers, Small business owners in the budget/mid-tier hospitality sector, Coastal residents in low-lying islands and settlements, women and youth, who are overrepresented in informal or service-sector jobs. Jobs dominating here are that of housekeeping, service and transport, small tour operations, and local souvenir markets that often lack job security, insurance, or formal protections, making communities highly economically fragile.

4.4 Wetlands: Nature's Kidney and Biodiversity Hotspots

Wetlands, often referred to as "nature's kidneys," are areas permanently or temporarily inundated with water, sourced primarily from aquifers, lakes, rivers, or even tidal surges. Functioning as ecotones, or transition zones, wetlands exhibit characteristics of both terrestrial and aquatic ecosystems. These environments support specialized vegetation, including hydrophytes such as mosses adapted to waterlogged soils, and robust trees thriving in shallow waters. Wetlands are globally distributed, occurring across diverse climatic regions on every continent except Antarctica. Their

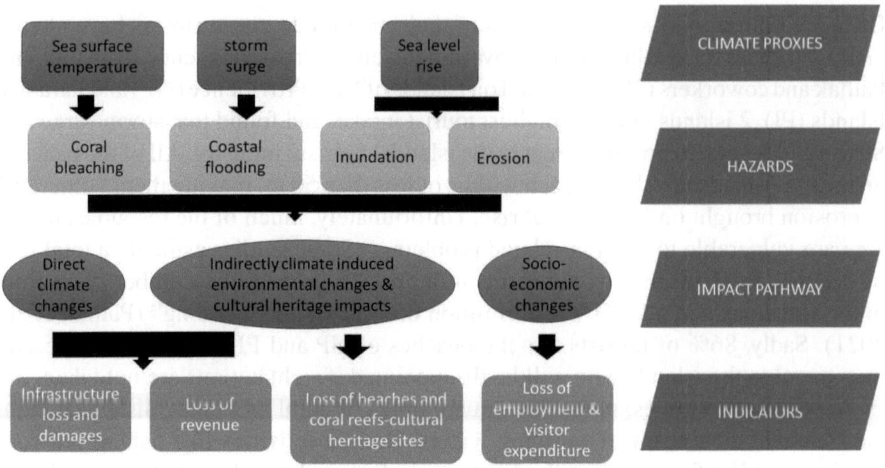

Fig. 4.6 Impact pathways through which climate change affects tourism in Bahamas and other SIDS. *Source* Adapted from Pathak et al. (2021)

ecological and socio-economic significance underscores their vital role in sustaining biodiversity and providing ecosystem services (Table 4.5).

4.4.1 Significance of Wetlands

Wetlands are essential to preserving the resilience and health of the environment. They act as natural water purifiers, effectively removing high levels of nutrients, sediments, and pollutants from surface waters. Their capacity to absorb excess rainfall and recharge groundwater reserves serves as a buffer against storms and floods. Coastal wetlands, including mangroves, coral reefs, and estuaries, provide critical protection against storm surges and tidal waves, reducing their destructive impact on coastal communities. Furthermore, wetlands are a primary source of freshwater, with approximately 95% of accessible freshwater stored in aquifers, making them indispensable for drinking water and irrigation. Ecologically, wetlands are biodiversity hotspots, offering a mosaic of habitats such as marshes, swamps, mangroves, and floodplains. These unique habitats support diverse flora and fauna, acting as critical transitional zones that benefit from the "ecotone effect", which enhances species richness by integrating terrestrial and aquatic communities. Wetlands also play a crucial role in avian ecology; migratory bird species rely on them as critical stopovers, feeding, and breeding grounds. Indian wetlands, for instance, form key nodes in the East Australasian and Central Asian Flyways. They also harbor numerous endemic and endangered species, such as the critically endangered brow-antlered deer (*Rucervus eldii*) in Keibul Lamjao National Park and the vulnerable

Table 4.5 List of World's most significant Ramsar sites

Ramsar site	Location	Area (hectares)
Camargue (Rhone Delta)	France	85,000
Wasur National Park	Indonesia	4,13,810
Everglades National Park	United States	6,10,500
Danube Delta	Romania	6,47,000
Lake Chad Wetlands	Chad	16,39,403
Kakadu National Park	Australia	19,79,766
Okavango Delta	Botswana	55,37,400
Sudd Wetland	South Sudan	57,00,000
Queen Maud Gulf	Canada	62,78,200
Ngiri-Tumba Maindombe	Democratic Republic of the Congo	65,69,624

Source Adapted from The List of Wetlands of International Importance (Ramsar Convention Secretariat. 2023. The List of Wetlands of International Importance. https://www.ramsar.org)

Great Indian Rhinoceros (*Rhinoceros unicornis*), over 70% of whose global population resides in Assam's Kaziranga National Park. Wetlands also contribute to socio-economic well-being through recreation and tourism, owing to their aesthetic and biological richness.

4.4.2 Climate Change and Wetland Vulnerability

Wetlands' integrity and functionality are seriously threatened by climate change, which also upsets the delicate ecological balance of these areas. Rising global temperatures have altered precipitation patterns, leading to hydrological events such as prolonged droughts and floods. Extreme hydrological events occur as a result of changing precipitation patterns brought on by rising global temperatures. These changes impair wetland hydrology, threatening water availability and ecosystem

Table 4.6 Summary of key factors on global wetland loss

Time period	Percentage loss	Primary causes	Source
Since 1700	21–35%	Human interventions such as land-use changes and drainage	Study finds global wetlands losses overestimated despite high rates
Since 1900	Approximately 55%	Urban development, pollution, land drainage, and climate change	Wetland conservation
1970–2015	Approximately 35%	Land-use change, agriculture, urbanization, and climate change	Global Wetland Outlook
1999–2019	Net loss of 4000 km^2	Tidal wetland loss due to sea-level rise, coastal development, and climate change	Landsat Reveals Dramatic Loss of Global Wetlands Over Past Two Decades

Global Wetlands Losses Overestimated Despite High Rates. Available at: https://phys.org/news/2023-02-global-wetlands-losses-overestimated-high.html
Ramsar Convention on Wetlands (2018). Global Wetland Outlook: State of the World's Wetlands and their Services to People. Ramsar Secretariat, Gland, Switzerland. https://www.global-wetland-outlook.ramsar.org/gwo-2018
Landsat Reveals Dramatic Loss of Global Wetlands Over Past Two Decades. Available at: https://landsat.gsfc.nasa.gov/article/landsat-revealsdramatic-loss-of-global-wetlands-over-past-two-decades

stability. Coastal wetlands are particularly vulnerable to sea-level rise, which results in inundation and salinization of freshwater systems, severely impacting the flora and fauna reliant on these habitats. Furthermore, temperature shifts and changes in water chemistry disrupt species composition, leading to biodiversity loss, particularly among endemic and sensitive species. Since they physically destroy ecosystems, extreme events such as hurricanes and cyclones worsen wetland degradation. Under warmer temperatures, peatlands and other carbon-rich wetlands run the risk of releasing stored greenhouse gasses, such as carbon dioxide and methane, contributing to a positive feedback loop that accelerates global climate change. These impacts are compounded by anthropogenic pressures, including drainage, pollution, and land-use changes, amplifying the vulnerability of wetlands. The cumulative effect of these stressors underscores the urgent need for targeted conservation strategies and climate-resilient management practices to safeguard wetlands and their ecosystem services. By integrating ecological, hydrological, and socio-economic perspectives, wetlands emerge as indispensable natural assets that require prioritized attention in global and regional environmental policies (Table 4.6).

4.4.2.1 Case Study 1

With a focus on the effects of storm surge and SLR, Mehvar and colleagues' (2019) study examines how climate change impacts the ecosystem services of coastal

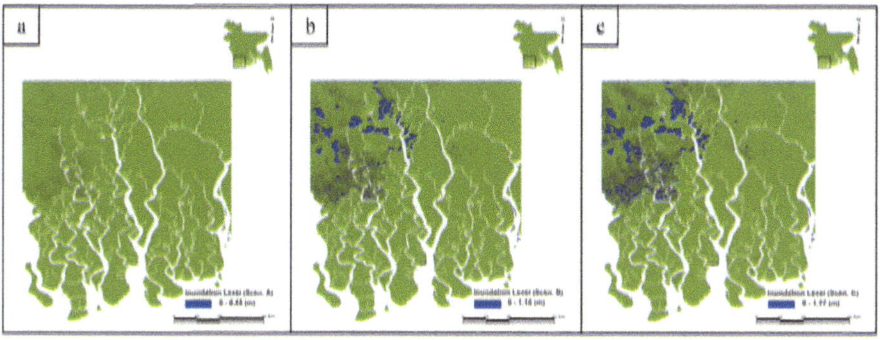

Fig. 4.7 Each scenario's inundation area extent in blue in 2100—There is a 0.25 m relative sea level rise (RSLR) in scenario A, 1.18 m in scenario B, and 1.77 m in scenario C. *Source* Adapted from Mehvar et al. (2019)

wetlands along Bangladesh's west coast. The authors estimate that 44% of the wetlands in the study area may be inundated with a one-meter rise in sea level, which would lead to a major loss of ecosystem services. Specifically, mangrove ecosystems, which provide critical storm protection, could reduce storm surge heights by up to 0.5 m, yet they are under threat of degradation. The loss of carbon sequestration capacity is estimated to release 0.48 million tons of CO_2 annually, further exacerbating climate change. Fisheries, another vital ecosystem service, could face a reduction of 30–50% in productivity due to habitat loss and saline intrusion. Figure 4.7 shows the extend of inundation area under different scenario. Economic valuation shows that the protective services provided by wetlands could save $300 million annually in avoided damages from storm surges. However, with continued degradation, this value is expected to decline significantly. The study also highlights that local livelihoods are at high risk, with 70% of the coastal population depending on wetland resources for income and food security. As livelihoods in agriculture and fisheries decline, many coastal residents may be forced to migrate inland or to urban areas. To mitigate these impacts, the authors recommend large-scale mangrove restoration, emphasizing that investing $1 in wetland conservation can yield benefits of $5–$6 in avoided economic losses. They advocate for integrated coastal zone management and community participation in policymaking. Flexible, locally informed governance structures are needed to maintain ecosystem services and strengthen coastal community resilience. These findings underline the critical role of coastal wetlands and the urgency of integrating their protection into climate adaptation policies.

4.4.2.2 Case Study 2

Mitchell (2013) examines the biological, social, and economic significance of wetlands in Sub-Saharan Africa, as well as their current state, dangers, and anticipated effects of climate change. Wetlands, which cover 1–2% of the region's land area, support diverse ecosystems and provide critical services like water purification, flood regulation, and carbon sequestration. They are home to unique species, including the shoebill stork and African manatee, and sustain over 50 million people who rely on them for agriculture, fishing, and water resources. However, wetlands face significant threats, with 70% under pressure from human activities such as agriculture, urbanization, and pollution. Figure 4.8 represents the number of dams built and the cumulative dam volume since 1800. Climate change exacerbates these challenges by altering wetland hydrology, leading to reduced water availability, prolonged droughts, and desertification, which could shrink wetland areas by 20–30% by 2050. The degradation of these ecosystems may release up to 0.8 gigatons of CO_2 annually, weakening their role as carbon sinks and contributing to global warming. Vulnerable systems like the Inner Niger Delta and Okavango Delta are already experiencing reduced water flows from climate variability and upstream water extraction. Conservation efforts face challenges, including weak governance, limited funding, and inadequate public awareness. The author suggests that wetland degradation in Sub-Saharan Africa can indirectly drive human migration by undermining local livelihoods dependent on wetlands. As agricultural productivity and water availability decline, rural communities may be forced to migrate to urban areas or across borders. This environmental migration can strain city infrastructure and deepen poverty. Strengthening wetland governance and conservation is crucial to reducing migration pressures. The paper underscores the need for integrated wetland management, combining scientific knowledge with community engagement and strengthened policy frameworks. Overall, the study calls for urgent action to protect Sub-Saharan wetlands and sustain their essential ecosystem services amid growing climate and human-induced pressures. Figure 4.9 shows the impact of water release from the Itezhi-tezhi Dam on Kafue flats.

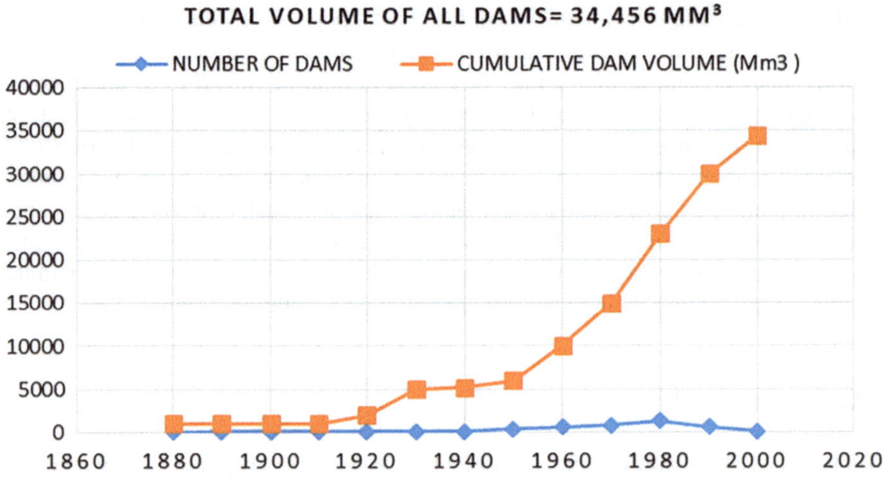

Fig. 4.8 Number of dams built and the cumulative dam volume since 1800 (from King & Pienaar, 2011)

Fig. 4.9 In late June 2010, the Zambezi Basin's Kafue Flats at high water demonstrate the impact of an environmental water release from Itezhi-tezhi Dam. Photo: Frances Bisschop. *Source* Adapted from Mitchell (2013)

4.4.2.3 Case Study 3

Lefebvre et al.'s (2019) study explores how vulnerable Mediterranean Basin wetlands that flood seasonally are to climate change. Using climate projections under two situations—RCP4.5 (stabilization) and RCP8.5 (high emissions)—researchers modeled future changes in water balance and wetland conditions for 2050 and 2100. Their analysis covered 229 localities, assessing the capacity of these ecosystems to sustain healthy conditions amid changing precipitation and temperature patterns. The study emphasizes how crucial wetlands are for ecosystem services, including water storage, habitat provision, and the availability of industrial resources, in addition to biodiversity. Key findings reveal significant increases in water deficits under both scenarios, with conditions worsening more rapidly under RCP8.5. By 2050, the proportion of wetlands able to maintain good ecological conditions is projected to drop from 97% currently to 81% under RCP4.5 and 68% under RCP8.5. By 2100, these Figures further decline to 52% and 27%, respectively. Countries such as Algeria, Morocco, Portugal, and Spain are identified as the most vulnerable to wetland degradation. However, the study notes the resilience of these wetlands to moderate reductions in precipitation, attributing it to their semi-permanent nature and ability to act as water reservoirs during dry periods. Figure 4.10 maps the seasonally flooded wetlands' anticipated habitat change in 2050 and 2100 under the RCP 4.5 and RCP 8.5 greenhouse emission scenarios. Authors stress the urgency of integrating climate change projections into wetland management strategies. They recommend proactive measures to mitigate non-climatic stressors such as land-use changes and water pollution. Maintaining the biodiversity and environmental services that Mediterranean wetlands offer requires doing these actions. Degradation of these habitats might have major ecological and socioeconomic repercussions throughout the region if nothing is done. This research advocates for ecosystem restoration through the re-establishment of natural water regimes and native vegetation to maintain wetland functionality. They emphasize the importance of sustainable land-use planning to reduce habitat fragmentation and mitigate the impacts of urban and agricultural expansion. The study also highlights the need for expanding protected areas and creating ecological corridors to secure critical habitats. Finally, engaging local communities and incorporating traditional knowledge are considered essential for developing adaptive, context-specific management strategies.

4.5 Conclusion

Invasion of exotic species in the Prairies, outbreak of destructive pests and subsequent wild fires in Canada, harmful algal blooms in US freshwaters, migration of fishes in the Mediterranean, Mountain hazards in China, and impacts on tourism sector in the Bahamas- different disasters in different corners of the planet. Consequences of these disasters occur in various dimensions. These include massive loss of ecosystems, high species extinction risk, weakening natural buffers, losses in global revenue, large scale human displacements, and exacerbating regional socio-economic inequalities. If you think this is a distant tragedy, affecting someone else, far away, and why you should care about a climate disaster that is not affecting you directly. You should care because climate change is affecting everyone in more ways than one can imagine. It is affecting the food on your plate, the money in your pocket, the sugar in your coffee, the tomato in your burger, and even the coffee in your cup. Climate extremes have become a new normal and are posing the biggest health threat of this century.

International frameworks like the Ramsar Convention and Kunming-Montreal Global Biodiversity Framework call for restoring degraded wetlands, conserving at least 30% of ecosystems, and tackling threats like pollution and invasive species by 2030. Sustainability in agriculture, aquaculture, and fisheries is vital to reducing wetland degradation. Integrating wetlands into national biodiversity strategies ensures their protection aligns with broader climate and development goals. Collaborative governance among governments, NGOs, and communities can mobilize resources and implement policies to protect wetlands. These efforts not only mitigate climate change but also enhance food and water security, benefitting nearly 4 billion people globally. Sustainable Development Goal 13 (SDG 13 or Global Goal 13), one of the 17 Sustainable Development Goals set by the UN General Assembly in 2015, is to take urgent action to combat climate change. This goal is also known as Global Goal 13 or SDG 13. Techniques for managing grasslands that can improve soil carbon sequestrations, maintain fuel breaks and build network of green infrastructures through reforestation with native species in forests, proper

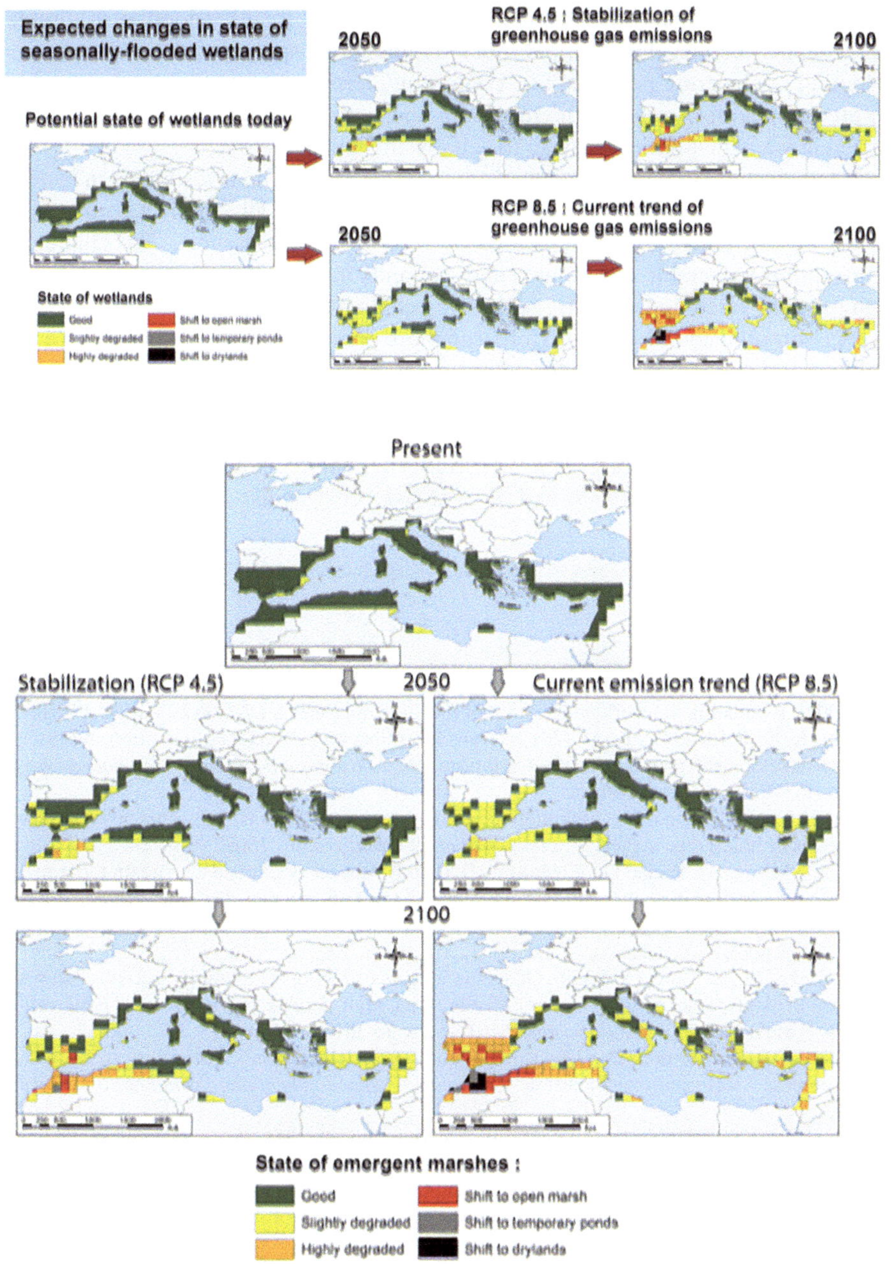

Fig. 4.10 Seasonally flooded wetlands' anticipated habitat change in 2050 and 2100 under the RCP 4.5 and RCP 8.5 greenhouse emission scenarios. *Source* Adapted from Lefebvre et al. (2019)

nutrient and watershed management to control algal blooms, adoption of circular bioeconomy and development of climate-resilient fish species in the marine sector, adopting engineering practices like contouring, terracing and spreading geotextiles to prevent runoff disaster in the mountain and sustainable tourism are some of the different ways that pave their way to achieve this goal.

Recently, governments and organizations have ramped up investments in restoration and conservation initiatives, such as the planned salt marsh nature reserve in the Severn Estuary, complemented by robust policy measures. The challenges confronting vulnerable ecosystems and wetlands in the face of climate change are complex and interconnected. By addressing ecological, socio-economic, and policy aspects simultaneously, global efforts can more effectively safeguard these essential habitats and strengthen resilience against ongoing climatic impacts. This integrated approach ensures that adaptation strategies are both environmentally sustainable and socially just, highlighting the critical need for ongoing research, international collaboration, and proactive conservation measures to counteract the multifaceted effects of climate change on wetlands and other at-risk ecosystems.

References

Agnew, D. J., Pearce, J., Pramod, G., Peatman, T., Watson, R., Beddington, J. R., & Pitcher, T. J. (2009). Estimating the worldwide extent of illegal fishing. *PLoS ONE, 4*(2), Article e4570.

Arkema, K. K., & Ruckelshaus, M. (2017). Transdisciplinary research for conservation and sustainable development planning in the Caribbean. In: *Conservation for the anthropocene ocean* (pp. 333–357). Academic Press.

Arnold, D. E. (1971). Ingestion, assimilation, survival, and reproduction by daphnia pulex fed seven species of blue-green algae 1, 2. *Limnology and Oceanography, 16*(6), 906–920.

Badh, A., Akyuz, A., Vocke, G., & Mullins, B. (2009). Impact of climate change on the growing seasons in select cities of North Dakota, United States of America. *The International Journal of Climate Change: Impacts and Responses, 1*(1), 105.

Bokar, H., Traoré, A. Z., Mariko, A., Diallo, T., Traoré, A., Sy, A., Soumaré, O., Dolo, A., Bamba, F., Sacko, M., & Touré, O. (2020). Geogenic influence and impact of mining activities on water soil and plants in surrounding areas of Morila Mine, Mali. *Journal of Geochemical Exploration, 209*, Article 106429. https://doi.org/10.1016/j.gexplo.2019.106429

Boojhawon, A., & Surroop, D. (2021). Impact of climate change on vulnerability of freshwater resources: A case study of Mauritius. *Environment, Development and Sustainability, 23*, 195–223.

Brandt, J. P. (2009). The extent of the north American boreal zone. *Environmental Reviews, 17*, 101–161. https://doi.org/10.1139/A09-004

Brandt, J. P., Flannigan, M. D., Maynard, D. G., Thompson, I. D., & Volney, W. J. A. (2013). An introduction to Canada's boreal zone: Ecosystem processes, health, sustainability, and environmental issues. *Environmental Reviews, 21*(4), 207–226. https://doi.org/10.1139/er-2013-0040

Calvo, E., Simó, R., Coma, R., Ribes, M., Pascual, J., Sabatés, A., Gili, J. M., & Pelejero, C. (2011). Effects of climate change on Mediterranean marine ecosystems: The case of the Catalan Sea. *Climate Research, 50*(1), 1–29.

Carvalhais, N., Forkel, M., Khomik, M., Bellarby, J., Jung, M., Migliavacca, M., Mu, M., Saatchi, S., Santoro, M., Thurner, M., Weber, U., Ahrens, B., Beer, C., Cescatti, A., Randerson, J. T., &

Reichstein, M. (2014). Global covariation of carbon turnover times with climate in terrestrial ecosystems. *Nature, 514*(7521), 213–217.

Central Intelligence Agency (CIA). (2018). *The world factbook: Bahamas, The.* Retrieved May 21, 2020, https://www.cia.gov/library/publications/the-worldfactbook/Geos/Bf.html

Chapra, S. C., Boehlert, B., Fant, C., Bierman, V. J., Jr., Henderson, J., Mills, D., Mas, D. M. L., Rennels, L., Jantarasami, L., Martinich, J., Strzepek, K. M., & Paerl, H. W. (2017). Climate change impacts on harmful algal blooms in US freshwaters: A screening-level assessment. *Environmental Science & Technology, 51*(16), 8933–8943.

Coll, M., Piroddi, C., Albouy, C., Ben Rais Lasram, F., Cheung, W. W., Christensen, V., Karpouzi, V.S., Guilhaumon, F., Mouillot, D., Paleczny, M., Palomares, M. L., Steenbeek, J., Trujillo, P., Watson, R. & Pauly, D. (2012). The Mediterranean Sea under siege: spatial overlap between marine biodiversity, cumulative threats and marine reserves. *Global Ecology and Biogeography, 21*(4), 465–480. https://doi.org/10.1111/j.1466-8238.2011.00697.x

Coma, R., Ribes, M., Serrano, E., Jiménez, E., Salat, J., & Pascual, J. (2009). Global warming-enhanced stratification and mass mortality events in the Mediterranean. *Proceedings of the National Academy of Sciences of the United States of America, 106*, 6176–6181.

Cook, J., Oreskes, N., Doran, P. T., Anderegg, W. R., Verheggen, B., Maibach, E. W., Carlton, J. S., Lewandowsky, S., Skuce, A. G., Green, S. A., Nuccitelli, D., Jacobs, P., Richardson, M., Winkler, B., Painting, R., & Rice, K. (2016). Consensus on consensus: A synthesis of consensus estimates on human-caused global warming. *Environmental Research Letters, 11*(4), Article 048002.

Crain, C. M., Halpern, B. S., Beck, M. W., & Kappel, C. V. (2009). Understanding and managing human threats to the coastal marine environment. *Annals of the New York Academy of Sciences, 1162*(1), 39–62.

DeKeyser, E. S., Dennhardt, L. A., & Hendrickson, J. (2015). Kentucky bluegrass (Poa pratensis) invasion in the northern Great Plains: A story of rapid dominance in an endangered ecosystem. *Invasive Plant Science and Management, 8*(3), 255–261.

DeMott, W. R. (1989). Optimal foraging theory as a predictor of chemically mediated food selection by suspension-feeding copepods. *Limnology and Oceanography, 34*(1), 140–154.

Deng, M., Chen, N., & Liu, M. (2017). Meteorological factors driving glacial till variation and the associated periglacial debris flows in Tianmo Valley, south-eastern Tibetan Plateau. *Natural Hazards and Earth System Sciences, 17*(3), 345–356.

Dial, R. J., Maher, C. T., Hewitt, R. E., & Sullivan, P. F. (2022). Sufficient conditions for rapid range expansion of a boreal conifer. *Nature, 608*(7923), 546–551. https://doi.org/10.1038/s41586-022-05093-2

Dusenge, M. E., Duarte, A. G., & Way, D. A. (2019). Plant carbon metabolism and climate change: Elevated CO_2 and temperature impacts on photosynthesis, photorespiration and respiration. *New Phytologist, 221*(1), 32–49.

Ellsworth, L. M. (2012). *Improved wildfire management in Megathyrsus maximus dominated ecosystems in Hawaii* (Doctoral dissertation, University of Hawai'i at Manoa).

Etter, A. G. (1951). How Kentucky bluegrass grows. *Annals of the Missouri Botanical Garden, 38*(3), 293–375.

FAO, Food and Agriculture Organization. (2009). Grasslands: Enabling their potential to contribute to greenhouse gas mitigation.

FAO, Food and Agriculture Organization. (2010). Fighting climate change with grasslands.

Fischlin, A., et al. (2007). Ecosystems, their properties, goods, and services. In M. L. Parry, O. F. Canziani, J. P. Palutifok, P. Van der Linden, & C. E. Hanson (Eds.) *Climate change 2007: Impacts, adaptation and vulnerability. Contribution of working group II to the fourth assessment report of the intergovernmental panel on climate change* (pp. 211–272). Cambridge University Press.

Fulton, R. S., III., & Paerl, H. W. (1987). Toxic and inhibitory effects of the blue-green alga Microcystis aeruginosa on herbivorous zooplankton. *Journal of Plankton Research, 9*(5), 837–855.

Gili, J. M., Fuentes, V., Atienza, D., & Lewinsky, I. (2010). *Report of the Medusa Project*. Tech Rep No. 8. Agencia Catalana de l'Aïgua, Generalitat de Catalunya.

Giorgi, F. (2006). Climate change hot-spots. *Geophysical Research Letters, 33*(8). https://doi.org/10.1029/2006GL025734

Global Wetland Outlook. (n.d.). Global Wetland Outlook. https://www.global-wetland-outlook.ramsar.org/

Global wetlands losses overestimated despite high rates. Available at: https://phys.org/news/2023-02-global-wetlands-losses-overestimated-high.html

Grêt-Regamey, A., Brunner, S. H., & Kienast, F. (2012). Mountain ecosystem services: Who cares? *Mountain Research and Development, 32*(S1). https://doi.org/10.1659/MRD-JOURNAL-D-10-00115.S1

Haynes, K. J., Allstadt, A. J., & Klimetzek, D. (2014). Forest defoliator outbreaks under climate change: Effects on the frequency and severity of outbreaks of five pine insect pests. *Global Change Biology, 20*(6), 2004–2018.

He, H., Kirkham, M. B., Lawlor, D. J., & Kanemasu, E. T. (1992). Photosynthesis and water relations of big bluestem (C4) and Kentucky bluegrass (C3) under high concentration carbon dioxide. *Transactions of the Kansas Academy of Science, 1903*, 139–152.

Hilborn, E. D., Roberts, V. A., Backer, L., DeConno, E., Egan, J. S., Hyde, J. B., Nicholas, D. C., Wiegert, E. J., Billing, L. M., DiOrio, M., Mohr, M. C., Joan Hardy, F., Wade, T. J., Yoder, J. S., & Hlavsa, M. C. (2014). Algal bloom-associated disease outbreaks among users of freshwater lakes—United States, 2009–2010. *MMWR. Morbidity and Mortality Weekly Report, 63*(1), 11–15.

Hlásny, T., König, L., Krokene, P., Lindner, M., Montagné-Huck, C., Müller, J., Qin, H., Raffa, K. F., Schelhaas, M.-J., Svoboda, M., Viiri, H., & Seidl, R. (2021). Bark beetle outbreaks in Europe: State of knowledge and ways forward for management. *Current Forestry Reports, 7*, 138–165.

Iglesias, V., Whitlock, C., Krause, T. R., & Baker, R. G. (2018). Past vegetation dynamics in the Yellowstone region highlight the vulnerability of mountain systems to climate change. *Journal of Biogeography, 45*(8), 1768–1780. https://doi.org/10.1111/jbi.13364

IPCC. (2014). Summary for policymakers. In C. B. Field, V. R. Barros, D. J. Dokken, K. J. Mach, M. D. Mastrandrea, T. E. Bilir, M. Chatterjee, K. L. Ebi, Y. O. Estrada, R. C. Genova, B. Girma, E. S. Kissel, A. N. Levy, S. MacCracken, P. R. Mastrandrea, & L. L. White (Eds.), *Climate change 2014: Impacts, adaptation, and vulnerability. Part A: global and sectoral aspects. Contribution of working group II to the fifth assessment report of the intergovernmental panel on climate change* (pp. 1–32). Cambridge University Press. ISBN 9781107058071.

Ives, J. D., Messerli, B., & Spiess, E. (1997). Mountains of the world-a global perspective. In *Mountains of the world: A global perspective* (pp. 313–335).

Ji, Z., Pei, T., Chen, Y., Wu, H., Hou, Q., Shi, F., Xie, B., & Zhang, J. (2022). The driving factors of grassland water use efficiency along degradation gradients on the Qinghai-Tibet Plateau, China. *Global Ecology and Conservation, 35*, Article e02090.

King, J., & Pienaar, H. (2011). Sustainable use of South Africa's inland waters: A situation assessment of resource directed measures 12 years after the 1998 national water act.

Landsat reveals dramatic loss of global wetlands over past two decades. Available at: https://landsat.gsfc.nasa.gov/article/landsat-reveals-dramatic-loss-of-global-wetlands-over-past-two-decades/

Lefebvre, G., Redmond, L., Germain, C., Palazzi, E., Terzago, S., Willm, L., & Poulin, B. (2019). Predicting the vulnerability of seasonally-flooded wetlands to climate change across the Mediterranean Basin. *Science of the Total Environment, 692*, 546–555.

Lenton, T. M., Held, H., Kriegler, E., Hall, J. W., Lucht, W., Rahmstorf, S., & Schellnhuber, H. J. (2008). Tipping elements in the Earth's climate system. *Proceedings of the National Academy of Sciences of the United States of America, 105*(6), 1786–1793. https://doi.org/10.1073/pnas.0705414105

Liu, L., Wang, Z., Wang, Y., Zhang, Y., Shen, J., Qin, D., & Li, S. (2019). Trade-off analyses of multiple mountain ecosystem services along elevation, vegetation cover and precipitation gradients: A case study in the Taihang Mountains. *Ecological Indicators, 103*, 94–104. https://doi.org/10.1016/j.ecolind.2019.03.034

Lucuik, G. S. (1984). Effect of climatic factors on post-diapause emergence and survival of spruce budworm larvae (Lepidoptera: Tortricidae). *The Canadian Entomologist, 116*(8), 1077–1083.

MacDougall, A. S. (2008). Herbivory, hunting, and long-term vegetation change in degraded savanna. *Biological Conservation, 141*(9), 2174–2183.

Mehvar, S., Filatova, T., Sarker, M. H., Dastgheib, A., & Ranasinghe, R. (2019). Climate change-driven losses in ecosystem services of coastal wetlands: A case study in the West coast of Bangladesh. *Ocean & Coastal Management, 169*, 273–283.

Mitchell, S. A. (2013). The status of wetlands, threats and the predicted effect of global climate change: The situation in Sub-Saharan Africa. *Aquatic Sciences, 75*(1), 95–112.

Mostofa, K. M., Liu, C. Q., Vione, D., Gao, K., & Ogawa, H. (2013). Sources, factors, mechanisms and possible solutions to pollutants in marine ecosystems. *Environmental Pollution, 182*, 461–478.

Navarro, L., Morin, H., Bergeron, Y., & Girona, M. M. (2018). Changes in spatiotemporal patterns of 20th century spruce budworm outbreaks in eastern Canadian boreal forests. *Frontiers in Plant Science, 9*, 1905.

Ogra, M. V., & Badola, R. (2015). Gender and climate change in the Indian Himalayas: Global threats, local vulnerabilities, and livelihood diversification at the Nanda Devi Biosphere Reserve. *Earth System Dynamics, 6*(2), 505–523.

Ohrtman, M. K., Clay, S. A., & Smart, A. J. (2015). Surface temperatures and durations associated with spring prescribed fires in eastern South Dakota tallgrass prairies. *The American Midland Naturalist, 173*(1), 88–98. https://doi.org/10.1674/0003-0031-173.1.88

Pathak, A., van Beynen, P. E., Akiwumi, F. A., & Lindeman, K. C. (2021). Impacts of climate change on the tourism sector of a Small Island Developing State: A case study for the Bahamas. *Environmental Development, 37*, 100556.

Perry, D. F., & Fleming, R. A. (1989). The timing of Erynia radicans resting spore germination in relation to mycosis of Choristoneura fumiferana. *Canadian Journal of Botany, 67*(6), 1657–1663.

Peters, M. K., Hemp, A., Appelhans, T., Becker, J. N., Behler, C., Classen, A., Detsch, F., Ensslin, A., Ferger, S. W., Frederiksen, S. B., Gebert, F., Gerschlauer, F., Gütlein, A., Helbig-Bonitz, M., Hemp, C., Kindeketa, W. J., Kühnel, A., Mayr, A. V., Mwangomo, E., … & Steffan-Dewenter, I. (2019). Climate–land-use interactions shape tropical mountain biodiversity and ecosystem functions. *Nature, 568*(7750), 88–92. https://doi.org/10.1038/s41586-019-1048-z

Rixen, M., Beckers, J. M., Levitus, S., Antonov, J., et al. (2005). The Western Mediterranean Deep Water: A proxy for climate change. *Geophysical Research Letters, 32*, L12608. https://doi.org/10.1029/2005GL022702

Ruiz, G. M., Fofonoff, P., Hines, A. H., & Grosholz, E. D. (1999). Non-indigenous species as stressors in estuarine and marine communities: Assessing invasion impacts and interactions. *Limnology and Oceanography*, *44*(3part2), 950–972.

Seymour, J. R., & McLellan, S. L. (2025). Climate change will amplify the impacts of harmful microorganisms in aquatic ecosystems. *Nature Microbiology*, 1–12.

Smadja, J., Aubriot, O., Puschiasis, O., Duplan, T., Grimaldi, J., Hugonnet, M., & Buchheit, P. (2015). Climate change and water resources in the Himalayas. Field study in four geographic units of the Koshi basin, Nepal. *Journal of Alpine Research|Revue de géographie alpine*, (103–2).

Smart, A. J., Scott, T. K., Clay, S. A., Clay, D. E., Ohrtman, M., & Mousel, E. M. (2013). Spring clipping, fire, and simulated increased atmospheric nitrogen deposition effects on tallgrass prairie vegetation. *Rangeland Ecology & Management, 66*(6), 680–687. https://doi.org/10.2111/REM-D-13-00054.1

Soto-Navarro, J., Jordá, G., Amores, Á., Cabos, W., Somot, S., Sevault, F., Macías, D., Djurdjevic, V., Sannino, G., Li, L., & Sein, D. (2020). Evolution of Mediterranean Sea water properties under climate change scenarios in the Med-CORDEX ensemble. *Climate Dynamics, 54*, 2135–2165. https://doi.org/10.1007/s00382-019-05105-4

Speedy, A., & Pugliese, P. L. (1992). *Legume trees and other fodder trees as protein sources for livestock* (No. 102, pp. xviii+-339). FAO.

Stocks, B. J. (1987). Fire potential in the spruce budworm-damaged forests of Ontario. *The Forestry Chronicle, 63*(1), 8–14.

Thiney, U., Banterng, P., Gonkhamdee, S., & Katawatin, R. (2019). Distributions of alien invasive weeds under climate change scenarios in mountainous Bhutan. *Agronomy, 9*(8), 442.

Thirkell, T. J., Pastok, D., & Field, K. J. (2020). Carbon for nutrient exchange between arbuscular mycorrhizal fungi and wheat varies according to cultivar and changes in atmospheric carbon dioxide concentration. *Global Change Biology, 26*(3), 1725–1738.

Touratier, F., & Goyet, C. (2011). Impact of the Eastern Mediterranean Transient on the distribution of anthropogenic CO_2 and first estimate of acidification for the Mediterranean Sea. *Deep-Sea Research I, 58*, 1–15.

USDA, N. (2018). *National resources inventory rangeland resource assessment*. USDA Natural Resources Conservation Service.

Vargas-Yáñez, M., García, M. J., Salat, J., García-Martínez, M. D. C., Pascual, J., & Moya, F. (2008). Warming trends and decadal variability in the Western Mediterranean shelf. *Global and Planetary Change, 63*(2–3), 177–184. https://doi.org/10.1016/j.gloplacha.2007.09.001

Wand, S. J., Midgley, G. F., Jones, M. H., & Curtis, P. S. (1999). Responses of wild C4 and C3 grass (Poaceae) species to elevated atmospheric CO_2 concentration: A meta-analytic test of current theories and perceptions. *Global Change Biology, 5*(6), 723–741.

Wang, J., Cui, Y., Choi, C. E., & Ng, C. W. (2019). The effect of climate change on alpine mountain hazards chain: a case study in Tianmo Ravine, Tibet, China. In *Proceedings of the 8th International Congress on Environmental Geotechnics Volume 3: Towards a Sustainable Geoenvironment 8th* (pp. 461–470). Springer Singapore.

Wang, Y., Hogg, E. H., Price, D. T., Edwards, J., & Williamson, T. (2014). Past and projected future changes in moisture conditions in the Canadian boreal forest. *Forestry Chronicle, 90*(5), 678–691. https://doi.org/10.5558/tfc2014-134

Wein, R. W. (1990). *The importance of wildfire to climate change: Hypotheses for the taiga* (pp. 185–190). Academic Press.

World Wetlands Day–2 February 2024. (n.d.). Convention on biological diversity. https://www.cbd.int/article/world-wetlands-day-2024.

WTTC, W. (2020). *2020 annual research: Key highlights Bahamas*. WTTC. Accessed May 15, 2020.

Yang, Y., Shi, Y., Sun, W., Chang, J., Zhu, J., Chen, L., Wang, X, Guo, Y., Zhang, H., Yu, L., Zhao, S., Xu, K., Zhu, J., Shen, H., Wang, Y., Peng, Y., Zhao, X., Wang, X., Hu, H., … & Fang, J. (2022). Terrestrial carbon sinks in China and around the world and their contribution to carbon neutrality. *Science China Life Sciences, 65*(5), 861–895.

Chapter 5
Patterns of Urban Growth and Green Space Decline: A Spatial Analysis of Built-Up Areas and Heat Islands in Guwahati

Abhijit Sarkar and Pranjit Kumar Sarma

Abstract Cities across the world are expanding at an unprecedented rate, especially in developing regions, where urban growth often comes at the cost of green spaces. This study takes a closer look at Guwahati, a swiftly urban growth city in Northeast India, to explore how land cover has experienced over the last thirty years. Using multi-temporal Landsat satellite imagery, we analyzed both the Normalized Difference Vegetation Index (NDVI) and Normalized Difference Built-Up Index (NDBI) for the years 1993, 2003, 2013, and 2023. The results show a dramatic shift-built-up areas expanded from just 21.09 km^2 in 1993 to 106.67 km^2 in 2023, while green spaces declined by nearly 39%. To assess the thermal consequences of these changes, we used Land Surface Temperature (LST) data from Landsat 8 for 2013 and 2023. Our analysis revealed that the area affected by Urban Heat Spots (UHS), zones of concentrated surface warmth, rose by 22.63%, mainly in dense, newly developed zones. This rise points to the growing influence of impervious surfaces and shrinking vegetation on local microclimates. Together, these insights highlight the importance of city growth approaches that champion ecological balance. The approaches and insights from this study provide a practical framework not just for Guwahati, but for other fast-growing cities grappling with similar challenges. As the world navigates climate risks and rapid urbanization, maintaining green cover is no longer optional; it's essential.

Keywords Urban expansion · Green space degradation · Land surface temperature · Remote sensing · Urban heat island · Climate resilience

---

A. Sarkar (✉)
Department of Geography, Gauhati University, Guwahati, Assam, India
e-mail: abhitackles@gmail.com

P. K. Sarma
Department of Geography, Mangaldai College, Darrang, Assam, India

U. Chatterjee et al. (eds.), *Eco-Resilience: Climate Change, Land Degradation and Sustainable Solutions*, Sustainability Solutions,
https://doi.org/10.1007/978-3-032-00708-7_5

5.1 Introduction

Urban expansion, fueled by rising populations and economic progress, has transformed landscapes globally, converting natural ecosystems into built environments (Wu, 2021; Zhao et al., 2020). According to a United Nations report, more than 50% of global population now inhabits in cities, with this share projected to grow to 66% by 2050, especially in emerging regions of Asia and Africa (Tian et al., 2021). Urbanization introduces numerous environmental challenges, such as forest clearing (Maria et al., 2017), contamination of water and air (Franco et al., 2018; Lim et al., 2020), urban waste accumulation (Lamboglia et al., 2018), disruption of ecosystem equilibrium (Tian et al., n.d.), deterioration of species diversity (Quintas-soriano et al., 2016), and the Urban Heat Island (UHI) effect, where built environments trap more heat than surrounding rural areas due to decline in green cover and increased impervious surfaces (Sheng et al., 2017; Zhou et al., 2019). These impacts are especially significant for urban centers like Guwahati, where rapid expansion has led to notable green space loss and intensified UHI effects.

In today's era, the earth has become an urban planet (Wigginton et al., 2016) and as a result, urban land use has influenced both local weather patterns as well as on regional climate conditions, primarily through the emergence of urban heat islands. Additionally, built-up surfaces in cities disrupt the natural flow of heat and moisture, and recent research also indicates that cities may play a key role in shaping precipitation regimes (Schneider et al., 2014). Guwahati has also seen accelerated urbanization, with India's urbanization rate increasing from 27.8 to 31.1% between 2001 and 2011 (Census of India, 2011), so this rapid urban growth tends to accelerate the expansion of impermeable surfaces, which in turn amplifies increases in land surface temperature (Climatic Effect of Urbanization, 2010; Yuan & Bauer, 2006). LST provides an effective measure of these temperature shifts, revealing where urban expansion has created heat-prone zones or "urban hotspots" (Jin, 2014). Monitoring LST also provides insight into vegetation stress and urban microclimates, guiding sustainable urban interventions (Anderson et al., 2007; Moran et al., 2000; Reginato & Pinter, 2019). Green spaces, which mitigate UHI effects by providing cooling benefits, are essential for environmental and public health (Nesbitt et al., 2019; Zhang et al., 2020). However, as Guwahati expands, green space degradation increases, impacting both the environment and resident well-being.

To explore these concerns at a local level, this study takes Guwahati, an increasingly urbanized city in Northeast India, as a case to examine how urban growth has influenced green space and surface temperature over time. By using satellite images from 1993, 2003, 2013, and 2023, we assess the shifts in green cover and built-up areas through NDVI and NDBI indices. We also calculate Land Surface Temperature (LST) to study heat patterns and identify urban heat spots (UHS), especially in the context of increasing built-up areas. While many studies focus on larger cities or broader regions, this research offers detailed, place-specific insights. The findings aim to support better urban planning by highlighting how fast-growing areas like

Guwahati are changing and what that might mean for the future in terms of heat, greenery, and liveability.

5.2 Study Area

In February 2003, Guwahati emerged as a prominent district headquarters, once renowned as Pragjyotishpur, meaning "light of the East". Its name derives from 'Guwa' for areca nut and 'Haat' for market in Assamese. According to a legend, the demon king Narakasura is said to have built this ancient city. Furthermore, Guwahati holds the ancient legacy of being 'Kamrup', the place where the part of the body of goddess Sati falls (Government of Assam, n.d.).

As the largest metropolis in Northeast India, Guwahati lies within the Indo-Burma Biodiversity Hotspot. Spanning 216.79 km^2, the city is home to nearly a million residents, resulting in a population density of 2,695.43 per km^2. Situated amidst eighteen hills, it sits on an undulating plain, with elevations vary from 49.5 to 55.5 m above sea level. Guwahati boasts eight Reserve Forests and two Wildlife Sanctuaries, including the internationally recognized Ramsar site, Deepor beel, enhancing its ecological significance. The Brahmaputra River meanders through its expanse, contributing to its rich biodiversity. With a tropical monsoon climate, Guwahati receives an annual rainfall of about 1600 mm and maintains a mean yearly temperature of 23 °C. Its diverse habitat includes forest fragments, shrublands, grasslands, plantations, wetlands, agricultural areas, urban settlements, and commercial areas, nurturing a thriving ecosystem. Recent assessments identify Guwahati as home to roughly 60 fish species, 25 amphibian species, 53 reptile species, 212 bird species, and 36 mammal species. Alarmingly, 33 species face extinction threats, and 62 species need evaluation (Government of Assam, n.d.) (Fig. 5.1).

5.3 Material and Methods

Among the diverse array of remote sensing datasets accessible, Landsat imagery, encompassing Multispectral Scanner (MSS), Thematic Mapper (TM)/Enhanced Thematic Mapper Plus (ETM+), and Operational Land Imager (OLI), is one among the most extensively utilized data sources for studying how urban areas develop and spread (Schneider & Mertes, 2014; Yin et al., 2011). For this research, Landsat 5 and Landsat 8 satellite imagery datasets were acquired to examine the evolution of urban landscapes and land cover changes over time. The Landsat 5 satellite imageries were acquired on the 30th of October, 1993, and the 13th of December, 2003. Similarly, Landsat 8 satellite imageries were obtained on the 22nd of November, 2013, and the 12th of December, 2023. These acquisition dates were chosen to ensure temporal consistency and comparability across the study period. The selected satellite imagery datasets have a spatial resolution of 30 m, providing detailed information on land

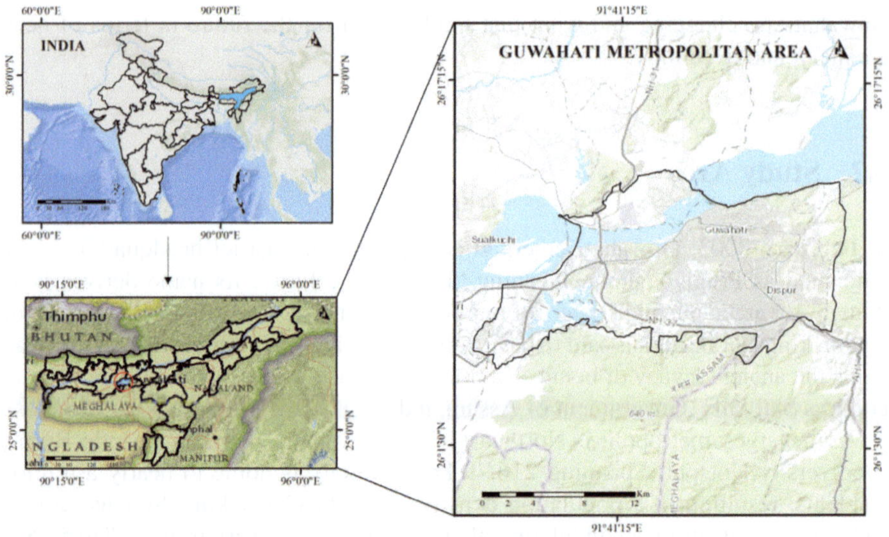

Fig. 5.1 Study area map of Guwahati Metropolitan Area

cover patterns and urban features within the research area. By leveraging Landsat 5 and Landsat 8 imagery datasets spanning multiple decades, this research seeks to explore the spatiotemporal dynamics of urban growth and land cover change, offering clear insights for a deeper understanding of environmental transformations in the research area (Table 5.1).

After acquiring Landsat 5 and Landsat 8 satellite imagery datasets for the years 1993, 2003, 2013, and 2023, the following steps involved calculating the Normalized Difference Vegetation Index (NDVI) and Normalized Difference Built-Up Index (NDBI) values for each year by utilizing the raster calculator function in ArcGIS 10.3.

Table 5.1 Satellite acquisition datasets

Satellite	Sensor	Year of acquisition	Path/ row	Spatial resolution (m)
Landsat 5	Thematic Mapper (TM)	30/10/1993	137/ 042	30
Landsat 5	Thematic Mapper (TM)	13/12/2003	137/ 042	30
Landsat 8	Operational Land Imager (OLI) and Thermal Infrared Sensor (TIRS)	22/11/2013	136/ 042	30
Landsat 8	Operational Land Imager (OLI) and Thermal Infrared Sensor (TIRS)	12/12/2023	137/ 042	30

Calculation of NDVI

The Normalized Difference Vegetation Index (NDVI) was derived using red and near-infrared bands, as these bands exhibit the highest electromagnetic energy absorption by chlorophyll (Xue & Su, 2017), making them well-suited for evaluating vegetation health and density in the research region.

$$\text{NDVI} = \frac{\text{NIR} - \text{Red}}{\text{NIR} + \text{Red}} \tag{5.1}$$

Calculation of NDBI

Another index utilized in this study, sensitive to built-up areas, is the Normalized Difference Built-Up Index (NDBI) (Zha, 2003), which is derived as follows:

$$\text{NDBI} = \frac{\text{SWIR} - \text{NIR}}{\text{SWIR} + \text{NIR}} \tag{5.2}$$

The NDVI values were computed to identify green cover areas, while the NDBI values were calculated to delineate built-up areas within the research area. These indices serve as proxies for vegetation health and urbanization, respectively, allowing for the quantification of land cover characteristics. Subsequently, the NDVI and NDBI values were reclassified individually for each year by utilizing a thresholding approach, where values below a certain threshold, like 0 and 1 were assigned to one class representing non-vegetated or non-built-up areas, while values above the threshold were assigned to another class representing vegetated or built-up areas. The reclassified NDVI and NDBI raster layers were then converted into polygon features using the raster to polygon conversion tool in Arc GIS. This conversion process facilitated the generation of vector polygons representing vegetated and built-up areas for each respective year. Finally, the area of green cover and built-up areas was measured for each year by summing the areas of the respective polygons. This approach allowed for the quantification of changes in green cover and built-up areas over time, offering key insights into the rate of urban growth and land cover changes within the research area.

Calculation of LST

The Land Surface Temperature (LST) data were derived from Band 10 of the Landsat 8 TIRS imagery for Guwahati city through the following steps:

(1) To convert OLI and TIRS band data to top-of-atmosphere spectral radiance, the radiance rescaling factors included in the metadata file are applied, using the following equation:

$$\text{TOA(L)} = \text{ML} * \text{Qcal} + \text{AL} \tag{5.3}$$

where,

ML The band-specific multiplicative scaling factor, obtained from the MTL metadata file under "RADIANCE_MULT_BAND_X."

QCAL The quantized and calibrated digital value of the band or its subset.

AL The band-specific additive scaling factor, found in the MTL metadata as "RADIANCE_ADD_BAND_X," where X denotes the band number.

(2) The TM thermal band data can be converted from spectral radiance to Brightness Temperature (BT), assuming a surface emissivity of 1.

$$\mathrm{BT} = (\mathrm{K2}/(\ln(\mathrm{K1}/\mathrm{L}) + 1)) - 273.15 \tag{5.4}$$

where,

BT Effective at-sensor brightness temperature in Kelvin

K1 Calibration constant 1 [W/(m^2 sr lm)] (607.76)

K2 Calibration constant 2 in K (1260.56)

L Spectral radiance at-sensor [W/(m^2 sr lm)].

(3) To calculate the NDVI (Normal Difference Vegetation Index), the bands 4 and 5 are used, applying the raster calculator, the equation is as follows:

$$\mathrm{NDVI} = (\mathrm{Band\ 5} - \mathrm{Band\ 4})/(\mathrm{Band\ 5} + \mathrm{Band\ 4})$$

(4) The data collected is crucial for calculating the proportion of vegetation (Pv) in the next step. This degree of vegetation coverage is computed using the following equation:

$$\mathrm{Pv} = \mathrm{Square}((\mathrm{NDVI} - \mathrm{NDVI_{min}})/(\mathrm{NDVI_{max}} - \mathrm{NDVI_{min}})) \tag{5.5}$$

where,

Square Corresponds to squaring the formula

$\mathrm{NDVI_{max}}$ Maximum values visible in the "table of contents"

$\mathrm{NDVI_{min}}$ Minimum values visible in the "table of contents".

(5) As a final step, emissivity values are derived using the raster calculatorby applying the following equation:

$$e = m\mathrm{Pv} + n \tag{5.6}$$

where,

m value of emissivity of vegetation, in this case 0.004 was used

Pv corresponds to the percentage of vegetation

n Soil emissivity value, in this case 0.986 was used

(6) Lastly, Land Surface Temperature (LST) is calculated using the following equation:

$$\mathrm{LST} = \mathrm{BT}/1 + w(\mathrm{BT}/p) * \mathrm{Ln}(\varepsilon) \tag{5.7}$$

where,

BT Brightness temperature (band 10 or 11 depending on the case)
w Length of the emitted radiation (band 10 or 11 as the case may be)
p Constant value obtained by the formula $h * c/s$ that when substituting the values is 1.438 * 10^− 34Js and results in 14,380 and ε is LSE obtained previously, this equation should be applied in bands 10 and 11 separately for later unify the results with the tool (cell statistics).

Delineation of UHI and UHS

The identification of Urban Heat Islands (UHI) and non-UHI areas was based on the range of LST derived using the following equations (Guerri et al., 2021; Liu et al., 2007)

For UHI areas:

$$\mathrm{LST} > \mu + 0.5 * \delta$$

where, μ and δ represent the mean and standard deviation of LST in the study area, respectively.

To identify Urban Heat Spots (UHS) in Guwahati, Land Surface Temperature (LST) values were employed, emphasizing ongoing monitoring of concentrated areas of intense heat. These zones, often arising within the Urban Heat Island (UHI) effect, create environments that are inhospitable for human habitation. Urban Heat Spots (UHS) were delineated using the following equation (Ahmed et al., 2023; Govil et al., 2019; Morabito et al., 2016; Swapan et al., 2021):

For UHS areas:

$$\mathrm{LST} > \mu + 2 * \delta$$

where again, μ and δ represent the mean and standard deviation of LST in the study area, respectively.

In the analysis of the Normalized Difference Built-Up Index (NDBI) map for 1993, the urbanized area spanned 21.09 km^2, while the non-built-up area encompassed 252.95 km^2. Over the subsequent decade, by 2003, the built-up area expanded to 63.60 km^2, with the non-built-up area shrinking to 210.42 km^2. This trend continued into 2013, where the urbanized area increased to 80.61 km^2, accompanied by a decline in the non-urbanized area to 193.39 km^2. By 2023, the built-up area further expanded to 106.67 km^2, while the non-built-up area is reduced to 167.34 km^2. Similarly, for the Normalized Difference Vegetation Index (NDVI) maps, in 1993, the vegetated area measured 211.68 km^2, with a non-vegetated area of 62.36 km^2. Subsequently,

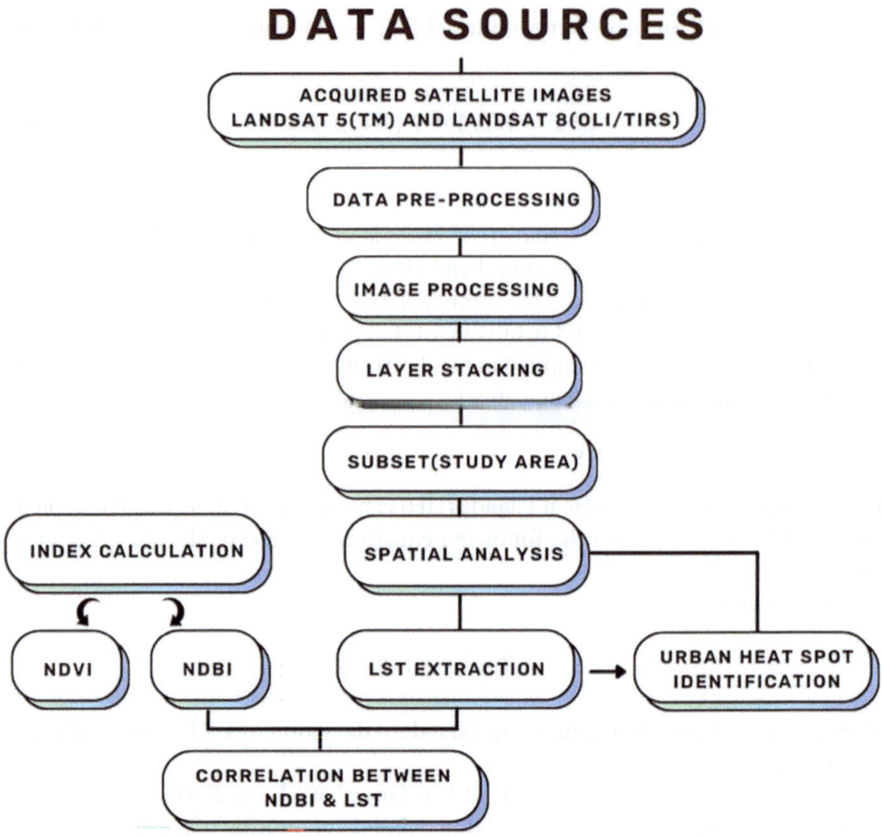

Fig. 5.2 Flowchart for the database and methodology

by 2003, the vegetated area decreased to 162.25 km^2, while the non-vegetated area increased to 111.76 km^2. This trend continued into 2013, where the vegetated area further declined to 157.24 km^2, accompanied by an increase in the non-vegetated area to 116.77 km^2. By 2023, the vegetated area experienced a significant reduction to 129.85 km^2, with the non-vegetated area expanding to 144.15 km^2 (Fig. 5.3).

5.4 Result and Discussion

Understanding the spatio-temporal patterns of urban development becomes easier by incorporating remote sensing data with landscape urban research (Jenerette & Wu, 2020; Wu et al., 2011). As the historical data captured by remote sensing over time are crucial for comprehending the progression of urban development (Deng et al.,

2009; Jensen & Cowen, 1999), so the analysis of the Normalized Difference Built-Up Index (NDBI) and Normalized Difference Vegetation Index (NDVI) maps for the years 1993, 2003, 2013, and 2023 has been prepared to provide key results into the spatio-temporal patterns of urban expansion and green cover change in Guwahati.

5.4.1 *Built-Up Area Expansion (NDBI Results)*

Urban development continues to reshape cities around the world, bringing with it both opportunities and challenges (Zhang et al., 2020). In the case of Guwahati, this transformation has become increasingly visible over the last thirty years. In 1993, the urbanized area covered just 21.09 km^2, but by 2023, it had expanded dramatically to 106.67 km^2. This steady rise reflects the city's rapid urban growth and the increasing demand for infrastructure and residential space. This consistent expansion reflects not only the city's response to growing population pressures but also its push toward economic modernization. Such a pronounced change in land cover directly ties into the main aim of this study, which is to trace the spatio-temporal patterns of urban expansion. At the same time, non-built-up areas shrank from 252.95 km^2 in 1993 to 167.34 km^2 by 2023, a clear sign of how natural and agricultural zones are being replaced by urban settlements.

Such patterns are not unique to Guwahati. Similar urban trajectories have been observed in other fast-growing South Asian cities, where uncontrolled expansion has often led to the encroachment of natural habitats and a significant reduction in ecosystem services (Balram & Dragicevic, 2012). Figure 5.3 offers a visual representation of this expansion, capturing the spatial extent of the built-up footprint over time. This clear land-use transformation emphasizes the necessity for more proactive and sustainable urban planning practices that integrate environmental considerations into policy and design.

5.4.2 *Green Space Degradation (NDVI Results)*

Over the last 30 years, Guwahati has witnessed a significant decline in its green cover. Based on the NDVI analysis, vegetated areas shrank from approximately 211.68 km^2 in 1993 to 129.85 km^2 by 2023, representing a 38.65% reduction. This reduction is largely associated with increasing urban encroachment, deforestation, and land use transformation patterns due to unregulated development (Nesbitt et al., 2019; Zhang et al., 2020). Figure 5.4 illustrates that the most substantial vegetation loss occurred in areas undergoing rapid urban transformation, particularly in the city core and expanding peripheral zones. Simultaneously, non-vegetated areas expanded significantly, from 62.36 km^2 in 1993 to 144.15 km^2 in 2023, highlighting how vegetated zones have been increasingly converted into impervious urban surface. This trend

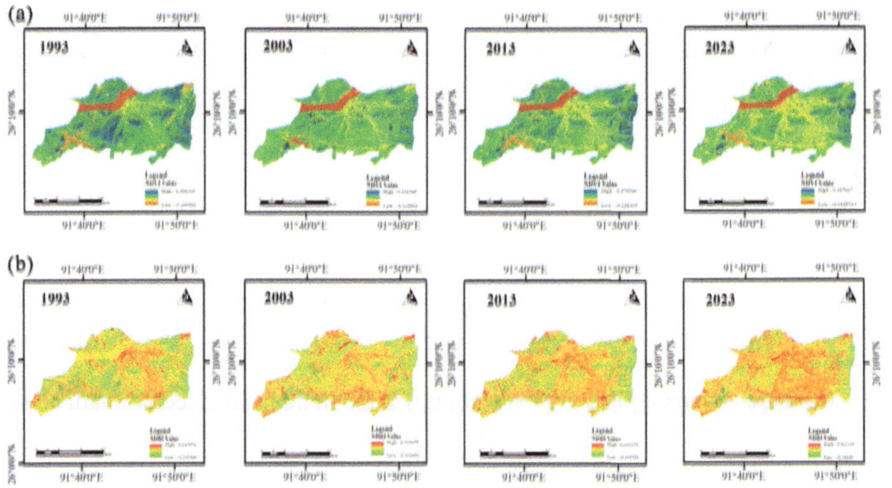

Fig. 5.3 **a** NDVI maps of Guwahati for the years 1993, 2003, 2013, 2023, **b** NDBI maps of Guwahati for the years 1993, 2003, 2013, 2023

reflects the ecological costs of urban growth, including the disruption of microclimates, the reduction of carbon sequestration, and habitat fragmentation (Bratman et al., 2019). The declining trend raises concerns about the city's environmental resilience and its ability to mitigate urban heat effects and support biodiversity (Cheng et al., 2021; Nowak et al., 2018). Furthermore, it emphasizes the urgency of integrating green infrastructure into future planning efforts to achieve a more sustainable urban environment (Fig. 5.5).

5.4.3 Land Surface Temperature (LST) and Urban Heat Island (UHI) Analysis

The assessment of Land Surface Temperature (LST) is essential for assessing the Urban Heat Island (UHI) effects linked to rapid urban growth and green space degradation (Anderson et al., 2007; Moran et al., 2000). LST data offer key insights into surface temperature variations across Guwahati, helping to identify Urban Heat Spots (UHS) where high temperatures are concentrated due to dense urbanization. This section presents LST data interpretation for the years 2013 and 2023, delineates UHS zones, and analyzes the correlation between urbanization and UHI intensity.

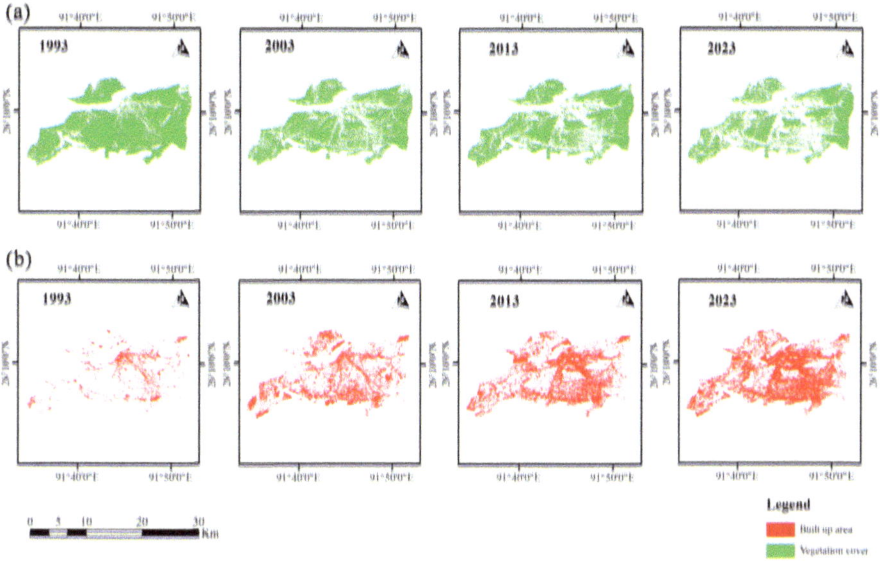

Fig. 5.4 **a** Green spaces of Guwahati for the years 1993, 2003, 2013, 2023, **b** Built areas of Guwahati for the years 1993, 2003, 2013, 2023

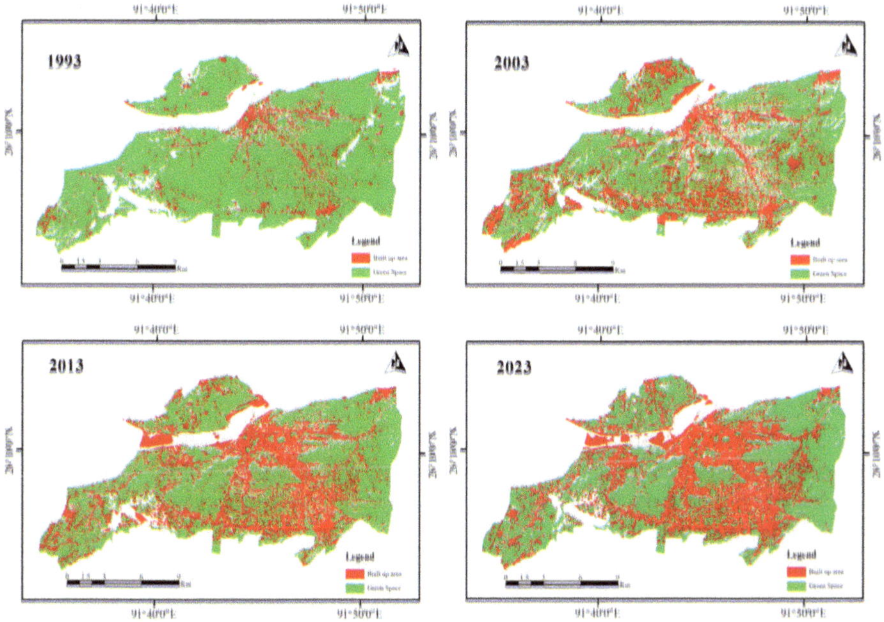

Fig. 5.5 Overlay maps of Green spaces and built up area for 1993, 2003, 2013, and 2023

5.4.3.1 LST for 2013 and 2023

This study uses Landsat 8's Thermal Infrared Sensor (TIRS) Band 10 to extract Land Surface Temperature (LST) data for the years 2013 and 2023. The raw digital numbers (DNs) were converted to spectral radiance, followed by Brightness Temperature (BT), and finally adjusted for land surface emissivity using NDVI-based correction, as suggested by Jime and Paolini (2004). This multi-step approach ensured reliable temperature estimates across varying land covers.

The spatial patterns derived from the LST maps (Fig. 5.6a) clearly demonstrate that urbanized areas consistently recorded more surface temperatures than green zones, reinforcing the presence of the Urban Heat Island (UHI) effect. Table 5.2 summarizes this contrast: in 2013, urban areas had a mean LST of 62.20 °C, slightly above the 61.29 °C in green zones, while in 2023, the urban mean dropped to 58.38 °C, compared to 57.64 °C in vegetated areas. Although both years show elevated temperatures in built-up regions, the slight reduction in overall values in 2023 may be influenced by differences in satellite overpass times or seasonal weather conditions during data acquisition (Anderson et al., 2007). Nevertheless, the persistent temperature gap between land covers highlights how impervious surfaces intensify localized warming, a critical element of the UHI phenomenon (Voogt & Oke, 2015). These findings directly support the study's objective of understanding the thermal consequences of land use change. As green spaces are replaced by built-up areas, the capacity of urban environments to regulate heat diminishes. This is inclined with global research emphasizing the role of vegetation in mitigating UHI effects (Zhou et al., 2019). Importantly, this evidence builds a foundation for developing spatial strategies to target UHI-prone areas. By identifying hotspots of thermal stress, city planners can prioritize regions for green infrastructure or reflective surface interventions, essential for building climate-resilient cities in rapidly urbanizing contexts like Guwahati.

5.4.4 Identification of Urban Heat Spots (UHS)

Urban Heat Spots (UHS) refer to specific areas within a city that experience noticeably higher surface temperatures compared to their surroundings. In this study, UHS were identified using Land Surface Temperature (LST) data derived from Landsat 8 imagery for the years 2013 and 2023 (Fig. 5.7). These zones represent clusters of elevated heat, often found in areas with high urban density, minimal vegetation, and expansive impervious surfaces, making them increasingly uncomfortable and, in some cases, unsuitable for human habitation (Ullah et al., 2023; Voogt & Oke, 2015). Between 2013 and 2023, the total area of UHS in Guwahati increased from 2.83 to 3.47 km^2, marking a significant 22.63% rise. This notable expansion reflects the ongoing urban growth and intensifying pressure on land resources. As concrete structures and asphalt replace natural landscapes, the land loses its ability to dissipate heat efficiently, resulting in concentrated warm zones (Weng, 2009; Zhang et al.,

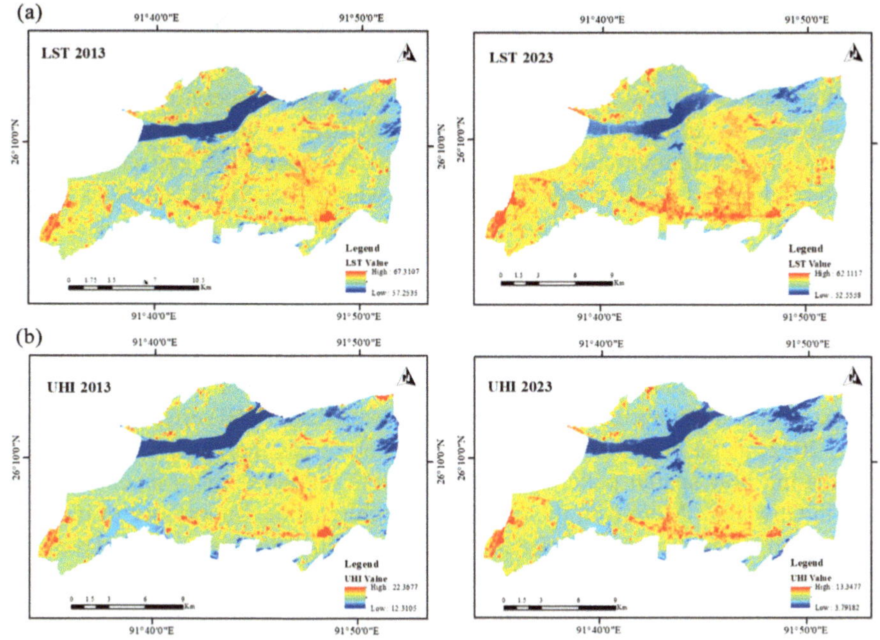

Fig. 5.6 **a** LST maps for 2013 and 2023, **b** UHI maps for 2013 and 2023

Table 5.2 Summary of LST for Urban and Green Zones (2013 and 2023)

Year	Zone type	Minimum LST (°C)	Mean LST (°C)	Maximum LST (°C)
2013	Urban	57.44	62.2	67.31
	Green	57.7	61.29	65.76
2023	Urban	52.6	58.38	62.11
	Green	53.93	57.64	61.41

2020). Spatial analysis showed that these heat pockets are primarily concentrated in the city's core, where built-up areas are densest. However, some peripheral zones, recently urbanized have also started to show signs of rising thermal intensity. This observation underscores the direct linkage between land use change and urban heat dynamics, further aligning with findings from (Gloria & Gnanasekaran, 2024), who reported similar spatial heat anomalies in other fast-growing urban centers. The identification of UHS contributes directly to the study's broader objective of analyzing spatial environmental stress caused by urbanization. By pinpointing where thermal stress is highest, this analysis not only emphasizes the impacts of green space degradation but also provides a foundation for climate-resilient planning strategies. Incorporating features, such as city parks, rooftop gardens, and tree-lined pathways could mitigate these warming effects and restore thermal comfort for residents (Cheng et al., 2021; Li et al., 2019; Nowak et al., 2018).

5.4.5 Correlation Between Built-Up Areas and LST

To better understand how built-up growth affects surface temperature, a correlation analysis was executed between the Normalized Difference Built-Up Index (NDBI) and Land Surface Temperature (LST) for the years 2013 and 2023. The results clearly show a pronounced positive relationship, areas with higher NDBI values also experienced elevated surface temperatures. This suggests that as more land is covered by impervious materials like concrete, asphalt, and rooftops, the natural ability of the surface to release heat is reduced, causing localized warming. The scatter plots in Fig. 5.8 further illustrate this trend, showing how NDBI values align with corresponding LST measurements across urban zones. In both years, densely built-up areas are strongly related with higher LST, validating the hypothesis that intensified urbanization contributes directly to the Urban Heat Island (UHI) effect (Alademomi et al., 2022; Zhou et al., 2019). This finding reinforces the study's broader goal of assessing how spatial patterns of urban growth influence environmental conditions in Guwahati. Moreover, the correlation highlights the importance of incorporating green infrastructure in high-density areas to reduce thermal stress and promote urban livability. These results are aligned with other studies across Asian cities that emphasize the thermal impacts of rapid urban expansion (Fig. 5.8).

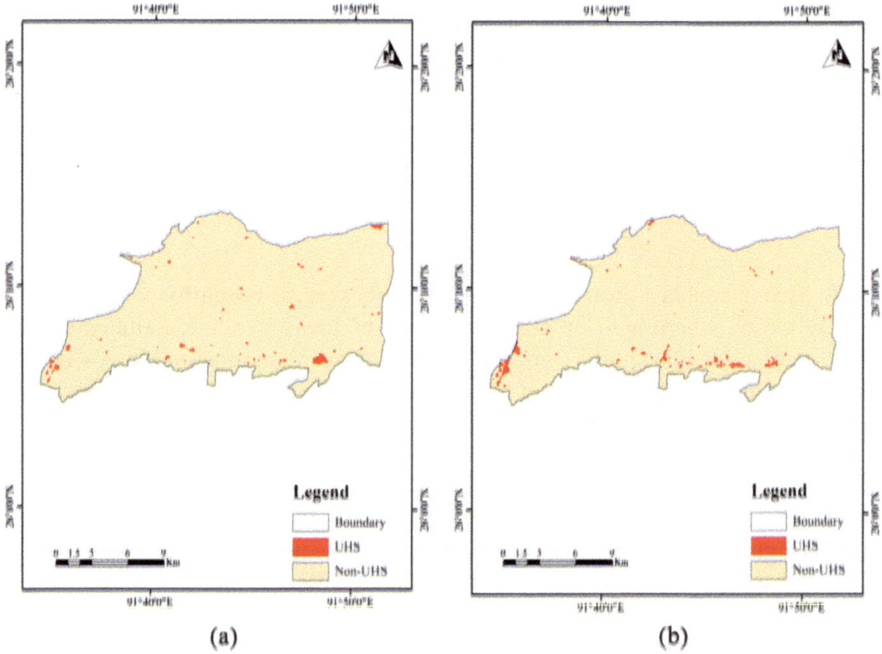

Fig. 5.7 **a** UHS map for 2013, **b** UHS map for 2023

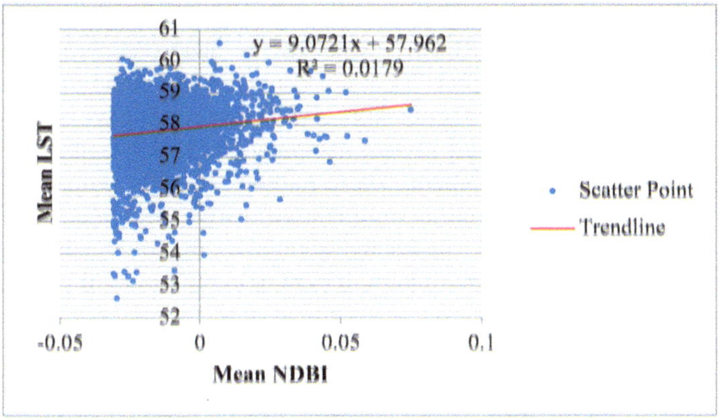

(a)

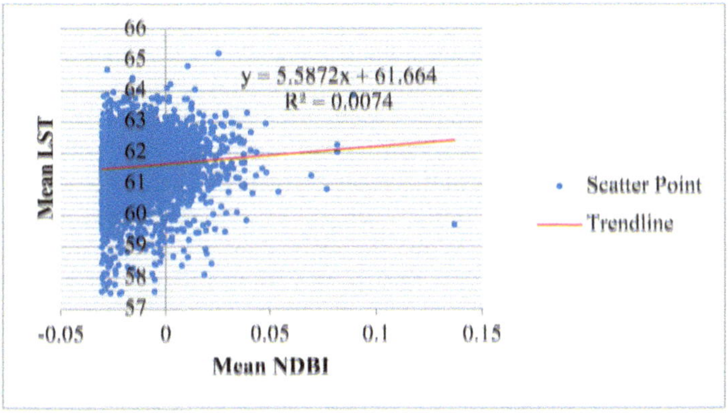

(b)

Fig. 5.8 **a** NDBI versus LST correlation, Guwahati (2013), **b** NDBI versus LST correlation, Guwahati (2023)

5.5 Conclusion

Forest ecosystems are integral to human civilization, offering critical ecological, social, and economic services (Çak et al., 2008). Within urban environments, green spaces act as essential buffers, providing not just ecological value but also supporting psychological wellness, physical health, and overall life satisfaction (Bratman et al., 2019; Sarkar et al., 2018). In this context, understanding the spatiotemporal dynamics of green space change becomes vital for supporting sustainable city planning and maintaining community resilience (Wu & Kyum, 2020). This study provides a comprehensive, data-driven evaluation of urban expansion and green space degradation in Guwahati between 1993 and 2023. Using integrated remote sensing indices—NDVI, NDBI, and LST, the research captures key land cover transitions, quantifies the loss of vegetated areas, and identifies the emergence of Urban Heat Islands (UHI) and Urban Heat Spots (UHS). Notably, the analysis reveals a sharp rise in built-up areas, particularly after 2003, corresponding to demographic pressures and accelerated urban growth. Simultaneously, green spaces have declined significantly, contributing to increased surface temperatures and stressing the need for adaptive environmental management. These findings strongly reinforce the necessity for protecting and expanding urban green infrastructure. Green spaces are not just passive landscape features, they actively regulate microclimates, absorb heat, and mitigate the negative effects of dense urbanization (Chen et al., 2023; Li et al., 2019). The research also underscores the critical need for resilient urban planning frameworks that embed ecological thinking at their core, treating green spaces as essential urban assets rather than aesthetic afterthoughts. In a broader context, this research advances to the growing global discourse on sustainable urban development. Guwahati, though a regional case, reflects the challenges faced by many cities across the Global South, where urbanization often outpaces ecological foresight. Although this study focuses on Guwahati, the patterns it reveals mirror broader urbanization challenges faced by many fast-growing cities across the Global South. These insights highlight the urgent need to incorporate ecological resilience into city planning, making this research relevant for both local decision-makers and global sustainability efforts. The results of this research offer a transferable model for monitoring and managing urban transitions, serving as a useful reference for policymakers, urban planners, and researchers. With climate resilience now a shared international priority, the integration of green infrastructure into city systems must be both a scientific and political imperative (Gill et al., 2007). To lessen the adverse impacts of urban heat islands and green space loss, future planning strategies should prioritize incorporating urban trees and other green infrastructure, vertical gardens, and permeable surfaces within city landscapes (Li et al., 2019). Moreover, continuous satellite-based monitoring and participatory urban design can further support adaptive city development and sustainable land management methods. These solutions, when localized and implemented effectively, can play a significant role in enhancing urban liveability and environmental resilience.

Acknowledgements We are deeply grateful to our institution for its constant support and encouragement throughout this research. A heartfelt thank you to Dr.Pranjit Kumar Sarma, whose guidance and insights have been invaluable in shaping this study. We also extend our appreciation to USGS Earth Explorer for providing the satellite imagery, which made our analysis possible. This research would not have been achievable without these contributions, and we are truly thankful.

Authors Contribution: All authors have contributed equally to this work.

Data Availability Statement:
The datasets during and/or analyzed during the current study available from the corresponding author on reasonable request.

Funding: No funding.

Potential conflicts of interest: No conflict of interest.

Informed Consent: Both authors secured informed consent for the study.

References

Ahmed, S., Bindajam, A., Waseem, M., Juhi, N., Horo, P., & Mallick, J. (2023). Response of soil moisture and vegetation conditions in seasonal variation of land surface temperature and surface urban heat island intensity in sub-tropical semi-arid cities. *Theoretical and Applied Climatology, 0123456789*.https://doi.org/10.1007/s00704-023-04477-2

Alademomi, A. S., Okolie, C. J., Daramola, O. E., & Akinnusi, S. A. (2022). The interrelationship between LST, NDVI, NDBI, and land cover change in a section of Lagos metropolis, Nigeria. *Applied Geomatics, June 2023*. https://doi.org/10.1007/s12518-022-00434-2

Anderson, M. C., Norman, J. M., Mecikalski, J. R., Otkin, J. A., & Kustas, W. P. (2007). A climatological study of evapotranspiration and moisture stress across the continental United States based on thermal remote sensing: 1. *Model Formulation, 112*, 1–17. https://doi.org/10.1029/2006JD007506

Balram, S., & Dragicevic, S. (2012). *Advances in geographic information science.*

Bratman, G. N., Anderson, C. B., Berman, M. G., Cochran, B., Lindahl, T., Meyer-lindenberg, A., Mitchell, R., Ouyang, Z., Roe, J., Scarlett, L., Smith, J. R., Van Den Bosch, M., Wheeler, B. W., White, M. P., Zheng, H., & Daily, G. C. (2019). *Nature and mental health: An ecosystem service perspective. July.*

Çak, G., Sivrikaya, F., & Kele, S. (2008). Forest cover change and fragmentation using Landsat data in Maçka State Forest Enterprise in Turkey. 51–66. https://doi.org/10.1007/s10661-007-9728-9

Chen, J., Yu, Z., Li, M., & Huang, X. (2023). Assessing the spatiotemporal dynamics of vegetation coverage in urban built-up areas.

Cheng, Y., Zhang, J., Wei, W., & Zhao, B. (2021). Landscape and Urban Planning Effects of urban parks on residents ' expressed happiness before and during the COVID-19 pandemic. *Landscape and Urban Planning, 212*(March), Article 104118. https://doi.org/10.1016/j.landurbplan.2021.104118

Deng, J. S., Wang, K., Hong, Y., & Qi, J. G. (2009). Landscape and Urban Planning Spatio-temporal dynamics and evolution of land use change and landscape pattern in response to rapid urbanization. *92*, 187–198. https://doi.org/10.1016/j.landurbplan.2009.05.001

Franco, S., Gaetano, V., & Gianni, T. (2018). Accepted manuscript.https://doi.org/10.1016/j.watres.2018.07.058.This

Gill, S., Forest, T. M., Ennos, R., & Pauleit, S. (2007). Adapting cities for climate change: The role of the green infrastructure. https://doi.org/10.2148/benv.33.1.115

Gloria, S. J., & Gnanasekaran, S. P. (2024). Impact of urban vegetation loss on urban heat islands: A case study of. 134–141.

Govil, H., Guha, S., Dey, A., & Gill, N. (2019). Heliyon seasonal evaluation of downscaled land surface temperature: A case study in a humid tropical city. *HLY, 5*(6), Article e01923. https://doi.org/10.1016/j.heliyon.2019.e01923

Guerri, G., Crisci, A., Messeri, A., Congedo, L., Munaf, M., & Morabito, M. (2021). Thermal summer diurnal hot-spot analysis: The role of local urban features layers. 1–29.

International Journal of An assessment of satellite remotely-sensed land cover parameters in quantitatively describing the climatic effect of urbanization (2010). 37–41.

Jenerette, G. D., & Wu, J. (2020). Analysis and simulation of land-use change in the central Arizona-Phoenix analysis and simulation of land-use change in the central Arizona—Phoenix region, USA. October 2001. 10.1023/A

Jensen, J. R., & Cowen, D. C. (1999). Remote sensing of urban / suburban infrastructure and socio-economic attributes. 29208.

Jime, J. C., & Paolini, L. (2004). Land surface temperature retrieval from LANDSAT TM 5. *90*, 434–440. https://doi.org/10.1016/j.rse.2004.02.003

Jin, M. (2014). Analysis of land skin. April 2004. https://doi.org/10.1175/BAMS-85-4-587

Lamboglia, R., Fiorentino, R., Mancini, D., & Garzella, S. (2018). From a garbage crisis to sustainability strategies: The case study of Naples ' waste collection firm. *Journal of Cleaner Production, 186*, 726–735. https://doi.org/10.1016/j.jclepro.2018.03.151

Li, X., Zhou, Y., Yu, S., Jia, G., Li, H., & Li, W. (2019). Urban heat island impacts on building energy consumption: A review of approaches and findings. *Energy, 515*. https://doi.org/10.1016/j.energy.2019.02.183

Lim, C., Ryu, J., Choi, Y., Woo, S., & Lee, W. (2020). Understanding global PM2. 5 concentrations and their drivers in recent decades (1998–2016). *Environment International, 144*(March), 106011. https://doi.org/10.1016/j.envint.2020.106011

Liu, J., Zhang, W., & Liu, M. (2007). Heat island intensity Assessing the effect of land use = land cover change on the change of urban heat island intensity. September 2014. https://doi.org/10.1007/s00704-006-0273-1

Maria, J., Prasad, S., Alexandre, J., & Diniz-filho, F. (2017). Agriculture on human welfare in the Brazilian Amazonia. *Land Use Policy, 65*(April), 135–142. https://doi.org/10.1016/j.landusepol.2017.04.003

Morabito, M., Crisci, A., Messeri, A., Orlandini, S., Raschi, A., Maracchi, G., & Munafò, M. (2016). Science of the Total Environment The impact of built-up surfaces on land surface temperatures in Italian urban areas. *Science of the Total Environment, 551–552*, 317–326. https://doi.org/10.1016/j.scitotenv.2016.02.029

Moran, M. S., Clarke, T. R., Inoue, Y., & Vidal, A. (2000). Estimating crop water deficit using the relation between surface-air temperature and spectral vegetation index. *263*(April 1994), 246–263.

Nesbitt, L., Meitner, M. J., Girling, C., Sheppard, S. R. J., & Lu, Y. (2019). Landscape and Urban Planning Who has access to urban vegetation ? A spatial analysis of distributional green equity in 10 US cities. *Landscape and Urban Planning, 181*(February 2018), 51–79. https://doi.org/10.1016/j.landurbplan.2018.08.007

Nowak, D. J., Hirabayashi, S., Doyle, M., Mcgovern, M., & Pasher, J. (2018). Urban Forestry & Urban Greening Air pollution removal by urban forests in Canada and its e ff ect on air quality and human health. *Urban Forestry & Urban Greening, 29*(September 2017), 40–48. https://doi.org/10.1016/j.ufug.2017.10.019

Quintas-soriano, C., Castro, A. J., Castro, H., & García-llorente, M. (2016). Land Use Policy Impacts of land use change on ecosystem services and implications for human well-being in Spanish drylands. *Land Use Policy, 54*, 534–548. https://doi.org/10.1016/j.landusepol.2016.03.011

Reginato, R. J., & Pinter, P. (2019). Canopy temperature as a crop water stress indicator. https://doi.org/10.1029/WR017i004p01133

Sarkar, C., Chris, P., Dsc, W., & Gallacher, P. J. (2018). Articles Residential greenness and prevalence of major depressive disorders : A cross-sectional, observational, associational study of 94 879 adult UK Biobank participants. *The Lancet Planetary Health, 2*(4), e162–e173. https://doi.org/10.1016/S2542-5196(18)30051-2

Schneider, A., Friedl, M. A., & Potere, D. (2014). A new map of global urban extent from MODIS satellite data. December 2009. https://doi.org/10.1088/1748-9326/4/4/044003

Schneider, A., & Mertes, C. M. (2014). Expansion and growth in Chinese cities, *9*. https://doi.org/10.1088/1748-9326/9/2/024008

Sheng, L., Tang, X., You, H., Gu, Q., & Hu, H. (2017). Comparison of the urban heat island intensity quantified by using air temperature and Landsat land surface temperature in Hangzhou. *Ecological Indicators, 72*, 738–746. https://doi.org/10.1016/j.ecolind.2016.09.009

Swapan, S., Mohd, T., Hoang, R., Hang, T., & Bhaskaran, S. (2021). Modelling urban heat island (UHI) and thermal field variation and their relationship with land use indices over Delhi and Mumbai metro cities. *Environment, Development and Sustainability, 0123456789*.https://doi.org/10.1007/s10668-021-01587-7

Tian, P., Cao, L., Li, J., Pu, R., Shi, X., & Wang, L. (n.d.). Landscape grain effect in Yancheng coastal wetland and its response to landscape changes. 1–20.

Tian, P., Li, J., Cao, L., Pu, R., Wang, Z., Zhang, H., Chen, H., & Gong, H. (2021). Assessing spatiotemporal characteristics of urban heat islands from the perspective of an urban expansion and green infrastructure. *Sustainable Cities and Society, 74*(June), Article 103208. https://doi.org/10.1016/j.scs.2021.103208

Ullah, N., Siddique, M. A., Ding, M., Grigoryan, S., Khan, I. A., Kang, Z., Tsou, S., Zhang, T., Hu, Y., & Zhang, Y. (2023). The impact of urbanization on urban heat island: Predictive approach using Google Earth Engine and CA-Markov modelling (2005–2050) of Tianjin City, China.

Voogt, J. A., & Oke, T. R. (2015). Thermal remote sensing of urban climates thermal remote sensing of urban climates. *4257*(December). https://doi.org/10.1016/S0034-4257(03)00079-8

Weng, Q. (2009). Thermal infrared remote sensing for urban climate and environmental studies: Methods, applications, and trends. *ISPRS Journal of Photogrammetry and Remote Sensing, 64*(4), 335–344. https://doi.org/10.1016/j.isprsjprs.2009.03.007

Wigginton, B. N. S., Fahrenkamp-uppenbrink, J., Wible, B., & Malakoff, D. (2016). Are the future. *352*(6288), 904–906.

Wu, J. (2021). Urban sustainability: An inevitable goal of landscape research Urban sustainability: An inevitable goal of landscape. December 2009. https://doi.org/10.1007/s10980-009-9444-7

Wu, J., Jenerette, G. D., Buyantuyev, A., & Redman, C. L. (2011). Quantifying spatiotemporal patterns of urbanization: The case of the two fastest growing metropolitan regions in the United States. *Ecological Complexity, 8*(1), 1–8. https://doi.org/10.1016/j.ecocom.2010.03.002

Wu, L., & Kyum, S. (2020). Science of the Total Environment Does socioeconomic development lead to more equal distribution of green space ? Evidence from Chinese cities. *Science of the Total Environment, Xxxx*, 143780. https://doi.org/10.1016/j.scitotenv.2020.143780

Xue, J., & Su, B. (2017). Significant remote sensing vegetation indices: A review of developments and applications.

Yin, J., Yin, Z., Zhong, H., Xu, S., Hu, X., & Wang, J. (2011). Monitoring urban expansion and land use/land cover changes of Shanghai metropolitan area during the transitional economy (1979–2009) in China. 609–621. https://doi.org/10.1007/s10661-010-1660-8

Yuan, F., & Bauer, M. E. (2006). Comparison of impervious surface area and normalized difference vegetation index as indicators of surface urban heat island effects in Landsat imagery. https://doi.org/10.1016/j.rse.2006.09.003

Zha, Y. (2003). Use of normalized di ff erence built-up index in automatically mapping urban areas from TM imagery. 583–594.

Zhang, J., Yu, Z., Cheng, Y., Chen, C., Wan, Y., Zhao, B., & Vejre, H. (2020). Evaluating the disparities in urban green space provision in communities with diverse built environments: The case of a rapidly urbanizing Chinese city. *Building and Environment, 183*(July), Article 107170. https://doi.org/10.1016/j.buildenv.2020.107170

Zhao, Q., Wen, Z., Chen, S., Ding, S., & Zhang, M. (2020). Quantifying land use/land cover and landscape pattern changes and impacts on ecosystem services.

Zhou, D., Xiao, J., Bonafoni, S., Berger, C., Deilami, K., Zhou, Y., Frolking, S., Yao, R., Qiao, Z., & Sobrino, J. A. (2019). Satellite remote sensing of surface urban heat islands: Progress, challenges, and perspectives. 1–36. https://doi.org/10.3390/rs11010048

Chapter 6
Climate-Land Cover Nexus on the Mbum Plateau of Northwest Region, Cameroon: Implications for Rural Livelihood Sustenance

Nfor Delphine Mbongsi and Suiven John Paul Tume

Abstract The land is a crucial natural resource that sustains agriculture and livelihoods in rural areas of Sub-Saharan Africa (SSA). Climate change and soil degradation are intricately linked, with detrimental effects on agriculture, environmental resilience, and livelihoods. This chapter is guided by the following specific objectives: (1) establish the climatic trend on the Mbum Plateau, (2) examine land cover dynamics, (3) assess the effects of climate-land cover nexus on crop production, (4) examine the livelihood implications of climate-land cover dynamics on the Mbum Plateau. Most researchers have not delved into the interconnectedness of agrarian soil degradation, climate change, and rural livelihoods. To address this knowledge gap, both quantitative and qualitative methods of data collection were used. This approach involved reviewing relevant literature, climatic data, and Landsat images spanning from the 1980s to 2024, along with an assessment of agricultural practices and soil degradation, to understand their collective impact on rural livelihoods. This was complemented by crop productivity data from 1990 to 2023. This chapter uncovered compelling evidence of the negative correlation between climatic changes, soil degradation, and the sustainability of rural livelihoods. These findings are groundbreaking as they demonstrate the interplay between climate change, soil degradation, and the sustenance of rural livelihoods. To ensure the continuity of rural livelihoods, it is imperative to foster a resilient environment free from unsustainable agricultural practices.

Keywords Agriculture · Climate change · Livelihood sustenance · Nature-based solutions

N. D. Mbongsi (✉) · S. J. P. Tume
Department of Geography and Planning, The University of Bamenda, Bamenda, Cameroon
e-mail: nfordelphinembongsi@gmail.com

U. Chatterjee et al. (eds.), *Eco-Resilience: Climate Change, Land Degradation and Sustainable Solutions*, Sustainability Solutions,
https://doi.org/10.1007/978-3-032-00708-7_6

6.1 Introduction

Climate variability and land cover changes represent significant global challenges, particularly affecting rural livelihoods in developing nations where communities heavily depend on natural resources for their sustenance. Recent studies have highlighted how these environmental transformations are reshaping traditional livelihood patterns and ecosystem services across various regions (Baffour-Ata et al., 2021; Lahai et al., 2022). The interplay between climatic factors and anthropogenic activities has led to substantial modifications in forest reserves, wetlands, and agricultural landscapes, creating complex challenges for sustainable development and food security (Msofe et al., 2014; Tume et al., 2020).

In Sub-Saharan Africa, these challenges are particularly pronounced due to the region's vulnerability to climate change and its heavy reliance on rain-fed agriculture. Research conducted across various African countries has demonstrated how changing rainfall patterns, increasing temperatures, and land degradation are compromising rural livelihoods and traditional farming systems (Lahai et al., 2022). Studies in Tanzania have shown that wetland ecosystems, crucial for local livelihoods, are experiencing significant transformations due to both climatic variations and human activities (Msofe et al., 2014). Similarly, research in Ghana has documented the impacts of forest reserve depletion on agricultural productivity and community resilience (Baffour-Ata et al., 2021).

Cameroon presents a unique case study due to its diverse ecological zones and complex socio-economic dynamics. The country's Northwest Region, particularly the Mbum Plateau, has emerged as a critical area for understanding the interactions between climate variability, land cover change, and rural livelihood sustainability. Recent investigations have revealed that environmental degradation, particularly soil impoverishment, has led to significant demographic shifts and changes in agricultural practices (Ndi & Amawa, 2014). The region's Mbororo communities, traditionally dependent on pastoral activities, have experienced substantial challenges due to climate variations, resulting in decreasing cattle numbers and the adoption of alternative livelihood strategies (Ngwani et al., 2024a, 2024b).

Research has identified various vulnerability hotspots, with hydrological systems, montane forests, and built-up areas facing significant threats due to population pressure and unsustainable land use practices on the Mbum Plateau (Tassah et al., 2020). While some positive trends have been observed in certain areas, such as improved land resource management (Tume et al., 2024), the overall trajectory of environmental change continues to pose substantial challenges for rural communities. The complex interactions between climate variability, land cover dynamics, and livelihood systems in this region necessitate a comprehensive understanding to inform effective policy interventions and adaptation strategies.

This chapter seeks to address these critical knowledge gaps by examining the climate-land cover nexus on the Mbum Plateau through a multi-temporal analysis spanning from 1980s to 2024. By integrating remote sensing techniques with ground-level observations, the chapter provides a detailed understanding of the spatiotemporal dynamics of land cover transformation and its implications for rural livelihood sustenance. The findings will contribute to the broader discourse on environmental change and sustainable development while offering practical insights for local stakeholders and policymakers in Cameroon and similar contexts across Sub-Saharan Africa. This chapter is guided by the following specific objectives: (1) establish the climatic trend on the Mbum Plateau, (2) examine land cover dynamics, (3) assess the effects of climate-land cover nexus on crop production, (4) examine the livelihood implications of climate-land cover dynamics on the Mbum Plateau.

6.2 Study Area and Methods

The Mbum Plateau lies between Longitudes 10° 50′ 48″ East of the Greenwich Meridian and Latitudes 6° 20′ 02″–6° 41′ 25″ North of the Equator, and Longitudes 10° 23′ 03″ and 11° 55′ 48″ east of the Greenwich Meridian. It constitutes the area covered by the Nkambe Central (which doubles as the Divisional headquarters for Donga-Mantung) and Ndu Sub-Divisions (Fig. 6.1). Mbum Plateau shares common borders to the West with the Misaje Sub-Division, to the North Ako Sub-Division, to the North-East Nwa Sub-Division, and the South, South-East/South-West Bui Division. The principal tribal group in the Mbum Plateau hailed from Tikari in the Adamawa Plateau of Cameroon. The Mbum Plateau comprises 41 communities, with a population of 120,781 inhabitants covering a total surface area of 2112.4 km^2 and a population density of 57.2 persons/km^2. Most inhabitants in the Nkambe Plateau are peasants with land as the lone source of livelihood.

The Mbum Plateau have two distinctive ecological subzones and consequently two climatic features. It has a tropical climate of Sudan Cameroon type, which has given rise to a cloud type of forest cover. The plateau is colder than the surrounding plains. The rainy season is eight months, longer than the dry season (four months), and runs from mid-March to mid-November. The maximum rainfall is in July, August, and September, with an annual maximum attaining 2332.8 mm, and a minimum of 1685.8 mm (Cameroon Tea Estate-CTE, Ndu Plantation Weather Station, 2015). Unlike previously when rainfall began in mid-March, the trends from 1995 up to date have shown that rainfall progressively starts close to the first half of January. The extent of the dry season varies because of the existence of the North East Trade Winds. The extent of the rainy season varies by the existence of the South West Monsoon, which carries moist air from the Atlantic Ocean into the hinterland. The rainy season generally begins in mid-March and ends in mid-November with rainfall amounts varying between 1300 and 1900 mm per annum. The mean annual range of temperature in Ndu hardly exceeds 5 °C. Annually, rainfall ranges from 1300 to 3000 mm, and a mean of 2000 mm. Ndu has the Cameroonian type of climate characterized by topographically induced rainfall. In the dry season, the area experiences cold mornings and hot afternoons, while the rainy seasons are generally warm. The

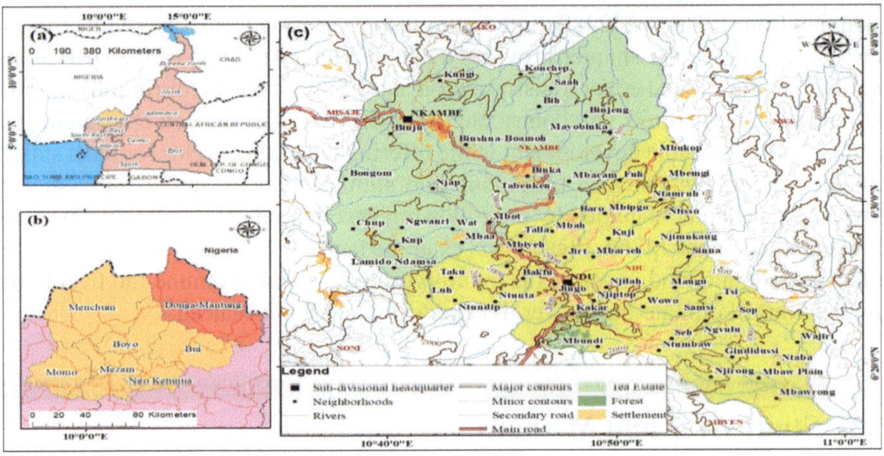

Fig. 6.1 Location of the Mbum Plateau. *Source* Adapted from the Administrative Map of Cameroon, 2016

relative humidity in the dry season is less than 40% while in the rainy season, relative humidity ranges from 80 to 95% especially from June to September. Livestock rearing is done on the hills where there is abundant pasture in the wet season, and intermittent streams and springs flow. During the dry season, most of the pasture dries up while intermittent and ephemeral streams disappear forcing livestock farmers to move down valleys where some crops are trampled upon by cattle causing farmer-grazier conflicts. Mbum Plateau has four vegetation types; grassland, savanna with trees and shrubs, forest, and wooded savanna. The dominant grass species *hyparrhenia* and *sporobus* that are good for grazing. There is a reduction in grass as one moves from valleys to slopes and hills. The grassland of Mbakop, Ntamru, Mbangong, Maka, Mayo Binka and Binshu could be considered plagioclimax. In the nineteenth century or so, cattle rearing was concentrated on hill slopes for fear of pasture scarcity and tsetse fly downslopes. Of recent, owing to the depletion of grazing land many cattle graze on the valleys, creating farmer-grazier conflicts.

6.3 Methodology

The methodology employed in this research adopted a comprehensive mixed-methods approach, integrating both quantitative and qualitative data collection techniques to examine the climate-land cover nexus on the Mbum Plateau. The study analysed climatic data and Landsat imagery spanning from the 1980s to 2024. The remote sensing analysis utilized a robust classification scheme comprising seven thematic classes: forest, savanna/cropland mosaic, landscaped green space, savanna, built structures, bare ground, and agricultural zone. Quality control was implemented through a confusion matrix for the 2023 classification, yielding a high Kappa Coefficient of 0.96%, which indicates strong reliability in the land cover classification. The temporal assessment was structured across four key periods (1990, 2000, 2010, and 2024), enabling systematic detection of land cover changes.

The climate analysis component incorporated detailed decadal characterizations of rainfall and temperature patterns, employing key metrics such as the Standardized Precipitation Index (SPI), rainfall Coefficient of Variation (CV), and reliability indices to quantify climatic variations. This methodological framework was strengthened by its multi-temporal approach and the integration of multiple data sources, though it could have benefited from more detailed descriptions of the household survey methodology and ground-truthing procedures. The systematic combination of remote sensing, climate data analysis, and household surveys provided a robust foundation for understanding the complex interactions between climate variability, land cover change, and rural livelihood sustainability in the study area.

Qualitative and quantitative methods were utilized in this study. Secondary data were sourced from various literature related to the topic, along with rainfall and temperature data for Ndu and Nkame, obtained from the Cameroon Tea Estates Plantation in Ndu, covering the period from 1981 to 2022. Additionally, crop production data were collected from the Regional Delegation of Agriculture and Rural Development in Bamenda, spanning from 1990 to 2023.

The analysis focused on mean annual, inter-annual, and seasonal variations, employing statistical measures such as standard deviations, variances, and specifically, the Coefficient of Variation (CV) to assess rainfall reliability. Decadal variations in climatic elements and the percentage changes in rainfall and temperature were also examined. All decadal variations were treated as anomalies to establish trends, which were illustrated through graphs (time series analysis), fitted with R^2 values and linear equations to demonstrate the percentage change of each climatic element.

Both qualitative and quantitative techniques were employed in the data analysis. The climatic elements, including rainfall and temperature, significantly influence the climate of the Mbum Plateau, impacting soil loss and rural livelihoods. To analyze these influences over time, we used monthly means, annual means, mean annual anomalies, running means, decadal means, and decadal anomalies to determine their effects on land cover change, agricultural output, and livelihoods.

Where: X_n = mean rainfall for month N.

R = mean annual rainfall

The Standardized Precipitation Index (SPI) was also calculated in this context. The SPI is a tool primarily developed to define and monitor drought conditions. It enables analysts to assess the rarity of a drought at various time scales for any rainfall station with historical data. McKee et al. (1993) utilized the SPI to establish criteria for defining drought intensities and events across different time scales. A drought event is identified when the SPI remains continuously negative and reaches an intensity of $-$ 1.0 or lower. The event concludes when the SPI rises above zero. Each drought event is characterized by its duration, which is defined by its start and end points, as well as the intensity for each month it persists. The cumulative sum of the SPI values for all months within a drought event can be referred to as the drought's magnitude (Table 6.1).

Rainfall is the only required input parameter for the analysis, which is effective in examining wet and dry cycles. To compute the Standardized Precipitation Index (SPI), consistent data sets are necessary. However, it is common for data sets to contain only 85–90% complete records. In this context, the analysis utilized unadjusted mean monthly rainfall data from 1981 to 2021. The SPI can also identify periods of unusually wet events. Conceptually, the SPI represents the number of standard deviations by which the recorded precipitation values deviate from the mean over specified periods. In statistical terms, the SPI is equivalent to the Z-score.

$\mathrm{Z-score} = x - \frac{\mu}{\delta}$ (McKee et al., 1993).

Where:

Z-score expresses the x score's distance from the mean (μ) in standard deviation (δ) units.

The Standardized Precipitation Index (SPI) is calculated based on the cumulative probability of a specific rainfall event occurring at a weather station. In the assessment of drought incidents on the Mbum Plateau, SPI was utilized effectively. All anomaly graphs generated from the data were fitted with trend lines and linear equations, which indicate whether there is an increase or decrease in the elements being studied. Additionally, other measures of central tendency for these weather elements included Standard Deviation (σ), Variance, and Coefficient of Variation (CV). The Coefficient of Variation is calculated as follows:

$$\sigma = \frac{\sqrt{\Sigma\left(Y - \overline{\Upsilon}\right)^{2}}}{N}$$

$$\mathrm{CV} = \sigma * \frac{100}{\overline{\Upsilon}}$$

where: $\bar{\Upsilon}$ = mean

Table 6.1 Drought probability of recurrence

SPI value	Category	Probability (%)	Freq. in 100 years	Severity of event
> 2.00	Extreme wet	2.3	100	1 in 1 year
1.5–1.99	Severely wet	4.4	70	1 in 1.1 years
1.00–1.49	Moderately wet	9.2	50	1 in 1.3 years
0–0.99	Mildly wet	34.1	45	1 in 1.5 years
− 0.1 to − 0.99	Mild dryness	34.1	33	1 in 3 years
− 1.00 to − 1.49	Moderate dryness	9.2	10	1 in 10 years
− 1.50 to − 1.99	Severe dryness	4.4	5	1 in 20 years
< − 2	Extreme dryness	2.3	2.5	1 in 50 years

Source McKee et al. (1993, 1995), World Meteorological Organization (2012)

N = sample size.

Based on these analyses, tables were created to summarize climatic characteristics. The drought vulnerability for this study is evaluated by reconstructing historical occurrences on a 12-month time scale (from January to December).

6.4 Results and Discussion

6.4.1 Climatic Trends

The climatological patterns observed on the Mbum Plateau between 1981 and 2022 reveal a complex interplay between rainfall variability and thermal dynamics. While precipitation exhibits pronounced oscillations characterized by intermittent peaks exceeding 300 mm in 1983, 2001, and 2019, the overall trend suggests minimal directional change. This precipitation regime manifests in a quasi-cyclical pattern with intervals of abundance and scarcity, potentially influencing the agricultural calendar and traditional farming wisdom that has historically guided cultivation practices in the region. The rainfall trend in the Mbum Plateau exhibits patterns similar to those found in other regions of the Bamenda Highlands of Cameroon and across the tropics. Data from Ndu and Nkambe show a slight decrease in inter-annual variation (Fig. 6.2).

In the same vein, the inter-annual pattern of the Standardized Precipitation Index (SPI) reveals the same pattern as inter-annual rainfall (Fig. 6.3).

The Standardized Precipitation Index (SPI) is an important indicator of changing climatic conditions, particularly concerning extreme weather events (Table 6.2). Persistently dry SPI episodes indicate a decline in rainfall, which is significant since the livelihoods in tropical regions primarily depend on rain-fed agriculture.

Extreme wet conditions were recorded in Ndu in 1982 and 2016, as well as in Nkambe in 1982. Severely wet conditions were observed at both locations in 2000. In Ndu, moderately wet rainfall episodes took place in 2015, 2017, 2020, and 2021, while in Nkambe, these episodes occurred in 2003, 2016, 2017, 2020, and 2021. Moreover, moderately wet rainfall events were noted in Ndu during the years 1981,

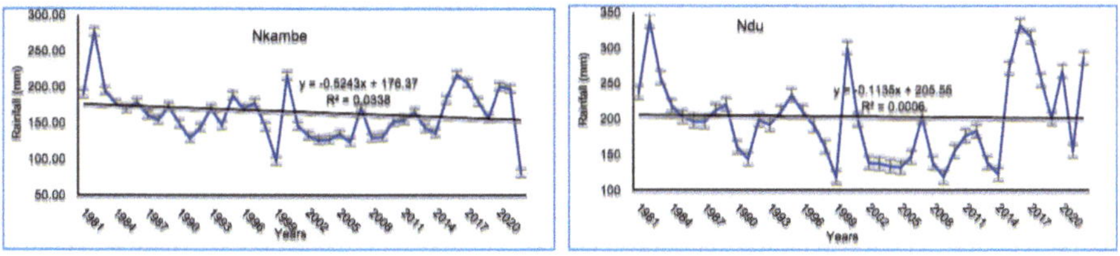

Fig. 6.2 Annual rainfall trend of Nkambe and Ndu (1981–2021). *Data source* Cameroon Tea Estates (CTE), Ndu Plantation

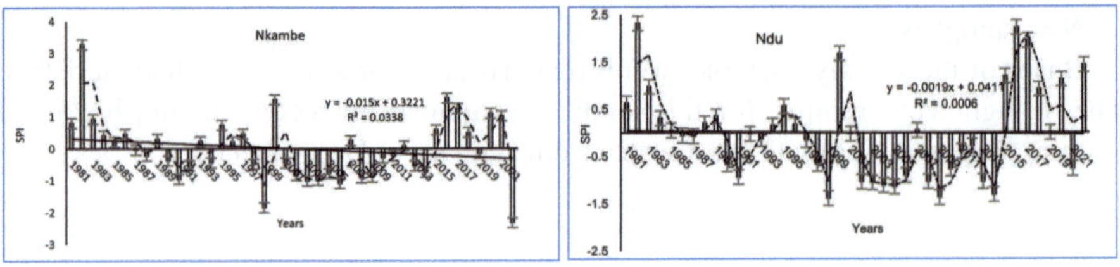

Fig. 6.3 Inter-annual SPI trend for Nkambe and Ndu (1981–2021). *Data source* Cameroon Tea Estates (CTE), Ndu Plantation

Table 6.2 Summary of SPI episodes in the Mbum Plateau

SPI value	SPI class	Years	№ of episodes	%
Ndu				
> 2.00	Extreme wet	1982, 2016	2	4.76
1.5–1.99	Severely wet	2000	1	2.38
1.00–1.49	Moderately wet	2015, 2017, 2020, 2022	4	9.52
0–0.99	Mildly wet	1981, 1983, 1984, 1985, 1988, 1989, 1994, 1995, 1996, 2007, 2018	11	26.19
− 0.1 to − 0.99	Mild dryness	1986, 1987, 1990, 1991, 1992, 1993, 1997, 1998, 2001, 2004, 2006, 2010, 2011, 2012, 2019, 2021	16	38.10
− 1.00 to − 1.49	Moderate dryness	1999, 2002, 2003, 2005, 2008, 2009, 2013, 2014	8	19.05
Total			**42**	**100**
Nkambe				
> 2.00	Extreme wet	1982	1	2.38
1.5–1.99	Severely wet	2000	1	2.38
1.00–1.49	Moderately wet	2003, 2016, 2017, 2020, 2021	5	11.90
0–0.99	Mildly wet	1981, 1983, 1984, 1985, 1986, 1989, 1993, 1995, 1996, 1997, 2007, 2012, 2015, 2018	14	33.33
− 0.1 to − 0.99	Mild dryness	1987, 1988, 1990, 1991, 1992, 1994, 1998, 2001, 2002, 2004, 2005, 2008, 2009, 2010, 2011, 2013, 2014, 2019	18	42.86
− 1.00 to − 1.49	Moderate dryness	1999, 2006	2	4.76
< − 2	Extreme dryness	2022	1	2.38
Total			**42**	**100**

Source Generated from Cameroon Tea Estates (CTE), Ndu Plantation rainfall data

1983, 1984, 1985, 1988, 1989, 1994, 1995, 1996, 2007, and 2018. In Nkambe, similar events were observed in 2003, 2016, 2017, 2020, and 2021.

Mildly wet conditions in Ndu were recorded in 1981, 1983, 1984, 1985, 1988, 1989, 1994, 1995, 1996, 2007, and 2018. For Nkambe, these incidents occurred in 1981, 1983, 1984, 1985, 1986, 1989, 1993, 1995, 1996, 1997, 2007, 2012, 2015, and 2018.

Regarding dry episodes, the distribution is as follows: mild dryness was recorded in Ndu during 1986, 1987, 1990, 1991, 1992, 1993, 1997, 1998, 2001, 2004, 2006, 2010, 2011, 2012, 2019, and 2021. In Nkambe, mild dryness occurred in 1987, 1988, 1990, 1991, 1992, 1994, 1998, 2001, 2002, 2004, 2005, 2008, 2009, 2010, 2011, 2013, 2014, and 2019.

For moderate dryness, Ndu experienced episodes in 1999, 2002, 2003, 2005, 2008, 2009, 2013, and 2014. Nkambe recorded moderate dryness in 1999 and 2006. An extreme dryness episode was solely noted in Nkambe in 2022.

These results revealed that seven of the eight SPI classes have been recorded in the Mbum Plateau (Fig. 6.4).

Between 1981 and 2022, the various Standardized Precipitation Index (SPI) episodes in the Mbum Plateau were recorded as follows:

- Extreme wet: 3 episodes (Ndu: 2, Nkambe: 1)
- Severely wet: 2 episodes (Ndu: 1, Nkambe: 1)
- Moderately wet: 9 episodes (Ndu: 4, Nkambe: 5)
- Mildly wet: 25 episodes (Ndu: 11, Nkambe: 14)
- Mild dryness: 34 episodes (Ndu: 16, Nkambe: 18)
- Moderate dryness: 10 episodes (Ndu: 8, Nkambe: 2)
- Extreme dryness: 1 episode.

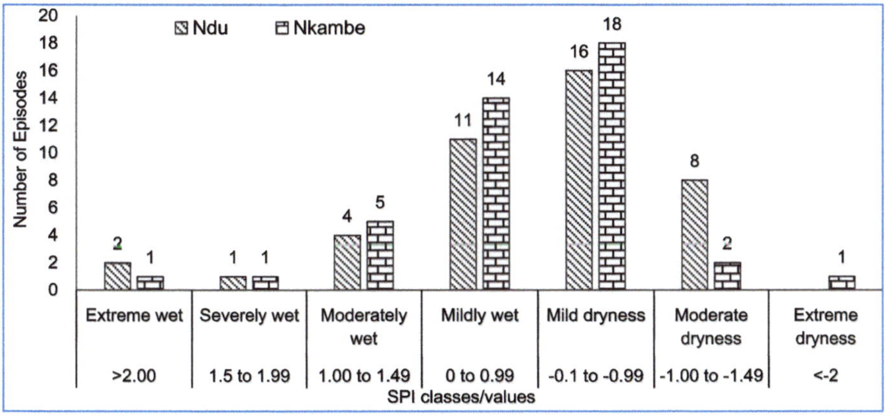

Fig. 6.4 Summary of SPI episodes for Ndu and Nkambe. *Source* Generated from Cameroon Tea Estates (CTE), Ndu Plantation rainfall data

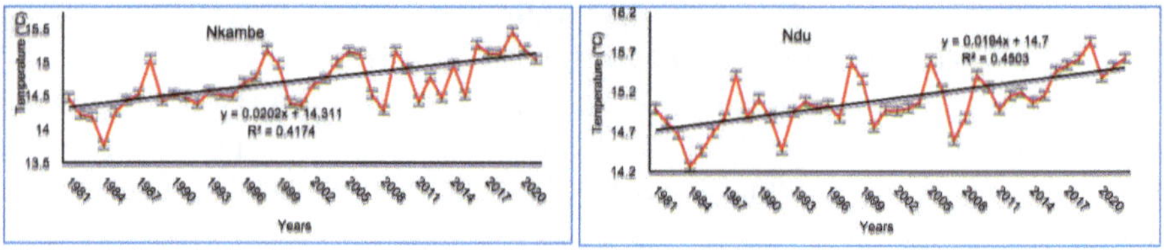

Fig. 6.5 Trend in minimum temperatures of Nkambe and Ndu. *Data source* Cameroon Tea Estates (CTE), Ndu Plantation

The thermal trajectory, conversely, demonstrates a more definitive pattern with a positive slope of 0.027 °C per year ($R^2 = 0.643$), translating to an approximate 1.5 °C increase over the study period, with notable thermal maxima occurring during the late 1990s and mid-2000s (Fig. 6.5). Minimum temperatures range between 14.5 and 15.5 °C, showing an increasing trend on the Mbum Plateau as indicated by data from Nkambe and Ndu (Fig. 6.5).

The minimum temperature trend in the plateau is on the rise, which has led to frequent agro-hydrological droughts negatively impacting both crops and livestock. This situation has also had adverse effects on animal and human health. Notable mean temperature anomalies occurred in 1983, 1997/1998, and 2013. The 1983 anomaly was attributed to widespread drought across West Africa, while the unusually high temperatures in 1997/1998 were caused by El Niño. The 2013 episode was also the result of drought. Overall, the mean annual temperature has been steadily increasing in Nkambe and Ndu (Fig. 6.6).

The trend in maximum annual temperatures for Ndu and Nkambe has been rising, showing a gradual increase. An anomaly was observed in 2005 due to a prolonged dry season characterized by exceptionally high temperatures (Fig. 6.7). The increase in temperatures is more pronounced in Nkambe than in Ndu, as indicated by the linear trend line.

The results showed that the mean monthly temperature trends in Nkambe and Ndu fluctuate at an increasing rate, as indicated by the linear trend line. The unusual temperature pattern resembles a zig-zag, with temperatures rising from January to March, declining until July, and then dropping further in December. These temperature variations impact cattle production by causing heat stress and increasing the

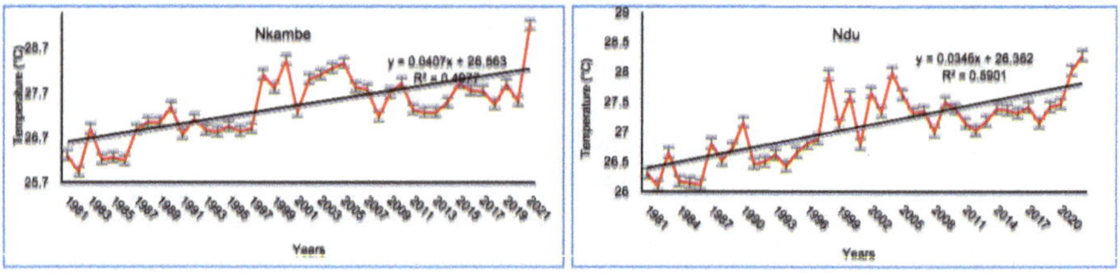

Fig. 6.6 Trend in maximum temperatures of Nkambe and Ndu. *Data source* Cameroon Tea Estates (CTE), Ndu Plantation

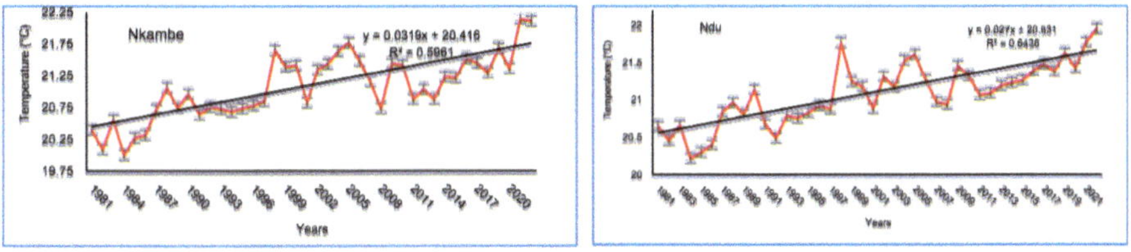

Fig. 6.7 Mean temperature trend of Nkambe. *Data source* Cameroon Tea Estates (CTE), Ndu Plantation

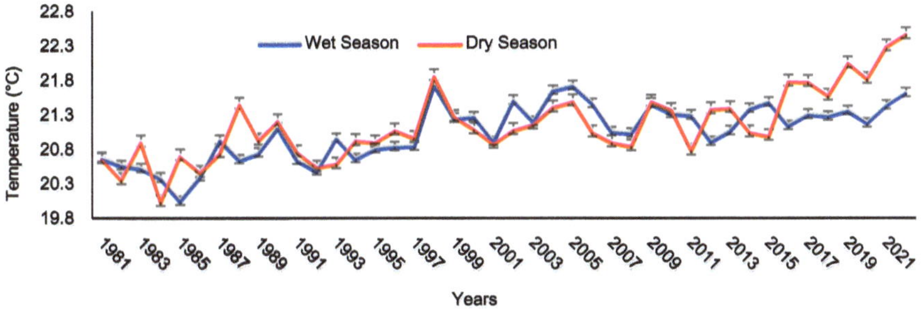

Fig. 6.8 Seasonal temperature pattern for the Mbum Plateau (1981–2022). *Data source* Cameroon Tea Estates (CTE), Ndu Plantation

likelihood of diseases. Higher temperatures also lead to increased humidity levels, which can adversely affect animal health. The lowest mean temperatures are recorded in April and May, while the highest occur in January, February, and December. Similarly, seasonal temperature fluctuations have comparable effects on both cattle production and human health (Fig. 6.8).

Seasonal temperature trends in the Mbum area show a rising trend during the wet season and a declining trend during the dry season, as indicated by a linear analysis. Recently, abnormal temperature patterns have emerged, with an increase from January to March, a decrease until July, and another rise from December to April. These unusual temperatures negatively affect livestock production by causing heat stress and facilitating the spread of diseases. Higher temperatures lead to increased humidity, which adversely impacts animal health. The lowest average temperatures are recorded in April and May, while the highest occur in January, February, and December.

These results indicate a decline in rainfall on the Mbum Plateau, with a total of 24 dry episodes recorded in Ndu (57.14%) and 21 in Nkambe (50%) amidst rising temperatures. Overall, there were 45 dry episodes, accounting for 53.57% of the total episodes across the plateau. A closer examination of the decadal climatic analysis in the Mbum Plateau reveals notable fluctuations in both rainfall and temperatures, with a general trend of rising temperatures and declining rainfall (Table 6.3). The climatic characteristics from 1981 to 2022 are summarized as follows:

Table 6.3 Summary of rainfall and temperature characteristics of the Mbum Plateau

Decades	Mean RF (mm)	Mean temp. (°C)	RF SD	RF CV (%)	Mean SPI	SPI class	RF reliability	RF trend	Temp. Trend
Ndu									
1981–1990	227.02	20.68	47.75	21.03	0.4	Mildly wet	Unreliable	Decrease	Increase
1991–2000	198.76	20.98	50.23	25.27	− 0.07	Mild dryness	Unreliable	Increase	Increase
2001–2011	155.77	21.26	27.94	17.94	− 0.8	Mild dryness	Reliable	Decrease	Stable
2012–2022	233.71	21.47	72.91	31.33	0.5	Mildly wet	Unreliable	Increase	Increase
Average	**203.82**	**21.10**	**49.71**	**23.89**	**0.01**	**Mildly wet**	**Unreliable**	**Decrease**	**Increase**
Nkambe									
1981–1990	185.4	20.54	36.23	19.54	0.59	Mildly wet	Reliable	Decrease	Increase
1991–2000	161.45	21.00	33.24	20.59	− 0.11	Mild dryness	Unreliable	Increase	Increase
2001–2011	141.18	21.32	14.81	10.42	− 0.66	Mild dryness	Reliable	Increase	Decrease
2012–2022	172.85	21.49	39.44	22.82	0.22	Mildly wet	Unreliable	Decrease	Increase
Average	**165.22**	**21.09**	**30.93**	**18.34**	**0.01**	**Mildly wet**	**Unreliable**	**Decrease**	**Increase**

RF SD rainfall standard deviation, *RF CV* rainfall coefficient of variation, *SPI* Standardized Precipitation Index
Source Generated from Cameroon Tea Estates (CTE), Ndu Plantation rainfall data

1980–1990

- Ndu: Mean rainfall was 227.02 mm, mean temperature was 20.68 °C, with a coefficient of variation (CV) of 21.02% (unreliable). The mean Standardized Precipitation Index (SPI) was 0.4 (mildly wet), indicating a trend of decreasing rainfall and increasing temperatures.
- Nkambe: Mean rainfall was 185.4 mm, mean temperature was 20.54 °C, with a CV of 19.59% (reliable). The mean SPI was 0.59 (mildly wet), suggesting a trend of decreasing rainfall and rising temperatures.

1991–2000

- Ndu: Mean rainfall decreased to 198.76 mm, while mean temperature increased to 20.98 °C, with a CV of 25.27% (unreliable). This period showed an increasing trend in both rainfall and temperature.

- Nkambe: Mean rainfall was 161.45 mm, mean temperature was 10 °C, with a CV of 20.59% (unreliable). There was an increasing trend in both rainfall and temperature during this decade.

2001–2011

- Ndu: Mean rainfall dropped to 155.77 mm, and mean temperature rose to 21.26 °C, with a CV of 17.94% (reliable). This indicated a trend of decreasing rainfall and stable temperatures.
- Nkambe: Mean rainfall was 141.18 mm, while mean temperature was 21.32 °C, with a CV of 10.42% (reliable). This period experienced an increasing trend in rainfall but a decreasing trend in temperature.

2012–2022

- Ndu: Mean rainfall increased to 233.71 mm, and mean temperature rose to 21.47 °C, with a CV of 31.33% (unreliable). This decade exhibited increasing trends in both rainfall and temperature.
- Nkambe: Mean rainfall was 172.85 mm, with a mean temperature of 21.49 °C, and a CV of 22.82% (unreliable). This period showed a decreasing trend in rainfall while temperatures continued to rise.

The juxtaposition of these climatic parameters presents intriguing implications for the agrarian landscape and ecosystem dynamics of the Mbum Plateau. The coupling of increasing temperatures with erratic rainfall patterns suggests a potential modification of soil moisture regimes, evapotranspiration rates, and consequently, the water budget available for both natural vegetation and cultivated crops. This climate-land cover nexus may be catalysing subtle shifts in the phenological cycles of local flora, altering pest dynamics, and potentially necessitating a reevaluation of traditional agricultural calendars that have historically anchored rural livelihoods in this region. The observed trends underscore the importance of developing adaptive strategies that can buffer against climate variability while maintaining the ecological integrity of agricultural landscapes. This analysis highlights the shifting climatic patterns in the Mbum Plateau, emphasising the complexities of rainfall and temperature trends over the decades.

More than 80% of the population in the study areas acknowledged that farm abandonment and water shortages are among the key challenges they face. The increase in population has led to insufficient farmland, prompting many farmers to move to the slopes, which have drier and less productive soil. This migration often encroaches on grazing lands, resulting in livestock straying into farms and damaging crops that have yet to be harvested. The situation creates conflict, with both farmers and herders blaming each other, frequently escalating to violent confrontations. The combined effects of dry, infertile soils and farmer-grazer conflicts have intensified crop failures and increased food insecurity, a concern expressed by farmers. This deepening crisis has left farmers searching for ways to sustain their livelihoods while also striving to maintain a resilient environment.

In Ndu, mean rainfall fluctuated considerably across decades, with the highest average (233.71 mm) recorded in 2012–2022 and the lowest (155.77 mm) in 2001–2011. The rainfall coefficient of variation (CV) indicates high variability, particularly in 2012–2022 (31.33%), suggesting increased precipitation unpredictability. Temperature shows a consistent warming trend, rising from 20.68 °C (1981–1990) to 21.47 °C (2012–2022). The SPI oscillated between mildly wet and mild dry conditions, with rainfall reliability generally classified as unreliable except for 2001–2011.

Nkambe exhibited slightly different patterns, with generally lower rainfall amounts compared to Ndu. The mean rainfall ranged from 141.18 mm (2001–2011) to 185.4 mm (1981–1990), showing less dramatic fluctuations than Ndu. Temperature trends paralleled those of Ndu, increasing from 20.54 to 21.49 °C. Notably, Nkambe demonstrated more stable rainfall patterns with lower CV values, particularly during 2001–2011 (10.42%), suggesting more predictable precipitation patterns during this period.

These findings reveal significant transformations in both climatic patterns and land cover dynamics on the Mbum Plateau, aligning with broader regional trends observed across Sub-Saharan Africa. The climatic analysis shows a pronounced warming trend with temperatures increasing by approximately 1.5 °C, which corresponds with findings by Morrison et al. who documented similar warming patterns in West African highlands. The rainfall patterns demonstrate high variability with notable fluctuations between wet and dry periods, particularly affecting agricultural planning and food security. This aligns with Baffour-Ata et al. (2021) observations in Ghana, where rainfall variability significantly impacted rural agricultural systems.

The decadal analysis reveals distinct spatial variations between Ndu and Nkambe, with Ndu experiencing more extreme rainfall fluctuations (CV up to 31.33%) compared to Nkambe's relatively stable patterns (CV as low as 10.42%). This spatial heterogeneity in climate patterns echoes findings by Msofe et al. (2014) in Tanzania, where localized climate variations significantly influenced ecosystem services and agricultural productivity. The temperature trends in both locations show consistent warming, with Ndu increasing from 20.68 to 21.47 °C and Nkambe from 20.54 to 21.49 °C, supporting the regional warming patterns documented by Shayan across similar ecological zones.

6.5 Land Cover Dynamics

The land cover analysis reveals significant temporal transformations in both localities. In Nkambe, forest cover declined dramatically from 46.73% in 1990 to 16.60% in 2010, before partially recovering to 27.23% in 2024 (Table 6.4).

This was accompanied by notable increases in agricultural zones and savanna/cropland mosaics, suggesting active land use conversion. The bare ground coverage fluctuated significantly, peaking at 18.29% in 2000, while built structures remained relatively stable around 2%. Ndu's landscape exhibited different dynamics, with

Table 6.4 Land cover change on the Mbum Plateau

Thematic classes	Area 1990 in (ha)	Area 1990 in (%)	Area 2000 in (ha)	Area 2000 in (%)	Area 2010 in (ha)	Area 2010 in (%)	Area 2024 in (ha)	Area 2024 in (%)
Nkambe								
Forest	21,957	46.73	15,971	33.99	7799	16.60	12,798	27.23
Savanna/ Cropland mosaic	7172	15.26	4422	9.41	7994	17.01	9924	21.12
Landscaped green space	0	0.00	0	0.00	0	0.00	0	0.00
Savanna	11,049	23.51	10,341	22.01	15,053	32.03	6800	14.47
Built structures	872	1.86	1961	4.17	975	2.07	983	2.09
Bare ground	1953	4.16	8596	18.29	5282	11.24	8492	18.07
Agricultural zone	3988	8.49	5702	12.13	9890	21.05	7996	17.02
Total	**46,992**	**100.00**	**46,992**	**100.00**	**46,992**	**100.00**	**46,992**	**100.00**
Ndu								
Forest	18,430	38.37	21,045	43.82	11,926	24.83	15,549	32.38
Savanna/ Cropland mosaic	5331	11.10	3203	6.67	6684	13.92	7497	15.61
Landscaped green space	741	1.54	757	1.58	750	1.56	654	1.36
Savanna	13,384	27.87	9635	20.06	15,365	31.99	7786	16.21
Built structures	987	2.06	1316	2.74	1117	2.33	742	1.54
Bare ground	4340	9.04	6529	13.59	5984	12.46	6125	12.75
Agricultural zone	4816	10.03	5544	11.54	6203	12.92	9674	20.14
Total	**48,028**	**100.00**	**48,028**	**100.00**	**48,028**	**100.00**	**48,028**	**100.00**

forest cover showing more resilience, fluctuating between 43.82% (2000) and 24.83% (2010), before stabilising at 32.38% in 2024. The agricultural zone expanded consistently from 10.03% in 1990 to 20.14% in 2024, indicating progressive agricultural intensification. Uniquely, Ndu maintained a small but consistent presence of landscaped green space (approximately 1.5%) throughout the study period, while savanna coverage showed considerable variability.

In 1990, the landscape was predominantly covered by dense, intact forest, which accounted for the majority of the land area (Fig. 6.9). Interspersed within the forest were pockets of savanna vegetation, primarily in the central and eastern portions of the region.

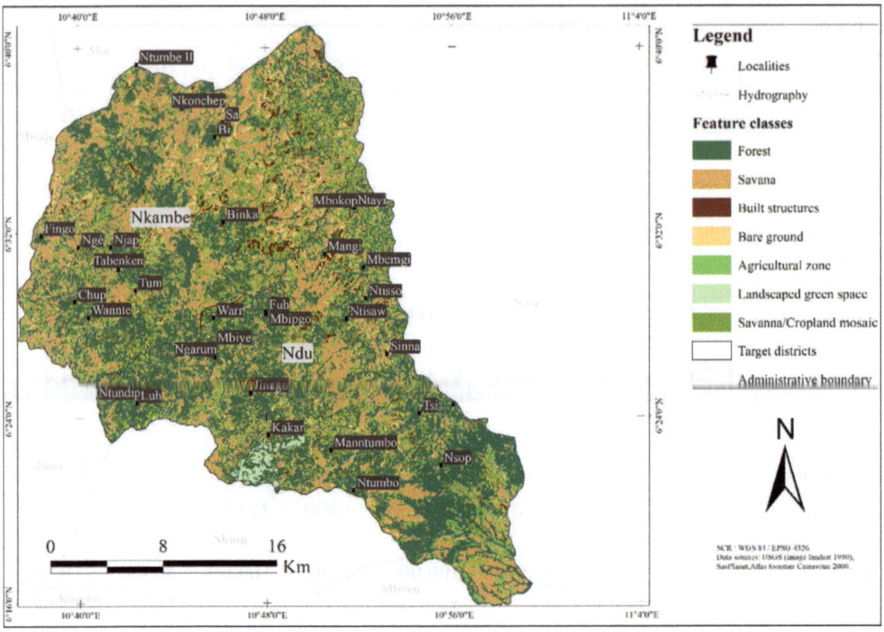

Fig. 6.9 Land cover on the Mbum Plateau in 1990

Agricultural land usage was relatively limited, confined to smaller, scattered parcels. Built-up areas and bare ground were also present, but occupied a small fraction of the landscape. The hydrology features, such as rivers, lakes, and streams, were delineated and unobstructed. Over the following decade, from 1990 to 2000, the Mbum Plateau began to undergo noticeable land cover changes. The forest cover started to fragment, with gradual encroachment and conversion of forested areas to other land uses (Fig. 6.10).

Agricultural expansion emerged as a key driver, with sizable increases in cultivated lands, particularly in the northern and central parts of the Mbum Plateau. Savanna vegetation also expanded, at the expense of some forest. While the built-up areas and bare ground increased compared to 1990, they still represented a relatively small portion of the landscape. The hydrology features remained largely intact, although some minor changes occurred. Accelerating into the 2000s and 2010s, the rate of land cover transformation intensified (Fig. 6.11).

By 2010, the forest cover had been reduced and fragmented, with large swaths converted to agricultural land, savanna, and built-up areas. The agricultural zone expanded considerably, occupying portions of the landscape, particularly in the central and northern regions of the Mbum Plateau. Savanna vegetation continued to spread, encroaching on the remaining forest patches. Built structures and bare ground also increased markedly, reflecting growing urbanization and infrastructure development. The hydrology network, while still present, showed signs of greater impact and disruption due to the extensive land use changes.

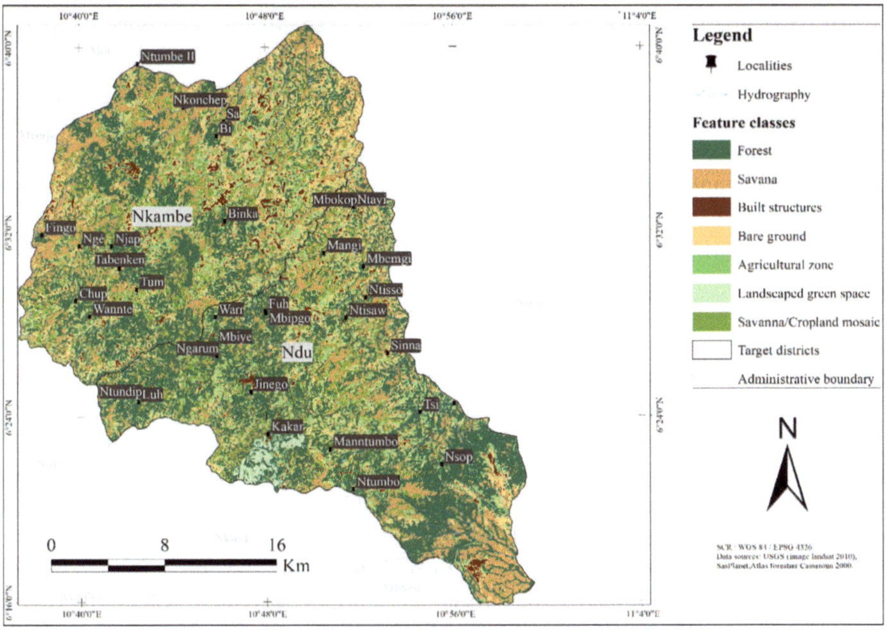

Fig. 6.10 Land cover on the Mbum Plateau in 2000

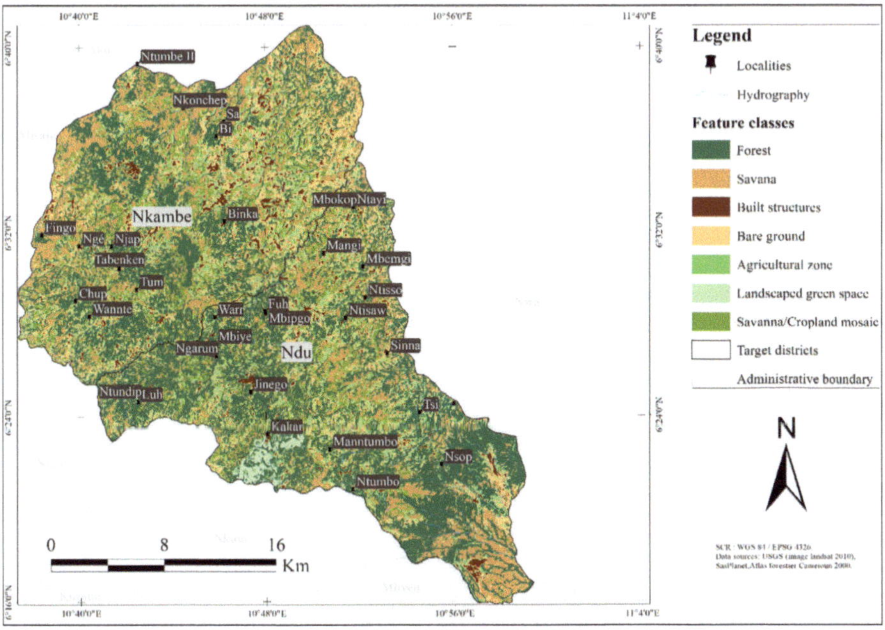

Fig. 6.11 Land cover on the Mbum Plateau in 2010

By 2024, the landscape had been transformed dramatically. The once-dominant forest cover had been heavily cleared and degraded, with only remnant patches remaining (Fig. 6.12). Agricultural land had become the predominant land use, occupying large expanses of the Mbum Plateau. Savanna vegetation had expanded significantly, replacing much of the former forest. Built-up areas and bare ground had also increased substantially, reflecting continued development and infrastructure growth. The hydrology features, while still visible, appeared to be under significant pressure from the intensified human activity and land use modifications across the Mbum Plateau.

Land cover changes demonstrate dramatic transformations, particularly in forest coverage. Nkambe experienced a severe forest decline from 46.73% in 1990 to 16.60% in 2010, before a partial recovery to 27.23% in 2024. This pattern of forest loss and partial recovery aligns with findings by Ndi and Amawa (2014) in their study of environmental degradation in Northwest Cameroon. Similarly, Ndu's forest cover fluctuated between 43.82 and 24.83%, showing more resilience but still indicating significant pressure on forest resources, comparable to patterns observed by Tassah et al. (2020) in their analysis of vulnerability hotspots in the region.

The expansion of agricultural zones in both locations (Nkambe: 8.49–17.02%; Ndu: 10.03–20.14%) indicates intensifying agricultural activities, potentially driven by population pressure and changing livelihood strategies. This trend corresponds with Ngwani et al. and's (2024a, 2024b) observations of shifting livelihood patterns among Mbororo communities in response to environmental changes. The increase in

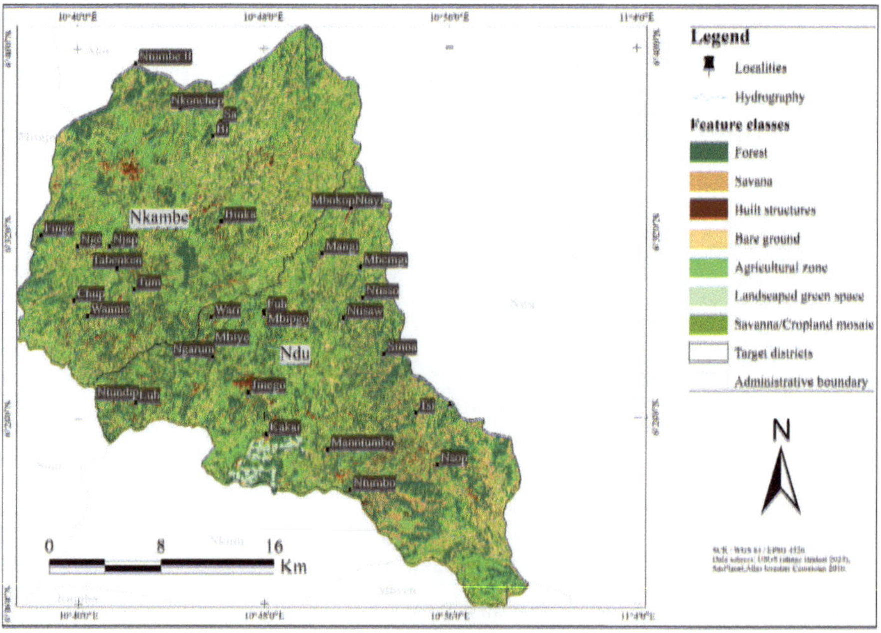

Fig. 6.12 Land cover on the Mbum Plateau in 2024

bare ground coverage, particularly in Nkambe (peaking at 18.29% in 2000), suggests accelerating land degradation, supporting Tume et al. (2024) on land climate and food security in the Kilum-Ijim Mountain Forest Communities.

The study's findings highlight the complex interplay between climate variability and land cover change, demonstrating how these factors collectively influence rural livelihood sustainability. The observed increases in temperature, coupled with unreliable rainfall patterns and significant land cover transformations, create challenging conditions for traditional agricultural practices. This supports the conclusions of Lahai et al. (2022), who revealed that rain-fed agriculture is highly vulnerable to climate change in Sub-Saharan Africa. The findings emphasize the need for adaptive strategies that can address both climatic uncertainties and land degradation while supporting sustainable rural livelihoods. In another dimension, Mbongsi et al. (2023) revealed that the remnants of montane forests on the Mbum Plateau of the Bamenda Highlands have the potential to offset and reverse the impacts of land degradation of Nature-based Solutions of indigenous conservation methods are reinforced.

6.6 Effects of Climate-Land Cover Nexus on Crop Production

The main crops that serve as subsistence and cash crops on the Mbum Plateau and elsewhere in the Bamenda Highlands of Cameroon are maize and beans. It has been established that rainfall is oscillating with a decreasing trend, while temperatures are rising. Agricultural land has also increased in the Mbum Plateau and has become the predominant land use. Despite this, maize and beans production have witnessed declines over time (Figs. 6.13 and 6.14).

These declines are attributed to climate change, land cover change, declining soil fertility, land degradation, and well as other factors of production. These have negative effects on livelihoods, thereby increasing the vulnerability of the rural masses to environmental change. These trends corroborate the findings of Mbongsi et al., (2024), who revealed that climatic shocks, land degradation, and other stresses on

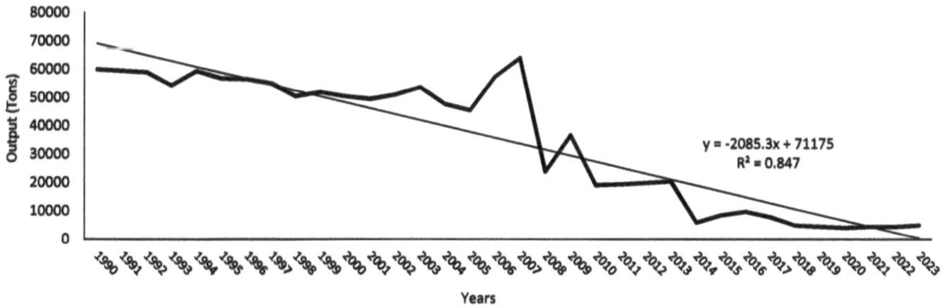

Fig. 6.13 Maize production trend (1990–2023)

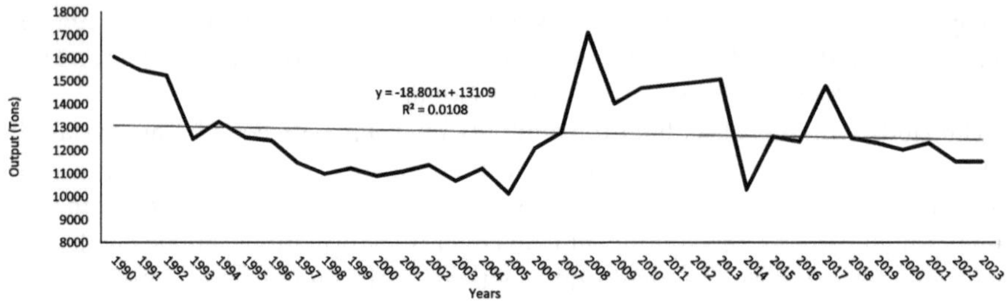

Fig. 6.14 Beans production trend (1990–2023)

the Mbum Plateau of the Bamenda Highlands of Cameroon have compelled rural women farmers, who are the most vulnerable, to embark on climate-smart agricultural initiatives to increase their resilience and improve livelihoods.

6.7 Livelihood Implications

The sedentary lifestyle of the inhabitants of the Mbum Plateau is distinctive, as their daily activities are heavily reliant on natural resources. These communities have successfully integrated both crop and animal production to sustain their livelihoods, melding with the traditional practices of Fulani pastoralists. Their activities are significantly influenced by seasonal variations in rainfall and the ecosystem services provided by rangelands, including pasture and water. Generally, the community's needs are met in adequate quantities and diverse locations, ensuring the survival of both their herds and the households that depend on them. However, with changing environmental conditions, sedentary farming practices have intensified. Coupled with climate fluctuations, the scarcity of drinking water in areas such as Wat, Binshu, Binka, Nkambe, Mbawngong, Ntisaw, Nkwintang, Mangu, and Ntubaw has diminished the quality and availability of pasture, placing additional pressure on these communities. This has resulted in alterations to the transhumance calendar, creating tensions between the Wimbum and Fulani communities. Consequently, cattle often trample farms, causing damage to crops and igniting conflicts between farmers and herders. The environmental factors contributing to these farmer-grazier conflicts include rugged terrain (20%), soil depletion, and the absence of clear boundaries between grazing and farming lands (40%), leasing of pastoral land (10%), and settlement on rangeland (10%). Limited land availability for both graziers and crop farmers inevitably leads to confrontations, with each party accusing the other of encroaching on their respective areas. The introduction of unclear boundaries has further intensified these conflicts, particularly in Dumbo, Misaje, and the Mbaw plain, especially during the dry season when competition for resources intensifies. While tensions between farmers and grazers have existed historically, they have become markedly more pronounced in recent years.

The emergence of mixed land use patterns, particularly the expansion of savanna/cropland mosaics (increasing to 21.12% in Nkambe and 15.61% in Ndu by 2024), suggests adaptive responses by local communities to changing environmental conditions. This transformation of traditional landscapes echoes Baffour-Ata et al. (2021) observations of community resilience and adaptation strategies in response to environmental change in West Africa. The findings underscore the importance of integrated approaches to environmental management that consider both climatic and land use dynamics in supporting sustainable rural development.

6.8 Conclusion

The study demonstrates the intricate relationship between climate variability, land cover changes, and rural livelihood sustainability on the Mbum Plateau of Northwest Cameroon. The findings reveal transformations in both climatic patterns and landscape composition, characterized by warming temperatures (1.5 °C increase over the study period), erratic rainfall patterns, substantial forest cover decline, and expanding agricultural zones (doubling in both localities). These changes have profound implications for rural livelihoods, as evidenced by the shifting land use patterns and adaptation strategies observed among local communities. The study's comprehensive analysis, supported by both quantitative data and household surveys, shows the urgent need for integrated environmental management approaches that address both climatic resilience and sustainable land use practices. Particularly concerning is the trend of forest degradation coupled with increasing bare ground coverage, suggesting accelerating environmental stress that could further compromise rural livelihood sustainability. These findings contribute significantly to the broader understanding of environmental change in Sub-Saharan Africa and emphasize the critical importance of developing climate-adaptive strategies and implementing Nature-based Solutions to ensure the long-term sustainability of rural livelihoods.

References

Baffour-Ata, F., Antwi-Agyei, P., & Nkiaka, E. (2021). Climate variability, land cover changes and livelihoods of communities on the fringes of Bobiri Forest Reserve, Ghana. *Forests*. https://doi.org/10.3390/F12030278

Lahai, M. K., Kabba, V. T., & Mansaray, L. R. (2022). Impacts of land-use and land-cover change on rural livelihoods: Evidence from eastern Sierra Leone. *Applied Geography*. https://doi.org/10.1016/j.apgeog.2022.102784

Mbongsi, N. D., Fogwe, Z. N., & Tume, S. J. P. (2023). Forest potentials and threat nexus on the Nkambe Plateau watershed. In Z. N. Fogwe, J. N. Kimengsi, T. E. Sunjo, S. S. Kometa, & C. A. Nkwemoh (Eds.), *Celebrating ten years of restless rest in landmark research and human resource development: Festschrift for professor Emeritus Cornelius Mbifung Lambi: Development and environmental dynamics* (pp. 364–377). The University of Bamenda Printing Press.

Mbongsi, N. D., Tume, S. J. P., & Moupou, M. (2024). Rural women in climate-smart agriculture for environmental resilience and livelihoods sustainability on the Mbum Plateau. In M. M. Nsangou, M. L. D. Kamgain, N. D. Mbongsi, & E. M. Fube (Eds.), *Femmes et Devéloppement Durable en Milieu Rural au Cameroun (Women and sustainable development in Cameroon)* (pp. 159–218). Editions de Midi.

McKee, T. B., Doesken, N. J., & Kleist, J. (1993). The relationship of drought frequency and duration to time scales. In *Proceedings of the Eighth Conference on Applied Climatology* (pp. 179–184). American Meteorological Society.

McKee, T. B., Doesken, N. J. & Kleist, J. (1995). Drought monitoring with multiple time scales. In *Proceedings of the Ninth Conference on Applied Climatology*. American Meteorological Society, Boston, 233–236.

Msofe, N. K., Lyimo, J. G., & Josephat, S. (2014). Anthropogenic drivers on land use/cover change and their implications to rural livelihood in Kilombero Wetlands, Tanzania. *Huria: Journal of the Open University of Tanzania, 16*, 61–74. https://doi.org/10.4314/huria.v16i0

Ndi, H. N., & Amawa, S. G. (2014). Environmental degradation and the emergence of agricultural frontiers in the North West of Cameroon. *Journal of Sustainable Development, 7*, 111. https://doi.org/10.5539/JSD.V7N5P111

Ngwani, A., Ngwa, F. G., Tume, S. J. P., Nkuh, Y. R., & Nformi, R. G. (2024a). Climatic variations and challenges to socio-economic development amongst mbororo communities in Mbum Plateau, North West Region of Cameroon. *Asian Journal of Geographical Research, 7*(2), 53–73. https://doi.org/10.9734/ajgr/2024/v7i2231

Ngwani, A., Tume, S. J. P., Ngwa, F. G., & Nkuh, Y. R. (2024b). The climate challenge to indigenous mbororo communities in the Mbum Plateau, North West Region, Cameroon. *JSM Environmental Science and Ecology, 12*(1), 1092(1–14). https://doi.org/10.47739/2333-7141/1092

Tassah, T. I., Becha, N. F., & Bomono, H. Y. (2020). Analysis of vulnerability hotspots in landcover for Momo Division, North West Cameroon.

Tume, S. J. P., Awazi, N. P., Tantoh, H. B., Kunsom, T. K., Kongso, M. E., Fokeng, R. M., Bailack, K. M., & Mairomi, W. H. (2024). Climate variability in the Kilum-Ijim mountain range of Cameroon: Implications for rural livelihoods and food security. In S. Tripathi, R. Bhadouria, & S. C. Garkoti (Eds.), *Warming mountains: Implications for livelihood and sustainability* (pp. 105–129). Springer. https://doi.org/10.1007/9783-031-62197-0_6

Tume, S. J. P., Zetem, C. C., Nulah, S. M., Ateh, E. N., Mbuh, B. K., Nyuyfoni, S. R., Ahfembombi, L. L., & Kwei, J. (2020). Climate change and food security in the Bamenda Highlands of Cameroon. In V. R. Squires, & M. K. Gaur, (Eds.), *Food security and land use change under conditions of climate variability: A multidimensional perspective* (pp. 107–124). Springer. https://doi.org/10.1007/978-3-030-36762-6_6

World Meteorological Organization. (2012). *Standardized precipitation index user guide*. WMO (1090), Geneva, 50.

Chapter 7
Linking Climate Stress with Vegetation and Wetland Degradation: A Geospatial Assessment of Eco-Resilience in the Barind Tract, Bangladesh

Md. Khayrul Islam Tuhin, Sanjida Rahman Mishu, Israt Jahan Sazuty, and Risat Ara Meem

Abstract The Barind Tract, a semi-arid landscape in northwestern Bangladesh, is undergoing a significant ecological transformation due to the increasing impacts of climate change. Wetlands and vegetation cover, crucial components to regional biodiversity and agricultural sustainability, are particularly vulnerable to transformative climatic stressors. With climate change becoming a global concern, understanding its impacts on vegetation and wetland dynamics in the Barind Tract, a droughtprone area in Bangladesh, is critical. While it applies an increase in climate stress with a strong correlation under stable stress (0.85) and moderate correlation under increasing stress (0.69) between climate stress and NDVI, it uses multi-temporal geospatial analysis and correlation modeling to identify significant links between climate stress and the loss of vegetation. Between 2020 and 2024, a significant increasing trend of recovery was observed, with areas of decreasing stress quadrupling, signifying potential for ecological recovery. The study also states that chronic environmental stress on the climate leads to land deterioration, putting regional diversity and sustainability at risk. The study thus recommends targeted interventions to enhance eco-resilience, including the conservation of wetlands, afforestation, and adaptive land-use planning. This research, by establishing the connection between climate stress and degradation patterns, will form the basis for sustainable strategies and policy responses that will support health and resilience over the long term in climate-vulnerable landscapes.

Md. K. I. Tuhin (✉)
Department of Geography and Environment, Dhaka College, Dhaka, Bangladesh
e-mail: khayrulgis@gmail.com

S. R. Mishu · I. J. Sazuty
Department of Geography and Environment, Eden Mohila College, Dhaka, Bangladesh

R. A. Meem
Department of Environmental Sciences, Jahangirnagar University, Dhaka, Bangladesh

U. Chatterjee et al. (eds.), *Eco-Resilience: Climate Change, Land Degradation and Sustainable Solutions*, Sustainability Solutions,
https://doi.org/10.1007/978-3-032-00708-7_7

Keywords Eco-resilience · Climate-vegetation interactions · Climate stress mapping · Barind tract · Geospatial modeling

7.1 Introduction

7.1.1 Background

Deforestation is a major global environmental issue with profound implications for climate change, biodiversity loss, and ecosystem disruptions. The clearance of forests hinders carbon sequestration, intensifying global warming and altering climate patterns. In Bangladesh, deforestation has been accelerating at an alarming rate due to a confluence of factors, such as rapid urbanization, agricultural expansion, and illegal logging. These drivers, compounded by the adverse effects of climate change, have led to a widespread degradation of forest and water ecosystems, posing considerable challenges to both environmental sustainability and socio-economic development. In general, deforestation refers to the decline of forest areas caused by human activities, resulting in environmental and social impacts that affect communities (Dewi et al., 2023). By releasing carbon dioxide, a key greenhouse gas, deforestation contributes to an intensive increase in global temperatures (Gu, 2023). As a result, ecological balance is disrupted as global warming indices increase by 15%, forest cover declines by 12%, and 15% of habitats are damaged (Sing et al., 2024). On the other hand, climate change refers to a shift in the climate that lasts for a long time, usually decades or more. It is distinguishable by variations in the mean and/or variability of its features. It describes how the climate changes over time, whether due to human action or natural fluctuation (IPCC, 2007). Deforestation both exacerbates and is exacerbated by climate change, demonstrating the close connection between these two processes.

Globally, deforestation is still a major cause of climate change. FAO Global Forest Resources Assessment (2020) estimates that between 1990 and 2020, 420 million hectares of forest were deforested (converted to other land uses); while the rate decreased over time, deforestation was still estimated to be 10 million hectares annually in 2015–2020 (approximately 0.25% annually). Furthermore, global forest coverage is expected to decline by 10.5% under "business-as-usual" conditions and 15.3% under "accelerated deforestation" by 2100 (Addae & Dragićević, 2023). Numerous factors contribute to this deforestation, including logging, growing populations, urbanization, grazing, acid rain, dam and reservoir construction, habitat fragmentation, slash-and-burn farming, wildfire, global warming, hydroelectric projects, quarrying, and mining (Maria et al., 2015). Herton, for instance, reported in 2020 that deforestation in the Brazilian Amazon had risen by at least 28% from the previous year. Climate change has an adverse effect on deforestation due to changes in the hydrological cycle and rising temperatures, which are prone to forest degradation.

This degradation has the potential to change carbon sources, increasing carbon emissions linked to deforestation and making ecological stability more difficult (Artaxo et al., 2021). Additionally, deforestation plays a crucial role in driving climate change by emitting greenhouse gases, mostly carbon dioxide (CO_2), which are responsible for around 17% human-induced annual emission (Gorte, 2010).

While deforestation is a worldwide issue, its effects differ from location to location. According to the Global Forest Resources Assessment report of FAO, 2020, South America witnessed the highest deforestation of 68 million hectares from 2000 to 2018, followed by Africa. The prime driver of this transition is the land-use conversion, losing 30.3% of moist forests and 25.2% of rainforests. This continuous deforestation results into rising surface temperature, while the Amazon going through a rise of approximately 0.4 °C, Southeast Asia 0.18 °C, and Africa 0.1 °C of temperature (Smith et al., 2020).

On the other hand, in the Northern Hemisphere, the disappearance of temperate and boreal forests causes the surface temperatures to significantly drop, especially in the spring and winter. In addition, it delays the retreat of snow cover in mid- and high-latitude regions, while short grass, with its greater albedo, reflects more sunlight and aids in cooling (Douville & Royer, 1996).

South Asia, with its dense population and rapidly growing economies, confronts particular problems with climate change and deforestation. Deforestation in South Asia is mostly caused by the conversion to commodity agriculture, specifically the dilation of the oil palm, rubber plantations, lumber, and pulp industries (Onder et al., 2021). India, for instance, has seen severe deforestation over the years; estimates place the yearly loss of forest at about 1.5 million hectares as a result of agricultural expansion, urbanization, illegal logging, and timber extraction (Haughan et al., 2022) while in China, urbanization consumed approximately 20.6% of natural water bodies over the 28-year period of 1990–2018 (Wu et al., 2024). Moreover, in Pakistan, deforestation is a pressing environmental challenge, specifically in the northwestern mountainous area where approximately 11 km^2 of forest cleared and transformed into barren land (Owais & Siddiqui, 2019). Therefore, these degradation primarily caused by governance failures such as weak community ownership and poor management, population growth, socio-economic pressure (fuelwood dependence), and low environmental awareness (Ullah et al., 2023).

As a tropical nation, Bangladesh has four main types of forests covering around 11% of its land with mixed-evergreen, deciduous, mangrove, and freshwater swamp forests (Reza & Hasan, 2019). Deforestation in Bangladesh, especially in the Chittagong Hill Tracts, is caused by illicit logging and neoliberal development projects, which seriously degrade the ecology (Nadiruzzaman et al., 2024). The Barind Tract, characterized by its ecological sensitivity and vulnerability to drought, is highly susceptible to land degradation due to climate change. The Sundarbans, the world's largest mangrove forest, has experienced severe deforestation and degradation due to a combination of natural and man-made factors, such as overexploitation of resources (Islam & Bhuiyan, 2018), expansion of shrimp farming (Khan et al., 2021), increasing urbanization (Saoum & Sarkar, 2024), exacerbating natural events (Islam et al., 2022), rising sea levels and salinity (Islam & Bhuiyan, 2018). Understanding the dynamics of vegetation and wetland degradation under climate-induced stress is essential for enhancing eco-resilience.

Extreme weather, habitat changes, and increased salt—all pose a threat to forest ecosystems, which lowers productivity and hampers biodiversity, especially in coastal regions that are already at risk (Chowdhury & Ndiaye et al., 2017). Local communities are adjusting planting schedules and increasing afforestation efforts, alongside other adaptation techniques, to deal with the effects of climate change (Chow et al., 2019; Miah et al., 2013).

Climate change has a variety of positive effects on deforestation in Bangladesh, many of which are entwined with larger environmental and socioeconomic dynamics. Even though deforestation is typically worsened by climate change, some adaptive techniques and changes in forest management have led to a better outcome. As an example, community-based strategies like mangrove afforestation and participatory forestry have successfully reduced deforestation while improving local lives (Rudra, 2016). According to Rudra (2016), agroforestry techniques that combine forestry and agriculture also support sustainability by lowering forest pressure and increasing land use efficiency. Therefore, the acknowledgment of ecosystem services has also improved forest management techniques, providing conservation and climate resistance priority. Furthermore, the government has been prompted by climate change to implement sustainable development programs, which have addressed deforestation and created economic possibilities (Kibria et al., 2011). But despite these adaptation measures, the overall issues caused by climate change still endanger forest ecosystems, therefore, Bangladesh must always work to maintain the balance between development and conservation. In Bangladesh, deforestation and land use issues are greatly worsened by climate change, especially in areas like the Barind Tract region, where human activity and climate-induced variables interact to worsen environmental degradation. Increasing rainfall, landslides, and temperature shifts are examples of irregular weather patterns that lead to soil erosion, forest degradation, and land conversion (Nadiruzzaman et al., 2024). Due to illegal logging, urbanization, and human pressures, Bangladesh lost over 39.1% of its forest cover between 1930 and 2014 (Islam & Sato, 2012; Reddy et al., 2016). Climate variability has increased the demand for agricultural land, leading to the conversion of forests to farmland and changes in the cultivation of salt and shrimp, which have exacerbated land degradation (Islam et al., 2021). Additionally, between 1980 to 2010, urban land increased by 201%, resulting in the loss of water bodies, wetlands, and ecosystem services, endangering biodiversity and environmental health (Al-Amin & Islam, 2023). The reduction in agricultural output poses a challenge to local livelihoods and food security (Islam et al., 2021; Sunderland et al., 2011). Bangladesh is still very concerned about the immediate effects of deforestation and changes in land usage, even if sustainable development programs and climate-resilient farming methods may lessen some adverse effects by encouraging reforestation and prudent land management.

The complicated issues raised by the interaction between deforestation, wetlands loss, and climate change demand immediate response. Even while local, regional, and global contexts show some adaptation techniques, the overall effects are still negative. The environment and the lives of millions of people in Bangladesh are at risk due to the combined effects of deforestation, waterbody shrinkage, and climate change. To overcome these obstacles, comprehensive policies that maintain a balance between economic growth and environmental preservation, as well as strong international collaboration and local governance, will be needed. Additionally, addressing anthropogenic environmental change through eco-resilience requires the generation and analysis of local climate stress impact data. Such data can be used for priority setting for sustainable land and water management and contribute to the development of a holistic approach.

7.1.2 Objectives

The primary objective of the study is to identify changes in vegetation and wetland from 1990 to 2024 in the Barind Tract to maintain its ecological sustainability. The specific objectives are:

- To assess the spatiotemporal patterns of climate stress across the Barind Tract region using multi-temporal geospatial data.
- To analyze the correlation between climate stress variations and vegetation degradation using NDVI and climatic variables.
- To quantify the extent of wetland loss in relation to increasing and stable climate stress zones over the past three decades.

7.2 Material and Methods

7.2.1 Study Area

The Barind Tract, locally known as Barind Bhumi, is the largest Pleistocene physiographic unit in the Bengal Basin, which covers 34,605.47 km^2 of northwest Bangladesh (Fig. 7.1). It is characterized by an elevated, undulating plateau that is distinctly differentiated from the adjacent floodplains due to its lateritic soil composition, limited groundwater availability, and semi-arid climatic conditions. The region spans around areas within Rajshahi and Rangpur divisions, encompassing multiple districts such as Rajshahi, Naogaon, Chapai Nawabganj, Joypurhat, Dinajpur, Thakurgaon, Panchagarh, Bogura, Natore, Pabna, Sirajganj, Gaibandha, Rangpur, Nilphamari, Kurigram, and Lalmonirhat. The region, with an estimated total population of approximately 37.96 million, has around 85% of the inhabitants residing in rural areas, and heavily dependent on agriculture for their livelihood. Geographically, it is situated between 23°48′–26°37′ N latitude and 88°01′–89°53′ E longitude, and is delineated by significant hydrological boundaries, including the Karatoya, Mahananda, Jamuna, and Padma rivers.

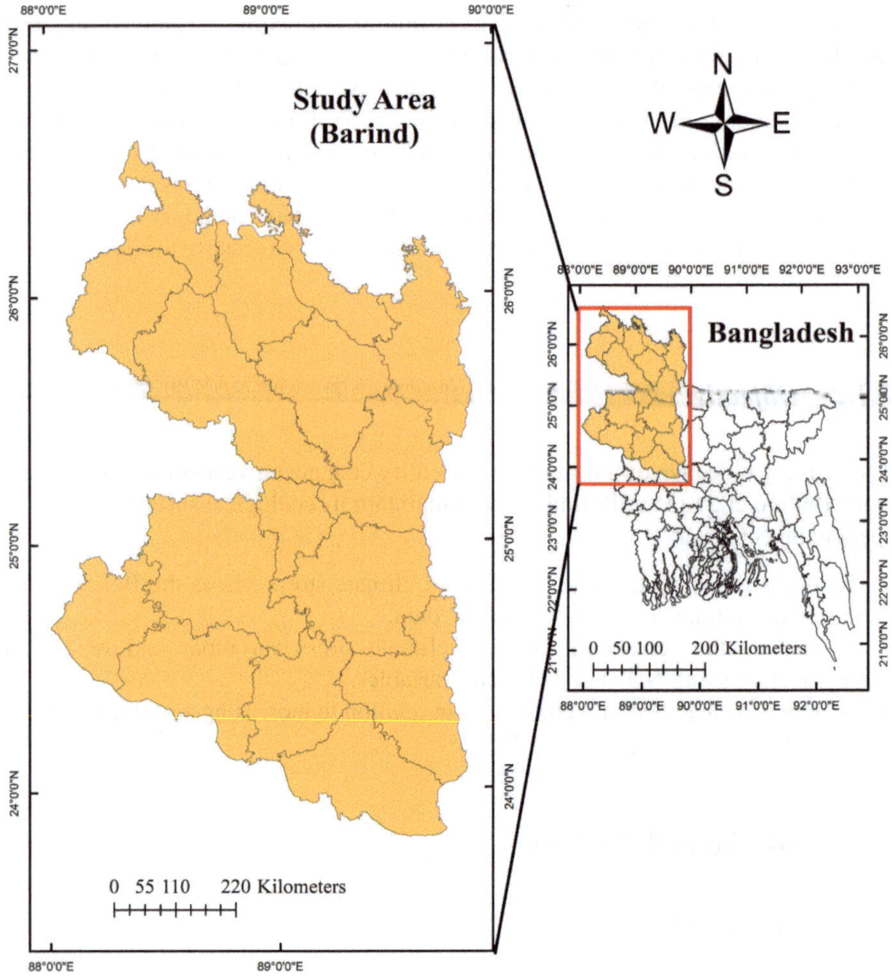

Fig. 7.1 Barind tract region: the study area

The Barind Tract experiences a warm and humid monsoon climate with four distinct seasons. Temperature fluctuations are considerable, ranging from 8 to 44 °C, while annual precipitation varies from 1200 to 1600 mm, contributing to desertification functions. Furthermore, the predominant soil type is reddish clay with low moisture retention capacity and supports crops like rice, wheat, maize, and mangoes.

The Barind Tract confronts severe climatic adversities, including deforestation, land degradation, and the depletion of wetland, primarily driven by rapid population growth and the expansion of agricultural activities. The overexploitation of groundwater resources and erratic monsoon patterns have exacerbated drought conditions, leading to contraction of the wetlands and a decline in biodiversity. Moreover, rising temperatures, prolonged dry spells, and shifting rainfall patterns are accelerating the onset of desertification-like conditions, making the region highly vulnerable to the adverse effects of climate change.

7.2.2 Methodology

This section outlines the methodological framework employed to identify the impact of climate change on vegetation and wetland dynamics in the Barind Tract of Bangladesh. The whole process is structured around three interconnected components, and those are: climate change assessment, vegetation change detection, and wetland dynamics analysis. An integrating approach combining Remote Sensing and Geographic Information System (GIS) technique has been utilized to analyze the spatial–temporal variations in climate stress, and its cascading ecological impact, particularly on vegetation cover and wetland areas (Fig. 7.2).

7.2.2.1 Data Collection and Sources

The study prioritized synergistic dataset combining climate variables and satellite images sourced from, credible platforms, Google Earth Engine, and TerraClimate to conduct the analysis thoroughly. The selected datasets are meticulously analyzed to ensure both temporal consistency, spatial accuracy, and reliability, ensuring the accuracy of the results across multiple time frames (1990, 1999, 2010, 2020, and 2024).

Climate Data

Climate data employed in the study are collected from the TerraClimate Monthly Data accessed through Google Earth Engine (GEE). The study primarily focuses on four key climatic parameters, which are maximum temperature, minimum temperature, precipitation, and soil moisture. These variables are significant for detecting and evaluating climate stress that affects the viability and functionality of vegetation and wetland. Additionally, the study covers five distinct time 1990, 1999, 2010, 2020, and 2024—allowing for a long-term, multi-decadal comparison.

Remote Sensing Data

Satellite images are obtained from Landsat missions (particularly Landsat 4–5 TM and Landsat 8–9 OLI/TIRS) to monitor and evaluate the vegetation and wetland changes across the selected timeline 1990–2024 (Table 7.1). The analysis used two key spectral indices:

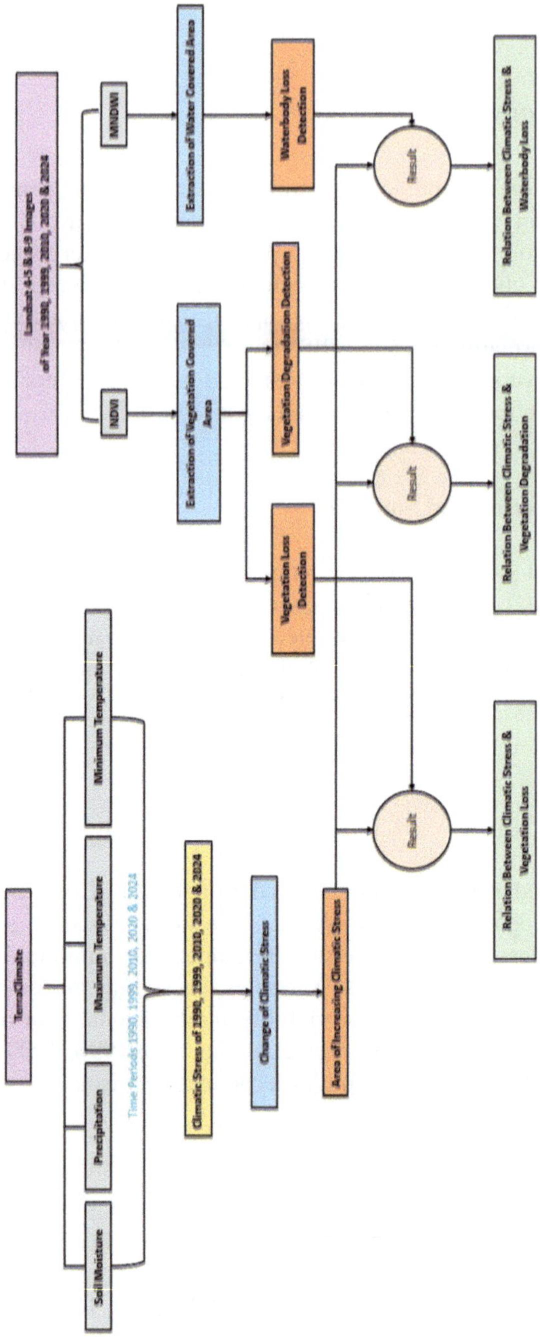

Fig. 7.2 Methodological flow chart

Table 7.1 Summary of data sources

Data type	Source	Parameters/Indices	Time periods
Climate data	Terra climate (GEE)	Temperature, precipitation, soil moisture	1990, 1999, 2010, 2020, 2024
Remote sensing dataset	Data landsat 4–5, 8–9	NDVI, MNDWI	1990, 1999, 2010, 2020, 2024

- **Normalized Difference Vegetation Index (NDVI)**: this index is used to analyze vegetation health and identify vegetation degradation and vegetation loss over the period (1990–2024).

$$\text{NDVI} = (\text{NIR} - \text{R})/(\text{NIR} + \text{R})$$

- **Modified Normalized Difference Water Index (MNDWI)**: this index is utilized to identify and quantify wetland and waterbody changes between the studied timeline.

$$\text{MNDWI} = (\text{Green} - \text{SWIR})/(\text{Green} + \text{SWIR})$$

The consistent use of these indices across five referred years (1990, 1999, 2010, 2020, and 2024) is to ensure that spatial and temporal comparisons are coherent and grounded in a scientifically established matrix.

7.2.3 Data Processing and Analysis

7.2.3.1 Climate Change Assessment

The assessment is executed with the collection and analysis of climate parameters such as temperature (maximum and minimum), precipitation, and soil moisture. These variables are anchored in Google Earth Engine (GEE), which enables efficient data handling and visualization. A weighted sum analysis is conducted to integrate the variables, emphasizing their cumulative impacts on environmental stress.

- **Climate Data Preprocessing**

The climate datasets are extracted from GEE for the years 1990, 1999, 2010, 2020, and 2024. To ensure the consistency and comparability over the study period, multi-year data compilation of maximum temperature, minimum temperature, precipitation, and soil moisture has been conducted. This process allows the development of a standardized dataset for climate stress analysis.

- **Weighted Sum Analysis**

The collected climate variables are analyzed to a normalized procedure to visualize the data in common scale and to quantify climate stress. The weighting system is applied to each variable based on its relative influence on climate stress, and the weighted sum approach is then utilized to generate climate stress maps, which highlight insights into the spatial distribution of climate-induced pressures over the period.

- **Climate Model**

The climate model (Fig. 7.3) operates through a systematic geospatial analysis framework that integrates four climate variables—soil moisture, precipitation, minimum temperature, and maximum temperature—for each selected year. Initially, each variable is reclassified into a uniform scale. These standardized layers are then used in a weighted overlay (Weighted Sum), reflecting their proportional impact on climate stress. The outputs represent climate stress maps for the years 1990, 1999, 2010, 2020, and 2024. Subsequently, these outputs undergo further reclassification to categorize the areas into distinct stress levels.

To analyze temporal trends, the reclassified climate stress maps from the studied years are compared using a combinational overlay tool (Combinatorial AND). This allows for the identification of changes in stress levels across four key time periods: 1990–1999, 1999–2010, 2010–2020, and 2020–2024.

- **Climate Stress Classification**

The climate stress maps are classified into five categories to identify the severe climatic stress-prone regions and relatively stable areas (Fig. 7.3). The classes are Very High Stress, High Stress, Moderate Stress, Low Stress, and Very Low Stress.

- **Change Analysis**

The shifts or conversions in the study area are detected by analyzing the degree of climate stress (whether it has increased, decreased or remained stable) in the study area using the combinational overlay tool across the reference time periods (1990–1999, 1999–2010, 2010–2020, and 2020–2024). It provides significant insights in long-term climate stress (Fig. 7.3).

7.2.3.2 Vegetation Change Detection

In this phase, with the assistant of Landsat satellite images (Landsat 4–5 and Landsat 8–9), the evolution of vegetation cover in response to changing climate conditions is detected. To ensure accuracy and flexibility, land use and land cover (LULC) classifications function. However, the prime purpose of this analysis is to identify areas of degradation and outright loss of vegetation across the decades (Fig. 7.4).

Fig. 7.3 Climate stress and change model

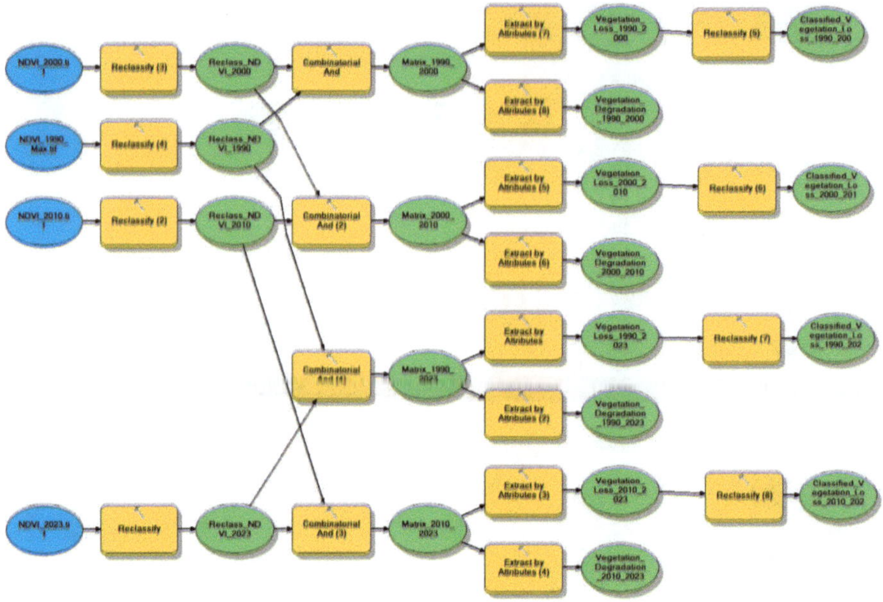

Fig. 7.4 Vegetation loss and degradation detection model

- **LULC Classification**

Supervised classification is primarily utilized to categorize the landscape into land cover types labeling as dense vegetation, moderate vegetation, sparse vegetation, water bodies, and built-up & bare soil areas. This classification results are validated using ground truth data and accuracy assessments while ensuring its reliability.

- **Vegetation Degradation Analysis**

This analysis involves tracking the transition of land cover from dense vegetation to moderate and sparse vegetation using NDVI data. Tools like Combinatorial AND and Con are applied to count and identify the degrading zones and quantify the severity of declining trend over the years (Fig. 7.4).

- **Vegetation Loss Analysis**

This phase focuses on the entirely vanished or disappeared vegetative areas, typically transitioning into water bodies and built-ups. The analysis is conducted for four-time intervals (i.e., 1990–2000, 2000–2010, 2010–2020, and 2020–2024), and to visualize the changing patterns within this time frame, a comparison calculation is done with the techniques of Combinatorial AND and Con techniques (Fig. 7.4). These changes are mapped to visually represent the pattern and extent of vegetation disappearance.

- **Statistical Analysis**

A zonal histogram analysis is conducted through ArcMap to correlate previously assessed climate stress levels and vegetation changes, which formed an interrelation between climate-induced stress variations and patterns of vegetation degradation and vegetation loss. As a result, it provides insights into the influencing patterns of climatic pressure on vegetation dynamics across various zones of Barind Tract.

7.2.3.3 Wetland Dynamics Assessment

This assessment is analyzed with the spectral index named MNDWI, which is specifically designed to detect water bodies. Its aim is to track the wetland changes and its associated climate stress variations.

- **Waterbody identification with MNDWI**

The Modified Normalized Difference Water Index (MNDWI) is calculated through the Landsat images (Landsat 4–5 and Landsat 8–9) for the years of 1990, 1999, 2010, 2020, and 2024. And a threshold technique is applied to accurately classify water features, generating maps for each intended years.

- **Wetland Loss Analysis**

To analyze the degree of wetland loss, MNDWI-derived data are processed, reclassified, and combined (with the tools named Combinatorial AND and Con) to compare among four periods (1990–1999, 1999–2010, 2010–2020, and 2020–2024). Areas where water bodies have alarmingly declined are identified and visualized into maps.

- **Correlation Analysis**

A zonal histogram analysis is conducted in ArcMap to determine the relationship between climate stress changes and wetland loss. It helps to detect the impacts of climate variations on wetland changes and provides significant insights of the vulnerable ecosystems of wetland.

This methodology assesses a systematic approach, while using remote sensing and GIS techniques to analyze the long-term impact of climate change on vegetation and wetland dynamics. The combination of climate data, vegetation indices, and wetland analysis provides a comprehensive understanding of environmental change in Barind Tract.

7.2.4 Accuracy Assessment

This study evaluates the accuracy of the classified outputs obtained from the remote sensing and geospatial methods, particularly focusing on vegetation and wetland

Table 7.2 Accuracy assessment table

Year	Overall accuracy (%)	Kappa coefficient (κ)
1990	85.7	0.81
1999	87.2	0.83
2010	89.5	0.86
2020	91.3	0.89
2024	92.1	0.9

Source Results from Landsat image analysis from https://earthexplorer.usgs.gov/

movement in Barind Tract. To validate the accuracy of the LULC classes, this assessment adopted the NDVI index to determine the land class, and MNDWI for water body detection across five (05) practiced time; 1990, 1999, 2010, 2020, and 2024.

Reference data are compiled from a variety of data sources, including historical topographic maps, high-resolution imagery from Google Earth, field observations (for recent years), and secondary literature. For earlier years (1990 and 1999), this validation is performed against archived satellite images and historical records. A stratified random sampling approach is employed to achieve adequate representation across each land cover type. For statistical robustness, at least 150–200 validation points are used per year to maintain statistical robustness.

A confusion matrix is produced to compare classified results with respect to the reference data. From this, the standard classification accuracy metrics are computed, including Overall Accuracy (OA), Producer's Accuracy (PA), User's Accuracy (UA), and Kappa Coefficient (κ). These metrics deliver information both about classification confidence and agreement above random chance. The findings indicated a trend towards increasing classification accuracy with time, attributed primarily to improvements in satellite image resolution, advancements in classification algorithms, and more accurate ground truth data (Table 7.2).

MNDWI drainage classifications for wetland mapping are assessed and validated using binary classification (presence/absence of water), with comparison to high-resolution satellite imagery and seasonal water maps. In most cases, overall accuracy was above 88%, suggesting high precision in water body identification.

To sum up, the accuracy assessment supports that the methods and datasets used in this study led to trustworthy outputs. Assuring confidence on subsequent evaluation of vegetation and wetland shifts attributed to climate and providing robustness for subsequent analyses on the spatial/temporal interpretation of environmental interactions surrounding vegetation dynamics in the Barind Tract.

7.3 Data and Discussion

Analyzing the transitional trend in vegetation and waterbody patterns over the selected time is critical for comprehending the environmental and anthropogenic pressures on ecosystem. In regions like Barind Tract, where rapid urbanization, agricultural expansion, rising temperature, and shifting rainfall patterns are the most prominent factors, acknowledging these drivers of deforestation and wetland loss is vital for implementing conservative measures. This research focuses on exploring the components that leverage the degree of forest and waterbody degradation in the Barind region over four time periods: 1990–1999, 1999–2010, 2010–2020, and 2020–2024. The study area is known for its semi-arid climate and monsoon-dependent agriculture where declining seasonal rainfall, groundwater depletion, and growing temperature rapidly alter the land use patterns. Therefore, the spatial patterns of climate stress, vegetation decline, and wetland degradation detected and identified in this study hightlight significant ecological vulnerabilities across the Barind Tract.

Through this analysis of vegetation loss, vegetation degradation, waterbody loss, and climate stress indicators, this study aims to identify the land use changing patterns and impact of climatic effects. Description of these analyses is illustrated below.

7.3.1 LULC

The analysis depicts a profound transformation among the land use categories, including waterbody, buildup & bare soil, sparse, moderate, and dense vegetation, primarily dominated by anthropogenic and ecological pressures (Fig. 7.5).

Buildup & bare soil areas have expanded approximately 101%, escalating from 7.66% (2,651.97 km^2) in 1990 to 15.4% (5,330.86 km^2) in 2024, highlighting rapid urbanization and land conversion (Table 7.3).

Furthermore, sparse vegetation demonstrated significant fluctuations over the study period, beginning at 22.89% (7,921.48 km^2) in 1990 and peaking at 54.15% (18,739.29 km^2) in 1999. Following this peak, the area experienced fluctuations, declining to 44.77% (15,492.51 km^2) in 2010, further dropping to 42.68% or 14,769.99 km^2 in 2020, and then slightly increasing to 45.3% (15,675.62 km^2) by 2024. In addition, moderate vegetation exhibited notable changes. It started at 47.41% (16,405.42 km^2) in 1990 and then experienced a sharp decline to 21.54% (7,452.61 km^2) by 1999, reflecting a notable plummet within a decade (Table 7.3).

However, moderate vegetation illustrated signs of growth in the following years, rising to 34.74% in 2010 and further to 36.12% (12,495.96 km^2) by 2020. In 2024, it stabilized at 32.33% (11,188.59 km^2), indicating partial restoration.

On the other hand, dense vegetation has declined dramatically over the decades, starting at 13.92% in 1990. Between 1990 and 1999, it has decreased to 8.25% (2,854.81 km^2) and continued to decline, reaching 3.62% (1,251.97 km^2) in 2010

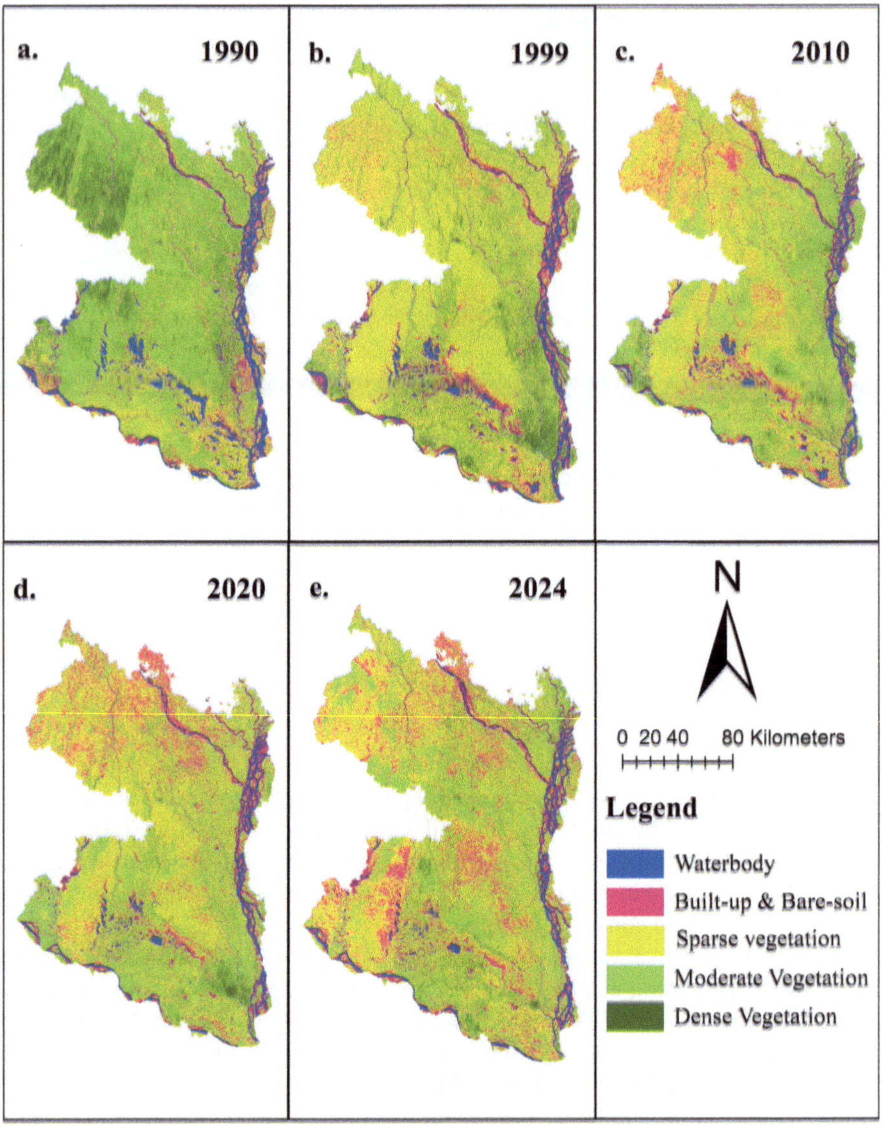

Fig. 7.5 Spatiotemporal changes in vegetation, built-up and waterbodies of barind tract region **a** 1990, **b** 1999, **c** 2010, **d** 2020, **e** 2024. *Source* Results from Landsat image analysis from https://earthexplorer.usgs.gov/

Table 7.3 Land use and land cover changes from 1990 to 2024

Land type	1990		1999		2010		2020		2024	
	Area km^2	Percent	Area km^2	Percent	Area km^2	Percent	Area km^2	Percent	Area km^2	Percent
Waterbody	2808.49	8.12	2034.81	5.88	1714.56	4.95	1553.79	4.49	1384.02	4
Buildup & bare soil	2651.96	7.66	3523.95	10.18	4125.57	11.92	5064.34	14.63	5330.86	15.4
Sparse vegetation	7921.47	22.89	18,739.3	54.15	15,492.5	44.77	14,770	42.68	15,675.6	45.3
Moderate vegetation	16,405.4	47.41	7452.6	21.54	12,020.8	34.74	12,496	36.12	11,188.6	32.33
Dense vegetation	4818.1	13.92	2854.8	8.25	12,51.97	3.62	721.38	2.08	1026.37	2.97

Source Results from Landsat image analysis from https://earthexplorer.usgs.gov/

and 2.08% (721.39 km^2) by 2020. However, between 2020 and 2024, dense vegetation showed a slight rise to 2.97% (Table 7.3). These observed correlations offer robust, evidence-based justification for prioritizing interventions aimed at enhancing ecological resilience.

7.3.2 Waterbody

The analysis of water bodies reveals a persistent and concerning pattern over the 34-year period. In 1990, water bodies approximately cover 2,808.49 km^2, accounting for 8.12% of the total land cover. By 1999, this trend dropped sharply to 2,034.82 km^2 (5.88%), followed by further diminutions to 1,714.56 km^2 or 4.95% in 2010 and 1,553.80 km^2 or 4.49 % in 2020. During 2020–2024, waterbody coverage dwindled to 1,384.02 km^2, representing just 4% of the total area—a net loss of 1,424.47 km^2 (50.7% of the 1990 baseline) (Table 7.3). This steady decline suggests long-term pressures such as unsustainable practices or anthropogenic modifications in the study area over the period.

7.3.3 Vegetation Loss in the Barind Tract

The trend of vegetation loss illustrates a continuous gradual ascent from 1990 to 2024. Compared to the total vegetation land in 1990, the vegetation loss amounts to 2,060.7291 km^2, representing a 7% decline. During this period, dense vegetation conversion is observed, with 23.7888 km^2 transitioning to water bodies and 183.4524 km^2 to built-up. Moderate vegetation contributes 112.7133 km^2 to water bodies and 626.1345 km^2 to built-up areas, while sparse vegetation accounts for 283.8204 km^2 and 830.8197 km^2, respectively. It rises to 8% (2336.952 km^2) during 1999–2010 where transitions from dense vegetation to water bodies and built-up areas are comparatively lower, with 32.9193 km^2 and 56.4282 km^2 respectively (Table 7.4).

Moderate vegetation transitions contribute 64.8657 km^2 to water bodies and 244.7838 km^2 to built-up areas, while sparse vegetation accounts for 169.6644 km^2 and 1,768.2903 km^2, respectively. However, the figure reached its peak in the next decade (1999–2010) hitting 3053.559 km^2 which is 10% of the total vegetative area (1990). Dense vegetation conversion to water bodies and built-up areas are recorded at 32.562 km^2 and 60.1758 km^2, respectively. Moderate vegetation transits to water bodies and built-up area are 119.4804 km^2 and 521.2503 km^2, while sparse vegetation accounts for 144.3375 km^2 and 2,175.7527 km^2 (Fig. 7.6). In addition, there is a slight decrease in the ongoing pattern for the last 4 years (2020–2024), and that is 2641.283 km^2 or 9%. Dense vegetation transition to water bodies and built-up areas dropped to 6.3873 km^2 and 39.7998 km^2. Moderate vegetation coverage contributes

Table 7.4 Vegetation loss in barind tract (1990–2024)

Timeline	Vegetation change	Area (km^2)	Total area (km^2)	(%)
1990–1999	Dense vegetation-water body	23.78	2060.72	7
	Dense vegetation-built up	183.45		
	Moderate vegetation-water body	112.71		
	Moderate vegetation-built up	626.13		
	Sparse vegetation-water body	283.82		
	Sparse vegetation-built up	830.81		
1999–2010	Dense vegetation-water body	32.91	2336.95	8
	Dense vegetation-built up	56.42		
	Moderate vegetation-water body	64.86		
	Moderate vegetation-built up	244.78		
	Sparse vegetation-water body	169.66		
	Sparse vegetation-built up	1768.29		
2010–2020	Dense vegetation-water body	32.56	3053.55	10
	Dense vegetation-built up	60.17		
	Moderate vegetation-water body	119.48		
	Moderate vegetation-built up	521.25		
	Sparse vegetation-water body	144.33		
	Sparse vegetation-built up	2175.75		
2020–2024	Dense vegetation-water body	6.38	2641.28	9
	Dense vegetation-built up	39.79		
	Moderate vegetation-water body	58.98		
	Moderate vegetation-built up	599.00		
	Sparse vegetation-water body	62.6		
	Sparse vegetation-built up	1874.43		
			10,092.52	

Source Results from Landsat image analysis from https://earthexplorer.usgs.gov/

58.9806 km^2 to water bodies and 599.0004 km^2 to built-up areas, while sparse vegetation accounts for 62.685 km^2 and 1,874.43 km^2, respectively (Table 7.4). These patterns suggest intensive anthropogenic activities and environmental pressure over the past three decades, with a marginal decline in recent years, possibly due to conservative measurements (Fig. 7.7).

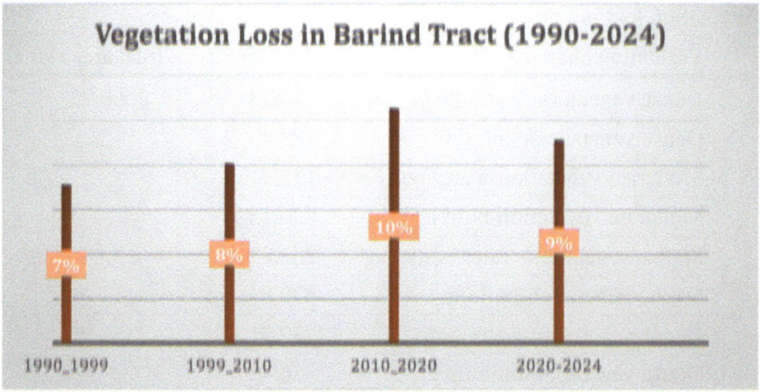

Fig. 7.6 Vegetation loss in barind tract (1990–2024)

7.3.4 Vegetation Degradation Patterns

Vegetation degradation has demonstrated significant fluctuations over the 34-year timeline (1990–2024). In terms of the total vegetation land area in 1990, during the first study period, from 1990 to 1999, vegetation degradation was comprehensive which covered a total of 14,362.4052 km^2. The major transition is notable in moderate to sparse vegetation (10,454.3361 km^2), dense to sparse vegetation (2,959.2378 km^2), and dense to moderate vegetation (948.8313 km^2). The decade is recorded as the highest degradation years, affecting 49 % of the total vegetative area. The figure witnesses a declined between 1999 and 2010 to 4,581.1512 km^2. Furthermore, the shifts of moderate to sparse vegetation contributed to 2,300.4639 km^2, followed by dense to moderate vegetation (1,953.9522 km^2) and dense to sparse vegetation (326.7351 km^2) (Table 7.5).

Degradation during this period accounts for 16% of the area. In the following decade, the figure rises slightly with 14,362.4052 km^2 or 17 % degradation, primarily due to the transitions from moderate to sparse vegetation (10,454.3361 km^2), dense to sparse vegetation (2,959.2378 km^2), and dense to moderate vegetation (948.8313 km^2). The most recent years (2020–2024) continue the declining trend accounting to 4,676.4297 km^2, with transitions dominated by moderate to sparse vegetation (4,188.9033 km^2), dense to moderate vegetation (310.3569 km^2), and dense to sparse vegetation (177.1695 km^2) (Table 7.5). Degradation during this period accounts for 16% (Fig. 7.8), highlighting continuous sustainable practices. To mitigate ongoing degradation and foster long-term resilience, targeted adaptation strategies are crucial. It includes agroecological practices (e.g., drought-resilient cropping systems and organic soil conservation techniques), restoration of degraded wetlands as natural buffers against climate variability, and nature-based solutions such as afforestation, watershed management, and climate-smart land-use planning (Fig. 7.9).

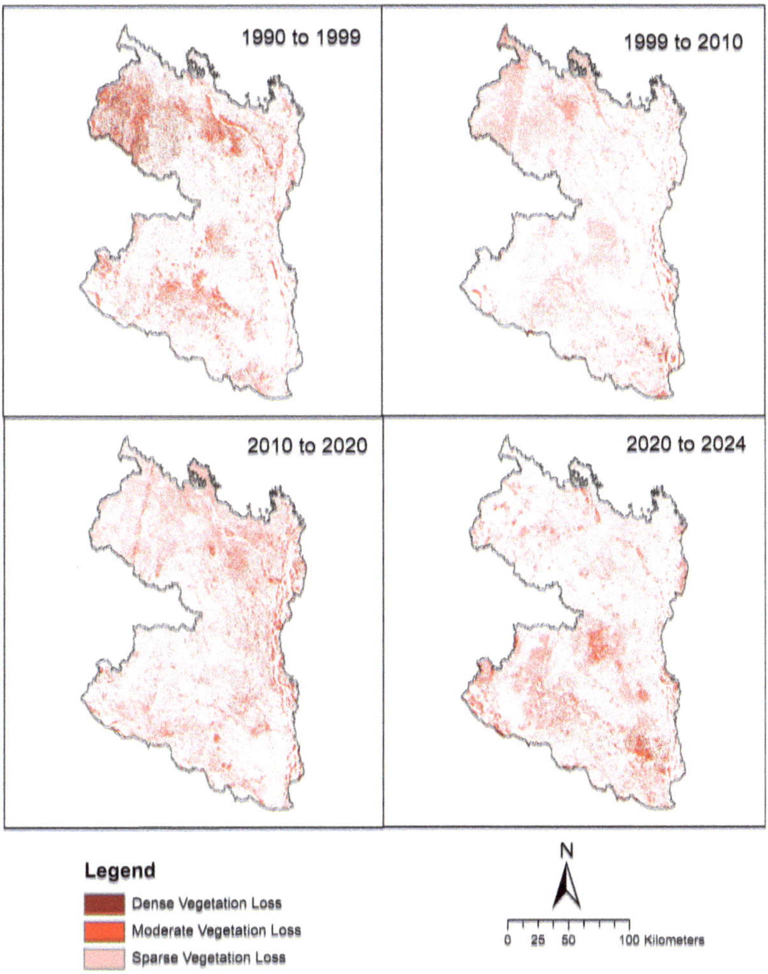

Fig. 7.7 Geospatial assessment of vegetation loss in the barind tract region (1990–2024). *Source* Results from Landsat image analysis from https://earthexplorer.usgs.gov/

7.3.5 Wetland Loss and Dynamics

In 1990, the total area covered by water bodies was recorded as 2808.4941 km^2 (Table 7.6). In comparison to this, the loss accounts for 65 % during 1990–1999, which covers 1837.7172 km^2 of the area and marking it as the greatest decline over the decades (Fig. 7.10).

This table dropped in the following decades, decreasing to 48% (1356.2064 km^2) for 1999–2010 and further to 42% (1172.2491 km^2) in 2010–2020 (Table 7.6). However, in recent timeline (2020–2024), waterbody loss has experienced a modest rebound, rising to 51 % or 1425.1581 km^2 (Table 7.6). This shows that despite some

Table 7.5 Vegetation degradation in barind tract (1990–2024)

Timeline	Vegetation change	Area (km^2)	Total area (km^2)	(%)
1990–1999	Dense vegetation-moderate vegetation	948.83	14,362.40	49
	Dense vegetation-sparse vegetation	2959.23		
	Moderate vegetation-sparse vegetation	10,454.33		
1999–2010	Dense vegetation-moderate vegetation	1953.95	4581.15	16
	Dense vegetation-sparse vegetation	326.73		
	Moderate vegetation-sparse vegetation	2300.46		
2010–2020	Dense vegetation-moderate vegetation	872.23	4934.96	17
	Dense vegetation-sparse vegetation	138.81		
	Moderate vegetation-sparse vegetation	3923.90		
2020–2024	Dense vegetation-moderate vegetation	310.35	4676.42	16
	Dense vegetation-sparse vegetation	177.16		
	Moderate vegetation-sparse vegetation	4188.90		
			28,554.95	

Source Results from Landsat image analysis from https://earthexplorer.usgs.gov/

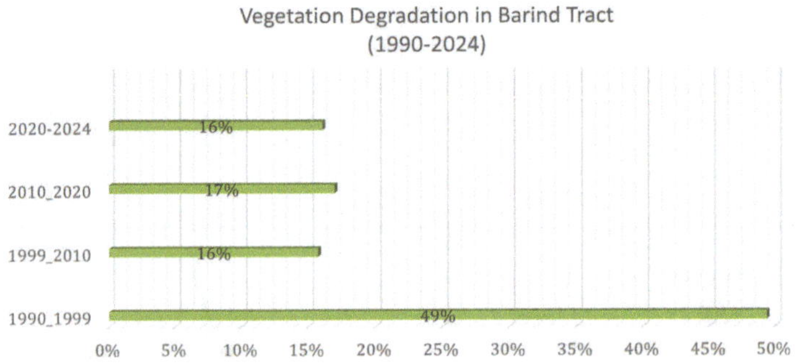

Fig. 7.8 Vegetation degradation in barind tract (1990–2024)

recovery in the early twenty-first century, waterbody loss remains an alarming issue (Fig. 7.11). It also suggests that the ecological recovery is possible under the right circumstances and management (according to the decreasing-stress zones we identified between 2020 and 2024, presented in Fig. 7.18). These recovery signals highlight notable opportunities for policymakers and land managers to replicate and scale-up effective resilience-building initiatives.

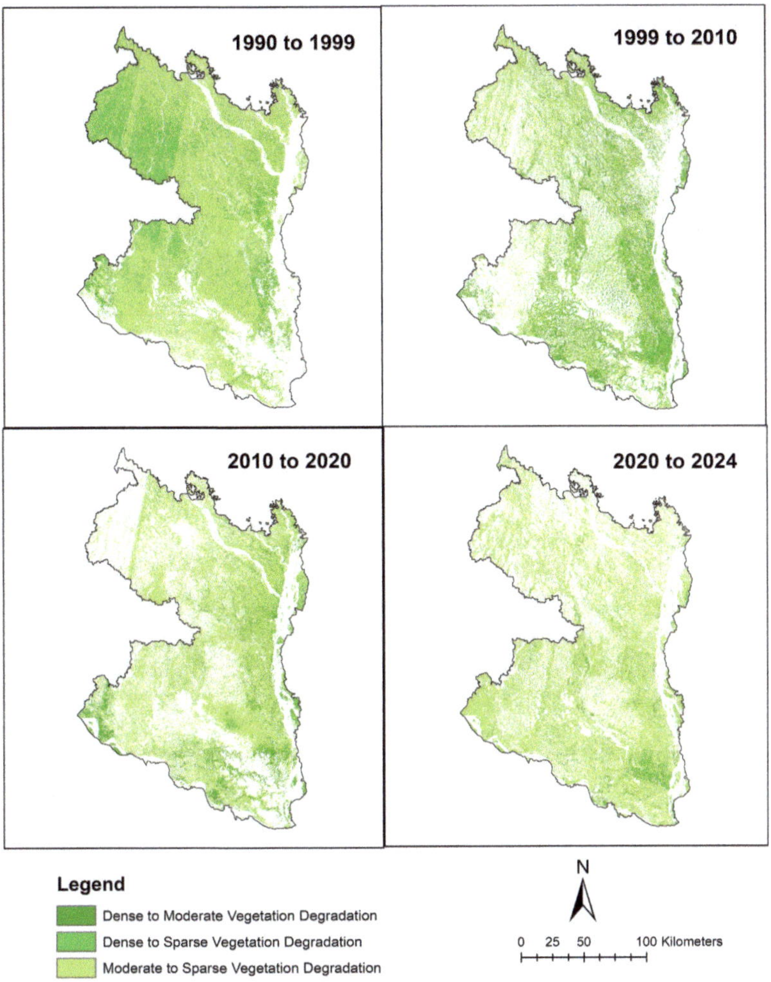

Fig. 7.9 Vegetation degradation patterns in the barind tract region (1990–2024)

Table 7.6 Waterbody loss from total water body

Timeline	Waterbody loss from total water body	
	km^2	%
1990–1999	1837.7172	65
1999–2010	1356.2064	48
2010–2020	1172.2491	42
2020–2024	1425.1581	51

Source Results from Landsat image analysis from https://earthexplorer.usgs.gov/

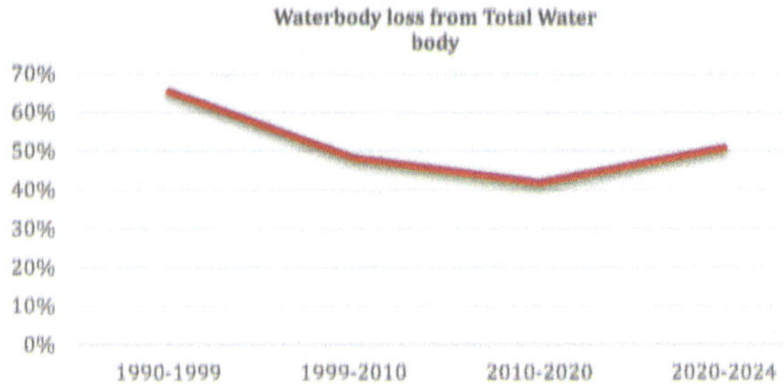

Fig. 7.10 Waterbody loss from total water body

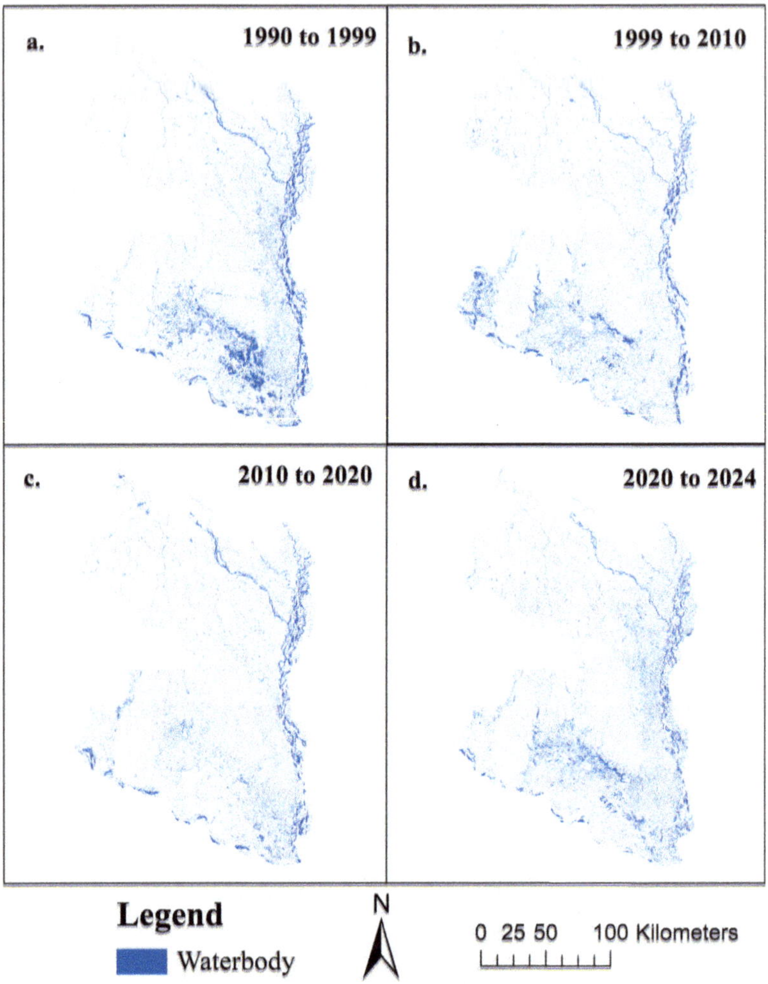

Fig. 7.11 Spatial distribution of wetland loss and dynamics in the barind tract region (1990–2024). *Source* Results from Landsat image analysis from https://earthexplorer.usgs.gov/

Table 7.7 Soil moisture

Soil moisture (in mm)		
Year	Low	High
1990	40.65	171.7
1999	45.85	175.55
2010	34	154.5
2020	106.125	174.558
2024	24.85	107.2

7.3.6 Climate Stress Indicators

7.3.6.1 Soil Moisture Dynamics and Drought Risks

The trends in soil moisture demonstrate a critical pattern of surging variability and intensifying drought risks. From 1990 to 1999, soil moisture levels reveal a slight surge, with low values increasing from 40.65 mm to 45.85 mm and high values rising marginally from 171.7 mm to 175.5 mm. Between 1999 and 2010, this pattern became more evident to drying conditions, with its values significantly dropping from 45.85 mm to 34 mm (low moisture value) and 175.55 mm to 154.5 mm (high value), respectively (Table 7.7).

However, in 2020, an unusual anomaly has been observed, with both low (106.125 mm) and high (174.558 mm) moisture values accelerating dramatically. This may indicate extreme weather events that disrupted the ongoing drying trend over the decades. Therefore, by 2020–2024, soil moisture level have declined to alarming lows, with low values recorded at 24.85 mm and high values at 107.2 mm (Table 7.7). This rate points to severe drying conditions, which suggests the possibility of desertification, accompanied by climatic effects (Fig. 7.12).

7.3.6.2 Precipitation Variability and Water Stress

Precipitation data reveals a similar concerning shift in climate conditions. From 1990 to 1999, high precipitation levels rose from 34.3 to 46.05 mm, indicating a favorable hydrologic period. Low values also experience a moderate rise, ranging between 8.7 and 9.6 mm (Table 7.8).

However, the following period (1999–2010) underwent an inversion of the ongoing pattern. High precipitation dropped to 36.7, while low level declined to 6.8. This pattern suggests a notable transformation in precipitation patterns. During 2010–2020, precipitation levels display a relative degree of stabilization, with high value decreasing to 28.8167 and low value plummeting to 12.425. Despite this, the last four years (2020–2024) witnessed a sharp decline in both high and low values, which are respectively 18.25 and 2.95 (Table 7.8 and Fig. 7.13). It marks this timeline

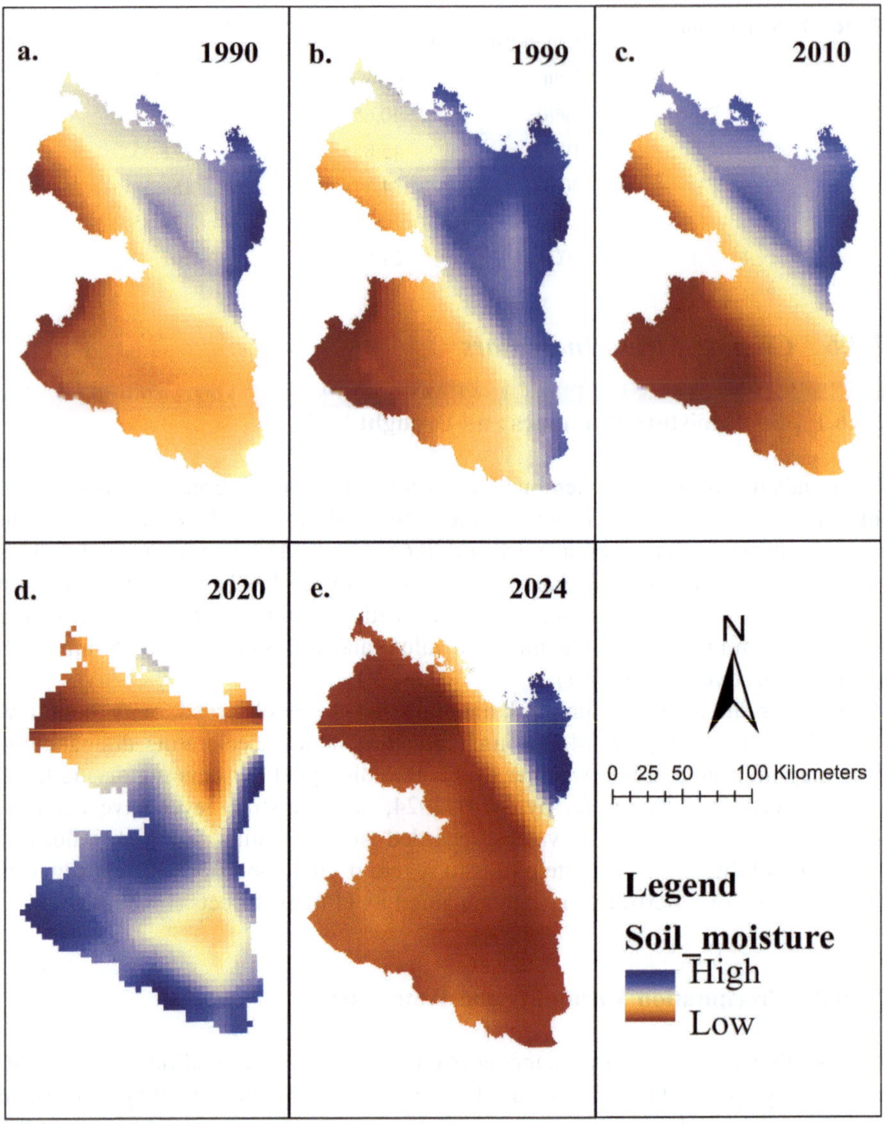

Fig. 7.12 Spatiotemporal analysis of soil moisture dynamics in the barind tract (1990–2024)

as the driest period over the decades, exacerbating the water stress across the region (Fig. 7.14).

Table 7.8 Precipitation

Precipitation (in mm)		
Year	Low	High
1990	8.7	34.3
1999	9.6	46.05
2010	6.8	36.7
2020	12.425	28.8167
2024	2.95	18.25

Source Results from Landsat image analysis from https://earthexplorer.usgs.gov/

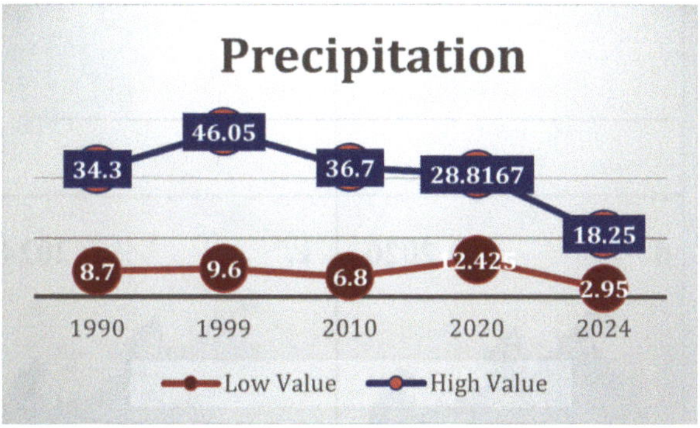

Fig. 7.13 Precipitation

7.3.6.3 Maximum Temperature Trends and Heat Stress

This trend also indicates the ongoing warming. Between 1990 and 1999, high value rose slightly from 35.65 to 35.85, and low value escalated marginally from 30.15 to 30.25. This upward trajectory continued in the following decade (1999–2010), with high maximum temperature value reaching 38.2 and low value climbing to 32.7, indicating a persistent warming pattern. However, an anomalous cooling pattern has been observed in 2010–2020 where the high value declined to 31.5 and low temperature dropped to 29.2417 (Table 7.9).

This anomaly can be the possible reason for transient climate events. Therefore, during 2020–2024, temperature has rebounded to around 2010 levels, with high values at 36.95 and low at 32.15, continuing the overall warming trend (Fig. 7.15). It even indicates the intensifying heat stress which might disrupt agricultural practices, ecosystem services, natural resources, and human health.

7.3.6.4 Minimum Temperature Irregularities

Minimum temperature data demonstrates a fluctuating pattern. Between 1990 and 1999, low values have slightly decreased from 21.9 to 21.8, and high values have

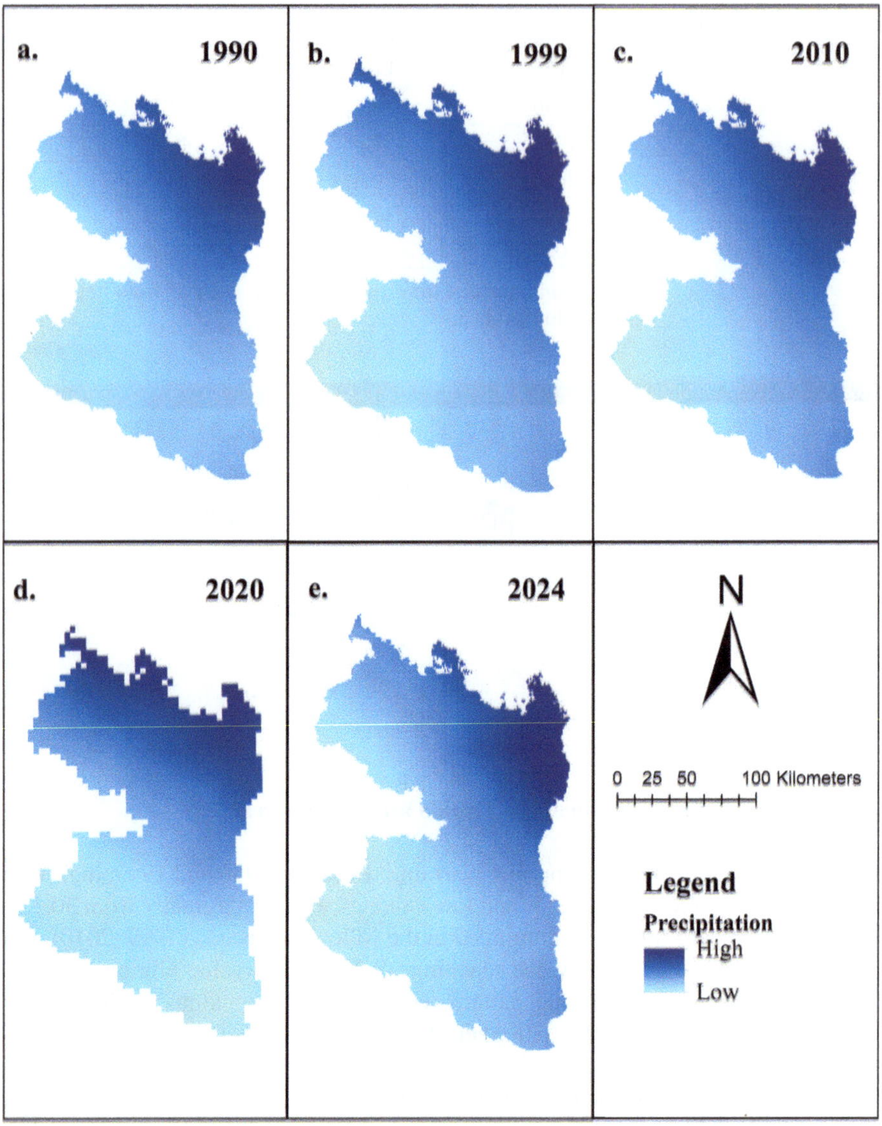

Fig. 7.14 Spatial Analysis of precipitation variability in the barind tract region (1990–2024)

dropped from 24.5 to 23.65 (Table 7.10). In the following decade (1999–2010), the temperature values increased, reflecting rising global warming. The high temperature increased from 23.65 to 25.2, and the low value reached from 21.8 to 23.1. However, the temperature again declined from the previous decade where low value fell to 19.1833 and high value dropped to 21.1167. By 2024, the temperature rose notably (Table 7.10).

Table 7.9 Maximum temperature (in °C)

Year	Low	High
1990	30.15	35.65
1999	30.25	35.85
2010	32.7	38.2
2020	29.2417	31.5
2024	32.15	36.95

Source Results from Landsat image analysis from https://earthexplorer.usgs.gov/

The high value reached 24.45 in the last selected four years (2020–2024), which is the recorded highest minimum temperature over the decades, and low value was elevated to 22.7 (Table 7.10). These fluctuations underscore the presence of climatic instability, thereby presenting potential risks to the viability of sensitive crops. This research shows that areas with such escalating or prolonged climate stress patterns are strongly associated with notable vegetation depletion and wetland shrinkage, indicating not only ecological susceptibleness but also possible permanent land damage unless proper corrective measures emerge. Regional climate adaptation continues to decline due to the critical ecosystem service that support local livelihoods, biodiversity conservation, and agricultural productivity while ensuring the adaptive capacity reduction of future climate shocks (Fig. 7.16).

In conclusion, integrated climate stress assessments with land-use planning can derive meaningful decisions towards achieving the objectives of eco-resilience, climate adaptation, and land degradation neutrality at Barind Tract and other environmentally vulnerable regions of the world.

7.4 Results and Findings

7.4.1 Condition of Climate Stress

In the initial phase of the observation period from 1990 to 1999, the region experienced minimal fluctuations in climatic patterns, the region has been dominated by stable climate stress, covering 17,722.38 km^2, which indicates a relatively stable condition (Table 7.11). Stable stress is concentrated in the northern, western, and central areas, while increasing stress (8454.93 km^2) is minimal and localized mostly in the northeastern and southeastern zones (Table 7.11).

Furthermore, decreasing stress is moderate and accounts for 8533.55 km^2, appearing in scattered patches across the northern and southern parts (Table 7.11). In the following decade (1999–2010), a significant shift has been revealed as areas experienced decreasing stress plummeting from 8,533.55 to 1,660.04 km^2, scattered predominantly throughout the region (Table 7.11).

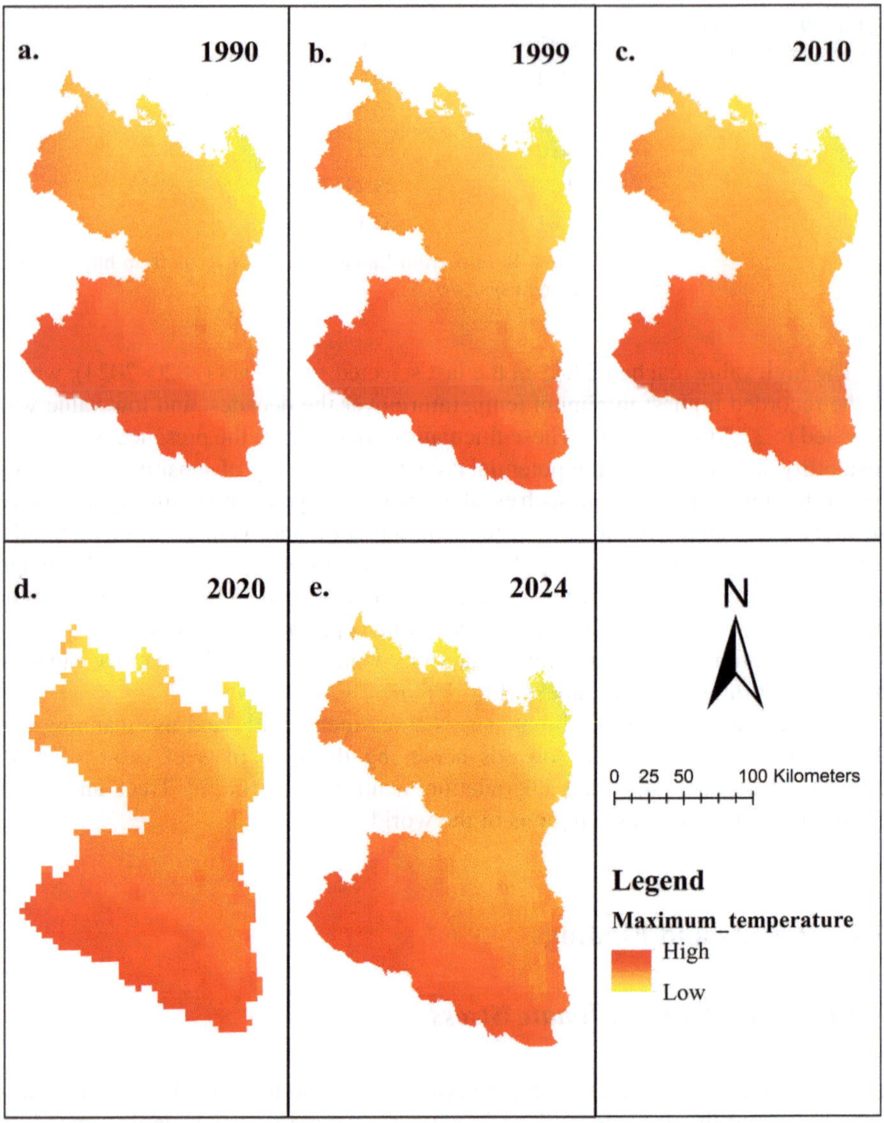

Fig. 7.15 Spatial analysis of maximum temperature trends in the barind tract region (1990–2024)

On the other hand, areas with increasing stress have almost doubled, covering 15,191.17 km^2, particularly in the southwest and northwest regions (Table 7.11). Although stable stress has began to decline, it still occupies a notable portion of the northeast and southeast regions. The next decade, 2010–2020, continues the trend of declination of stable stress, which covered 13,953.68 km^2, predominantly in the northern and southeastern zones (Table 7.11 and Fig. 7.17). Increasing stress

Table 7.10 Minimum temperature (in °C)

Year	Low	High
1990	21.9	24.05
1999	21.8	23.65
2010	23.1	25.2
2020	19.1833	21.1167
2024	22.7	24.45

Source Results from Landsat image analysis from https://earthexplorer.usgs.gov/

remained dominant, particularly in central and southern regions. However, its growth had slowed down compared to the previous decade. And the decreasing stress demonstrated slight increase, appearing more prominently in the northern areas.

In the last four years (2020–2024) a remarkable reverse has been observed. The decreasing stress multiplied nearly fourfold, expanding to 21,746.21 km^2, with the most significant improvement in the southern, central, and northeast regions. Meanwhile, the increasing stress experienced a dramatic decline, decreasing to 3,774.26 km^2 and occupying only a small area in the north. Stable stress continued its downward trend, which covered 9,190.39 km^2 and mostly occupying the northern regions in scatters (Table 7.11).

In summary, stable climate stress consistently declined over the decades, reflecting that the region faced more climate change over time. While the decreasing stress showed fluctuations before significantly improving in recent years (Fig. 7.17). However, in spite of starting with a surge, increasing stress dwindled at the end of the study period.

7.4.2 Correlation Between Climate Stress and Wetland Changes

The Barind Tract has experienced a significant regional shift in climate stress over the decades, directly affecting the wetland loss patterns. From 1990 to 1999, most regions of the study area have faced severe increasing climate stress, particularly in the northwestern zone (Fig. 7.18). This decade can be marked for the highest wetland loss, with 1,824.62 km^2 (Table 7.12).

The dominance of stable stress (1005.24 km^2) across the majority of the region suggests widespread environmental degradation, while the decreasing climate stress (8454.93 km^2) is mostly seen in southern regions, and increasing climate stress (1005.24 km^2) affects majorly the northwestern zone (Table 7.12). In the following decade (1999–2010), a gradual stabilization (699.94 km^2) mostly in the central parts of the region emerged, with fragmented decreasing stress (96.83 km^2) throughout the region.

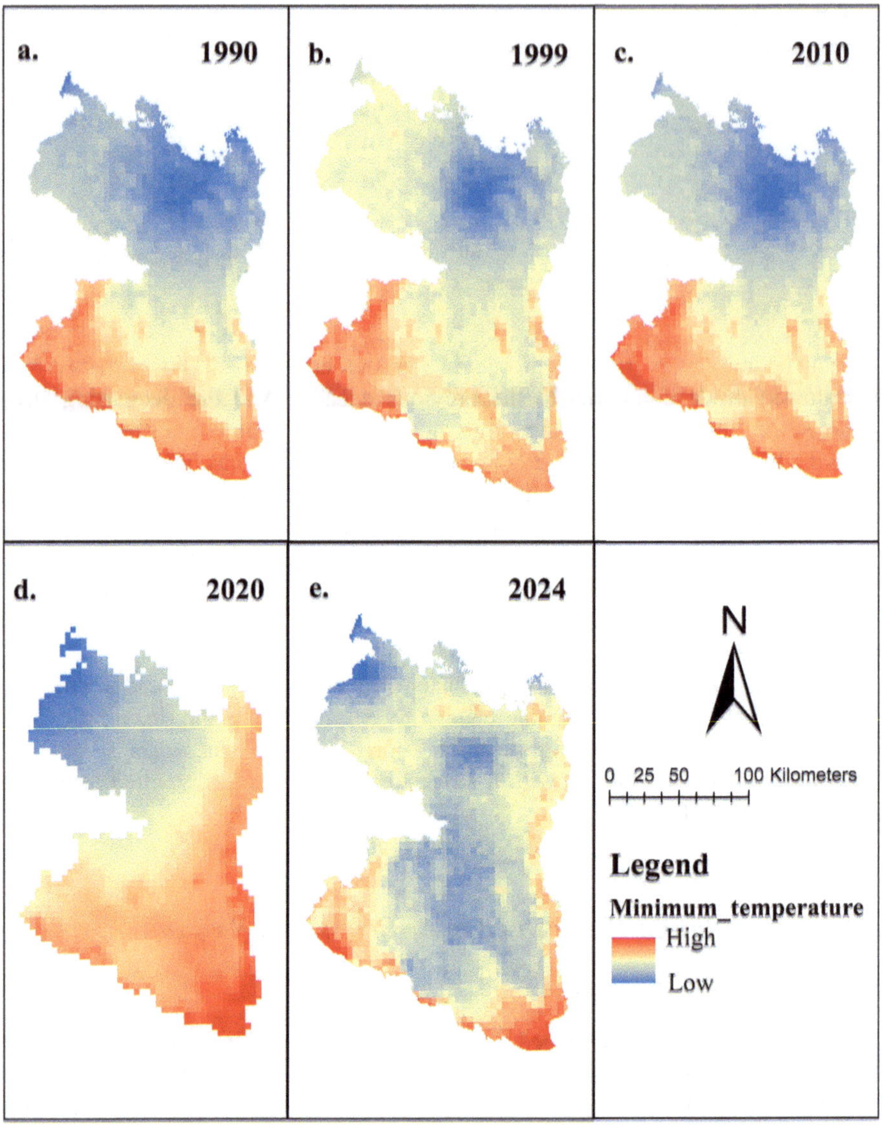

Fig. 7.16 Spatial analysis of minimum temperature irregularities in the barind tract region (1990–2024)

However, increasing stress (545.83 km^2) has influenced the northern and southern areas, continuing the drivable wetland degradation, which accounts for 1342.61 km^2 in total (Table 7.12). Furthermore, from 2010 to 2020, the wetland loss continued its downward trend, reaching 1121.92 km^2 (Table 7.12). Northern parts of the study area illustrate a decreasing stress (Fig. 7.18), marking a positive shift in environmental conditions, which is 96.83 in km^2. And the southern and southwestern region faces

Table 7.11 Year wise climate stress change

Types of climate stress change	1990–1999	1999–2010	2010–2020	2020–2024
	Area_km^2	Area_km^2	Area_km^2	Area_km^2
Decreasing stress	8533.55	1660.04	5224.69	21,746.21
Stable stress	17,722.38	17,859.64	13,953.68	9190.39
Increasing stress	8454.93	15,191.17	15,532.48	3774.26
Total area	34,710.86	34,710.86	34,710.86	34,710.86

Source Results from Landsat image analysis from https://earthexplorer.usgs.gov/

an increasing stress (405.91 km^2), which also declined from the previous decade (Table 7.12). However, the quantity of stable climate stress continued its declining pattern hitting 3053.559 km^2, which is mostly observed in northern and central zones (Table 7.12).

In the most recent period (2020–2024), the decreasing stress areas have soared dramatically, particularly in the southern, central, and western parts of the Barind Tract (Fig. 7.18). These conversions suggest the practice of sustainable activities. Meanwhile, the northern and northeastern parts depicted a combination of stable (334.54 km^2) and increasing stress (124.12 km^2), highlighting environmental challenges that need targeted interventions (Table 7.12).

7.4.3 Correlation Between Climate Stress and Vegetation Loss

Table 7.13 shows that the analysis of vegetation loss and climate stress from 1990 to 2024 displayed a remarkable temporal shift across sparse, moderate, and dense vegetation categories. For sparse vegetation loss, stable stress predominated initially (589.29 km^2 in 1990–1999) which indicates that areas experienced prolonged loss without significant recovery, however between 1999 and 2010, the intensification of both stable Stress (1,006.76 km^2) and increasing stress (823.35 km^2) suggests that the region faced increasing pressure, causing further loss of vegetation and environmental systems, while decreasing stress plummeting to 97.76 km^2 indicates limited recovery and restoration of the land during this period. However, the trend recovered slightly in 2010–2020, with decreasing stress rising to 445.35 km^2, although stable stress-maintained elevation (1,120.48 km^2). In the last four years (2020–2024) decreasing stress dramatically surged to 1,312.46 km^2 indicates a positive shift towards land recovery by sustainable land management practices, alongside a decline in increasing stress (126.88 km^2) and stable stress (451.87 km^2) points to a reduction in ongoing degradation and environmental pressure, allowing ecosystems to stabilize and begin regenerating. Therefore, moderate vegetation loss exhibited stable stress initially (1990–1999: stable stress 358.53 km^2), followed by a decline across all categories by 1999–2010 (e.g., decreasing stress: 24.98 km^2). Stress resurged in

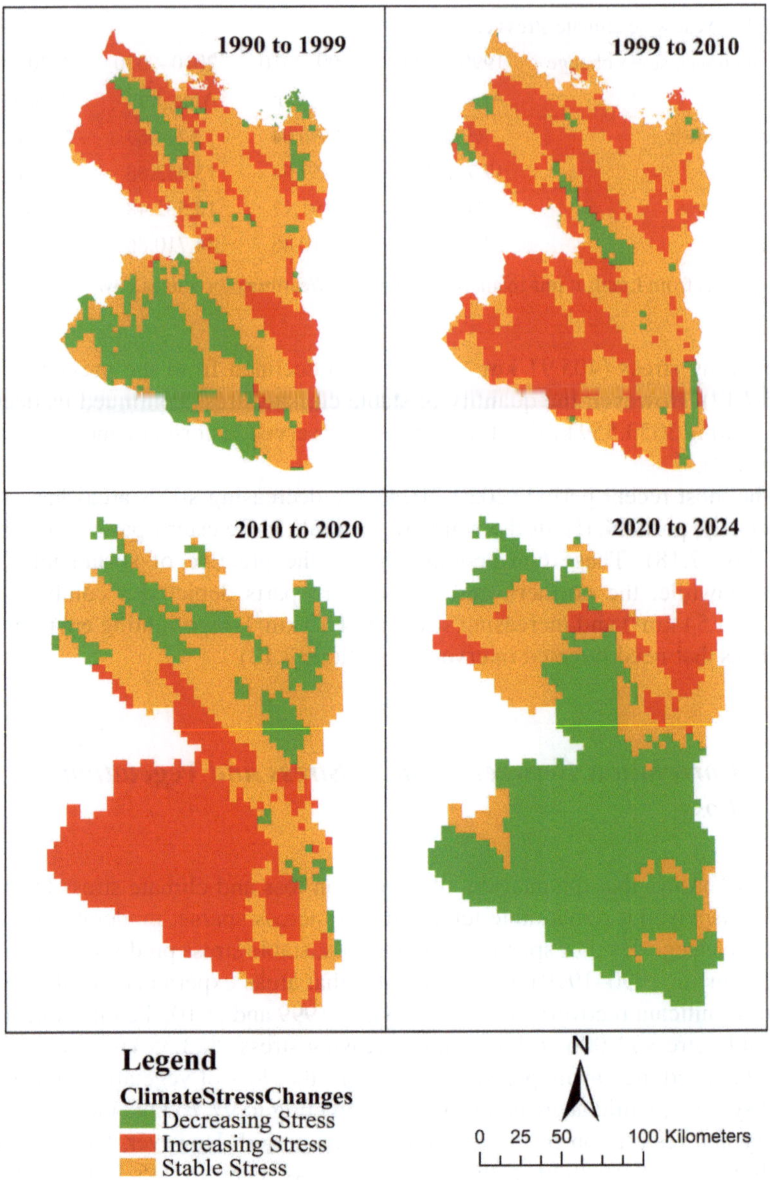

Fig. 7.17 Spatial analysis of climate stress changes in the barind tract region (1990–2024)

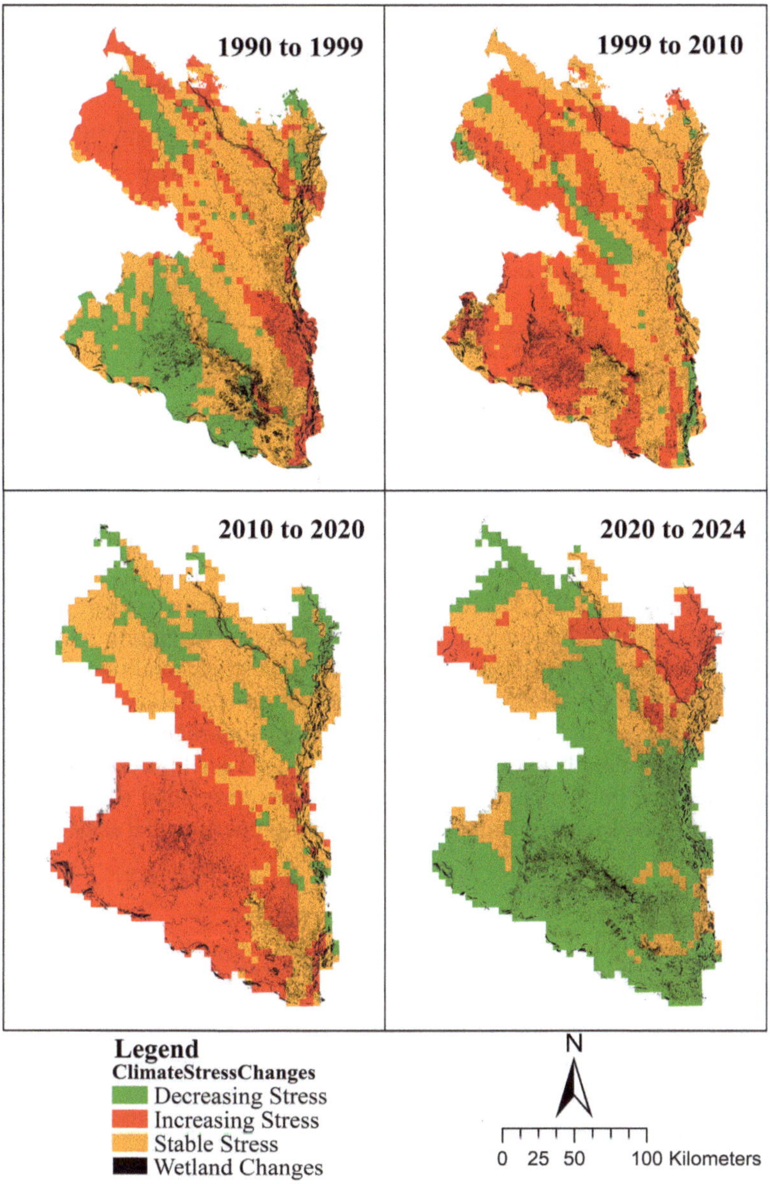

Fig. 7.18 Geospatial analysis of climate stress and wetland dynamics in the barind tract region (1990–2024)

Table 7.12 Year wise wetland loss in various climate stress area

Year	Wetland loss in decreasing climate stress area (km^2)	Wetland loss in stable climate stress area (km^2)	Wetland loss in increasing climate stress area (km^2)	Total wetland loss
1990–1999	365.41	1005.24	453.97	1824.62
1999–2010	96.83	699.94	545.83	1342.61
2010–2020	173.81	542.21	405.91	1121.92
2020–2024	922.86	334.54	124.12	1381.51

2010–2020 (increasing stress: 221.75 km^2) highlights a period of intensified environmental pressure, further exacerbating vegetation loss and contributing to land degradation, however, in 2020–2024, decreasing stress has risen to 460.26 km^2, with stable and increasing stress dropping to 121.56 km^2 and 41.17 km^2, respectively.

Dense vegetation loss, initially predominated by increasing stress (118.63 km^2 in 1990–1999), with a dwindle in increasing stress by 1999–2010 (17.47 km^2), signaling early signs of environmental degradation. A combination trend followed in 2010–2020, but by 2020–2024, increasing stress accounted for 1.61 km^2, with a modest rise in decreasing stress (33.57 km^2) and minimized stable stress (8.42 km^2).

Figure 7.19 shows that the correlational analysis of vegetation loss and stress variations provides the key insights into how different stress conditions interact with vegetation degradation. A strong positive correlation (0.85) is observed between vegetation loss types and stable climate stress, indicating regions with severe vegetation loss are more likely to experience more sustained stable stress conditions over the study periods. Furthermore, a moderate positive correlation (0.69) is observed between vegetation loss type and increasing stress, suggesting that as vegetation loss

Table 7.13 Vegetation loss and climate stress over time (in km^2)

Year	Label	Decreasing stress	Stable stress	Increasing stress
1990–1999	Sparse vegetation loss	280.81	589.29	236.53
	Moderate vegetation loss	187.43	358.53	190.33
	Dense vegetation loss	24.91	63.08	118.63
1999–2010	Sparse vegetation loss	97.76	1006.76	823.35
	Moderate vegetation loss	24.98	183.59	99.76
	Dense vegetation loss	19.1	52.24	17.47
2010–2020	Sparse vegetation loss	445.35	1120.48	646.49
	Moderate vegetation loss	97.63	302.75	221.75
	Dense vegetation loss	12.32	49.99	26.47
2020–2024	Sparse vegetation loss	1312.46	451.87	126.88
	Moderate vegetation loss	460.26	121.56	41.17
	Dense vegetation loss	33.57	8.42	1.61

Source Results from Landsat image analysis from https://earthexplorer.usgs.gov/

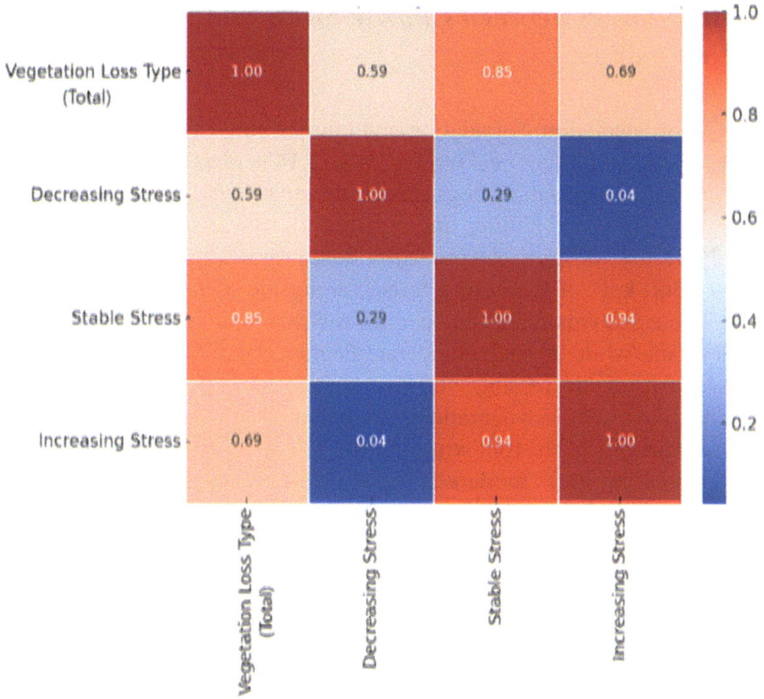

Fig. 7.19 Correlation between vegetation loss and stress factors

intensifies—progressing from sparse to moderate to dense—the degree of increasing stress also rises, although it is not as significant as stable stress. In contrast, the correlation between vegetation loss type and decreasing stress is relatively weak (0.59), indicating that different vegetation loss types do not significantly influence the decreasing stress level.

Moreover, A notably strong correlation (0.94) is observed between stable stress and increasing stress, revealing that regions experiencing stable stress are highly prone to future increases in stress. On the other hand, the correlation between decreasing stress and increasing stress is negligible (0.04), suggesting that areas undergoing decreasing stress depict almost no relationship with increasing stress, and indicate a distinct stress pattern.

These findings suggest that areas with sparse vegetation loss tend to endure high stable stress, which may indicate long-term stress persistence with limited recovery. Meanwhile, regions experiencing dense vegetation loss demonstrate a strong association with increasing stress, underscoring the link between extreme vegetation degradation and worsening environmental conditions. The weak relationship between decreasing stress and vegetation loss type implies that vegetation recovery is not strongly influenced by the specific type of vegetation loss. However, effective land management strategies should consider these stress patterns to mitigate environmental degradation and promote sustainability.

7.4.4 *Correlation Between Climate Stress and Vegetation Degradation*

The analysis of vegetation degradation and climate stress from 1990 to 2024 indicates complicated and categorical trends. Table 7.14 depicts that during 1990–1999, moderate to sparse vegetation degradation dominated the most, with 5335 km^2, 2894 km^2 and 2183 km^2 falling under stable, decreasing, and increasing stress, respectively. Dense to Sparse Degradation demonstrates high ecological pressure, marked by 1306 km^2 of increasing Stress, alongside 459 km^2 of decreasing and 1184 km^2 stable Stress indicates intensified environmental pressures and long-term ecological strain, while dense to moderate degradation maintains comparatively a lower level (167 km^2 decreasing, 166 km^2 stable, and 311 km^2 increasing climate stress), indicating less severe degradation and greater potential for land recovery and improved sustainability. The following decade (1999–2010) haswitnessed a dramatic decline in decreasing stress for dense to sparse degradation (16 km^2 decreased from 459), indicating some recovery in vegetation, continued with a rise in stable stress for dense to moderate degradation (1093 km^2 increased from 466) signifying persistent environmental pressure. Continuing the trend, moderate to sparse degradation withheld its dominance but with declining increasing stress (969 km^2) and increased stable stress (1217 km^2). In 2010–2020, moderate to sparse degradation reached its peak in increasing stress (1669 km^2), reflecting heightened environmental pressures that further intensified vegetation loss, along with 1521 km^2 stable and 662 km^2 decreasing stress suggested some localized recovery potential.

Furthermore, dense to moderate degradation depicts the rise in increasing stress (394 km^2) and stable stress (315 km^2), while dense to sparse degradation stabilizes (64 km^2 in increasing, 47 km^2 in stable, 23 km^2 in decreasing). The most recent years (2020–2024) highlights a dramatic reversal where moderate to sparse degradation experienced a rise in decreasing stress (2879 km^2, has risen from 662 km^2) and dropped in increasing stress (230 km^2 went down from 1669 km^2). Similarly, dense to sparse and dense to moderate degradation has witnessed near-elimination near-annihilation of increasing stress (1 km^2 and 3 km^2, respectively), with decreasing stress escalating sharply for dense to moderate (232 km^2) and dense to sparse (134 km^2). Stable Stress dwindled across all categories, notably for dense to sparse (1184 in 1990–1999 plummet to 38 km^2).

Figure 7.20 shows that the correlation analysis findings reveal significant relationships between vegetation degradation and different stress types. A moderate correlation (0.59) is identified between vegetation degradation type and decreasing stress, indicating that areas with higher degree of vegetation degradation tend to endure increasing rate of decreasing stress. Similarly, a moderate correlation between vegetation degradation type and stable stress (0.52) indicates that stable stress levels are associated with the degree of vegetation degradation. Additionally, a moderate correlation (0.53) between vegetation degradation type and increasing stress highlights that changes in vegetation degradation changes are accomplished by the variations of increasing stress.

Table 7.14 Vegetation degradation and climate stress over time (in km^2)

Year	Vegetation degradation type	Decreasing stress	Stable stress	Increasing stress
1990–1999	Dense to sparse vegetation degradation	459	1184	1306
	Dense to moderate vegetation degradation	167	466	311
	Moderate to sparse vegetation degradation	2894	5335	2183
1999–2010	Dense to sparse vegetation degradation	16	198	111
	Dense to moderate vegetation degradation	75	1093	777
	Moderate to sparse vegetation degradation	97	1217	969
2010–2020	Dense to sparse vegetation degradation	23	47	64
	Dense to moderate vegetation degradation	136	315	394
	Moderate to sparse vegetation degradation	662	1521	1669
2020–2024	Dense to sparse vegetation degradation	134	38	1
	Dense to moderate vegetation degradation	232	64	3
	Moderate to sparse vegetation degradation	2879	976	230

Source Results from Landsat image analysis from https://earthexplorer.usgs.gov/

Moreover, a strong correlation (0.86) between stable stress and increasing stress suggests that areas with higher stability are also prone to higher increasing stress. In addition, another strong correlation (0.71) between decreasing stress and stable stress indicates that these two factors often follow a similar pattern, implying that regions experiencing decreasing stress are often related to stable stress conditions. These findings underscoring the need for effective management to mitigate degradation and promote sustainability.

7.4.5 *Correlation Between Climate Stress and Wetland Degradation*

Analysis of wetland loss and climate stress from 1990 to 2024 reveals the distinct temporal shifts in stress dynamics. Table 7.15 shows that during 1990–1999, stable stress has dominated (1,005.24 km^2), accompanied by moderate increasing stress

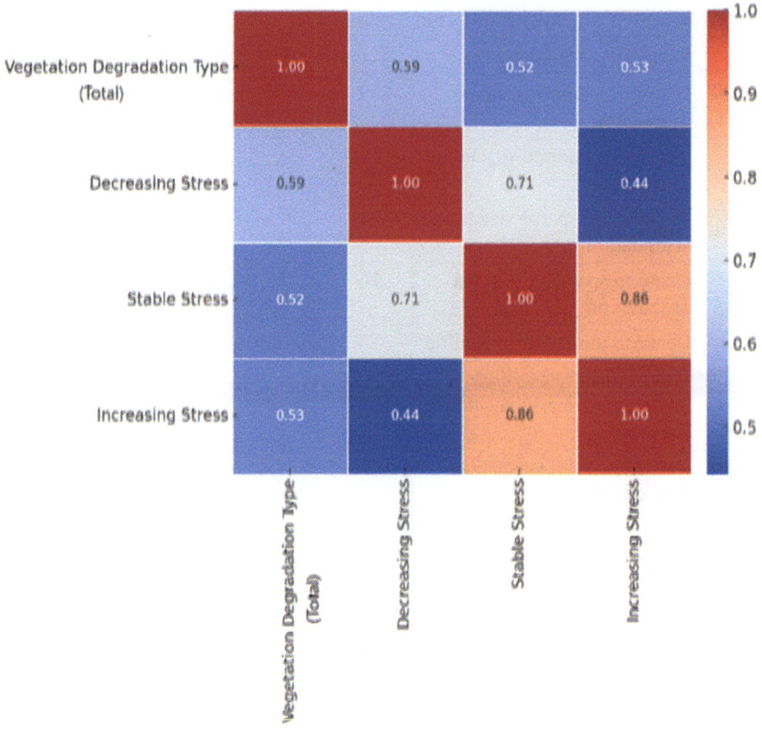

Fig. 7.20 Correlation between vegetation degradation and stress factors

(453.97 km^2) and decreasing stress (365.41 km^2), highlighting baseline ecological pressures with limited mitigation. By 1999–2010, the increasing stress escalated to 545.83 km^2, while decreasing stress plummeted to 96.83km^2 and stable stress declined to 699.94 km^2, indicating intensified degradation during this period, which undermines ecological stability. The 2010–2020 interval reveals partial recovery, with decreasing stress rebounded by 173.81 km^2, increasing stress decreased by 405.91 km^2 and stable stress fell further to 542.21 km^2, indicating the emerging mitigation efforts. The most dramatic shift occurred in 2020–2024, with decreasing stress rose dramatically by 922.86 km^2, increasing stress plummeting to 124.12 km^2, and stable stress declining to 334.54 km^2, highlighting a worsening situation.

The correlational analysis between various stress categories and wetland loss provides significant insights into environmental patterns. The relationship between decreasing stress and stable stress is weakly negative, suggesting that wetland loss in these areas is influenced by distinct environmental factors. This further implies that intensification of degradation in one category does not necessarily correspond with similar trends in the other, likely due to variations in local climatic conditions, land use, and ecological responses (Fig. 7.21).

Similarly, the negative correlation between decreasing stress and increasing stress highlights that regions experiencing declining stress demonstrate different wetland

Table 7.15 Wetland loss and climate stress over time (in km^2)

Year	Decreasing stress	Stable stress	Increasing stress
1990–1999	365.41	1005.24	453.97
1999–2010	96.83	699.94	545.83
2010–2020	173.81	542.21	405.91
2020–2024	922.86	334.54	124.12

Source Results from Landsat image analysis from https://earthexplorer.usgs.gov/

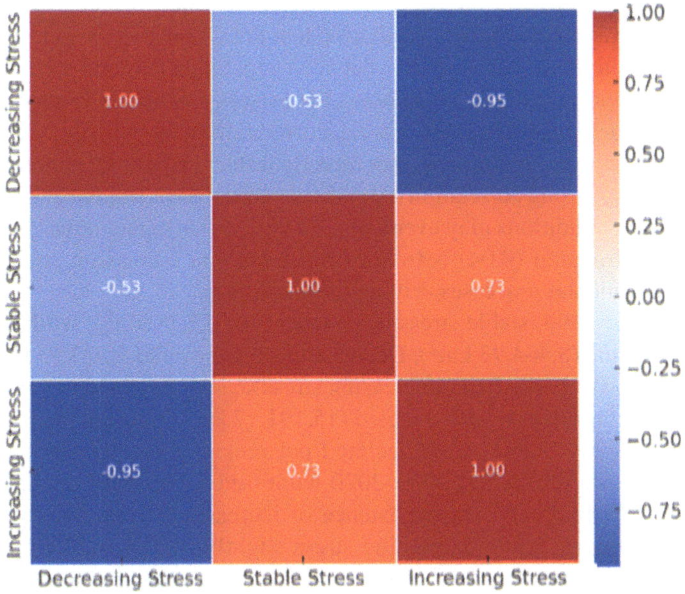

Fig. 7.21 Correlation between wetland loss and stress factors

loss patterns compared to those under increasing stress. This difference may be attributed to factors such as shifting climate conditions, land-use changes, or localized environmental policies that aim wetland conservation (Fig. 7.21).

In contrast, a strong positive correlation between stable stress and increasing stress indicates a pattern where wetland loss in areas with stable stress often coincides with loss in regions experiencing increasing stress. These findings suggest that broader climatic or environmental trends are driving wetland degradation across multiple stress categories, highlighting systemic ecological decline, widespread land degradation, and significant challenges to long-term sustainability.

7.4.6 Findings

The correlation analysis reveals key relationships between wetland loss, vegetation degradation, and climate stress conditions. A moderate correlation exists between vegetation degradation and decreasing stress (0.59), stable stress (0.52), and increasing stress (0.53), indicating that all stress types are influenced by vegetation degradation levels. Profound correlations are observed between stable and increasing stress (0.86) and between decreasing and stable stress (0.71), indicating interconnected stress patterns.

A notable strong correlation (0.85) is also discovered between vegetation loss and stable stress, implying that severe vegetation loss areas tend to have persistent stress. A moderate positive correlation (0.69) between vegetation loss and increasing stress suggests that with intensifying vegetation loss, stress conditions worsen. However, vegetation loss demonstrated solely a weak correlation (0.59) with decreasing stress, indicating that recovery patterns are not directly influenced by vegetation loss types. In addition, a very strong correlation (0.94) between stable and increasing stress highlights that stable conditions may eventually transition into higher stress levels, while a negligible correlation (0.04) between decreasing and increasing stress suggests distinct environmental processes driving these changes.

From 1990 to 1999, stable stress dominated (17,722.38 km^2), while increasing stress was minimal (8,454.93 km^2). Decreasing stress covered 8,533.55 km^2 in fragmented patches. By 1999–2010, decreasing stress sharply declined to 1,660.04 km^2, whereas increasing stress nearly doubled (15,191.17 km^2), particularly in the southwest and northwest. Stable stress declined but remained significant in the northeast and southeast. The period 2010–2020 witnessed a continued decline in stable stress (13,953.68 km^2) and the persistence of increasing stress, mainly in central and southern regions, while decreasing stress slightly rose in northern areas. The 2020–2024 period witnessed a dramatic fourfold increase in decreasing stress (21,746.21 km^2), marking a notable recovery, while increasing stress fell sharply to 3,774.26 km^2, suggesting a decline in stress-driven wetland loss. Stable stress further declined to 9,190.39 km^2, predominantly in northern areas, highlighting a shift in environmental stress patterns.

Overall, stable stress has consistently declined over the decades, while decreasing stress fluctuated but displayed a significant recovery in recent years. Increasing stress initially surged but dropped substantially at the end of the study period, indicating potential shifts in environmental conditions. These findings provide the critical need for targeted conservation strategies that focus on mitigating stress impacts, restoring degraded wetlands, and enhancing long-term sustainability. Strengthening eco-resilience through these interventions is essential to support the adaptive capacity of vulnerable landscapes under changing climatic conditions.

7.4.7 Implications for Environmental Management

The findings from this research have significant implications for environmental management and wetland conservation. The strong correlation between vegetation degradation and stress levels underscores the urgent need for proactive land-use planning and conservation efforts. Policymakers should prioritize reforestation and wetland restoration projects in the regions that are experiencing stable or increasing stress to prevent long-term ecosystem degradation.

The high correlation between stable and increasing stress (0.94) suggests that areas with stable conditions may eventually transit to more severe stress. This highlights the need for early intervention strategies, such as adaptive management practices, sustainable land-use planning, and climate mitigation measures, to slow down or mitigate the stress progression.

The recent escalation in decreasing stress (2020–2024) provides an opportunity to reinforce ecological resilience through targeted conservation programs. Governments and environmental organizations should focus on this positive trend by strengthening wetland protection policies, improving water resource management, and promoting afforestation initiatives.

Furthermore, the negligible correlation (0.04) between decreasing and increasing stress suggests that distinct mechanisms have driven these changes. This calls for region-specific strategies rather than a one-size-fits-all approach. In regions with historically high increasing stress, emphasis should be placed on climate adaptation measures, controlled water drainage, and improved land management practices to mitigate further wetland loss.

Overall, the findings emphasize the need for combined climate adaptation and wetland conservation strategies that address the complex interactions between vegetation loss, climate stress, and wetland degradation. Strengthening policy frameworks, investing in ecosystem restoration, and adopting sustainable land-use practices will be crucial for ensuring long-term environmental stability.

7.5 Conclusions

7.5.1 Final Remarks

The Barind Tract has undergone notable climatic shifts from 1990 to 2024, resulting in significant impacts on vegetation cover and wetland dynamics. Increased temperatures, unusual rainfall patterns, and uneven soil moisture have prompted processes including deforestation, land use change, and loss of wetland; climate stress peaked from 1999 through 2010 and has gradually decreased since 2020. In light of the recent decreases in climate stress, especially across the southern, central, and northeastern zones, there is an opportunity for repair and rehabilitation of ecosystems.

Yet the region's real challenges lie in its ecological fragility. The reductions in shrubland, native forest, and wetland area (especially from 1990 to 2010) are generally associated with climate-induced stress and direct human impacts, e.g., illegal agricultural expansion and urban development. These changes are putting biodiversity, local agricultural productivity, and local livelihoods at risk and emphasize the need for urgent integrated responses.

This study shows the role of climate modeling through remote sensing and GIS to evaluate land degradation patterns. The knowledge generated provides a scientific basis to promote eco-resilience in the Barind Tract through sustainable land management, tree planting, afforestation, climate-resilient agriculture, and wetland restoration interventions. To achieve sustainable resilience, environmental policies should first and foremost be guided by nature-based solutions and a system of optimal water resource management to be weighed against zones with a malfunctioning climate. Integrating such adaptive strategies with local environmental realities will be crucial in reducing degradation, maintaining ecosystem services, and enabling sustainable development in this climate-sensitive region.

7.5.2 Knowledge Contributions, Eco-Resilience Implications, and Future Research

This study highlights the complex interrelationships among climate-related stress, vegetation degradation, and wetland loss contribute to a holistic understanding of climate change and land degradation in the Barind Tract, one of the ecologically fragile regions stressed to the extent of being a victim of climate change and land degradation. Utilizing spatial analysis and correlation modeling, the study identifies hotspots of environmental pressure and degradation, informing contextualized science-based adaptive land and resource management.

A major knowledge contribution is the identified strong association between stable and increasing climate stress zones, implying that areas under seemingly stable conditions can be at risk for future degradation. This is at odds with expectations of stability and illustrates the immediate need for ongoing environmental monitoring and dynamic, proactive conservation planning. Moreover, the low correlation for decreased and increased stress suggests that localized environmental conditions and anthropogenic pressures are important modulators of land degradation patterns, especially in wetland ecosystems (Rountree et al., 2020).

Such knowledge bears significant implications for eco-resilience and sustainability. The spatial identification of climate-stressed zones provides a roadmap for targeted interventions, including: Sustainable Land Stewardship—Community-Based Resource Management; Restoring vegetation and keeping agricultural productivity through afforestation and climate-smart agriculture; Wetland conservation and restoration for improved water regulation and biodiversity; & Local context-specific fantastic solutions and integrated water resource management.

The prerogative for delivery of resilience-building interventions at scale has never been stronger; the dip in climate stress reported for 2020–24 suggests early opportunities for ecosystem restoration, and rumbustious growth for climate resilience. Incorporating these results into policy frameworks can enable planners and decision-makers to mitigate both short-term threats and long-term vulnerabilities.

Future work combining lower-scale land use data and more detailed hydrological modeling and accounting for anthropogenic drivers will clarify wetland and vegetation loss mechanisms. Integration of remote sensing, machine learning, and climate projection models could contribute to more effective early warning systems and predictive capacity for environmental risk assessment (Moe et al., 2024). Moreover, broadening the temporal range of the study beyond 2024 and including socioeconomic variables as well—like effects on local livelihoods, food security, and biodiversity—will be important to the development of cohesive, holistic plans for climate adaptation and ecosystem resilience.

The advancement of these research directions contributes to the broader aim of land degradation neutrality and eco-resilient landscapes as part of national adaptation plans and global sustainability targets (IPCC, 2014).

7.5.3 AI Disclosure

This chapter is the result of the authors' original research, analysis, and critical interpretation. All data collection, geospatial modeling, correlation analysis, and scientific insights were conducted entirely by the authors. To support the clarity, flow, and language consistency of the text, AI-powered tools such as ChatGPT (OpenAI), DeepSeek, and GrokAI were used selectively for grammar refinement, paraphrasing, and formatting assistance. These tools did not influence the intellectual content, research design, or conclusions. The final manuscript fully represents the authors' expertise, analytical reasoning, and scholarly contribution.

References

Addae, B., & Dragićević, S. (2023). Modelling global deforestation using spherical geographic automata approach. *ISPRS International Journal of Geo-Information, 12*(8), 306.

Al-Amin, M., & Islam, G. T. (2023). Impact of land use change on ecosystem services in a coastal town of Bangladesh. In *Transcending Humanitarian Engineering Strategies for Sustainable Futures* (pp. 1–19). IGI Global.

Artaxo, P., Machado, L., Franco, M. A., Albuquerque, I. M., & Rizzo, L. (2021, December). Deforestation and climate change: The multiple pressures over Amazonian forests. In *AGU fall meeting abstracts* (vol. 2021, pp. B41C-04).

Chow, J., Khanom, T., Hossain, R., & Khadim, J. (2019). Forest management for climate change adaptation in Bangladesh. In *Confronting climate change in Bangladesh: Policy strategies for adaptation and resilience* (pp. 39–50).

Chowdhury, M. R., & Ndiaye, O. (2017). Climate change and variability impacts on the forests of Bangladesh: A diagnostic discussion based on CMIP5 GCMs and ENSO. *International Journal of Climatology, 37*(14), 4768–4782.

Dewi, R. U. I., Zuhaiery, A., Sa'diyah, N. H., Rizqiya, T. F., & Nurpratiwi, H. (2023). Harmonisasi masyarakat dan pemerintah untuk mengatasi deforestasi di Selatan Tulungagung. *Dewantara: Jurnal Pendidikan Sosial Humaniora, 2*(2), 149–157.

Douville, H., & Royer, J. F. (1996). Influence of the temperate and boreal forests on the Northern Hemisphere climate in the Météo-France climate model. *Climate Dynamics, 13*(1), 57–74.

FAO. (2020). Global forest resources assessment 2020. *Food and Agriculture Organization*. http://www.fao.org/forest-resources-assessment

Gorte, R. W., & Sheikh, P. A. (2010, March). Deforestation and climate change.

Gu, S. (2023). The impact of increasing forest loss areas on the global temperature and tourism industry. *Asian Journal of Research in Agriculture and Forestry, 9*(3), 42–55.

Haughan, A. E., Pettorelli, N., Potts, S. G., & Senapathi, D. (2022). The role of climate in past forest loss in an ecologically important region of South Asia. *Global Change Biology, 28*(12), 3883–3901.

Intergovernmental Panel on Climate Change (IPCC). (2007) Climate change 2007: Synthesis report. Contribution of Working Groups I, II and III to the Fourth Assessment Report of the Intergovernmental Panel on Climate Change (Core Writing Team, R. K. Pachauri, & A. Reisinger [Eds.]). IPCC. https://www.ipcc.ch/report/ar4/syr/

Intergovernmental Panel on Climate Change (IPCC). (2014). Climate change 2014: Synthesis report. In R. K. Pachauri & L. A. Meyer, (Eds.), Contribution of Working Groups I, II and III to the Fifth Assessment Report of the Intergovernmental Panel on Climate Change. IPCC. https://www.ipcc.ch/report/ar5/syr/

Islam, K., & Sato, N. (2012). Deforestation, land conversion, and illegal logging in Bangladesh: The case of the Sal (Shorea robusta) forests. *iForest—Biogeosciences and Forestry, 5*(3), 171.

Islam, M. D., Di, L., Mia, M. R., & Sithi, M. S. (2022, July). Deforestation mapping of Sundarbans using multi-temporal Sentinel-2 data & transfer learning. In *2022 10th International Conference on Agro-geoinformatics (Agro-Geoinformatics)* (pp. 1–4). IEEE.

Islam, S. D. U., & Bhuiyan, M. A. H. (2018). Sundarbans mangrove forest of Bangladesh: Causes of degradation and sustainable management options. *Environmental Sustainability, 1*(2), 113–131.

Islam, S., Ma, M., Hossain, M. N., Ganguli, S., & Sarker, M. N. I. (2021). Temporal evaluation of climate change on land use and land cover changes in the southeastern region of Bangladesh from 2001 to 2016. In *Climate vulnerability and resilience in the global south: Human adaptations for sustainable futures* (pp. 509–525).

Khan, M. S., Abdullah, S., Salam, M. A., Mandal, T. R., & Hossain, M. R. (2021). Review assessment of biodiversity loss of Sundarbans forest: Highlights on causes and impacts. *Indonesian Journal of Forestry Research, 8*(1), 85–97.

Kibria, M. G., Rahman, S. A., Imtiaj, A., & Sunderland, T. (2011). Extent and consequences of tropical forest degradation: Successive policy options for Bangladesh. *Journal of Agricultural Science and Technology, 1*(1).

Maria, S. P. (2015). Deforestation: Causes and consequences. *The International Journal of Social Sciences and Humanities Invention, 2*(3).

Miah, M. G., Ahmed, M., & Afroz, T. (2013). Climate change and adaptation: Evidence from a forest-dependent community in Bangladesh. *Journal of Forest and Livelihood, 11*(1), 94–108.

Moe SJ et al (2024) Integrating climate model projections into environmental risk assessment. *Integr Environ Assess Manage 20*(2): 367–381. OUP Academic

Nadiruzzaman, M., Shewly, H. J., Rashid, B., Dutta, O., & Mukul, S. A. (2024). Environmental degradation, neoliberal development, and deforestation in the Chittagong Hill Tracts of Bangladesh: A critical analysis in the context of the ongoing climate crisis.

Ondr, S., Erbaugh, J. T., & Kosmidou-Bradley, G. C. (2021). Economic issues related to Asian deforestation. In *Oxford research encyclopedia of environmental science*.

Owais, S. M., & Siddiqui, S. (2019). Appraisal of deforestation and forest degradation in District Swat, Pakistan. *International Journal of Economic and Environmental Geology (IJEEG), Published by SEGMITE, 10*(3), 7–15. https://doi.org/10.46660/IJEEG.VOL10.ISS3.2019.301

Reddy, C. S., Pasha, S. V., Jha, C. S., Diwakar, P. G., & Dadhwal, V. K. (2016). Development of national database on long-term deforestation (1930–2014) in Bangladesh. *Global and Planetary Change, 139*, 173–182.

Reza, A. A., & Hasan, M. K. (2019). Forest biodiversity and deforestation in Bangladesh: The latest update. In *Forest degradation around the world* (pp. 1–19).

Rowntree, J. E., Stanley, P. L., Maciel, I. C. F., Thorbecke, M., Rosenzweig, S. T., Hancock, D. W., Guzman, A., & Raven, M. R. (2020). Ecosystem impacts and productive capacity of a multi-species pastured livestock system. *Frontiers in Sustainable Food Systems, 4*, 544984. https://doi.org/10.3389/fsufs.2020.544984

Rudra, A. K. (2016). Forest management for enhancing ecosystem services in the climate change scenario of Bangladesh. *Energy and Environmental Engineering, 4*, 21–29.

Saoum, M. R., & Sarkar, S. K. (2024). Monitoring mangrove forest change and its impacts on the environment. *Ecological Indicators, 159*, Article 111666.

Singh B, Reddy VM, Reddy KR, Sharma P, Nautiyal M, Bhatnagar S, Islam R, Alabdeli H (2024) Modeling the impact of deforestation on global warming using system dynamics. E3S Web of Conferences, 581:01023. https://doi.org/10.1051/e3sconf/202458101023

Smith, C., Spracklen, D. V., & Baker, J. C. A. (2020). *Impacts of tropical deforestation on local climate.* 11712. https://doi.org/10.5194/EGUSPHERE-EGU2020-1171

Sunderland, T. C., Kibria, M. G., Rahman, S. A., & Imtiaj, A. (2011). Depleting tropical forests at a landscape scale: Finding solutions for Bangladesh.

Ullah, S., Wu, Y., & Khan, A. (2023). Evaluating the socioeconomic factors on deforestation in Northern Pakistan: A study on existing economic incentive tools for reducing deforestation. *Sustainability, 15*(7), 5894. https://doi.org/10.3390/su15075894

Wu, D., Zheng, L., Wang, Y., Gong, J., Li, J., & Chen, Q. (2024). Characteristics of urban expansion in megacities and its impact on water-related ecosystem services: A comparative study of Chengdu and Wuhan, China. *Ecological*

Chapter 8
Estimating Soil Erosion Using Revised Universal Soil Loss Equation (RUSLE) Model in the Upper Bharathapuzha

D. Monika, P. Masilamani, J. Subhashini, J. Killivalavan, and R. Thanuja Krishnan

Abstract The Contemporary chapter aims to investigate Land depletion in Upper Bharathapuzha Basin, Tamil Nadu, using the Revised Universal Soil Loss Equation (RUSLE). By means of an unified approach of geospatial records and Geographic data platform, various aspects contributing to soil erosion, such as precipitation-induced erosion force, soil fragility, inclination of land, land use, and conservation efforts are incorporated for the analysis. The study identifies areas with steeper slopes, lower vegetation density, and higher land use intensity as more susceptible to erosion. Furthermore, the impact of soil conservation practices like terracing and afforestation is assessed for their effectiveness in reducing erosion. The current methodology offers a predictive framework for identifying regions vulnerable to landslides and floods, enhancing early warning systems for the affected areas, and managing the disaster. The insights obtained from the study onto design targeted soil conservation strategies. By implementing these strategies, the integrity of soil resources, water quality, and the ecosystem can be better preserved in this vital watershed area. These findings also support the decision-making for sustainable land use and socio-economic well-being in communities that rely on the Upper Bharathapuzha Basin. The study's multidisciplinary approach aligns with broader efforts to address environmental challenges, promote resilience, and foster sustainable development in water-sensitive areas. The study supports the eco-resilience by addressing climate change, land degradation, and also proposes sustainable solutions.

D. Monika
Department of Natural Resources and Environmental Studies, National Dong Hwa University, Hualien, Taiwan

P. Masilamani (✉) · J. Killivalavan · R. Thanuja Krishnan
Department of Geography, Bharathidasan University, Tiruchirappalli, Tamil Nadu, India
e-mail: masilamani@bdu.ac.in

J. Subhashini
Department of Geography, Sri Vijay Vidyalaya College of Arts and Science, Dharmapuri, Tamil Nadu, India

U. Chatterjee et al. (eds.), *Eco-Resilience: Climate Change, Land Degradation and Sustainable Solutions*, Sustainability Solutions,
https://doi.org/10.1007/978-3-032-00708-7_8

Keywords Soil erosion · RUSLE · Upper bharathapuzha · Land degradation · GIS and remote sensing · Erosion risk mapping

8.1 Introduction

Soil Erosion is the phenomenon bringing forth denudation of the top layer of soil. This process leads to the loss of fertility in the soil and brings a combined bad outcomes to the environment (Prasannakumar et al., 2012). It's estimated that about 35.9 petagrams of soil are lost from land each year (Borrelli et al., 2017). It is a significant conservation-related issue that is hazardous to the production of cultivation, the stability of water resources, and land sustainability. This study provides insights into erosion-prone zones, enabling support for sustainable management practices. Soil erosion is highly concerning nature and community health risk facing human society (Bhat et al., 2019). It is the core mechanism by which soil loses fertility under agriculture across the globe. The Universal Soil Loss Equation, originally paper framework, it was made digital and improved by scientists (Renard et al., 1997), and the updated was called as Revised Universal Soil Loss Equation (RUSLE). The significance of the issue has grown, including harmful effects on farming, water quality, and infrastructure affecting both the site of erosion and surrounding areas (Pimentel, 2006). It also leads to flooding because it can create thick layers of sediment obstruct streams and drainage flowing steadily and flooding. Erosion is triggered by both natural phenomenon and human interventions. The natural factors that affect it include things like the type of soil, rainfall intensity, climate, inclination of the land, soil's probability to erosion, and the presence of vegetation. Soil erosion can also be caused by human activities like deforestation, slash-and-burn farming, intensive agricultural practices, and overgrazing. The findings of (Ananda & Herath, 2003) uncover that harmful outcomes of technical change, irrelevent government regulations, and poor institutions are more reliable for the continuous soil erosion in developing countries. The continuous overexploitation of the land resources combined with the climatic factors tends to remove the good top layer of the soil (Lin et al., 2019). The impacts of erosion are significant, not only due loss of fertility, but also because of a series of residual effects, such as sedimentation, siltation, eutrophication of waterways, and increased flooding (Borrelli et al., 2017), which is considered as one of the big issues in the Western Ghats (Thomas et al., 2018). Soil erosion rates are increasing significantly now-a-days due to the individual activities such as stripping of nature, especially deforestation practices for farming (Verheijen et al., 2009). Soil erosion framework proves a measurable and steady procedure to assessment of erosion under a numerous range of conditions and to the comprehensive development of controlling measures of soil erosion. The main environmental benefits of soil erosion controlling includes reducing the run-off velocity, maintaining the integrity of soil and controlling the pollutants and maintaining a healthy biodiversity, and maintaining the ecosystem. This RUSLE is a way to find out the amount of soil eroded every year and uses five factors: rainfall (R), soil erodibility (K), length slope (LS), cover management (C), and conservation practices (P). This study employs the RUSLE method because this method is impactful with respect to cost for data provisions, model parameterization, and modelling the soil erosion. Mountain areas are mostly affected by soil erosion. This is due to the heavy downpour, strong winds, and slope gradient. High-yield farming and socio-economic stress for land made soil erosion on inclining lands (Shi et al., 2004). In Bharat, about 91% of the zone belongs to five erosion categories varying from <5 to 40 t ha^{-1} yr^{-1}, multiple methods are needed to protect these (Sharda et al., 2013). The RUSLE model has been identified as the after identifying suitable models for each parameter, tailored to the study area's needs, GIS and remote sensing technologies are employed to analyze each parameter. This method provides a time-efficient and cost-effective alternative to traditional field-based approaches, streamlining soil erosion estimation (Ghosal & Das Bhattacharya, 2020). Major types of soil erosion include (i) Splash erosion removes soil particles by raindrop impact, while (ii) sheet erosion uniformly removes topsoil during runoff, (iii) Rill erosion forms small channels, (iv) gully erosion creates deep grooves, (v) ravines are advanced gullies, (vi) landslides occur on steep slopes, and (vii) stream bank erosion affects riverbanks during high flows (Telkar et al., 2015). The upper Bharathapuzha watershed, located in the southern Indian state of Kerala, encompasses diverse landforms, including hills, plateaus, and plains. This region is characterized by intensive agricultural activities, primarily rice cultivation, as well as urbanization and industrialization. The geographic features, coupled with anthropogenic pressures, make the upper Bharathapuzha watershed particularly vulnerable to erosion and land degradation. Degradation of soil and water resources poses risk agricultural sustainability and the daily life of local communities. The total area of the upper Bharathapuzha is 2700 km^2. The novelty of this study lies in integrating GIS techniques with RUSLE to generate high resolution erosion maps. The main objectives of this study are to quantify potential loss of soil and identify the erosion-prone areas, and to guide sustainable management strategies. Soil erosion is a significant risk to land sustainability and agricultural productivity in the Upper Bharathapuzha. By applying the RUSLE model, this study aims to enhance our understanding of erosion processes, identify vulnerable areas, and inform soil conservation strategies. Through this interdisciplinary study, it researches towards safeguarding the soil and water resources of the upper Bharathapuzha region for future generations. Main objectives of this study are to (i) estimate the soil erosion rates (ii) identify key factors influencing soil erosion (iii) to assess the impact of landuse changes (iv) develop sustainable management practice. The objectives of this study are connected as they work together to identify soil erosion by considering factors like rainfall, soil type, and land use changes, the study identifies areas most affected by erosion. Based on the findings, the study recommends sustainable practices to reduce erosion and ensure long-term environmental protection. This connection ensures the research offers both an understanding of the problem and practical solutions for the future.

8.2 Study Area

The Bharathapuzha river is river in (Kerala and Tamil Nadu) in south India. It is the largest river basin in Kerala and is located between 10°26′ to 11°13′ N latitudes and 75°33′ to 77°13′ E longitudes. The southwest and northeast monsoons bring more rain to the windward and leeward sides of the river (John et al., 2022). Daily highest and lowest temperature ranges between 15.5 °C and 39 °C (Nair & Jayaprakash, 2021). The study area comes under the administrative limit of Palakkad district and covers the entire Palakkad Taluk, Chittur Taluk, Alathur Taluk and Eastern part of Ottappalam Taluk and Southern part of Mannarkkad Taluk. This area was selected because of its unique as well as diverse physiographic and climatic features, different from the Lower Bharathapuzha Basin and also the rest of Kerala. It is situated within the Western Ghats Mountain range, a biodiversity hotspot renowned for its rich flora and fauna. The river originates in the western ghats, at an elevation of 1100 m above mean sea level, this river is nourished by its tributaries: Kalpathypuzha, Gayathripuzha, Thootha, and Chitturpuzha. These tributaries traverse diverse geological and geomorphological landscapes across the state of Kerala (Magesh et al., 2013). The watershed extends from the highlands of the Anamalai and Nilgiri hills in the east to the coastal plains of the Arabian Sea in the west. In terms of relief, the basin can be divided into three clearly defined zones. These zones include the coastal plains, characterized by elevations below 7.5 m; the midland area, which encompasses elevations ranging from 7.5 to 75 m; and the highland area, with elevations exceeding 75 m (John et al., 2022). Despite its natural beauty and ecological significance, the Upper Bharathapuzha watershed faces several environmental challenges, like deforestation, soil erosion, water pollution, and habitat degradation. Soil erosion in Upper Bharathapuzha is primarily driven by two factors- natural factors and anthropogenic factors. Naturally, the soil erosion occurs due to the heavy downpour that occurs during the southwest monsoon. The precipitation of the area is mainly contributed from the southwest monsoon) and also by northeast monsoon with an average annual rainfall of 2924.4 mm. The humidity is higher during monsoon months from June to October (Magesh et al., 2013). These natural factors contribute to soil erosion by dislocating and washing away the soil particles, causing surface runoff. The steep slopes and fragile soil types induce soil erosion by supporting the natural factors. It affects the soil fertility, water quality, and agriculture productivity of the upper Bharathapuzha watershed. The main challenges to address soil erosion are due to limited technical capacity and competing land use priorities. Collaborative efforts among government agencies, research institutions, and local communities are indeed for developing solutions to Soil Erosion. By understanding these causes, impacts, and spatial distribution of soil erosion can implement conservation measures and its risk. The study area is mapped using the ArcGIS software, and it is represented as (Fig. 8.1: Study area).

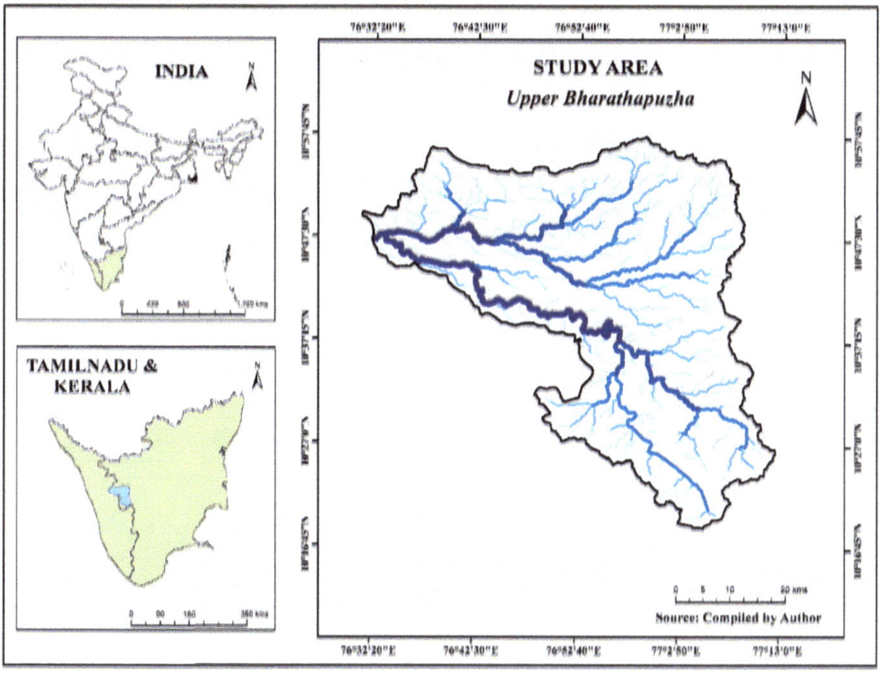

Fig. 8.1 Study area—Upper Bharathapuzha

8.3 Methodology

The USLE was improved and replaced by the revised RUSLE with additional data and recent research results of the last 40 years to predict erosion better, as it was updated by a group of scientists (Renard et al., 1997). The effects of precipitation, soil, topographic, land utilization, and conservation practices were considered to estimate the soil erosion (Kebede et al., 2021). These models are essential in erosion modelling within hydrological basins, effectively supporting assessments that span from localized areas to global applications (Patriche, 2017). The equation of the model is

$$A = R * K * L * S * C * P.$$

Where A is the soil loss in t ha^{-1} $year^{-1}$; K is the eroding factor, R is the rainfall runoff erosivity factor; S is the slope steepness factor; L is the slope length factor, and C is the cover and management factor; P is the conservation practices determinant. The L, S, C, and P values do not have any units. Above mentioned factors (R * K * LS * C * P) were blended into GIS field for creation of erosion prediction map. By integrating this formula in the research can greatly help in estimating the soil erosion of an area (Balasubramani et al., 2015). The methodology is represented in a flowchart type

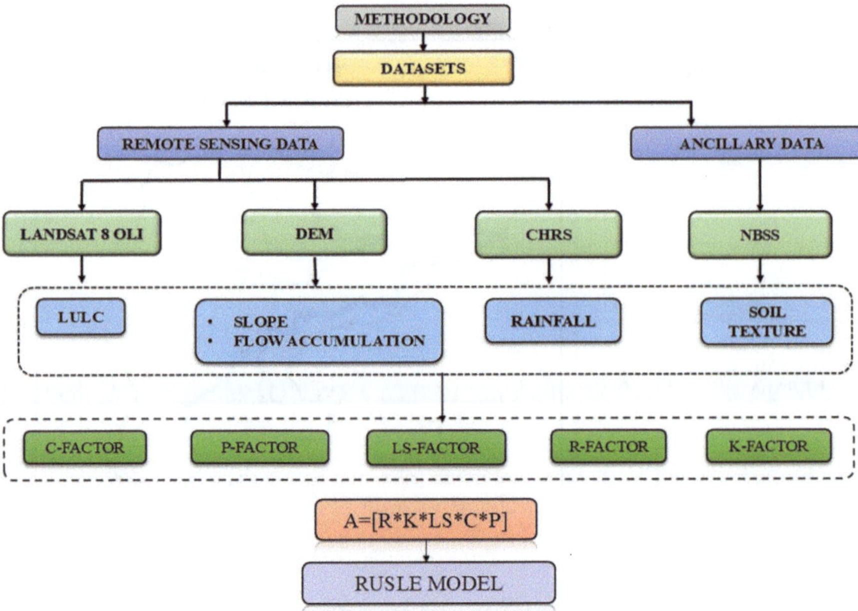

Fig. 8.2 Methodological flowchart of the research

in (Fig. 8.2). Revised Universal Soil Loss Equation is one of the most popular and widely used soil erosion models throughout the world (Udayakumara et al., 2010). Through this methodology, the study intends to find out the areas with more soil loss, especially along the foothills and river paths. This investigation highlights the urgent need to manage the land more carefully and sustainably. The results help us better understand how land is getting damaged over time. It also shows why it is important to bring together ideas like climate change adaptation, eco-resilience, and sustainable practices to protect the environment for the future.

8.4 Rainfall and Landuse/Landcover

Rainwater is one of the principal contributor for soil erosion due to surface runoff, so measuring rainfall is an influential factor for assessment soil erosion. Average rainfall for upper Bharathapuzha has been calculated for the past 20 years (2002–2022). The data has been obtained for the CHRS webportal. Human activities/Anthropogenic activities, land use change are one of the primary causes which leads to accelerated soil erosion (Borrelli et al., 2017). Landuse and land cover are one of the most significant determinants which is utilized for estimating erosion of an area. Landsat 8 image has been used to map the landuse/landcover of the region.

8.5 Rainfall Erosivity—R-factor

Rainfall is a major factor in soil erosion, and the connection between the two is measured by a factor called rainfall erosivity (Pruski & Nearing, 2002). The R factor in Revised Universal Soil Loss Equation represents the effect of precipitation rates, which leads to soil erosion. R is the cumulative rainfall with combined moving potential of particles and the highest downpour strength in 30 mm per hour (Wischmeier & Smith, 1978). It is estimated by the R factor, which leads to runoff and other raindrop impacts, with an escalation in the runoff process rising to erosion peak first and then reducing (Ran et al., 2012). The role of R determinent is very vital in the RUSLE for estimating topsoil degradation as it reflects the erosive forces that detach and transport soil particles. Higher R values indicate more erosive rainfall conditions, leading to increased soil erosion. The rainfall data of Upper Bharathapuzha, taken from the rainfall erosivity factor (R) assesses the consequences of downpour on soil erosion by considering the shifting potential associated with precipitation events. Additionally, it predicts the rate and volume of runoff generated, which is closely linked to the intensity and duration of the rainfall event (Wischmeier & Smith, 1957).

$$R = 0.0483 * P\ 1.61 \tag{8.1}$$

8.6 K-Factor

K-factor-RUSLE model is a parameter assessing the soil erodibility, indicating the proneness of a soil type to erosion. It examines soil properties comprising texture, structure, proportion of biological compounds, and permeability. An area with greater K values is highly prone to erosion reason of minimal biological compounds and coarse texture. It is basically the capacity of the top layer to withstand against erosion. K represents the amount of soil loss per unit. K factor is mapped by Tamil Nadu NBSS data. It has only two classification, and they are (i) Clay loam (ii) Sandy clay. Both soil has different characteristics and texture. Sandy clay soil tends to erode more easily due to the larger particle size, and it has lower cohesion leading to faster runoff and loss of soil. But Clay loam soil has smaller particles and higher cohesion, which can make it more resistant to erosion. Here the soil map is prepared by the National Bureau of Soil Survey (NBSS) with reference to the (Table 8.1: Soil types and their K values).

Table 8.1 Soil types and their K values

Soiltype	K value
Clay loam	0.24
Sandy clay	0.28

8.7 LS-Factor

LS-Factor in RUSLE model combines slope length and steepness to assess their impact on soil erosion. K factor offers a great opportunity to monitor soil characteristic changes and soil erodability (Jakab, 2004). These are obtained from the Digital Elevation Model. Flow accumulation can affect the effective slope length by influencing the path and distance over which water flows downslope. 30 m High resolution SRTM Digital Elevation Model has been utilized to calculate this factor as the study of (Panagos et al., 2015a, 2015b). suggests that the high-resolution DEM helps with more precision, accuracy. Estimating soil erosion with more accuracy. Areas with high flow accumulation tend to have longer effective slope length, which may result in enhanced soil erosion potential. The soil erosion extent is determined by the topographical character of the area (Fayas et al., 2019). It results in Sheet and Rill erosion (Balasubramani et al., 2015). Planting oil seeds, soybean, barley, and winter wheat can minimize the erosion risks due to their high density, and they can be preserved for thousands of years (Basic et al., 2004). Proper management of the land use factors is essential for sustainable land utilization and conservation.

$$LS = \{FA * (\text{cellsize}/23.13)\}^{0.4} * \{\text{Sin (slope of DEM } 0.017545)/0.09\} * 1.6 \tag{8.2}$$

8.8 C-Factor

C-Factor in RUSLE evaluates the type and condition of vegetation cover and land management efforts. The C factor exhibits conservation plans will affect the average yearly soil loss potential and will be dispersed during construction period, crop rotation method in agriculture or any other management activities. This can be calculated from Land utilization and cover classification (Ward & Pulido-Velazquez 2008). The land utilization and cover play an irreplaceable role in measuring soil erosion. The soil loss is highly dependent upon the area and type of vegetation cover (Benkobi et al., 1994). When compared with other factors, the C factor is ore influenced by the people intervention (Ayalew et al., 2020). Different land areas have different characteristics, while the vegetation controls the erosion by safe-guarding the soil surface from the effects of rainfall, runoff. Dense forest or grassland with good groundcover typically have low C value while the baren or sparsely vegetation area have high values. Dry scrub leaves protect soil from downpour and minimize the water running in slope (Ranzi et al., 2012). The C determinent represents the cropping system which influence the erosion, and is important for developing conservation plans. This factor is based on current land condition to the bare and empty ploughed field (Renard et al., 2017).

8.9 P-Factor

Soil loss is estimated based on the particular land use during cultivation from upslope to downslope. It is termed as conservation practice, a support determinent (Renard et al., 2017). This P relates to practices that limit the flow and minimize its effects of soil loss (Balasubramani et al., 2015). Its values are determined by slope-specific farming techniques (Shi et al., 2004). It represents the domination of the conservation and minimizes the intensity of the runoff and erosion. This factor exhibits the proportion of erosion with practice to the corresponding tillage along the slope (Dabral et al., 2008; Wischmeier & Smith, 1957). Usual support practices consist cross-slope farming, contour farming, and strip cropping, terracing, and grassed waterways. In this study, the P determinent map was created from the land utilization/ land cover and support determinents. These support methods reduce soil loss and are especially effective in dry and drought-prone areas (Ebabu et al., 2022).

$$P = 0.2 + 0.03 * S \tag{8.3}$$

Were,

S = Slope gradient %.

These are the parameters and their formula used for this soil erosion study. By incorporating these parameters, the RUSLE model enables researchers and land managers to assess soil erosion risks, prioritize conservation efforts, and develop sustainable land management strategies.

8.10 Results and Discussion

8.10.1 Rainfall

The value of the rainfall has been classified into 5 major classifications and they are very high, high, moderate, low, and very low and they are indicated in five different colours- Yellow, Teal, Green, Blue, and Dark blue. By interpreting the average rainfall map for the past 20 years, we can clarify that the rainfall slowly decreases in the east direction when nearing Tamil Nadu. The Kerala in the western part, like Malampuzha, Palakad, and Mundur, has top precipitation average of 1555–1600 mm. High rainfall in regions like Chittur, Kanjikad region receives rainfall around 1520–1560 mm. Moderate rainfall value lies between 1480 and 1520 mm, and the region that receives moderate rainfall are Walayar, Ettimadai, and Velanthavalam. The low rainfall areas of average rainfall of 1460–1490 mm are in the regions of Othakalmandapam, Pollachi, Anaimalai, and Aliyar reserve forest areas. Very low rainfall receiving areas, according to the study, are Thirumoorthy, Eastern part of

Anaimalai Tiger Reserve, and Seeposalai regions. They receive rainfall of 1440–1465 mm. The rainfall is mapped using ArcGIS software and represented as (Fig. 8.3: Rainfall map).

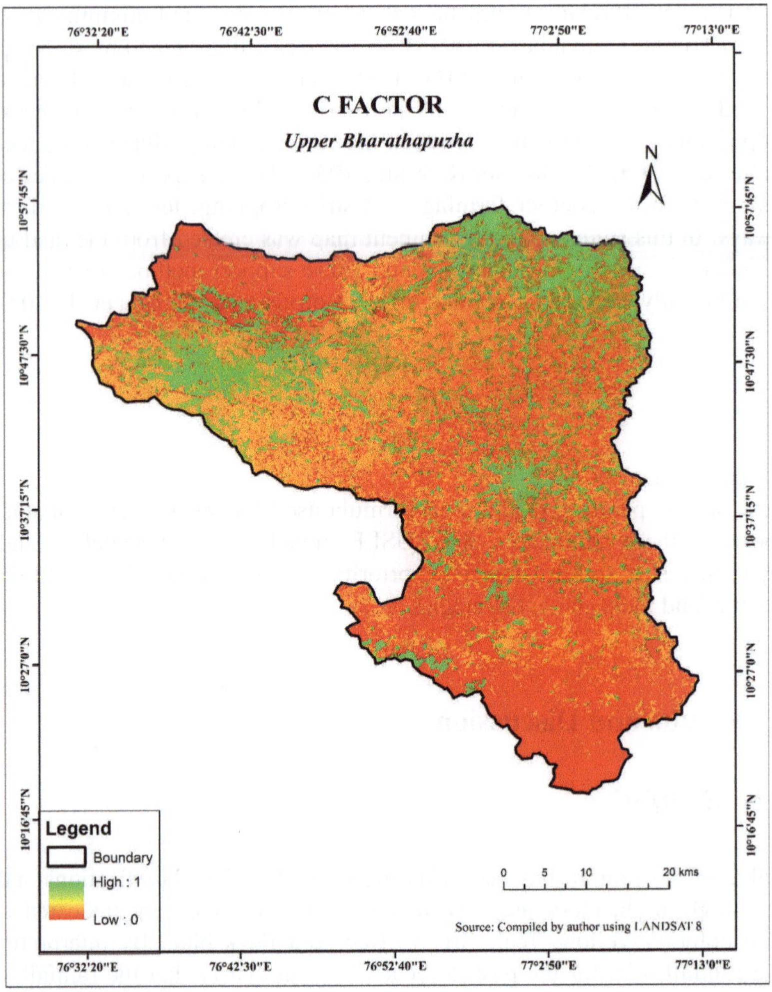

Fig. 8.3 Rainfall map

Table 8.2 Classification of land use and its percentage

S. No.	Land use	Total area in km^2	Area in percentage
1	Waterbody	40.14	1.47
2	Vegetation	915.67	33.70
3	Agricultural land	1133.08	41.71
4	Built-ups	388.96	14.31
5	Barren land	238.52	8.78

8.11 Landuse/Landcover

Different land uses exhibit varying surface characteristics such as vegetation layer, soil type, and slope, which significantly influence erosive forces acting on the soil. For instance, forests and grasslands tend to provide better soil protection compared to bare land or agricultural fields. So, studying land utilization and land cover is very important to study erosion. The land utilization and cover of the Upper Bharathapuzha is tabulated below as (Table 8.2: Classification of Landuse and its percentage).

The entire upper Bharathapuzha area is 2727.77 km^2. From the above interpretation we can clearly understand that nearly 40% of the area is coated by farming land. Next to agriculture land, the vegetation cover covers up to 33% of the total area, Buildups cover up to 14%, Barren land covers 8% of the entire area, and Waterbody cover the least area of 1.47% in the 2727.77 km^2, which is also represented as a map in (Fig. 8.4: Landuse/Landcover map).

8.12 Rainfall Erosivity (R)

The highest rainfall of the research location happens in Palakkad and Mazhampuzha, Kerala, due to the Western Ghats with its range of 800 mm. The amount of rainfall decreases when it moves towards Tamil Nadu. Pollachi, Madhukarai, Kinathukadavu, and Anaimalai regions receive moderate rainfall. Regions of low rainfall include Udukkampalayam, Kedimedu, and Erisinampatti because these regions are situated in the lee side or Rain shadow of the Western Ghats Mountain range, which can be clearly noted in Fig. 8.5.

8.13 Soil Erosivity Factor (K)

The K-factor of Upper Bharathapuzha has the majority of Clay loam soil type, which has the value of 0.24. Most of the regions have Clay loam soil which can resist erosion, Pollachi, Vaalparai, Madhukarai, have Clay soil and Kerala have clay loam soil in the regions of Parali, Palakkad, have clay loam soil which is visually represented

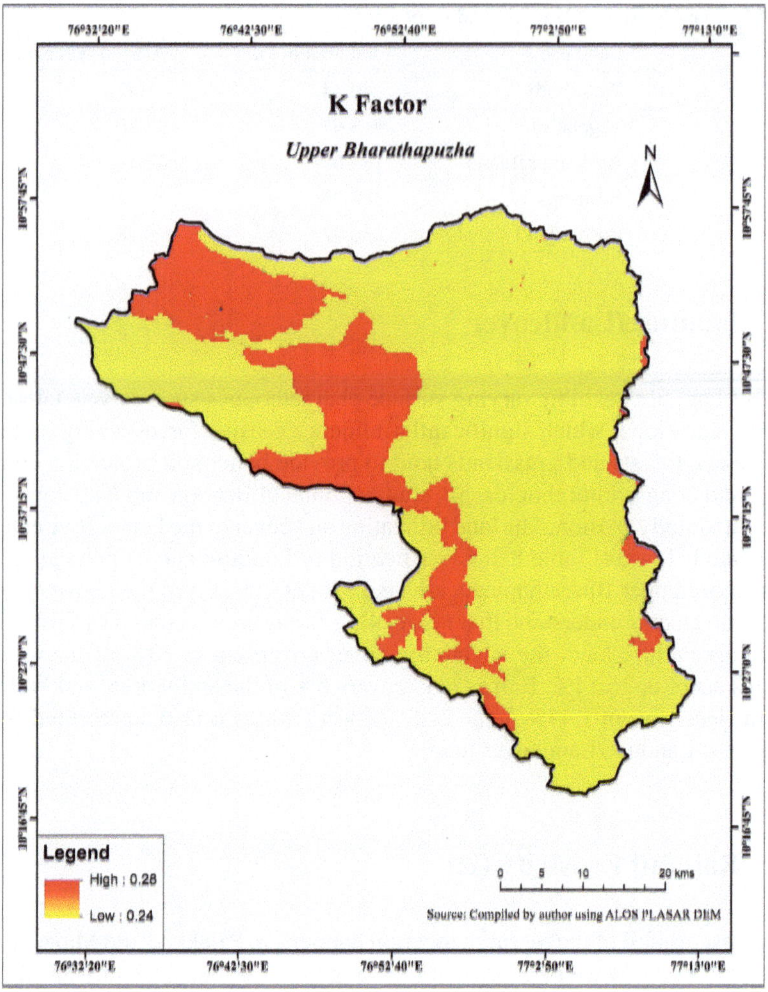

Fig. 8.4 Landuse/landcover map

as (Fig. 8.6). Clay loam soil is found in areas with moderate to good drainage, such as agricultural fields and grassland. Value 0.28 represents that Sandy clay soil is commonly present in poorer drainage areas, flood plains. In some mountainous regions, this soil can be present on slopes where weathering and sediment transport sediment deposit it on the lower slope. In the area, Sandy Clay soil is present in the foothills of Western Ghats in Malampuzha, Kozhhinjampara, Kottur, Sethumadai on the Eastern side. In the western side of the Watershed Thirumoorthy, Udukkampalayam, and Kedimedu have sandy clay soil type. The areas with Sandy clay soil type are more suspectable to erosion. Higher slope values have higher erosion rates. The other regions of the watershed have no steep slopes or mountain ranges.

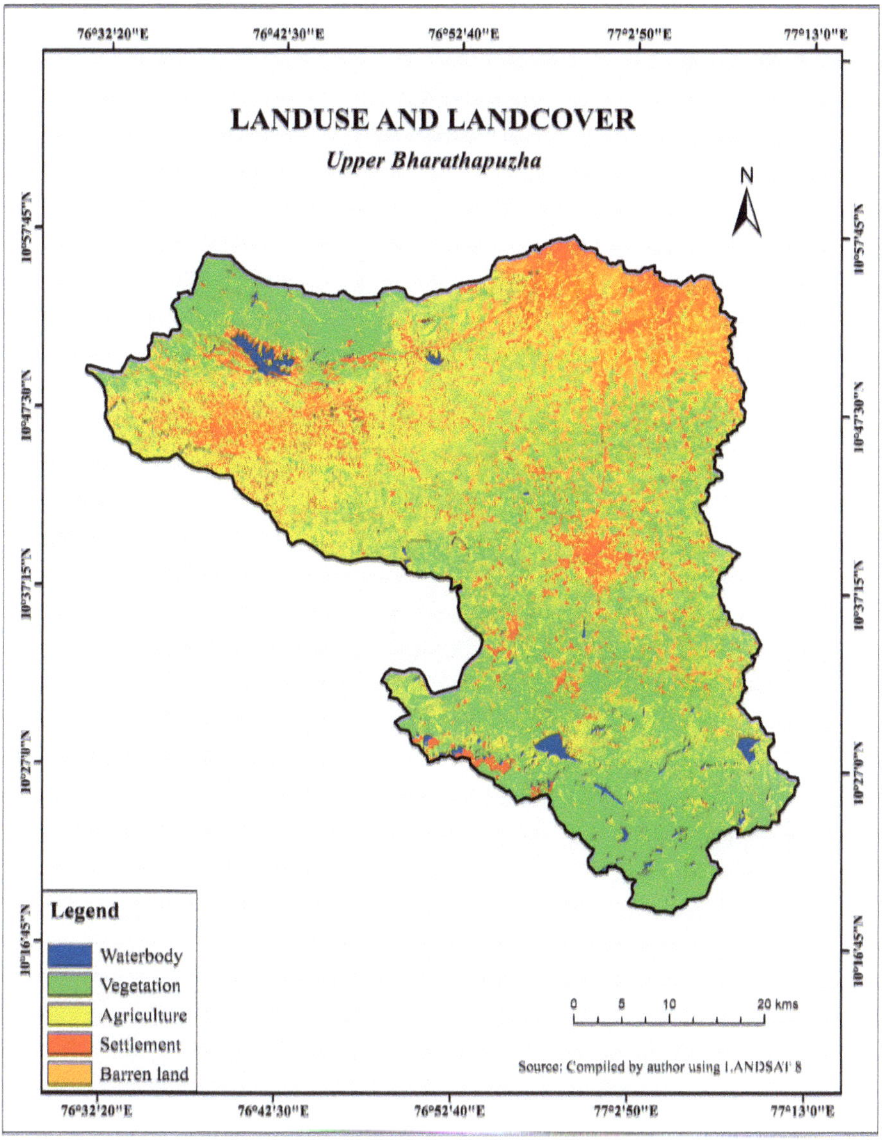

Fig. 8.5 Rainfall erosivity

8.14 Conservation Practice Factor (P)

P factor values vary from 0 to 1 Areas without conservation practices, such as open grasslands, are classified into 1, while built-up and barren lands are classified as 0. Slope map so prepared by processing the Digital Elevation Model. The slope ranges

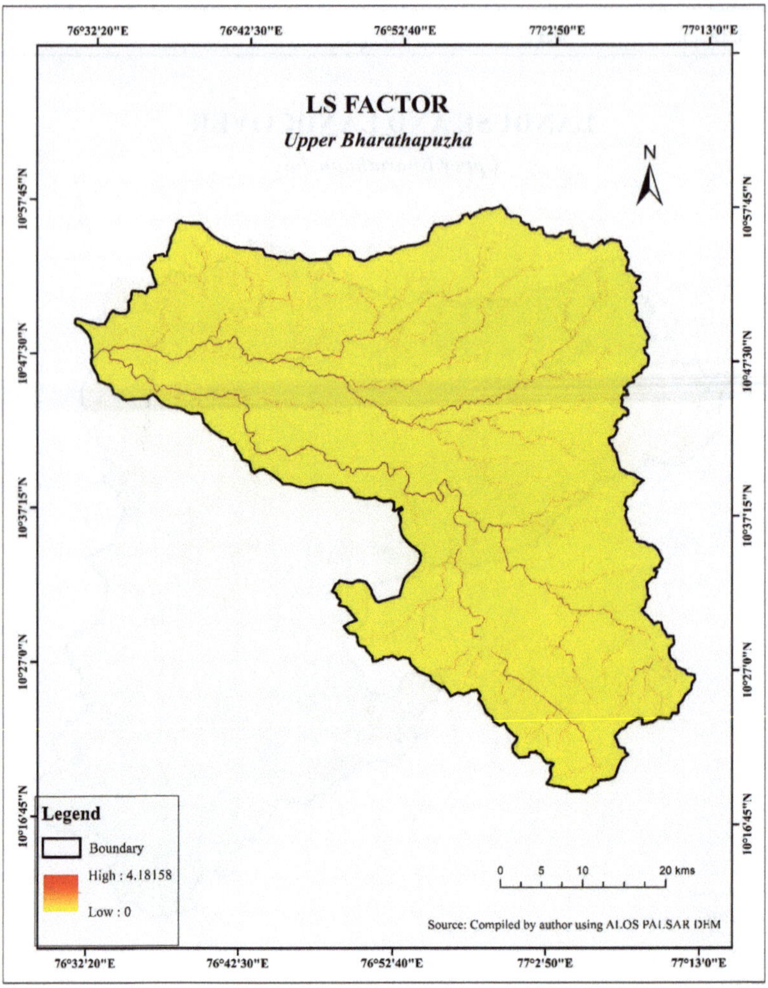

Fig. 8.6 Soil erosivity

from 0 to 89.98. These values are converted into slope gradient. P- determinent is calculated from the slope gradient. c. Longer and steeper slopes contribute more to soil erosion. In the P- factor of Upper Bharathapuzha Watershed. The high slope values are present in the mountain areas of Anaimalai Reserve Forest, Aliyar Reserve Forest in Tamil Nadu. The presence of the Western Ghats in Kerala, the slopes of these mountains are very steep. The southern area of study area and the northwestern part have a slope. The high slope values of the area indicate that the steep slope areas are easily susceptible to erosion. The interior part of the watershed won't be the susceptible region of soil erosion, and the visual representation is as Fig. 8.7.

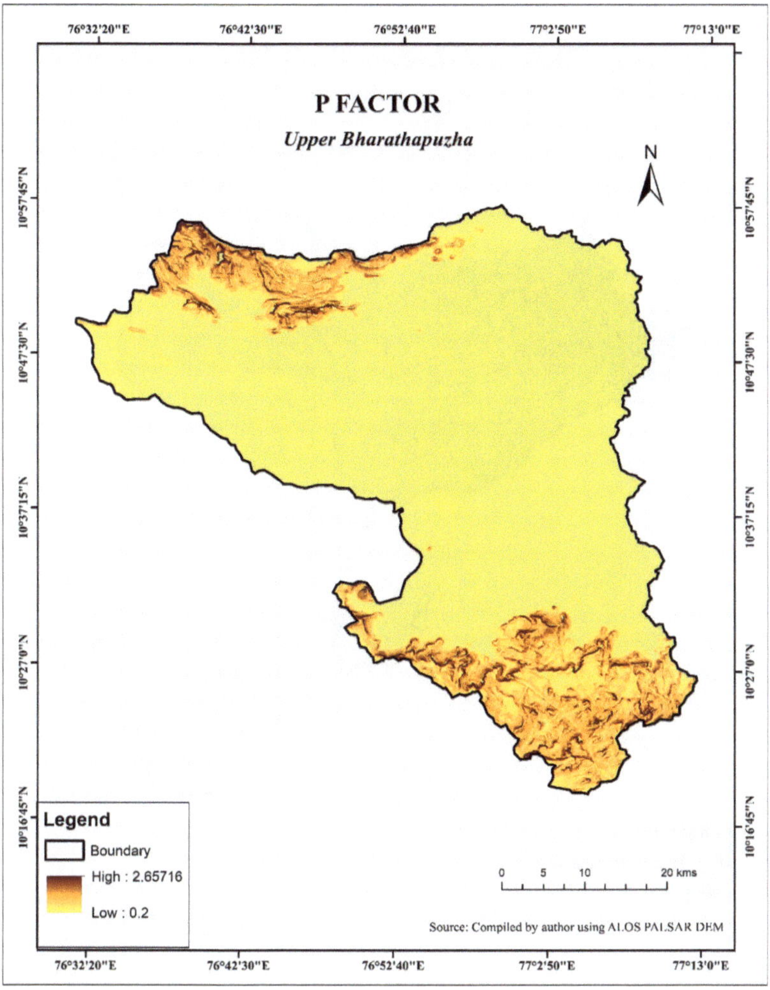

Fig. 8.7 Conservation practice factor

8.15 Slope Length and Steepness Factor (LS) and Crop Management Factor (C)

Here, flow of the river and its tributaries are clearly noticed, and the flowing direction or pathway is more inducing erosion. The value of the LS factor is high as 4.18. The mountain region is having high steep and slope length, which results in higher flow pathways down the slope, which can provide a great cause for soil erosion. The main stream flowing through the Bharathapuzha transports sediments in large scale and by which the erosion is induced and it is shown in Fig. 8.8.

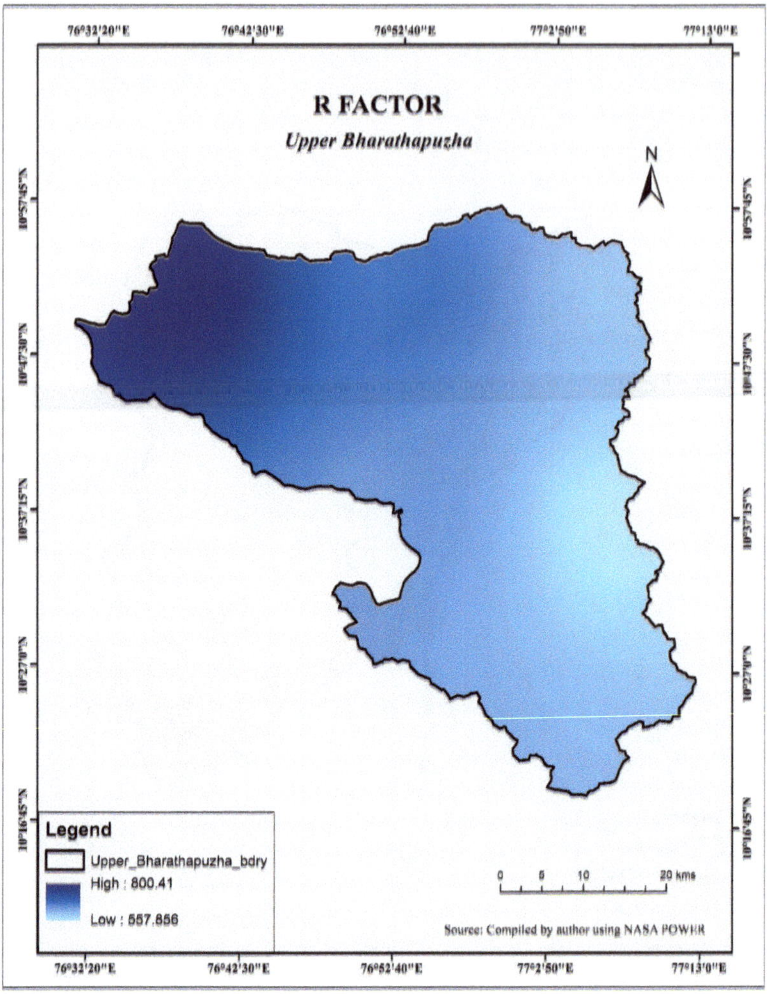

Fig. 8.8 Slope length and steepness

The study of C factor of the Bharathapuzha have a combination of both high and low values. The high vegetation areas are the densely vegetated areas like Aliyar and Anaimalai reserved forest in the southern region have a low C factor, and in the northern direction, that is Kerala's western Ghats. The high values of C factor have a high probability of erosion occurrences. These areas are mostly barren lands. The interior regions, like villages of the watershed, have a moderate erosion profile. The places with dense forest have low C values, while the barren land causes erosion because of high C values (Table 8.3 represents the Land use types and its C factor value, which is extracted from the Crop management factor map represented as Fig. 8.9).

Table 8.3 Land use and their C values

Landuse	C-Factor
Build up	1.0
Agricultural land	0.28
Barren land	0.18
Vegetation	0.004
Waterbody	0.00

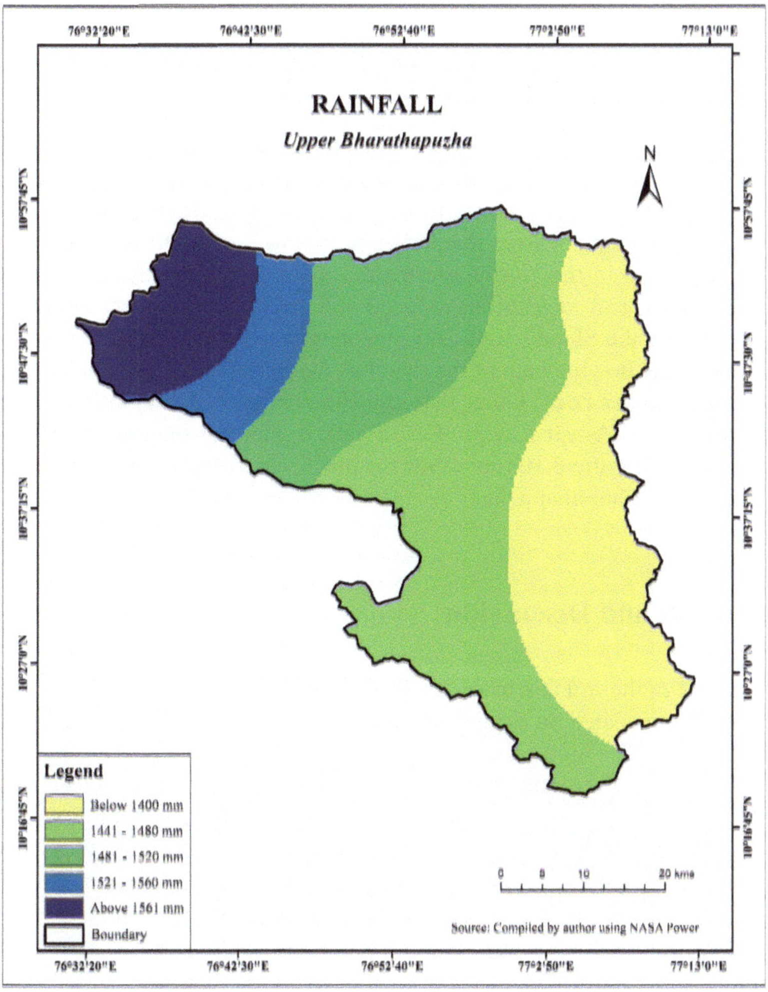

Fig. 8.9 Crop management

8.16 Soil Loss (A = R * K * L * S * C * P)

The soil loss values which are obtained using the RUSLE, depend on 5 factors, and it can change significantly because of varying climate conditions. Typical annual soil erosion was organized by different scales as exposed by Dabral et al. (2008) for Indian conditions Where A is the annual soil loss (t ha^{-1} $year^{-1}$); R is the rainfall Erosivity determinant (MJ mm h^1 ha^{-1} $year^{-1}$); K is soil Erodibility determinant [Mg ha^{-1} MJ^{-1} mm −1]; LS, C and P factors are dimensionless or they have no units. All the RUSLE determinants were multiplied using the formula, and the erosion was mapped. By analysing the final map, the majority of the regions in the Upper Bharathapuzha watershed are very less affected by soil erosion. In the regions of Anaimalai Reserve Forest and Aliyar Reserve Forest, there is no soil erosion due to dense vegetation. The vegetation significantly reduces the chance of erosion by stabilizing soil, slowing down runoff, and providing protective cover. The base of Western ghats and the barren land in the Palakkad and Mazhampuzha regions of Kerala, and Barren land in areas like Pollachi, foot hills of Anaimalai, and the larger barren area present in the Northeastern side of the study area have moderate soil erosion due to lack of vegetation, reduced soil stability, increased surface runoff. When comparing the Slope length and Steepness determinent and Soil loss map, it concludes that the flowing path of the river has the runoff and as a result of this event has erosion in higher rates. Other than the Flowing path the regions like foothills of mountains of the Aliyar Reserve Forest have higher erosion rates which is up to 1566.1 t h^{-1} y^{-1}. Figure 8.10 represents the areas which are more vulnerable to soil loss in the Upper Bharathapuzha.

8.17 Result and Discussion

The projection of the soil loss using the Revised Universal Soil Loss Equation model in this area has revealed an extensive variety of soil loss, which is maximum upto 1566.1 t h^{-1} y^{-1}. The high erosion rates are found in foothills of the western ghats, Anaimalai mostly. The slope gradient and surface runoff in these areas contributed to the soil loss rate. Lower erosion rates are found in the agricultural areas, areas with dense vegetation cover, where there is better land management. Areas receiving high rainfall correspond to high soil loss, during high rainfall events, the particles are transported downhill. The high rainfall of the area is 1600 mm, and the lowest is 1460 mm. The Land use and land cover characteristics play a crucial role in determining the intensity and distribution of erosion. Their result indicates that forest areas show the lowest soil loss, which acts as a natural shield, and barren lands contribute to indirect runoffs. RUSLE, factors exhibited a significant influence in this study. Rainfall erosivity- R factor, with higher rainfall intensities continued by greater soil detachment. K factor or Soil Erodability revealed that clay soil with k value of 0.24 is found in exceeding 80% of the research area, and sandy soil

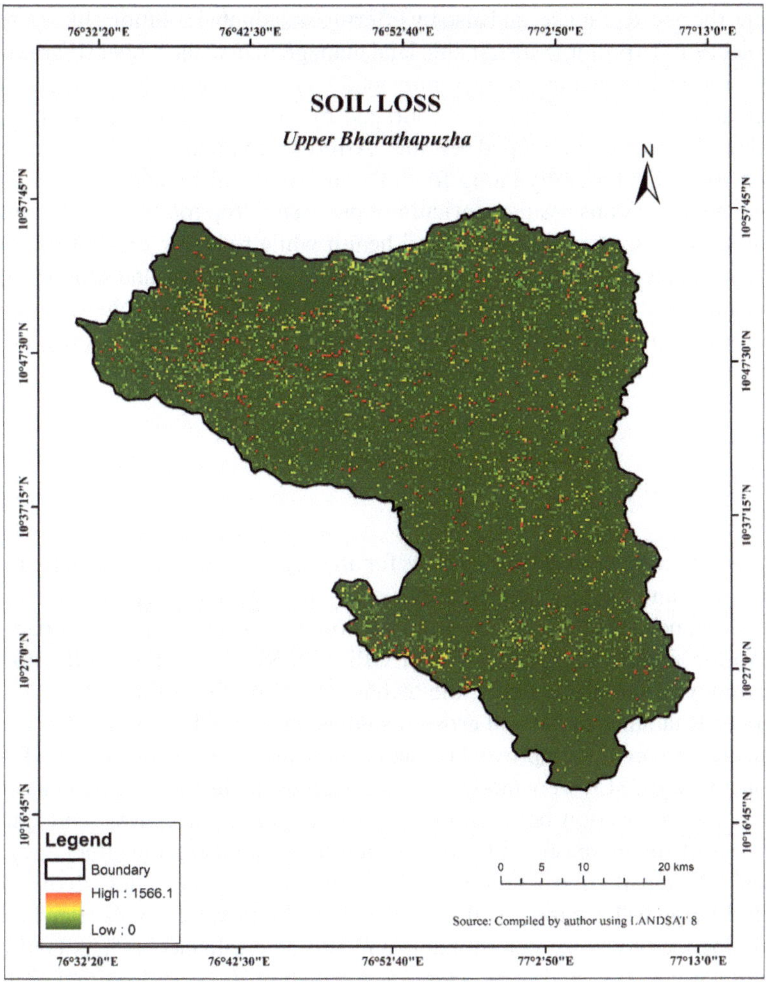

Fig. 8.10 Soil loss

with 0.24–0.28 is found in high-elevation and slope areas. LS factor is the dominant driver in areas with steep slopes and rough terrain, especially along foothills and hilly landscapes. C factor, in dense forest areas and high vegetation, shows lower c values, indicating better protection of erosion, while barren lands with high erosion rates with high c values. P factor reflects the soil conservation measures that can reduce the water runoff to reduce soil erosion. Value 1 indicates low practice with buildup, and 0 indicates high practice, like grassland and vegetation. The result finds that soil erosion in the upper Bharathapuzha region is majorly influenced by the above-mentioned factors. The study directly contributes to the book's core theme by demonstrating how soil erosion, which affects the land-by-land degradation, climate variability, which threatens the eco-resilience, and also by offering sustainable solutions like restoration and protection. To promote sustainable land management in the Upper Bharathapuzha region, practical measures are recommended based on the study findings. Contour farming and terracing should be encouraged by the slow down surface runoff and minimize soil erosion on sloped terrains. Afforestation and vegetative barriers can stabilize the soil, especially along foothills and riverbanks where erosion risks are high and adopting conservation agriculture practices, crop rotation, and maintaining continuous soil cover will improve soil health while reducing erosion risks. These strategies together can build eco-resilience, reduce degradation, and support the long-term sustainability of the region. The study emphasises climate change, especially in the form of increased rainfall. The linkage of the conservation practice factor exacerbates land management strategies.

8.18 Conclusion

Soil is one of the most important sources for all organisms to live, especially for food, maintenance, and the functioning of the ecosystem. Management of soil decreases soil erosion and surface runoff, so tracking soil erosion risk is a crucial part of soil conservation practices. Combination of GIS with RUSLE improved the efficiency and accuracy of the erosion risk; the technology also allows for wider study areas like Upper Bharathauzha in soil erosion studies and provides the necessary tools to analyse these in order to improve the management and conservation of soil. Different components of RUSLE were modelled using various mathematical formulae and GIS to assess the connection between erosion susceptibility, steepness, and Land Use maps. Loss of soil increases with slope, which shows total erosion after heavy downpour with slope particularly to greater slopes and rainfall intensities, so the foothills of the hilly regions have more soil erosion risk. C factor varies to place and it is very much influenced by some determinants and sub-determinants such as land utilization, surface cover, surface roughness, and moisture. It is advised that soil losses of up to 25 t ha^{-1} yr^{-1} are possibly considered tolerable due to the natural ability of these ecosystems to regenerate soil over time. Support practices are essential for controlling soil erosion, especially in places with steep angles and high erosion risks. Managing the cover and crops, represented by the C factor, can effectively limit soil erosion in farmland, helping to retain nutrients. Measures like expanding plant layer margins, maintaining stone walls, and practicing relatable farming can substantially lower soil loss in cultivated lands. Future studies can consider focusing on investigating areas suggested for conservation purposes, Applying ustep by step changes to make the model better and enhance the quality of recommendations as it leads to restoration of topsoil, which can increase the productivity of agriculture, forest, and grassland ecosystems. The utilization of the Revised Universal Soil Loss Equation (RUSLE) in assessing erosion in the Upper Bharathapuzha Basin has provided valuable insights into erosion dynamics. By considering factors that include the rainfall erosivity, soil erodibility, slope length, management practices, conservation efforts, the study mapped soil erosion susceptibility across various landscapes. Additionally, the analysis suggests that factors such as slope length and steepness influence erosion rates, with higher rates observed along the flowing path of the river and in mountainous regions. This data assists and supports planners dominate the decline in soil productivity, minimize expenses caused by erosion, and optimize the benefits of precipitation. The study about erosion and data about it helps the planners and researches to control the degradation of productive soils, optimizing it and decreasing expense. Soil erosion has been a very critical situation in the countries like China, Philippines, Costa Rica according the study of (Ananda, & Herath, 2003), so India is also a very developing country in every field so it also result in losses of certain resources and conservation practices should be addressed and followed to manage the land resources and sustainability. Also, the population growth can promote more intensive agricultural practices and poor land management, which can induce more erosion. So the land management is a key challenge that needs to be noted in the study area. These insights underscore the need for targeted conservation measures, which are eco-friendly and long-term land care practices to mitigate soil risks and protect soil resources at the Upper Bharathapuzha Basin.

References

Ananda, J., & Herath, G. (2003). Soil erosion in developing countries: A socio-economic appraisal. *Journal of Environmental Management, 68*(4), 343–353.

Ayalew, D. A., Deumlich, D., Šarapatka, B., & Doktor, D. (2020). Quantifying the sensitivity of NDVI-based C factor estimation and potential soil erosion prediction using Spaceborne earth observation data. *Remote Sensing, 12*(7), 1136.

Balasubramani, K., Veena, M., Kumaraswamy, K., & Saravanabavan, V. (2015). Estimation of soil erosion in a semi-arid watershed of Tamil Nadu (India) using revised universal soil loss equation (rusle) model through GIS. *Modeling Earth Systems and Environment, 1*, 1–17.

Basic, F., Kisic, I., Mesic, M., Nestroy, O., & Butorac, A. (2004). Tillage and crop management effects on soil erosion in central Croatia. *Soil and Tillage Research, 78*(2), 197–206.

Benkobi, L., Trlica, M. J., & Smith, J. L. (1994). Evaluation of a refined surface cover subfactor for use in RUSLE. *Rangeland Ecology & Management/Journal of Range Management Archives, 47*(1), 74–78.

Bhat, S. A., Dar, M. U. D., & Meena, R. S. (2019). Soil erosion and management strategies. Sustainable Management of Soil and Environment, 73–122.

Borrelli, P., Robinson, D. A., Fleischer, L. R., Lugato, E., Ballabio, C., Alewell, C., Panagos, P., et al. (2017). An assessment of the global impact of 21st century land use change on soil erosion. *Nature Communications, 8*(1), 1–13.

Dabral, P. P., Baithuri, N., & Pandey, A. (2008). Soil erosion assessment in a hilly catchment of North Eastern India using USLE, GIS and remote sensing. *Water Resources Management, 22*, 1783–1798.

Ebabu, K., Tsunekawa, A., Haregeweyn, N., Tsubo, M., Adgo, E., Fenta, A. A., Meshesha, D. T., Berihun, M. L., Sultan, D., Vanmaercke, M., & Poesen, J. (2022). Global analysis of cover management and support practice factors that control soil erosion and conservation. International *Soil and Water Conservation Research, 10*(2), 161–176.

Fayas, C. M., Abeysingha, N. S., Nirmanee, K. G. S., Samaratunga, D., & Mallawatantri, A. (2019). Soil loss estimation using ruslemodel to prioritize erosion control in KELANI river basin in Sri Lanka. *International Soil and Water Conservation Research, 7*(2), 130–137, ISSN 2095–6339.

Ghosal, K., & Das Bhattacharya, S. (2020). A review of RUSLE model. *Journal of the Indian Society of Remote Sensing, 48*, 689–707.

Jakab, Á. K. T. H. G. (2004). The effect of soil physical parameters on soil erosion. *Geographical Bulletin, 53*(1–2), 77–84.

John, J., Rosamma, C. N., & Thampi, S. G. (2022). Assessment and prediction of soil erosion and its impact on the storage capacity of reservoirs in the Bharathapuzha River Basin, India. *Environmental Modeling & Assessment, 27*(1), 77–103.

Kebede, Y. S., Endalamaw, N. T., Sinshaw, B. G., & Atinkut, H. B. (2021). Modeling soil erosion using RUSLE and GIS at watershed level in the upper beles, Ethiopia. *Environmental Challenges, 2*, Article 100009.

Lin, B. S., Chen, C. K., Thomas, K., Hsu, C. K., & Ho, H. C. (2019). Improvement of the K-factor of USLE and soil erosion estimation in Shihmen reservoir watershed. *Sustainability, 11*(2), 355.

Magesh, N. S., Jitheshlal, K. V., Chandrasekar, N., et al. (2013). Geographical information system-based morphometric analysis of Bharathapuzha river basin,Kerala, India. *Applied Water Science, 3*, 467–477.

Nair, G. S., & Jayaprakash, T. (2021). Land use and climate change impacts on watershed hydrology of Bharathapuzha River Basin: A swat model application, India. *International Journal of Electrical Engineering and Technology (IJEET), 12*(1), 182–208.

Panagos, P., Borrelli, P., & Meusburger, K. (2015a). A new European slope length and steepness factor (LS-Factor) for modeling soil erosion by water. *Geosciences, 5*(2), 117–126.

Panagos, P., Borrelli, P., Meusburger, K., Alewell, C., Lugato, E., & Montanarella, L. (2015b). Land use policy estimating the soil erosion cover-management factor at the European scale. *Land Use Policy, 48*, 38–50.

Patriche, C. V. (2017). Soil erosion modelling. Landform Dynamics and Evolution in Romania, 397–423.

Pimentel, D. (2006). Soil erosion: A food and environmental threat. *Environment, Development and Sustainability, 8*, 119–137.

Prasannakumar, V., Vijith, H., Abinod, S., & Geetha, N. J. G. F. (2012). Estimation of soil erosion risk within a small mountainous sub-watershed in Kerala, India, using Revised Universal Soil Loss Equation (RUSLE) and geo-information technology. *Geoscience Frontiers, 3*(2), 209–215.

Pruski, F. F., & Nearing, M. A. (2002). Runoff and soil-loss responses to changes in precipitation: A computer simulation study. *Journal of Soil and Water Conservation, 57*(1), 7–16.

Ran, Q., Su, D., Li, P., & He, Z. (2012). Experimental study of the impact of rainfall characteristics on runoff generation and soil erosion. *Journal of Hydrology, 424*, 99–111.

Ranzi, R., Le, T. H., & Rulli, M. C. (2012). A RUSLE approach to model suspended sediment load in the Lo river (Vietnam): Effects of reservoirs and land use changes. *Journal of Hydrology, 422*, 17–29.

Renard, K. G., Foster, G., Weesies, G., McCool, D., & Yoder, D. (1997). *Predicting soil erosion by water: A guide to conservation planning with the revised universal soil loss equation (Rusle)* (vol. 703). United States Department of Agriculture.

Renard, K. G., Laflen, J. M., Foster, G. R., & McCool, D. K. (2017). The revised universal soil loss equation. In *Soil erosion research methods* (pp.105–126). Routledge.

Sharda, V. N., Mandai, D., & Ojasvi, P. R. (2013). Identification of soil erosion risk areas for conservation planning in different states of India. *Journal of Environmental Biology, 34*(2), 219.

Shi, Z. H., Cai, C. F., Ding, S. W., Wang, T. W., & Chow, T. L. (2004). Soil conservation planning at the small watershed level using RUSLE with GIS: A case study in the Three Gorge Area of China. *CATENA, 55*, 33–48.

Telkar, S. G., Solanki, S. P. S., Dey, J. K., & Kant, K. (2015). Soil erosion: Types and their mechanism. *International Journal of Economic Plants, 2*(4), 178–180.

Thomas, J., Joseph, S., & Thrivikramji, K. P. (2018). Assessment of soil erosion in a tropical mountain river basin of the southern Western Ghats, India using RUSLE and GIS. *Geoscience Frontiers, 9*(3), 893–906.

Udayakumara, E. P. N., Shrestha, R. P., Samarakoon, L., & Schmidt-Vogt, D. (2010). People's perception and socioeconomic determinants of soil erosion: A case study of Samanalawewa watershed, Srilanka. *International Journal of Sediment Research, 25*(4), 323–339.

Verheijen, F. G., Jones, R. J., Rickson, R. J., & Smith, C. J. (2009). Tolerable versus actual soil erosion rates in Europe. *Earth-Science Reviews, 94*(1–4), 23–38.

Ward, F. A., & Pulido-Velazquez, M. (2008). Water conservation in irrigation can increase water use. *Proceedings of the National Academy of Sciences, 105*(47), 18215–18220.

Wischmeier, W. H., & Smith, D. D. (1978). Predicting rainfall erosion losses: a guide to conservation planning (No. 537). Department of Agriculture, Science and Education Administration.

Wischmeier, W., & Smith, D. (1957). Factors affecting sheet and rill erosion. *Transactions. American Geophysical Union, 38*(6), 889–896.

Chapter 9
Land Degradation-Climate Change Nexus in the Rural–Urban Front of Bamenda, Cameroon

Mildred Endam Acha and Suiven John Paul Tume

Abstract The global population has surpassed 8 billion, with projections that 70% will reside in urban areas by 2050. The unchecked urbanization and unsustainable development practices are leading to significant land degradation and contributing to climate change. Development actors often focus on enhancing infrastructure to accommodate the growing urban population, which unintentionally intensifies land degradation. Despite the importance of this issue, the land degradation and climate change nexus is rarely explored. To address this gap, this chapter investigates development actions contributing to land degradation, the connection between land degradation and climate change, the associated challenges, and potential measures to combat land degradation. A mixed research design was employed, using both qualitative and quantitative methods. A total of 200 questionnaires were distributed using purposive sampling techniques, supplemented by structured interviews and focus group discussions to further explore the connections. The findings reveal that intensified human activities and changing consumption patterns significantly contribute to land degradation. Additionally, there is a notable relationship between ecological dynamics—evidenced by the loss of forests and wetlands—and seasonal as well as inter-annual climate variability, which leads to reduced water availability and agricultural productivity. The chapter recommends rigorous monitoring of human activities and encourages a shift toward sustainable practices. It advocates for promoting green consumption and green agriculture, implementing landscape restoration initiatives (such as planting fruit trees in residential areas), and establishing buffer zones around wetlands to mitigate climate change and environmental degradation.

Keywords Land degradation · Climate change · Development actions · Green practices · Nature-based solutions · Vulnerability

M. E. Acha (✉) · S. J. P. Tume
Department of Geography and Planning, The University of Bamenda, Bamenda, Cameroon
e-mail: yivisiendam@gmail.com

U. Chatterjee et al. (eds.), *Eco-Resilience: Climate Change, Land Degradation and Sustainable Solutions*, Sustainability Solutions,
https://doi.org/10.1007/978-3-032-00708-7_9

9.1 Introduction

Land degradation is a widespread problem, yet key uncertainties remain concerning its extent, severity, and linkages to climate change (Reed & Stringer, 2016). Despite the problems of measuring the extent and severity of land degradation, given its complex nature, climate change and land degradation remain some of the biggest and most urgent challenges for humanity (Phoenix, 2025). The current global extent, severity, and rates of land degradation are not well quantified. There is no single method by which land degradation can be measured objectively and consistently over large areas because it is such a complex and value-laden concept. Existing scientific and locally based approaches, including the use of indigenous and local knowledge, can assess different aspects of land degradation or provide proxies as well as remote sensing data (International Network on Appropriate Technology, 2025; Olsson et al., 2019).

Land degradation is a function of human-induced actions, with significant implications on biodiversity, soil fertility, and overall health conditions (OSS, 2024). This accelerating degradation compromises food insecurity, posing a serious threat to the growing global population (IPBES, 2019; OSS, 2024). Land degradation is a problem of global dimensions and affects ecosystems (IPBES, 2019). The degradation of the Earth's land surface negatively impacts the well-being of at least 3.2 billion people, pushing the planet towards a sixth mass species extinction, and costing more than 10% of the annual global loss of biodiversity and climate change (IPBES, 2019). This warrants urgent and concerted action, without which land degradation will worsen in the face of population growth, unprecedented consumption, and an increasingly climate change and globalised economy (IBPES, 2018). Land degradation impacts transcend globally and influence other factors of degradation, such as rainfall patterns and the occurrence of extreme weather events such as droughts or floods (OSS, 2024). Land is of great importance, providing a variety of needs in our ecosystem, such as water, food, shelter, clean air, and biodiversity conservation (Filho et al., 2021). Human-induced negative impacts on land are apparent and have led to land degradation (OSS, 2024; Umoru, 2023). Although several natural drivers might be root causes, anthropogenic factors significantly trigger this phenomenon. Unsustainable agriculture, urbanisation, deforestation, pollution, and overexploitation of water resources lead to land degradation (OSS, 2024).

Over half of the world's population lives in urban areas, and is predicted to increase to 70% by the middle of the 21st Century (Phoenix, 2025; United Nations, 2015). This is a threat to land degradation and ecosystem services (Setoa et al., 2012). This is worsened by the fact that as cities expand, the management of ecosystem services to prevent land degradation is always neglected in the planning processes (Olsson et al., 2019). On the contrary, economic development and urban expansion are prioritised (Phoenix, 2025). The growth in built-up space has increased impervious surfaces; as such, 40–83% of rainfall becomes surface water runoff (Pataki et al., 2011).

Rainfall intensity is predicted to increase in many parts of the world under climate change (Nishioka, 2024; Royal Society, 2016), associated with increased runoff, which will get worse. Urbanisation, land degradation, and climate change are therefore strongly interlinked, suggesting the need for common solutions (Reed & Stringer, 2016). Without environmental education and protection of susceptible areas, environmental degradation will continue (Gimah & Bodo, 2019a, 2019b).

Measures to reverse land degradation are available, but economic, political, institutional, legal, and socio-cultural barriers, including lack of access to resources and knowledge, restrict their implementation (Olsson et al., 2019). Proven strategies to reverse land degradation include tenure reform, tax incentives, payments for ecosystem services, participatory integrated land-use planning, farmer networks, and rural advisory services. Deferred action increases the costs of addressing land degradation and can lead to biophysical and human losses (Olsson et al., 2019). Since the 1850s, about 35% of human-caused CO_2 emissions to the atmosphere emanate from land as a combined effect of land degradation and land-use change (Foley et al., 2005), and about 38% of the Earth's land area has been converted to agriculture (Foley et al., 2011). Combatting land degradation is an urgent priority to protect the biodiversity and ecosystem services that are vital to all life on Earth and to ensure human well-being. Land degradation represents an economic loss in the order of 10% of annual global gross product (IBPES, 2018; IPBES, 2019). Restoring degraded lands is essential to enhance water and food security, increase employment, avoidance of conflict, improve gender equality, migration, and meet the global sustainable development goals (IPBES, 2019).

Emerging and developing economies are increasingly witnessing rising consumption, which is dominantly driving land degradation (IBPES, 2018; IPBES, 2019). While the unsustainable management of arable land is currently the most extensive direct driver of land degradation, climate change exacerbates the impacts, limiting options for addressing land degradation (IPBES, 2019; OSS, 2024). Land degradation has been recognised for over 100 years in Africa, with concerns of Sahel desertification becoming prominent in the 1970s (IBPES, 2018; OSS, 2024).

During the twentieth century, the average temperature on the continents increased by 1.5 °C compared to the pre-industrial era, exceeding the global average of 1.1 °C (Nishioka, 2024). This increase from 1991 to 2020 affected Sub-Saharan Africa, with hikes of extreme heat in North Africa. World Bank projections reveal that Sub-Saharan Africa will continue to experience higher temperatures (Nishioka, 2024; OSS, 2024), leading to land degradation and inefficiency of ecosystem services. The traditional societies in Africa before the 1950s believed in the preservation of the environment against all forms of adulteration and misuse (Nduka & Orisakwe, 2009). Most forests were inaccessible or preserved for special purposes, as some were called evil forests or reserves for their gods to scare people from using the forest resources. The streams and rivers were also protected against all forms of misuse during the period. Farmers practised shifting cultivation to allow the use of land to regain soil nutrients before returning after many years (Bodo & Gimah, 2018; Gimah & Bodo, 2019a, 2019b). Unfortunately, there is a considerable shift in our present society, with natural resources being used up without any consideration for preservation or sustainable usage (Nishioka, 2024). In recent years, this balance has been seriously threatened by excessive demand for fuel, exceeding carrying capacity in arable land, industrial mining, and housing facilities, resulting in devastating costs like oil spills, overheating, landslides, flooding, and degrading land (Patnaik, 2018).

Rapid urban growth and unsustainable consumption practices are ravaging Cameroon's urban areas. Bamenda doubles as the regional capital for the Northwest and the Mezam Division, harbouring a good number of the regional delegations and major educational institutions (both secondary and higher institutions), and is the focal point for business in the region (Acha & Fombe, 2023). Given these opportunities offered by the Bamenda Municipality and its proximity to most villages of the North West Region, it has witnessed an influx of population and increased socio-economic activities. The intensification of urban functions within the city core and urban fringe over time results in land use change and degradation. Development proponents are more focused on socio-economic development to enhance the welfare of the increasing population, as such development actions undermine land protection, consequently influencing climate change. This accounts for only a negligible fraction in the literature, with previous studies focused more on land use dynamics and wetland degradation (Acha & Fombe, 2023; Al-Kindy, 2025; Kimengsi et al., 2017; Kometa & Ndi, 2012); land use dynamics and variations in sprawl (Fombe & Acha, 2020); gentrification patterns and development implications (Acha & Fombe, 2023), and farm-based climate adaptation dynamics insights from the vegetable sector (Kimengsi et al., 2022). Little has been done to establish the nexus between land degradation and climate change, which constitutes the focus of this chapter.

9.2 Rationale

Climate change and land degradation are recurrent phenomena that characterise major cities of the world, increasing the vulnerability of urban landscapes, producing a negative ecological footprint, and having extensive and frequently irreversible implications on the environment and climate (Webb et al., 2017). The twentieth century is marked by rapid urban growth, wherein the natural landscapes are pressured and rapidly degrade, creating brown cities that impact the climate negatively. The recurrent influx of population and rapid transformation within the urban and peri-urban zone of Bamenda are accompanied by climate-incompatible developments and land degradation, increasingly threatening liveability within the urban landscape. The growing population is increasingly adopting unsustainable consumption patterns and community practices contrary to SDG 12. These warrant planning options and policy actions that counteract the negative effects of urban development, including mitigating climate change (SDG13), promoting compact development patterns to facilitate land conservation (SDG 15), reducing hunger (SDG 2) promoting the well-being of the population in line with SDG 11 Agenda 21. This will help achieve Cameroon's Vision 2035, challenge number 4, focused on environmental protection. The chapter aims to examine development actions that degrade land, the nexus between land degradation and climate change, implications, and measures to facilitate land conservation and minimise climate change. It is guided by the following specific questions: (1) What are the development actions that degrade land in Bamenda?, (2) What is the relationship between climate change and land degradation?, (3) What are the associated implications and potential measures to address land degradation?

9.3 Study Area and Methods

9.3.1 Study Area

This study was carried out in Bamenda, the administrative seat of the Northwest Region of Cameroon. It is made up of the Bamenda I, II, and III municipalities, located between Latitude 5°94″ and 5° 98″ North of the Equator and Longitude 10.15° and 10.18° East of the Greenwich Meridian (Fig. 9.1), and at an elevation 1258 m above sea level (Master Plan of Bamenda City Council, 2011–2027).

Bamenda is bounded to the west and southwest by Momo Division and Bali, respectively. To the north, it is flanked by Bafut, to the northeast by Tubah, and to the south by Santa.

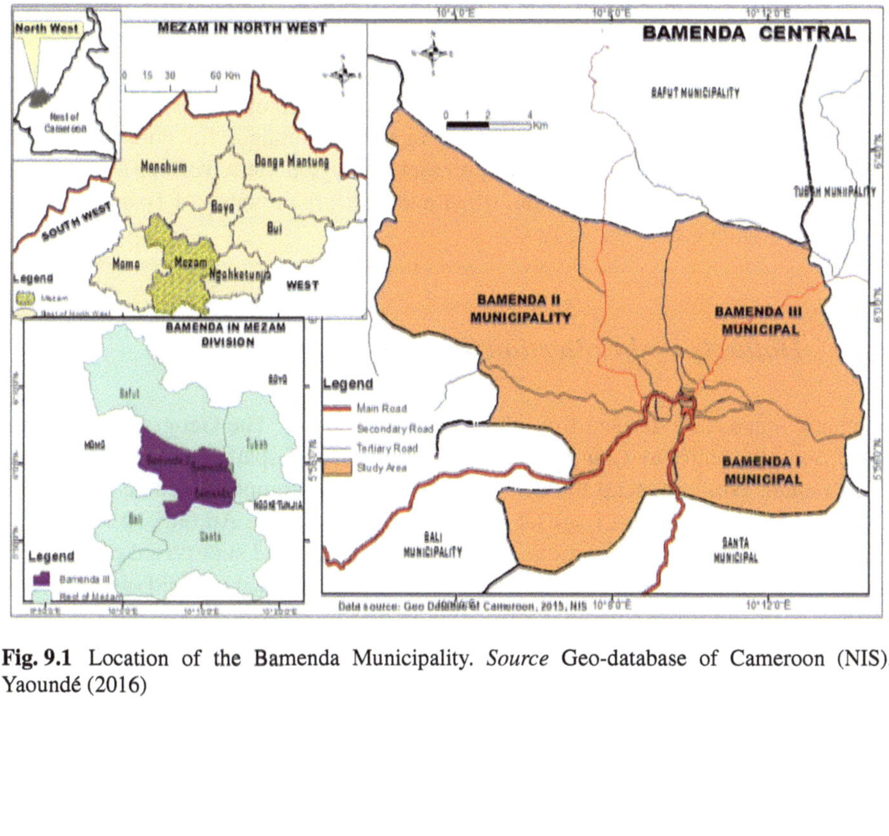

Fig. 9.1 Location of the Bamenda Municipality. *Source* Geo-database of Cameroon (NIS), Yaoundé (2016)

This study employed the survey, descriptive, and correlational designs to unravel development actions that degrade land and establish the nexus with climate change. Recurrent field surveys within the urban core and fringe zone revealed development practices that degrade land and climate outcomes. Focus group discussions and interviews with farmers and urban dwellers were conducted to uncover their actions in land degradation and efforts in enhancing land conservation to minimise climate change. Purposive and random sampling techniques were employed in questionnaire administration. The snowball technique was used to identify key focal points selected for this study, among which were church premises where people from different neighbourhoods converge and meeting houses.

9.3.2 Data Sources

Secondary data from Council Development Plans and project records of institutions provided the basis for a literature review on development practices that promote land degradation and interfere with climate-resilient city growth. Farmers were asked to reveal the frequency of fertiliser and pesticide application and trends for common crop types cultivated. Land cover change maps were produced to show the spatio-temporal dynamics of different land uses as an indicator of land degradation. Published papers from diverse scientific journals and dissertations regarding land degradation and climate change were vital. The exploratory design consists of the exploration and understanding of past phenomena from available data and Landsat images. It enabled the researcher to make an inventory of the demographic change and dynamics in the urban landscape concerning evolution in Landsat images for landmark years, valuable to this study. The study explored meteorological data from 1971 to date to present variations in climate in the urban and peri-urban zones of Bamenda. This was complemented by interviews.

9.3.3 Datasets and Collection Tools

Thematic maps were derived from the secondary sources. The Geographic Position System was the major tool used to detect changes to enable spatial landscape mapping in Bamenda. GIS enhanced the detection of the spatio-temporal density of vegetal loss and climate outcomes. Landsat images were classified in ArcGIS 10.5 using the supervised classification technique, using training areas and supervised pixel-based image classification. Land cover classes were attributed based on field observations, and real-time high-resolution Google Earth images of Bamenda were used to assess the accuracy of the attributed land covers. Diachronic data for land cover analysis from 1971 to the present with Landsat images were utilised. The GPS instrument was used to calculate the spatial–temporal dynamics of identified land uses. The trend in crop yield was analysed for specific portions of land to infer soil quality change over time.

9.4 Results

9.4.1 Development Actions that Degrade Land

This section focuses on the anthropogenic actions that degrade land in Bamenda. A large array of human actions triggers land degradation in Bamenda (Fig. 9.2 and Table 9.1).

Over the years, the urban and peri-urban zones of Bamenda have witnessed rapid urbanisation and sprawl with the intensification of socio-economic activities, exacerbating deforestation, pollution, and land degradation.

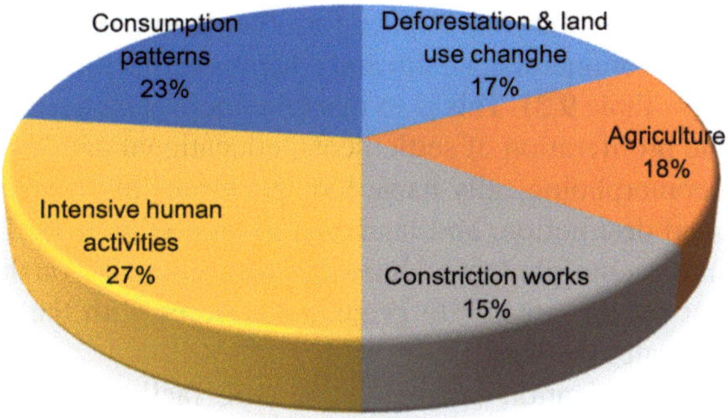

Fig. 9.2 Development actions that degrade land in Bamenda

Table 9.1 Development actions that degrade the environment

Area	Human activity	Environmental consequence
Intensified human activities	Construction activities; changing consumption patterns and increasing solid waste generation; deforestation and gallery forest loss; commercial expansion	Growing construction waste; Increased disposal of garbage and household waste; loss of biodiversity/dwindling water sources and climate variation; Land use change; Importation of containerised products and chemicals
Agriculture	Overgrazing, over-cultivation, heavy application of chemical fertilisers and agrochemicals	Soil compactness, contamination and poisoning of the soil/surface and groundwater sources; Loss of soil fertility
Commercial growth	Construction of commercial structures	Increased commercial waste

9.4.1.1 Intensified Socio-Economic Activities

The Urban and peri-urban zones of Bamenda are witnessing growing pressure upon their environment linked to the growing influx of people in their urban space, indicative of urbanisation and urban sprawl. A majority of the population (27%) indicates that the intensification of socio-economic activities is the main trigger of land depreciation. Demographic surge in Bamenda I, II, III municipalities (Fig. 9.2), significantly fosters intensification of urban activities; settlement, commercial growth, construction, changing consumption patterns, and increased land pollution. Figure 9.2 reveals a significant surge in population across the Municipality of Bamenda from 15,000 inhabitants in 1972 to 629,817 inhabitants projected for 2024. Implying an increase of 614,817 inhabitants within 52 years. The influx has been consistent within the various sub-divisions dominated by Bamenda II, with an increase in population from 28,253 in 1976 to 299,843 inhabitants projected for 2024, followed by Bamenda III with an increase in population from 8,190 in 1976 to a projection of 260,975 for 2024. Bamenda I, though, has the least growth, comparatively, has equally recorded a significant increase from 2,708 in 1976 to 68,999 in 2024. This unprecedented population growth implies more space for other land uses, among them are: farmland, built-up land, commercial, health and educational facilities, and other socio-economic facilities (Table 9.3). This is to meet the increasing needs of the population (Fig. 9.3) and the proliferation of settlements, educational and leisure facilities into wetlands and geomorphologically hazardous environments with resultant ecological loss, watershed destruction, and land degradation (Fig. 9.4). The proliferation of settlements into geomorphological hazard zones, especially on the Bamenda escarpment, carved out for the green city project, interferes with the ongoing nurturing projects and degrades the land.

Densification of socio-economic facilities (educational, health facilities, and markets) is most in Bamenda II, followed by Bamenda III, and least in Bamenda I. This implies land degradation because of the impact created by waste from the occupants and users of all these facilities. The reckless disposal of waste from these different sources into wetlands, valleys, farmland, and almost everywhere degrades the land.

Table 9.3 Distribution of socio-economic facilities in the sub-divisions of Bamenda

Sub-division	Secondary schools	Health facilities	Markets	Structures	Hotels	Car parks	Travel agencies
Bamenda I	6	5	1	1343.7	1	1	1
Bamenda II	39	57	5	3116.7	41	4	8
Bamenda III	22	31	4	2607.21	14	3	5
Total	67*	93#	10	–	66**	8	13

Source *DDSE statistics, (2021/2022), #Bamenda Health District Annual Report, (2021), **CDP of Bamenda II, III (2020 & 2021)

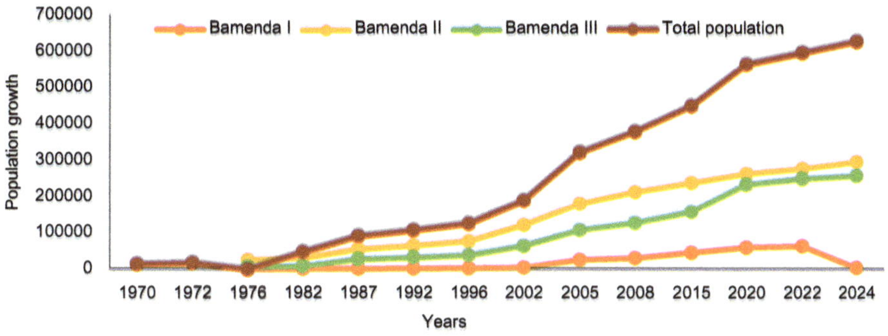

Fig. 9.3 Evolution of population of the Bamenda municipality from 1970 to 2024. *Source* National Census Data BUCREP (2005), CDP of Bamenda I, II & III

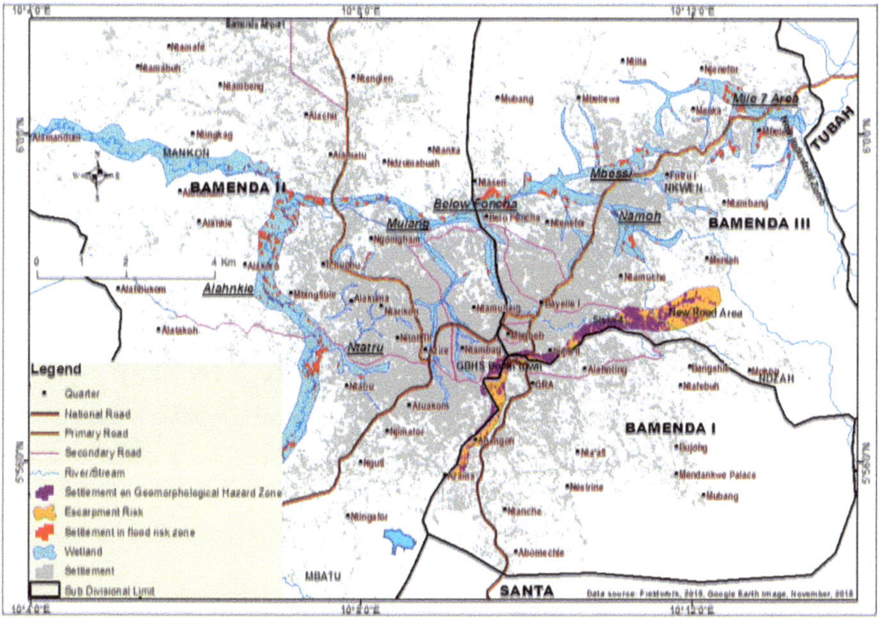

Fig. 9.4 Settlement proliferation on geomorphological hazards and flood-prone zones. *Source* Base Map NIC, Updated with Fieldwork (2024)

9.4.1.2 Urbanisation-High Consumption Patterns and Waste Generation

Urbanisation in Bamenda is amplified by continued population growth, associated with increased consumption levels, and changing taste and fashion. 23% of the population noted that a change in consumption patterns from local to modern imported products increases plastic and other forms of waste in the city. The situation in the peri-urban zone of Bamenda is becoming more problematic, compounded by the fact that a greater percentage of input becomes output that is disposed of recklessly in open dumpsites, degrading the land. Very little has been done to reduce waste generation until recently, when a few NGOs are into waste repurposing. The continuous influx of population in the urban and peri-urban zone of Bamenda is associated with increased construction activities for settlement, recreation, education, health, commercial, and transportation. 15% of the population reveals that increased construction waste is linked to the material type, which is often non-local and non-degradable (Fig. 9.5). Poor management of building waste degrades land.

Urbanisation has propelled land use change in Bamenda to the detriment of farmland and forestland, leading to land degradation (Fig. 9.6).

Multiple changes are observable in the land use pattern in Bamenda, most significant being agricultural land to settlement land (45.16%) and the least being forest to farmland (9.68%). The majority of these land uses are causing global deforestation and land degradation (Fig. 7a, b and Table 9.4).

Out of the different land uses other than the lake, built-up was the least land use type (1972) with 588.96 hectares, comparatively for the whole Municipality occupying

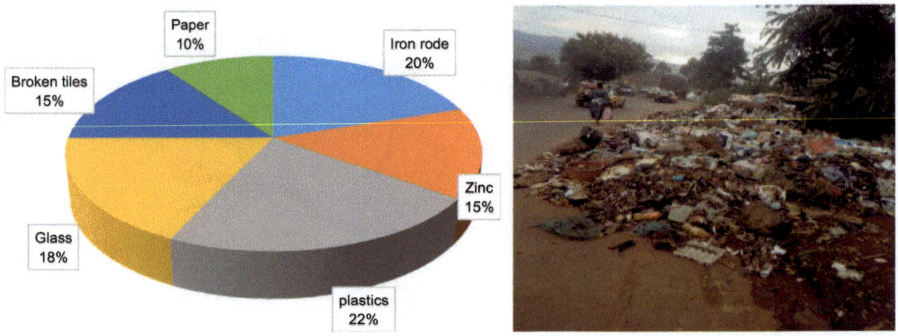

Fig. 9.5 Types of waste generated in Bamenda

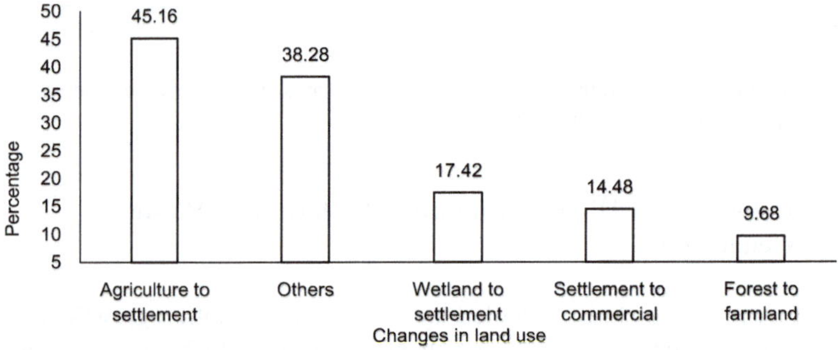

Fig. 9.6 Pattern of land use changes

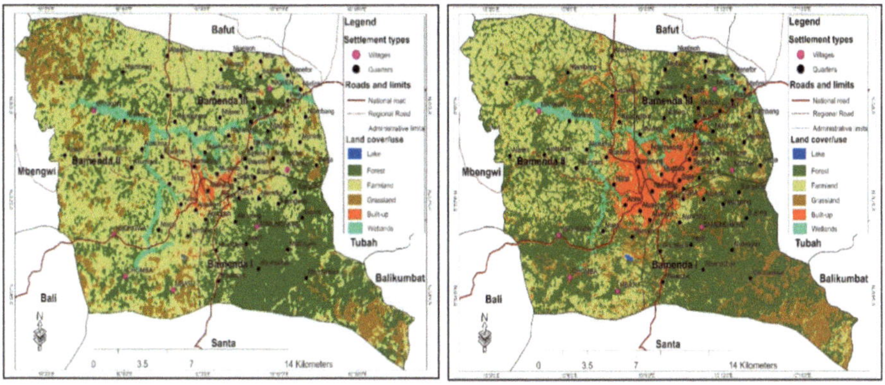

(a)

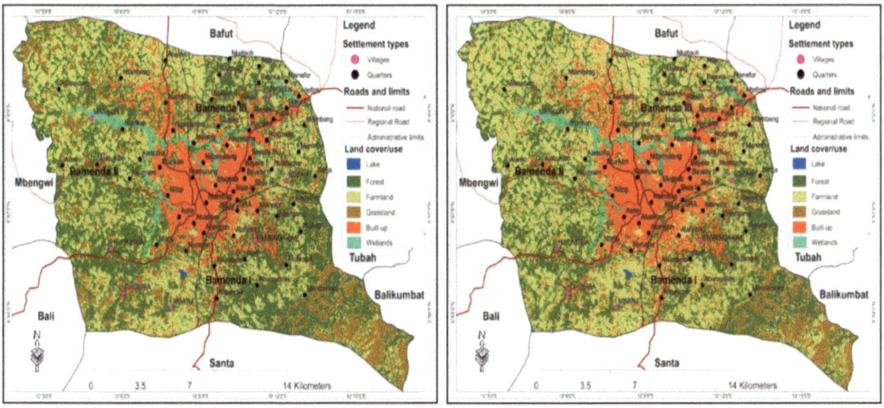

(b)

Fig. 9.7 **a** State of land cover of municipality of Bamenda 1972 and 1996. *Source* Landsat 3 (NASA, 1972/01/1)) and Landsat 5 (NASA,1996/12/18). **b** State of land cover transformation of the Municipality of Bamenda, 2008 & 2022. *Source* Landsat 7 (NASA, 2008/02/15) Landsat 9 (NASA, 2022/12/21

1.79%, while farmland constituted the highest in spatial extent (16,329.58 hectares), occupying 49.65%. Over the years, built-up has been on a consistent increase, from 588.96 in 1972 to 2865.15 by 1996, to 5252.94 in 2008, and 7656.84 by 2022. This implies that within 50 years, built-up has grown by 7067.88 hectares (from 1.79% to 21.49%) and farmland has recorded a corresponding decrease from 16,329.58 to 8268.3 hectares with a change in surface coverage by -8061.28 hectares. Forest and wetlands have equally decreased in spatial extent from 33% and 2.67% respectively, to 29.09% and 1.95% by 2022. These trends point directly to land degradation in Bamenda. The dwindling of farmlands is evident in growing food prices, intensifying urban poverty, and compounding urban liveability. To meet up with the dwindling cultivable land and food shortages, most farmers have adopted unsustainable agricultural practices, further degrading the land.

Table 9.4 Land cover change from 1972 to 2022 in the Bamenda Municipality

Land cover/use	1972 Surface coverage (ha)	% (1972)	1996 Surface coverage (ha)	2008 Surface coverage (ha)	2022 Surface coverage (ha)	% (2022)	Areal change (1972–2022) (ha)	% change (1972–2022)
Built-up	588.96	1.79	2865.15	5252.94	7656.84	23.28	7067.88	21.49
Farmland	16,329.58	49.65	10,326.69	11,099.06	8268.3	25.13	−8061.28	−24.52
Forest	11,170.82	33.96	13,773.96	11,567.7	9544.65	29.09	−1626.17	−4.87
Grassland	3904.92	11.87	5275.26	4303.08	6761.4	20.55	2856.48	8.75
Lake	15.86	0.05	16.29	14.31	16.47	0.05	0.62	–
Wetlands	879.28	2.67	632.07	652.33	641.76	1.95	−237.52	−0.67
Total	32,889.42	100	32,889.42	32,889.42	32,889.42	100	–	–

Source Calculated from land cover maps of 1972, 1996, 2008, 2022

Table 9.5 Fertilizer and pesticide use in Bamenda

Crop/Vegetable	Period before maturity (months)	Frequency of fertiliser application	Trend in the quantity of applications over time
Maize	4	3	Increasing
Yam	8	2	Increasing
Cocoyam	7–10	2	Increasing
Cassava	9–24	2	Increasing
Vegetables	2	4	Increasing
Tomatoes	3	4	Increasing

9.4.1.3 Agriculture

The City of Bamenda, over the years, has relied heavily on agriculture from within and out of the city. The growing population and rising food demand have triggered significant modifications in farming practices in recent decades. Food crop production in Bamenda has witnessed increased application of chemical fertilisers to manage the growing food demand. This is the case of vegetable and crop production where a greater percentage of farmers recount in the interview how they cannot do without fertilisers. "*without fertiliser, we toil in vain, the yield will not even be commensurate with the labour and inputs, as such no profit can be attained. Again, when we grow crops without fertilisers, the output can not attract a market because customers like healthy crops*" (Table 9.5). The use of pesticides is significantly increasing due to the proliferation of pests with a changing climate; this creates a cycle of unsustainable agricultural practices and consequent degradation.

All crops are cultivated with the use of fertilizer, applied more than once, based on the type of crop. The trend over time has been increasing, leading to land degradation. Land degradation is a product of intensive agricultural systems involving high use of agrochemicals, leading to eutrophication and toxic effects of pesticides on the environment. Another aspect of agriculture is overgrazing, a common activity on the highland of Bamenda, which extends even to valleys because this activity is not enclosed in ranches; as such, the level of trampling cannot be easily controlled. Interviews reveal that grazing has intensified following the growing population. This, together with poor farming practices like slash and burn, and recurrent burning of hills in dry seasons, intensifies deforestation and land degradation.

9.4.2 Land Degradation and Climate Change Nexus

Climate change is directly and indirectly driven by land degradation. To understand the nexus, the trend of climatic elements (rainfall and temperature) and variables indicative of land degradation (vegetal loss, farmland indicative of productivity) were assessed. Trends in mean temperature (MT) and mean rainfall (MR) from 1971 to 2023 show considerable variations (Table 9.6).

Rainfall has been decreasing over time (Fig. 8a). This is associated with decreasing forest coverage over the years (Fig. 9.9), implying that as forest coverage dwindles, carbon sinks are lost, giving rise to increasing temperatures from 18.68 in 1971 to 22.97 in 2023 (Fig. 8b). This shows that a direct relationship exists between forest coverage and rainfall. The decline in vegetal cover is directly linked to land degradation, reflected in the decrease in farmland from 16,329.58 hectares in 1972 to 8268.3 hectares in 2022. Increasing temperature trends (Fig. 8b) are directly associated with increasing grassland (Fig. 9.9), indicative of overgrazing and land degradation. Land degradation negatively impacts climate as it consequently leads to rising temperatures and decreasing rainfall, intensifying greenhouse effects and global warming.

On the other hand, it is established that climate equally negatively impacts land degradation in Bamenda. Incidents of recurrent heavy rainfall result in significant erosion in upland areas and flash floods in lowland areas, degrading the land. Climate change, evident by variations in temperature and rainfall, exacerbates the magnitude of several ongoing land degradation processes in Bamenda, producing new degradational patterns. Human-induced global warming has already caused observed changes in two drivers of land degradation: decreasing rainfall amounts and increased heat stress. Climate and land interact directly over time and space, with increasing rainfall intensity exacerbating soil erosion and nutrient depletion. Indirectly, climate change impacts land degradation in that declining agricultural productivity caused by climate change drives an intensification of agriculture through fertiliser use, further leading to land degradation.

Table 9.6 Rainfall and temperature characteristics for Bamenda (1971–2023)

Periods	MR	SD	CV	MSPI	SPI class	RT	Reliability	MT	TT
1971–1980	190.02	19.69	5.36	0.59	Mildly wet	Increase	Reliable	18.68	Increase
1981–1990	187.94	12.6	6.7	0.55	Mildly wet	Decrease	Reliable	18.56	Decrease
1991–2001	184.11	14.04	7.62	0.49	Mildly wet	Decrease	Reliable	19.39	Increase
2002–2012	169.65	21.71	12.8	0.24	Mildly wet	Decrease	Reliable	21.27	Increase
2013–2023	143.92	32.93	22.88	-0.19	Mild dryness	Decrease	Unreliable	22.97	Increase
Average	175.13	20.19	12.07	0.34	Mildly Wet	Decrease	Reliable	20.17	Increase

MR mean rainfall (mm); *SD* standard deviation; *CV* coefficient of variation (%); *MSPI* mean standardised precipitation index; *RT* rainfall trend; *MT* mean temperature (°C), *TT* temperature trend

Data Source NW regional meteorological service, Bamenda (2023)

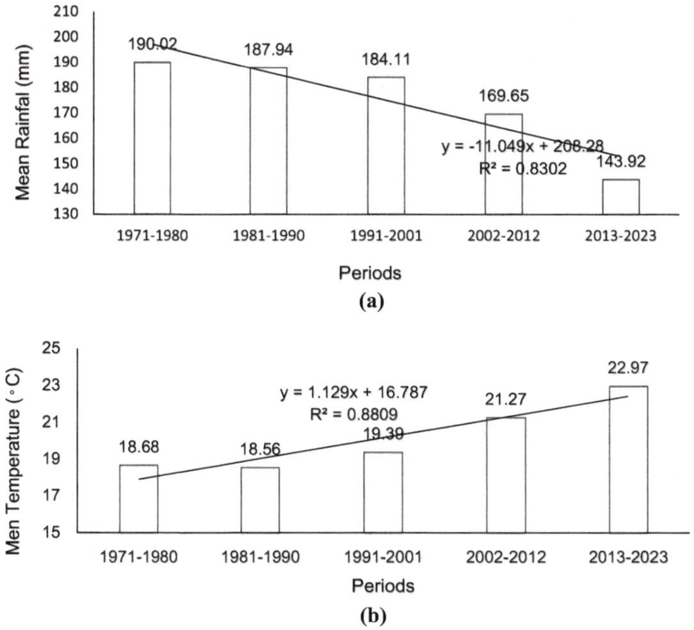

Fig. 9.8 **a** Mean rainfall trends in Bamenda from 1971 to 2023. **b** Mean temperature trends in Bamenda from 1971 to 2023. *Source* NW Regional Meteorological Service, Bamenda (2023)

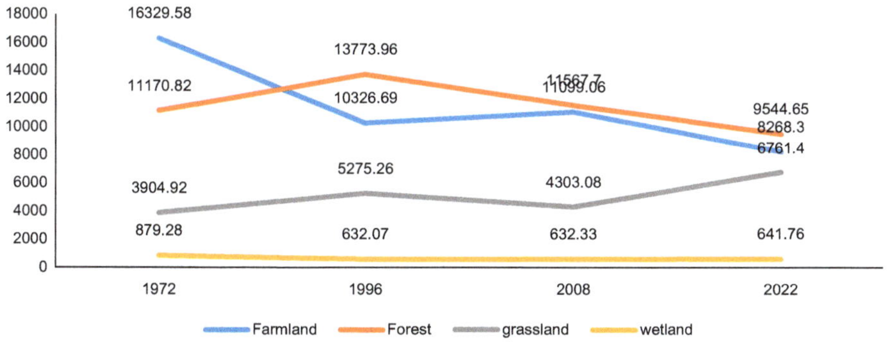

Fig. 9.9 Trend of farmland, forest, grassland, and wetland in Bamenda from 1970 to 2022. *Source* Land cover maps of 1972. 1996, 2008, and 2022

Decreasing rainfall trends and increasing temperature trends coincide with decreasing forest coverage and increasing grassland (Fig. 9.9) and built surfaces, indicative of land degradation, as revealed by statistics from the land use maps. The nexus between climate change and land degradation is direct in that land degradation directly contributes to climate change by triggering temperature increases and diminishing rainfall averages, consequently changing the climate. This is because the deforestation process that drives land degradation implies an increase greenhouse effect, leading increase in average temperatures at the Earth's surface. Climate change, on its part, indicated by rising temperature, contributes to land degradation by exacerbating evaporation rates, producing drier soils, which are more vulnerable to water erosion. Seasonal changes are characterised by short phases of drought that stress vegetation and enhance land degradation in Bamenda. An interview with one of the neighbourhood leaders of Bamenda I revealed that "*several tree species are threatened and some are at risk of extinction due to deforestation and climate change. The Vepris bali tree that was in the Ngemba Forest Reserve is possibly extinct, while the Vepris onanae is critically endangered*".

9.5 Challenges Mitigating Climate Change and Land Degradation in Bamenda

Land degradation has significant ramifications, such as loss of ecosystem services. This comes from institutional lapses and limited awareness of the population that land degradation is a problem.

9.5.1 Institutional Lapses and Non-compliant Population

Government policies to address land degradation are often reactive, and not proactive, thus failing to address the ultimate causes of degradation. Multiple actions are ongoing within the city of Bamenda that trigger land degradation and contracts existing policy, yet the culprits are rarely penalised. The Master plan of Bamenda (2011–2027) categorises land into developable and undevelopable land. The invasion and possession of building permits by urban dwellers into undevelopable lands (wetlands, river courses, flood plains, water catchment areas, steep slopes) points to institutional lapses and weak enforcement of environmental regulations. Failure to penalise defaulters of the law in many instances implies weak policy enforcement. As more people surge into Bamenda, the demand for land in the urban fringe increases, in the face of limited land, these marginal lands (riverine/marshy areas and steep slopes) are procured and occupied, in violation of Law N° 96/12 of 5th August 1996 relating to environmental management regulating land degradation. The reclamation and construction of houses in proximity to riverbanks, non-respect of "*Stream Setbacks of 15 m from small rivers, 10 m from seasonal streams and 5 m from a gully or runoff channel*" (Consultant HTR of the master plan of Bamenda city 2011–2027) is evident of week enforcement of regulation driving land degradation.

- Law No.2004/003 of 21 April 2004 prohibits construction on any land in danger of natural hazards, ecologically protected areas, as well as land likely to jeopardise public health or local cultural values that are unfit for habitation.

- Law No. 89/27 of 29 December 1989 relating to toxic and dangerous wastes
- Law No.1-323 of 1993 on Land Reclamation, notwithstanding that the process continues.

Regardless of the laws, the population are non-compliant. This is evident by the growing magnitude of land reclamation and the invasion of wetlands, degrading them.

9.5.2 Limited Awareness of Land Degradation as a Problem

The implementation of land management policies is strongly influenced by the perceptions of human–environment relationships. Land degradation is an unintended consequence of economic development, with unprioritized actions to mitigate the consequences. Challenges posed by land degradation are undermined, and unsustainable actions progress unrestricted. An interview with a farmer revealed that "*I do not care if the land is degrading or not, all I want is land, and once it is available, I can apply as much fertiliser as possible to get my desired output*". In another interview, a farmer revealed that "*everybody uses methods suitable to them, I have equally devised mine, that is, applying fertilisers as many times as possible to get my desired output*". The ignorance of their actions significantly triggers land degradation. Other challenges include growing population pressure and over-exploitation of land-based resources (Eisenack, 2024).

9.5.3 Measures to Address Land Degradation and Climate Change

Land degradation can be avoided, reduced, or reversed by implementing sustainable land management, restoration, and rehabilitation practices (Li & Elumalai, 2023). This provides many benefits, including adaptation to and mitigation of climate change (Tume et al., 2024). Delaying the application of proven activities to combat land degradation will make the necessary steps increasingly difficult and costly (Eisenack, 2024; Li & Elumalai, 2023). Sustainable land management and land restoration practices in Bamenda are essential for climate change mitigation and adaptation; without them, viability in the face of climate change may be compromised (Tume et al., 2024). Sustainable land management involves a comprehensive array of technologies and enabling conditions that have proven effective in addressing land degradation at multiple landscape scales, from local farms to entire watersheds (FAO, 2024).

Sustainable forest management in Bamenda is crucial for preventing deforestation, preserving and enhancing carbon sinks, and contributing to the reduction of greenhouse gas emissions. The gallery forest, forest reserves, and other man-made vegetation in the city are facing significant decline. Implementing sustainable forest management practices is essential for generating socio-economic benefits while providing the necessary fibre, timber, and biomass to meet the growing demands of the local population. Decision-makers and land managers need relevant information to enhance land sustainability and optimise natural resource usage. Effective monitoring strategies, restrictions on forest encroachment, and a focus on sustainable farming practices that primarily use organic fertilisers are vital. This includes imposing penalties on non-compliant farmers while incentivising those who adhere to these sustainable practices, which is paramount for reducing land degradation and conserving biodiversity. Key stakeholders and local communities play essential roles in designing and implementing more sustainable land management practices. Local land leaders in Bamenda should oversee land use activities and report any unauthorised practices to the appropriate authorities. Additionally, creating buffer zones around wetlands, forest reserves, stream banks, and watersheds is necessary to safeguard these areas from encroachment and degradation. Moreover, as part of the green city initiative, there will be a renewed emphasis on tree planting, particularly fruit trees along streets and within residential areas. This practice aims to minimise evaporation and reduce the greenhouse effect, ultimately helping to moderate local temperatures.

Environmental literacy is essential for all members of society to ensure the responsible use of our natural resources without harming the environment. Promoting environmental adult education in the municipality of Bamenda will inform the community about the consequences of their actions on the environment. Numerous organisations in Bamenda are dedicated to environmental education and sustainability, which can lead to a shift away from unsustainable land use practices and help mitigate land degradation. Enhancing environmental adult literacy can empower residents to understand, appreciate, and address the challenges of environmental degradation and the complexities of environmental transformations. It is crucial to educate farmers about sustainable cultivation practices and the risks associated with over-fertilisation, ranching, and unsustainable agricultural methods. There is no universal solution for sustainable land management; successful approaches depend on adapting methods to suit specific biophysical, social, economic, and political contexts. This may involve a variety of sustainable farming practices, substituting chemical fertilisers with organic alternatives, managing forests responsibly, and promoting urban designs that utilise local materials to minimise waste and support sustainable consumption patterns, along with leveraging indigenous knowledge systems. To effectively reduce the effects of degradation and enhance the resilience of both ecosystems and livelihoods in Bamenda, it is necessary to incorporate various practices into landscape planning. This includes local-level sustainable finance and business strategies to alleviate human pressure on land and decrease degradation (Table 9.7).

Table 9.7 Pathways to addressing land degradation in Bamenda

Ambition (Goal)	Strategy
Safeguarding biodiversity	Significantly protect biodiversity through enlarged and more effective protected areas (wetlands and reserves), halting the conversion of natural land through limiting invasion for settlement or other activities
Consumption lifestyles	Promote local consumption patterns, including the adoption of less land-degrading diets, such as more vegetable-based diets, and low- and renewable-energy-based housing, energy-efficient transportation and industrial systems
Minimise global human population growth rate	Women's empowerment and moving towards improved access to education, voluntary family planning, and social welfare for ageing populations discourages family allowance and early marriages
Zero-waste economy	Minimise waste through circular economy practices, and maximise resource efficiency, improve sanitation management systems, refurbish, repair and reuse materials, especially for buildings and other infrastructure
Sustainable production systems and resource management	Optimise land use, minimise external inputs and maximise local inputs, energy efficiency, reusing water, and depend highly on renewable energy, for production systems for food and other commodities
Holistic approach to land management	Protection and improving soil health, croplands, rangelands, forestry, water systems, human settlements, and surrounding landscapes, to reverse land degradation

Adapted from IBPES, 2018

9.6 Discussion

Climate change and land degradation have become the concern of most governments worldwide as rapid socio-economic changes and population growth are unavoidable (Olsson et al., 2019). This chapter is focused on establishing the nexus between land degradation and climate change, which aligns with previous studies, like Reed and Stringer (2016) on land degradation, desertification, and climate change. This is interlinked with the focal theme of this book, '*Eco-Resilience, Climate Change, Land Degradation and Sustainable Solutions*', in that the findings of this chapter uncover the triggers to land degradation, the nexus between land degradation and climate change, challenges, and pathways to reduce and restore land degradation, to attain sustainability. It anchors squarely on the theme of this book as it reveals the contemporary developmental actions that drive land degradation and the interplay with climate change, emphasising policy actions that can be adopted to reduce land degradation, without which it will become very costly for humanity and future generations. This aligns with the findings of Tume (2022), who revealed that increasing temperatures and variable rainfall are clear indicators of climate change in Bamenda, resulting in land degradation and other environmental shocks. These results also indicate that intensified human activities and changing consumption patterns significantly contribute to land degradation. Multiple human activities, including construction, agriculture, overgrazing, and deforestation, linked to population growth, trigger degradation. This aligns with the findings of (OSS, 2024) that land degradation is a function of human-induced actions, with significant implications on biodiversity, soil fertility, and overall health conditions (OSS, 2024). This accelerating degradation compromises food and water security, posing a serious threat to the growing global population (Tume, 2022; Tume & Kongso, 2022). This is also in line with the findings of Ndikebeng et al., (2023), who revealed that land cover change has serious implications on land degradation and water resources, leading to threats to water security (Veryee et al., 2023).

Human overpopulation is among the most pressing environmental issues, driving global warming, environmental pollution, habitat loss, the mass extinction of biodiversity, unsustainable agricultural practices, and unsustainable consumption patterns, at faster than the rate of regeneration (Nishioka, 2024; OSS, 2024). The unprecedented population growth in the Municipality of Bamenda, with unsustainable patterns of consumption, is the major force behind land degradation and climate change, with observable developmental actions that undermine environmental sustainability. This corroborates the ideas of Fombe and Balgah (2012) that as the urban population is rising quite rapidly, the economies are changing rapidly in terms of technological development, consequently, consumption patterns have led to a massive generation of solid waste, degrading land. There has been growing dependence on fertiliser application to meet growing food demand. A similar idea is shared by IPBES (2018, 2019) that a continued increase in nutrient and chemical inputs and a shift towards industrialised livestock production systems, with pesticide and fertiliser use, is expected to double by 2050, further exacerbating land degradation. The rate of fertiliser application in Bamenda is on a growing trend to meet the current food shortage and growing prices.

There is a nexus between ecological dynamics, such as the loss of forests and wetlands, and seasonal and inter-annual variability in climate, resulting in water scarcity for agriculture and domestic chores. A direct relationship exists between land degradation and climate change in and vice versa. This is similar to the findings of Olsson et al., (2019) and International Network on Appropriate Technology (2025) that land degradation impacts climate change. Global warming will exacerbate ongoing land degradation processes through increasing floods, drought frequency and severity, intensified cyclones, with outcomes being modulated by land management (Olsson et al., 2019).

Curbing land degradation in Bamenda is inhibited by institutional lapses and a non-compliant population. The policies on land degradation, like the environmental law of 1996, restrict human habitation from wetlands and areas vulnerable to hazards, notwithstanding, these areas are densely occupied by settlements. There is a gross violation of policy, and defaulters are hardly penalised. This is not different from the ideas (IBPES, 2018) that the widespread failure of policies and institutions to enforce and incentivise sustainable practices and internalise the long-term economic costs of unsustainable production has meant that the exploitation of natural resources typically leads to greater levels of land degradation.

The chapter reveals that effective monitoring, environmental education, and sustainable forest management strategies like creating buffer zones around the green spaces in the city of Bamenda can significantly curb the level of land degradation in Bamenda. In the same vein, Gimah and Bodo (2019a, 2019b), Olsson et al., (2019) noted that without environmental adult education and a strong plan to create terrestrial and marine protected areas, the environment will continue to suffer degradation. Nasi et al., (2011), in the same light, noted that the successful implementation of sustainable forest management requires well-established and functional governance, monitoring, and enforcement mechanisms to eliminate deforestation, illegal logging, arson, and other activities that are inconsistent with sustainable forest management principles.

9.7 Conclusion

The Anthropocene is characterised by the dominance of human activities, which often involve indiscriminate resource use and a diminishing respect for ecological balance with other living beings. Every effort made by humanity to exploit natural resources for comfort and development impacts the environment. Bamenda is witnessing challenges related to land degradation and climate change. This region, known for its tropical savannah climate and prolonged wet season, displays a rich diversity of human activities and dynamic interactions. Over the decades, Bamenda has experienced significant urbanisation, resulting in sprawl and changes in land use, particularly in peri-urban areas. The encroachment upon and transformation of the natural landscape through human settlements—invading wetlands and highlands—coupled with over-cultivation, excessive use of fertilisers on dwindling agricultural lands, seasonal livestock grazing, and increasing waste generation, obstruct sustainable land management and disrupt the landscape. In recent years, there has been a notable rise in human-induced pressures, evidenced by the intensification of human activities, expansion of built-up areas, deforestation, and increased loss of ecosystem services. These factors, combined with the region's tropical climate, significantly contribute to land degradation. Meteorological data collected since the 1970s shows a consistent increase in annual temperatures alongside a decline in rainfall. Climate models project a warming trend, predicting an average rise in annual temperatures by 3–5 °C by 2080, while rainfall is expected to decrease. This overarching warming trend is associated with ecological loss, declining crop yields, and worsening urban poverty. The problem of land degradation has not only reduced agricultural productivity but has also intensified the reliance on rural areas for food production, further expanding the ecological footprint and driving up food prices, which exacerbates urban poverty. Unless urgent and coordinated action is taken, land degradation will continue to escalate alongside ongoing population growth. Addressing land degradation requires systemic changes on a macroeconomic level, including concerted efforts to improve the sustainability of both production systems and consumer lifestyles. At the same time, it is essential to foster a socioeconomic environment that encourages lower population growth rates and per capita consumption. Strengthening institutional capacities can enhance the effectiveness of policy instruments aimed at avoiding, reducing, and reversing land degradation.

References

Acha, M. E. & Fombe, L. F. (2023). Gentrification patterns, trends and development implications of the Bamenda Metropolis from 1972–2022. *International Journal of Social Sciences and Management Review, 6*, 139–153. https://doi.org/10.37602/IJSSMR.2023.6312

Al-Kindy, S. N. M. (2025). Climate change initiatives and achievement of SDGS: A paradoxical analysis. *American Journal of Climatic Studies, 5*(1), 1–9. http://www.ajpo.org/

Bodo, T., & Gimah, B. G. (2018). Government programmes in checking the occurrence of habitat loss and their implications for maintaining a sustainable environment in Ogoni, Rivers State, Nigeria. European. *Journal of Biomedical and Pharmaceutical Sciences, 5*, 64–71.

Eisenack, K. (2024). Why local governments set climate targets: effects of city size and political costs. *Environmental and Resource Economics, 87*(11), 2935–2965. https://doi.org/10.1007/s10640-024-00919-1

FAO. (2024). Terminal evaluation of the project "Climate Adaptation in Wetland Areas in the Lao People's Democratic Republic". Project Evaluation Series, 19/2024. Rome. https://doi.org/10.4060/cd0869en

Fombe, L.F., & Balgah, S. N. (2012). The urbanization Process in Cameroon: Patterns, Implications and Prospects. *Nova Science Inc.* New York; USA, 213 p

Filho, W. L., Azul, A. M., Brandli, L., Salvia, A. L., & Wall, T. (Eds.). (2021). *Life on land* (p. 1117). Springer.

Foley, J. A., Defries, R., Asner, G. P., Bardford, C., Bonan, G. S., Carpenter, S. R., Chapin, F. S., Coe, M. T., Daily G. C., Gibbs, H. K., Helkowski, J. H., Holloway, T., Howard, E. A., Kucharik, C. J., Monfreda, C., Patz, J. A., Prentice, C., Ramankutty, N., & Sndyer, P. K. (2005). Global consequences of land use. *Science, 309*, 570–574, https://doi.org/10.1126/science.1111772

Foley, J. A., Ramankutty, N., & Zaks, D. P. M. (2011). Solutions for a cultivated planet. *Nature, 478*, 337–342. https://doi.org/10.1038/nature10452

Fombe, L. F., & Acha, M. E. (2020). Land use dynamics and variations in sprawl across sub-divisions in Bamenda urban-scape 1996–2018. *Journal of Sustainable Development, 13*(4), 224. https://doi.org/10.5539/jsd.v13n4p224-234

Gimah, B. G. & Bodo, T. (2019a). Curbing human activities that degrade the environment: The relevance of environmental adult education. *Earth and Environmental Science Research and Reviews, 2*(5), 1–7. ISSN: 2639-7455

Gimah, B. G., & Bodo, T. (2019b). Creation of awareness through environmental adult education as a solution to the problem of habitat loss in Ogoni, Rivers State, Nigeria. *International Journal of Advanced Research and Publications, 3*, 22–28.

IBPES. (2018). Summary for Policy makers of the assessment report on land degradation and restoration of the intergovernmental science-policy platform on biodiversity and ecosystem services. In R. Scholees, L. Montainarella, A. Brainich, N. Barger B. ten Brink, M. Cantele,

B. Erasmus. J. Fisher, T. Gardener, T.G. Holland, F. Kohler, J. S. Kotiano, G. Von Maltitz, G. Nangendo, R. Pandit, L. J. Parrotta, M. Dpotts, S. Prince, M. Sankaran & L. Willemen (Eds.), IPBES Secretariat, Bonn, Germany, p. 44.

International Network on Appropriate Technology. (2025). www.appropriatetech.net. Accessed on 8 May, 2025.

IPBES. (2019). Global assessment report on biodiversity and ecosystem services of the intergovernmental science-policy platform on biodiversity and ecosystem services. In E. S. Brondizio, J. Settele, S. Díaz, & H. T. Ngo, (Eds.), *IPBES secretariat*. Bonn, Germany, p. 1148. https://doi.org/10.5281/zenodo.3831673

Kimengsi, J. N., Akumbo, M. C., Balgah, R. A., Tingum, E. N., Tume, J. P. S., & Akhere, S. N. (2022). Farm-based climate adaptation dynamics: Insights from the vegetable sector in the Western Highlands of Cameroon, *Cogent Social Sciences, 8*, 2126452, 1–22.

Kimengsi, J. N., Balgah, S. N., & Achia, S. N. (2017). Peri-Urban land use dynamics and development implications in the Bamenda III municipality of Cameroon. *Sustainability in Environment, 2*(3), 273–288. https://doi.org/10.22158/se.v2n3p273

Kometa, S. S., & Ndi, R. A. (2012). The Hydro-geomorphological Implications of Urbanisation in Bamenda, Cameroon. *Journal of Sustainable Development, 5*(6), 64–73. https://doi.org/10.5539/jsd.v5n6p64

Li, P., & Elumalai, V. (Eds.) *Recent advances in environmental sustainability*. Springer, Cham, p. 465.

Nasi, R., Putz, F. E., Pacheco, P., Wunder, S., & Anta, S. (2011). Sustainable forest management and carbon in tropical Latin America: The case for REDD+. *Forests, 2*, 200–217. https://doi.org/10.3390/f2010200

Ndikebeng, K. R., Forba, C. F., Tume, S. J. P., Banseka, J. Y., & Kimengsi, J. N. (2023). Land cover dynamics and implications on water resources in Bamenda III sub-division, North West Region, Cameroon. *International Journal of Global Sustainability, 7*(1), 37–52. https://doi.org/10.5296/ijgs.v7i1.21031

Nduka, J. K., & Orisakwe, O. E. (2009). Effects of effluents from warri refinery petrochemical company WRPC on water and soil qualities of "Contiguous Host" and "Impacted on Communities" of Delta State, Nigeria. *The Open Environmental Pollution & Toxicology Journal, 1*, 11–17.

Nishioka, M. (2024). Changes in temperature and CO_2 in the atmosphere at various latitudes. *Current Research in Environmental Science and Ecology Letters, 1*(1), 01–09.

Olsson, L., Barbosa, H., Bhadwal, S., Cowie, A., Delusca, K., Flores-Renteria, D., Hermans, K., Jobbagy, E., Kurz, W., Li, D., Sonwa, D.J., & Stringer, L. (2019). Land degradation. In P. R. Shukla, J. Skea, E. Calvo Buendia, V. Masson-Delmotte, H.-O. Pörtner, D. C. Roberts, P. Zhai, R. Slade, S. Connors, R. van Diemen, M. Ferrat, E. Haughey, S. Luz, S. Neogi, M. Pathak, J. Petzold, J. Portugal Pereira, P. Vyas, E. Huntley, K. Kissick, M. Belkacemi, J. Malley, (eds.), *Climate change and land: An IPCC special report on climate change, desertification, land degradation, sustainable land management, food security, and greenhouse gas fluxes in terrestrial ecosystems*. https://doi.org/10.1017/9781009157988.006

OSS. (2024). African land: The degradation and the absolute requirement of sustainable management. OSS-Livre Teress-dAfrique-En (1).

Pataki, D. E., Carreoiro, M. M., Cherrier, J., Grulke, N. E., Jennings, V., Pincetl, S., Pouyat, R. V., Whitlow, T. H., & Zipperer, W. C. (2011). Coupling biogeochemical cycles in urban environments: Ecosystem services, green solutions, and misconceptions. *Frontiers in Ecology and the Environment, 9*, 27–36. https://doi.org/10.1890/090220

Patnaik, R. (2018). Impact of industrialisation on the environment and sustainable solutions-reflections from South Indian Region. *IOP Conference Series: Earth and Environmental Science, 120*.

Phoenix, K. (2025). Four ways Cambridge is advancing climate education. core-cms.prod.aop.cambridge.org. Accessed on May 7, 2025.

Reed, M. S., & Stringer, L. C. (2016). *Land degradation, desertification and climate change: Anticipating, assessing and adapting to future change* (p. 198). Routledge.

Royal Society. (2016). *Resilience to extreme weather* (p. 124). The Royal Society.

Setoa, K. C., Reenbergb, A., Boonec, C. G., Fragkiasd, M., Haasee, D., Langankef, T., Marcotulliog, P., Munroeh, D. K., Olahi, B., & Simon, D. (2012). Urban land teleconnections and sustainability. *Proceedings of the National Academy of Sciences, 109*, 7687–7692. https://doi.org/10.1073/pnas.1117622109

Tume, S. J. P. (2021). Impact of climate change on domestic water accessibility in Bamenda III Sub-Division, North West Region, Cameroon. *Journal of the Cameroon Academy of Sciences, 17*(2), 131–145. https://doi.org/10.4314/jcas.v17i4.3

Tume, S. J. P. (2022). Standardized precipitation index valuation of climate change in Bamenda. *Geology, Earth and Marine Sciences, 4*(1), 1–6. https://doi.org/10.31038/GEMS.2022414

Tume, S. J. P., & Kongnso, M. E. (2022). Standardized precipitation index valuation of the impact of climate variability and change on domestic water accessibility in Bamenda III Municipality, North West Region, Cameroon. *Journal of Science Frontier Research (H), Environment and Earth Science, 22*(8), 39–55.

Tume, S. J. P., Nyuyse, Y. B., Foka, S. K., Kuyar, C., Ngoran, T. L., Ngoran, H. A., Lueong, L. A., Ngala, E. N., Tambe, A. E. M., Fanka, I. B., Kiseem, B. F., & Mbinkar, V. J. (2024). Pathway for net zero policies: aligning nationally determined contributions (NDCs) for holistic sustainable development in Cameroon. *Global Scientific Journals, 12*(9), 1279–1300.

Umoru, K. (2023). Major drivers and effects of land degradation in Nigeria: A review. *International Journal of Scientific Research in Biological Science, 10*(6), 46–50, www.isroset.org

United Nations. (2015). *World urbanisation prospects: The 2014 revision* (p. 517). Department of Economic and Social Affairs.

Veryee, L., Fombe, L. F., & Kometa, S. S. (2023). State of urban household drinking water security situation in Bamenda, North West Region, Cameroon. *Environmental Management and Sustainable Development, 12*(2), 130–145. https://doi.org/10.5296/emsd.v12i2.21459

Webb, N. P., Marshall, N. A., Stringer, L. C., Reed, M. S., Chappell, A., & Herrick, J. E. (2017). Land degradation and climate change: Building climate resilience in agriculture. *Frontiers in Ecology and the Environment, 15*(8), 450–459. https://doi.org/10.1002/fee.1530

Chapter 10
Assessment of Land Surface Phenology and Its Response to Climate Change in the Tawa River Basin, India

Pradeep Kumar Rajput

Abstract This study analyzes the vegetation phenology using time-ordered data remote sensing data such as MODIS data 250 m, and 16-day Synthesis NDVI data from 2002 to 2021 to understand seasonal and interannual variations. Key phenological parameter start and end of the developing season, season length, baseline, and amplitude were extracted using TIMESAT software. An adaptive Savitzky-Golay smoothing technique was employed to the NDVI time-ordered data to improve the clarity and accuracy of seasonal pattern detection. Over 19 growing seasons, the longest season occurred in 2010, indicating significant variability. The Identifying Shifts in Additive Seasonal and Trend Patterns was also employed to identify structural changes and Transitions in Perpetual vegetation trends. The combined use of TIMESAT and BFAST enabled a comprehensive assessment of both seasonal dynamics and long-term vegetation changes. These insights support ecological monitoring, informing land use, management, and conservation strategies by providing a robust framework for tracking phenological responses to environmental changes across landscapes.

Keywords NDVI · TIMESAT · Phenology · Season · BFAST · Climate change

10.1 Introduction

Change of climate, the study of phenology to understand the relationships between organisms and their surrounding environment, and changes in the environment may impact the timing of these events. Phenology data can be gathered through ground-based methods, such as Phenocam or remote sensing data, and is used in many fields such as agriculture, ecology, and climate change research. Phenology is the study of the life cycle of plants and biological events in the annual cycle time period of plants

P. K. Rajput (✉)
Department of Geography, School Basic Science, CSJMU Kanpur University, Kanpur, Uttar Pradesh, India
e-mail: pkrseven@gmail.com

U. Chatterjee et al. (eds.), *Eco-Resilience: Climate Change, Land Degradation and Sustainable Solutions*, Sustainability Solutions,
https://doi.org/10.1007/978-3-032-00708-7_10

and animals, such as start of season, leafing, flowering, fruiting, and migration (Lieth, 2013). It is a sensitive aspect of plant and animal biology, as these events are influenced by environmental factors such as temperature, precipitation, and photoperiod (day length). Timesat is a mostly used software for analysing time ordered variable, such vegetation index particularly in the field of remote sensing and vegetation studies. It provides a comprehensive set of algorithms and functions for extracting and analyzing temporal information from satellite imagery, making it a valuable tool for studying phenology, land cover dynamics, and other temporal patterns. It provides a variety of methods for identifying and describing significant phenological rhythm. These algorithms are built upon tried-and-true techniques like the Savitzky-Golay filter, threshold-based strategies, and statistical modelling methods. Timesat can get useful insights into the dynamics and behaviour of vegetation throughout time by applying these algorithms to data. Jönsson and Eklundh (2004). The ability of Timesat to handle multi-temporal satellite imagery, such as information from the MODIS sensor, is one of its important characteristics. It can analyse a lot of satellite data and extract useful data at various temporal and spatial resolutions. When comprehensive coverage is needed, regional and worldwide research benefit most from this capability. To assist in the interpretation and dissemination of data, it offers a variety of visualisation tools, such as time series plots, phenological diagrams, and land cover maps. Additionally, the software offers batch processing, enabling users to execute numerous time series dataset analyses concurrently. Timesat has been extensively utilised in numerous scientific applications and studies. It has been used to extraction of season the phenology of the vegetation in a variety of environments, including woods, grasslands, and croplands. This package has also been widely utilized for monitoring land cover changes, ranging from localized studies to broad-scale assessments. Multiple studies have validated its reliability and precision in capturing temporal trends and patterns from remote sensing data. It is an invaluable tool for researchers and practitioners working on phenology analysis, land cover change detection, and other temporal analyses because of its broad range of algorithms and functionalities and user-friendly interface. It can be utilised for many different purposes, including crop classification, crop yield estimation, vegetation monitoring, and assessment of vegetation productivity (Bendix et al., 2002). The processing of long-term time series of satellite data is possible thanks to Timesat ability to handle massive amounts of data, which is one of its primary advantages. The derived biophysical parameters use of tools for visualisation are also included for the representation and analysis of the (Bendix et al., 1999).

Land surface phenology (LSP), means which refers and understand the Rhythm of seasonal Natural events such as leaf growing, greening, and senescence, and serves as a key indicator of terrestrial ecosystem responses to climate change variability and environmental change (Richardson et al., 2013; White et al., 2009). With advances in Earth Observation technologies, phenological patterns can now be monitored consistently over broad spatial and temporal scales, providing meaningful understanding of vegetation behaviour and patterns across various landscapes (Friedl et al., 2006; Zhang et al., 2003). The use of remote sensing satellite-derived vegetation indices data, such as the Normalized Difference Vegetation Index (NDVI) and Enhanced

Vegetation Index (EVI), has enabled the study of LSP from global to local scales (Didan, 2015). Globally, shifts in phenological events are important and the most delicate biological adaptations in response to shifting climate patterns. Numerous studies across continents have documented earlier onset of spring, delayed autumn senescence, and increases growing seasons, particularly in temperate and boreal ecosystems (Jeong et al., 2011; Parmesan & Yohe, 2003; Piao et al., 2019). These changes have profound implications for carbon cycling, biodiversity, agriculture, and ecosystem services (Cleland et al., 2007; Richardson et al., 2018). At the regional level, particularly in the monsoon-influenced landscapes of South Asia, phenological responses are modulated by complex interactions between precipitation patterns, temperature anomalies, and land use dynamics (Ganguly et al., 2010; Joshi et al., 2020). India, with its diverse agro-climatic zones and dependency on sectors vulnerable to climate variability, including agriculture and forestry, is especially vulnerable to phenological disruptions. However, studies on LSP in central Indian river basins such as the Tawa River Basin are limited, despite their ecological and socio-economic importance. This research aims to assess the slengthily phenological trends and breakpoints in the Tawa river basin using MODIS product of NDVI time ordered data (2002–2021) and to examine how LSP responds to climatic variations over time. By employing time-series decomposition, break detection methods like BFAST, and phenological metrics derived from TIMESAT and Phenofit, the research offers a multi-scalar perspective from global patterns to regional specificity, filling critical knowledge gaps in climate-phenology interactions in tropical semi-humid environments. This research is not only relevant for local environmental management but also contributes to the global discourse on ecosystem sensitivity to climate variability, making it of interest to international readers concerned with land system science, climate change adaptation, and satellite remote sensing applications.

10.2 Study Area

The Tawa River Basin, situated in the Indian state of Madhya Pradesh, encompasses the river, a major tributary of the Narmada River, extending over a length of about 200 km (Fig. 10.1). Basin is an important source of water for irrigation, drinking water, and hydropower generation. The area is known for its fertile agricultural land and is one of the most important agricultural regions in India. This Basin covers an area of approximately 4500 km^2 and has a population of around two million people. The river basin is characterized by a semi-arid climate region with high temperatures and minimum rainfall, making efficient water management a crucial issue. The river basin is also facing challenges such as rapid urbanization, industrialization, and increased water demand, which are putting pressure on the water resources and ecosystem of the region. Therefore, proper water management and sustainable development strategies are necessary to ensure to maintain the ecological health of the Basin. The River is a major tributary of the Narmada River and has its origin in the Satpura Range of Madhya Pradesh. The river basin is characterized by its tropical climate, with monsoon rains providing most of the annual precipitation. The basin is home to a diverse range of vegetation, including tropical deciduous forests, scrub forests, and grasslands. The river Basin is also home to a number of important wildlife species, including tigers, leopards, and many species of birds. The River Basin is a Significant source of water for Hydration, agriculture land and Consuming water supply in the region. The basin is also important for hydroelectric power generation, with a number of hydropower projects setup along the Tawa River and its tributaries. The Tawa River Basin has also been identified as a potential site for irrigation and flood control projects. In conclusion, the Tawa River Basin is an important region in Madhya Pradesh, providing crucial water resources for the surrounding communities and serving as a vital habitat for a variety of wildlife species are there.

10.3 Material and Methods

10.3.1 Data Used

The Vegetation Indices temporal data products are a series of remote sensing datasets sourced from the Moderate Resolution Imaging Spectroradiometer (MODIS) sensor, carried by NASA's Terra and Aqua satellite missions. These products are characterized by an observation frequency of 16 days' composite and a spatial resolution of 250 m, making them highly suitable for large-scale study and long-term monitoring of terrestrial ecosystems. The Normalized Difference Vegetation Index (NDVI) stands out as one of the most commonly utilized indicators for evaluating vegetation health, monitoring productivity, and tracking seasonal changes in plant dynamics. NDVI is calculated using the Bounce-back values from the red and near-infrared (NIR) bands of the MODIS sensor data (Fig. 10.3). Due to its sensitivity to vegetation biomass and canopy structure, NDVI has become an essential tool in environmental monitoring, land cover classification, drought assessment, agricultural planning, and phenological studies. It provides critical insights into how vegetation responds to climatic variations and anthropogenic pressures over time. In this study, NDVI was utilized to examine and monitor vegetation phenology, focusing on the rhythm of plant growing cycle events and their responses to seasonal and interannual variations in climate and environmental conditions. Phenological events like the onset of the developing season, peak vegetation activity, and senescence serve as vital indicators of ecosystem health and productivity. Tracking these changes over a long period allows researchers to identify trends, shifts, or anomalies that could signal ecological disturbances or changes in land recourses use and climate changes. To extract and analyze NDVI data over a 20-year period, from 2002 to 2021. I was able to systematically extract NDVI values for the study region across the entire 20-year period at 16-day intervals. This resulted in a comprehensive time series dataset with consistent spatial and temporal resolution, ensuring accurate analysis of vegetation changes and phenological transitions.

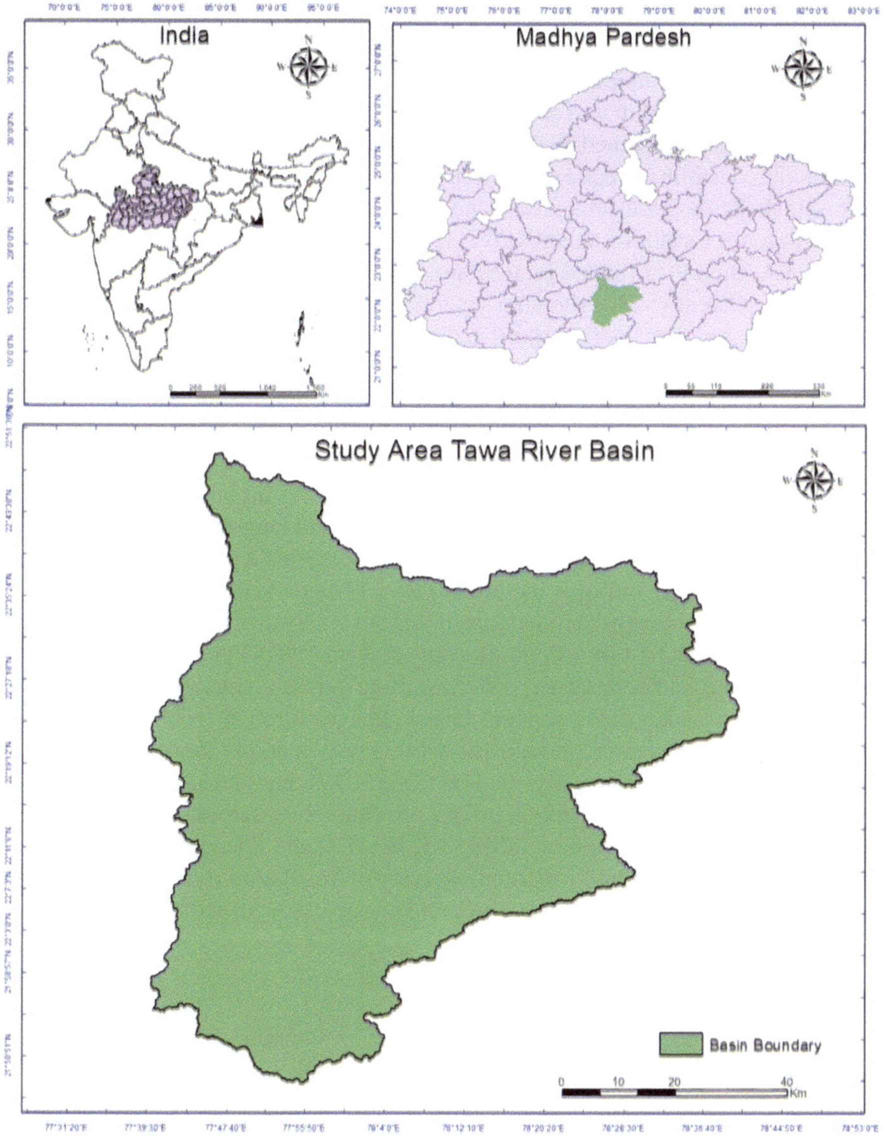

Fig. 10.1 The study area map

10.3.2 Software Package and Analysis

The study employs a comprehensive methodology to analyze vegetation phenology using NDVI time series data obtained from satellite observations covering the period 2002 to 2022, focusing on forested areas. TIMESAT software plays a central role in this analysis by extracting a key indicator of which is shown in figure (Fig. 10.2). To ensure data accuracy, the NDVI time series undergoes a two-step filtering process first smoothed with a median filter to reduce noise and outliers, subsequently, the Savitzky-Golay algorithm was applied to reduce the noise in data is caused by atmospheric noises. The smoothed NDVI curves are then fitted using an asymmetric Gaussian function, enabling precise extraction of phenological metrics. TIMESAT operates on stacked time series image data and facilitates pixel-based processing for identifying seasonal variations using NDVI, EVI, and biophysical parameters like LAI and FAPAR. Binary formatted images are processed with Seasonal Trend Decomposition using Loess (STL), allowing separation of trends from seasonal patterns. At the same time, the Breaks for Additive Seasonal and Trend (BFAST) algorithm is utilized to find structural changes within NDVI time series data. This method breaks down the series into its trend, seasonal, and residual elements, and then applies the OLS-MOSUM test to pinpoint significant breakpoints. These breakpoints highlight sudden or progressive changes in vegetation patterns, often triggered by events like droughts, land-use alterations, or climate variability. As an unsupervised method, BFAST effectively detects both abrupt disturbances and long-term ecological shifts. Additionally, the Phenofit R package is employed to refine daily-scale phenological signal extraction using multiple methods, TRS identifies SOS and EOS by tracking when NDVI values cross predefined thresholds but is sensitive to noise and threshold selection. DER uses first or second derivatives of the NDVI curve to capture peak rates of greening and senescence, offering more precise temporal estimates. The Inflection Method identifies curvature-based changes through local maximum and minimum in rate of change, though its stability varies across landscapes. The Gu Method fits a double logistic curve method to the NDVI data and identifies phenological events based on curve inflection points, proving robust across diverse vegetation types. Overall, the integration of TIMESAT, BFAST, and Phenofit allows for high-resolution (Table 10.1), multi-method analysis of vegetation dynamics over time, supporting ecosystem monitoring, climate change assessment, and land management strategies.

Adaptive Savitzky-Golay filtering

The adaptive Savitzky-Golay filter reduces noise by smoothing data using the least-squares fitting approach to a second-degree polynomial.

$$f(t) = C_1 + C_2 t + C_3 t^2 \tag{10.1}$$

where the weighted least-square technique is used to calculate $C_1 \ldots C_2$ and C_3, and the window's width regulates how smooth the data are. The original value is then replaced by the estimated figure in the data point. When there are significant increases or drops near the data points, TIMESAT tightens the window size to account for the quick change in the data. Because of this, the approach is particularly adaptable and useful for portraying abrupt seasonal shifts.

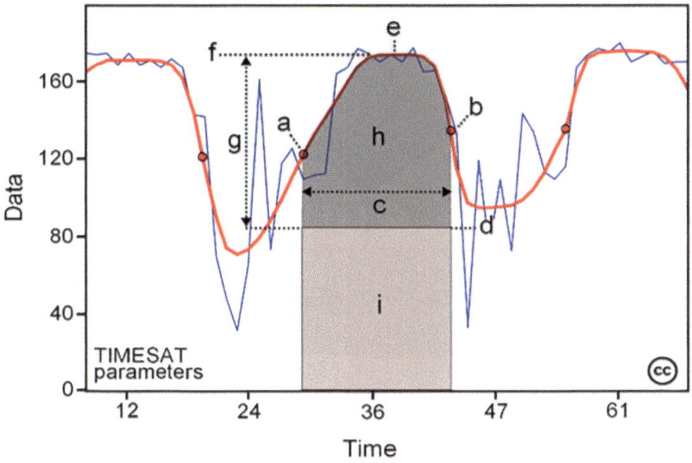

Fig. 10.2 Key seasonal metrics produced by TIMESAT: **a** start of the growing season, **b** conclusion of the growing season, **c** duration of the growing season, **d** baseline value. *Source* Authors (Eklundh & Jönsson, 2016)

Chronological Pattern Dissection

The time series y_t modeled as:

$$y_t = T_t + S_t + R_t \tag{10.2}$$

where

T_t **Trend component** (long-term changes).
S_t **Seasonal component** (periodic fluctuations).
R_t **Remainder component** (noise or random errors).

This decomposition separates the systematic behaviour of the time series into interpretable parts. BFAST identifies breakpoints (structural changes) in both T_t (trend) and S_t (seasonality) by segmenting the time series.

Seasonal and Trend Decomposition

The seasonal and trend components are modelled separately:

Trend (T_t) is estimated using a regression model:

$$T_t = B_0 + B_1 t + \epsilon_t \tag{10.3}$$

$$s_t = \sum_{k=1}^{k} \left(a_k \cos\left(\frac{2\pi kt}{p}\right) + b_k \sin\left(\frac{2\pi kt}{P}\right)\right) \tag{10.4}$$

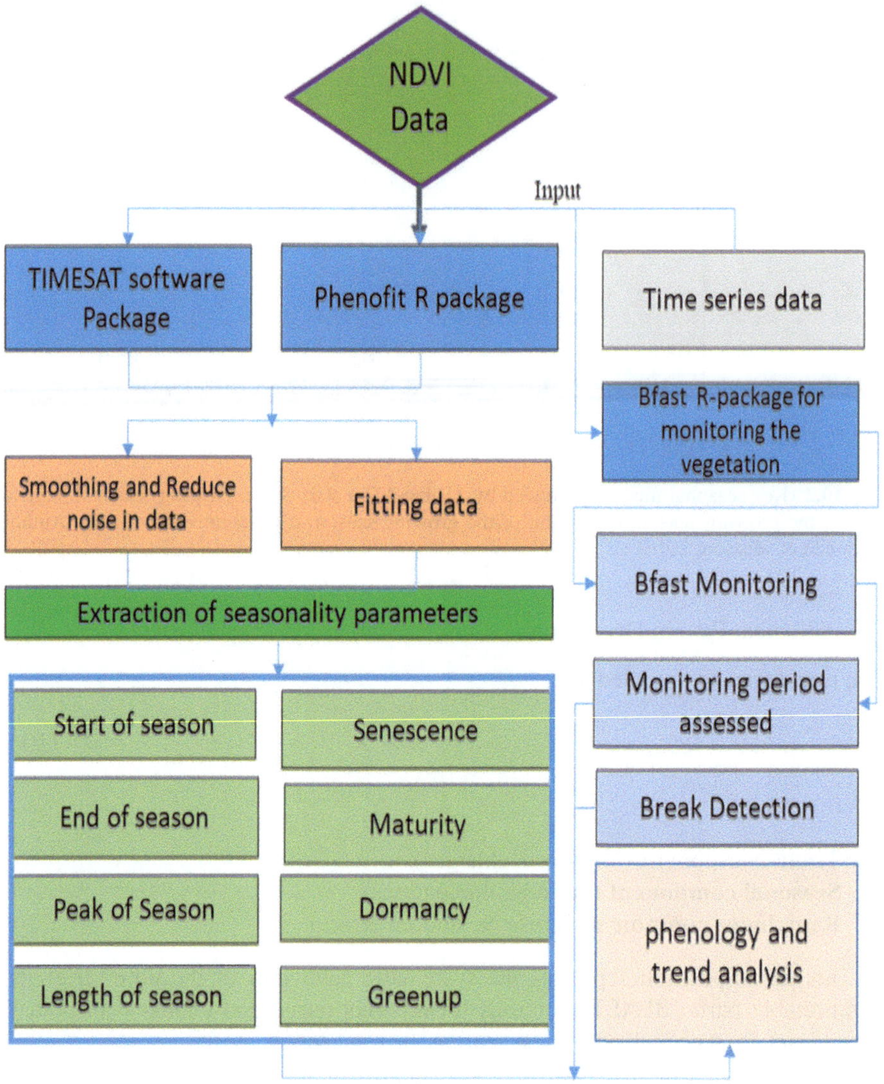

Fig. 10.3 Methodology flow chart

Breakpoint Detection

BFAST identifies breakpoints (structural changes) in both Tt (trend) and St (seasonality) by segmenting the time series.

$$T_t^{(i)} = \beta_0^{(i)} + \beta_1^{(i)} t \tag{10.5}$$

Table 10.1 Parameter for fine fitting curve

t_0	Transition point between the vibrant bloom of spring and the crisp decline of autumn
mn	Background value
mx	Maximum value
sos	Beginning of the growth period
eos	Conclusion of the growth cycle
rsp	Speed of springtime greening
rap	Rate of autumn senescence

Source Authors (Kong et al. 2022)

Breakpoints are identified by reducing the residual sum of squares (RSS) to its minimum across all potential segment configurations.

$$\min \sum_{\{i=1\}}^{m} \sum_{\{t=t_{\{i-1\}}+1\}}^{\{t_i\}} \left[Y_t - T_t^i - S_t^i\right]^2 \tag{10.6}$$

Monitoring Changes with BFAST Monitor

For near-real-time applications, the BFAST monitor compares a new observation Y_t against the fitted components T_t and S_t to detect anomalies.

The monitoring equation is (Fig. 10.3):

$$y_t - (T_t + S_t) = \delta_t \tag{10.7}$$

10.4 Results and Discussion

10.4.1 Extraction Phenological Parameter and Fine Fitting Curve

The Phenofit R package offers robust functions for fine curve fitting of phenological data. Here, T represents the temporal sequence corresponding to y (e.g., a Sequential data of Vegetation Index, VI). The parameters mn and mx define the lowest and highest values of y, respectively. The terms sos and eos indicate the start and end of the developing season, respectively. The parameters rsp and rau describe the rate of spring green-up and autumn senescence, respectively. t_0 marks the inflection point between vegetation green-up and senescence, while a_3 and a_5 serve as morphological parameter for the Asymmetric Gaussian (AG) model. The provided dataset represents the values of different variables related to vegetation dynamics over a period of 20 years (Table 10.2). The variables include mn (minimum value of NDVI), mx (maximum value of NDVI), rsp (rate of spring Greenup), and rau (rate of autumn senescence). NDVI is a commonly used index derived from remote sensing data that provides an indication of vegetation health and density (Figs. 10.4 and 10.5). Analyzing the dataset, it is observed that the minimum and maximum values of NDVI (mn and mx) vary across different years. The range of variation suggests temporal fluctuations in vegetation density and health within the spatial area. The values of mn range from 0.157 to 0.301, indicating the lowest level of vegetation cover or health, while the values of mx range from 0.712 to 0.81, representing the highest level of vegetation cover or health. These rates provide insights into the timing and speed at which vegetation transitions from dormant or senescent to active growth and vice versa. The values of rsp range from 0.00779 to 0.0466, indicating the rate at which vegetation experiences spring green up. However, the other hand, the values of rau, a measure of the rate of autumn senescence, range from 0.00535 to 0.0112. The rates of spring green up and fall senescence show variation after 20 years of data analysis, indicating changes in the occurrence and length of these phenological occurrences (Figs. 10.6, 10.7 and 10.8). Climate, temperature, precipitation, and ecological conditions unique to the research area could all have an impact on these variations. In order to measure ecosystem health, follow plant responses to environmental changes, and examine the Consequences of land use patterns and climatic fluctuations on the dynamics of floral growth and spatial distribution, it is crucial to comprehend the patterns and dynamics of NDVI and phenological rates, as indicated by rsp and rau. Such analyses can provide valuable information for ecological studies, biodiversity conservation, and land management practices. Further analysis of the dataset, in conjunction with additional environmental data and digital data through satellite imagery, could provide a more comprehensive understanding of the factors driving the observed variations in NDVI and phenological rates. This could involve exploring correlations with climate variables, land use changes, and other environmental factors to investigate potential drivers of forest dynamics in the area of interest. Further research and analysis are warranted to fully understand the underlying mechanisms driving these variations and to inform effective planning strategies for sustainable land use and ecosystem conservation.

10.4.2 Phenology Parameters Extraction

The analysis of vegetation phenology using long-term remote sensing data provides critical insights into seasonal vegetation dynamics, ecological responses, and environmental change. In this research, MODIS product, which is 250 m spatial resolution, 16-day composite NDVI as indicator was utilized to assess seasonal phenological parameters over a nineteen-year period, from 2002 to 2021. The use of MODIS data allows for consistent, high-temporal-resolution observations of vegetative patterns across large landscapes, offering a powerful tool for monitoring changes over time. Specifically, the TIMESAT software was employed to process and analyze

Table 10.2 Phenological parameter by Phenofit

Year	t_0	mn	mx	rsp	a_3	rau	a_5
2000	250	0.248	0.722	0.0112	2	0.00728	2.87
2001	641	0.157	0.731	0.0121	2	0.00743	2
2002	1012	0.252	0.721	0.0114	2	0.00771	3.31
2003	1420	0.23	0.736	0.00779	2	0.0106	2
2004	1745	0.214	0.746	0.00798	2	0.00894	2
2005	2097	0.271	0.784	0.0164	2.8	0.0075	2
2006	2470	0.28	0.755	0.0258	3.54	0.00824	3.04
2007	2867	0.291	0.719	0.012	2	0.0112	3.52
2008	3175	0.291	0.729	0.022	6	0.00787	3.31
2009	3586	0.301	0.715	0.0127	2	0.00817	5.07
2010	3931	0.273	0.765	0.00971	2	0.00786	3.74
2011	4298	0.267	0.77	0.0127	2	0.00856	2.3
2012	4681	0.247	0.712	0.00879	2	0.00804	2
2013	5005	0.249	0.722	0.0408	6	0.00535	2.25
2014	5416	0.234	0.757	0.019	6	0.00748	2
2015	5751	0.281	0.784	0.00905	2.01	0.0088	2.53
2016	6137	0.234	0.757	0.0179	6	0.0078	2
2017	6452	0.294	0.77	0.0466	5.13	0.00713	3.43
2018	6834	0.201	0.785	0.0466	4.38	0.00673	2
2019	7238	0.274	0.81	0.0232	6	0.00793	2.81

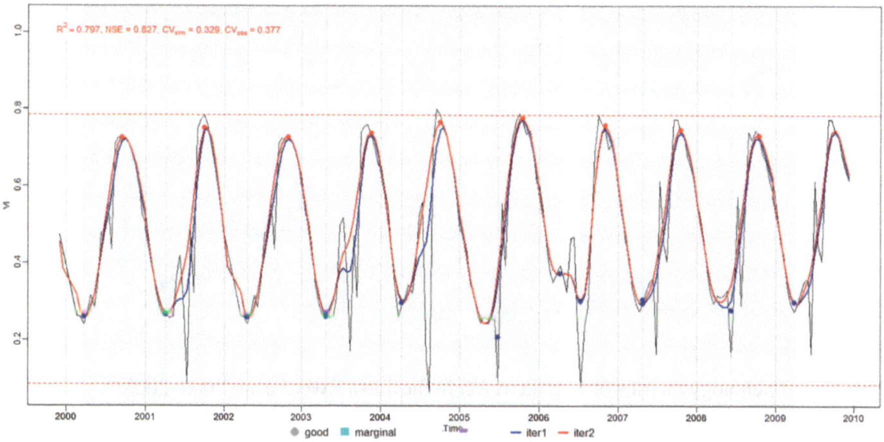

Fig. 10.4 Curve fitting and seasonal growth segmentation for the Tawa River Basin from January 2010 to December 2016, utilizing MOD13Q1 Collection 6 16-day NDVI data as the primary input

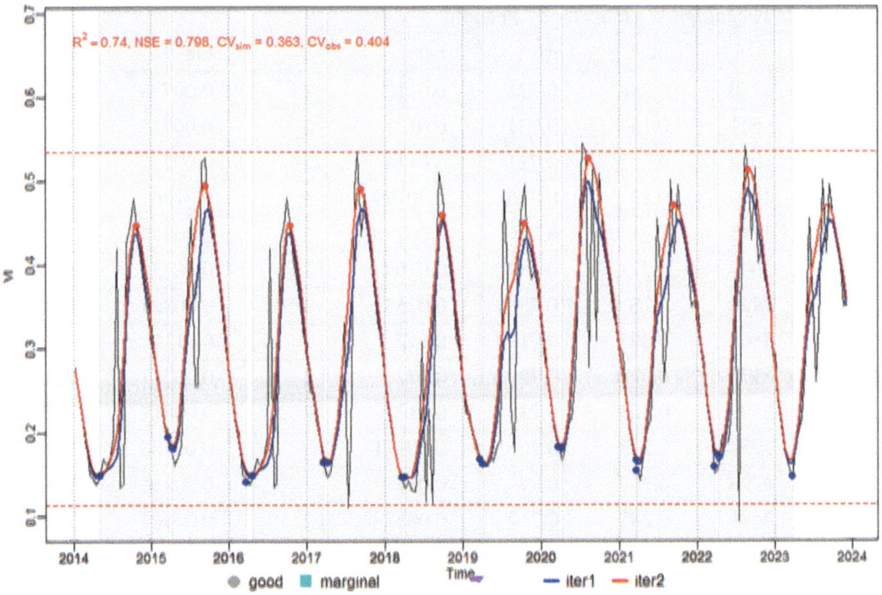

Fig. 10.5 Vegetation dynamics and seasonal segmentation in the Tawa River Basin (Jan 2014–Dec 2022) using MOD13Q1 C6 16-day NDVI data

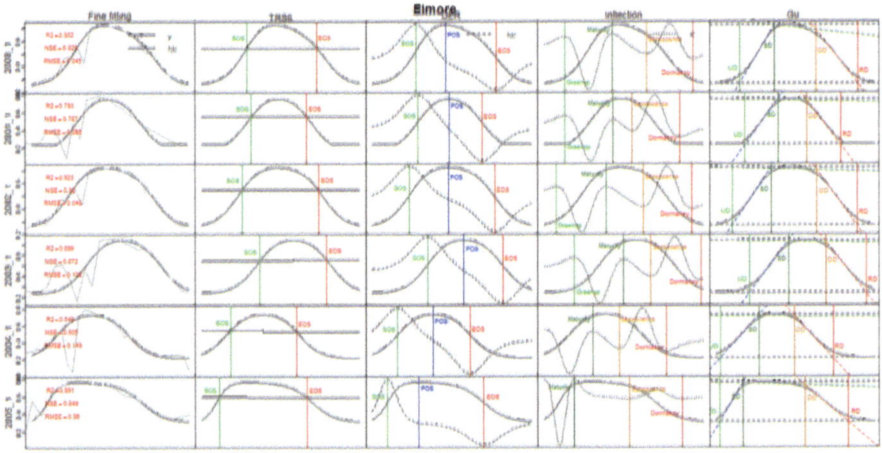

Fig. 10.6 Phenology metrics of fine-fitted AG in the Tawa River basin (2000–2005)

the vegetation time ordered, enabling the extraction of vital seasonality parameters. The primary seasonality parameters extracted through TIMESAT, base level, and the integral of the season (a cumulative measure of vegetation productivity over the season). These parameters collectively provide a comprehensive picture of vegetation dynamics, allowing researchers to identify when vegetation starts to grow,

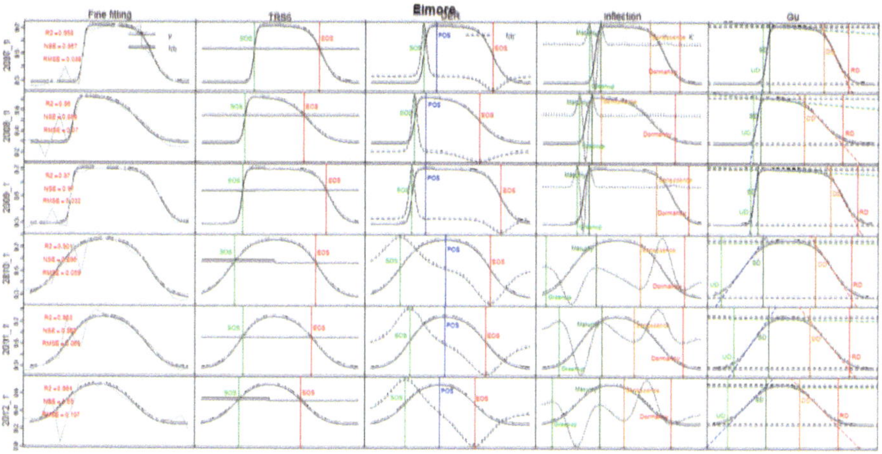

Fig. 10.7 Phenology metrics from fine fitting AG in the Tawa River basin (2007–2012)

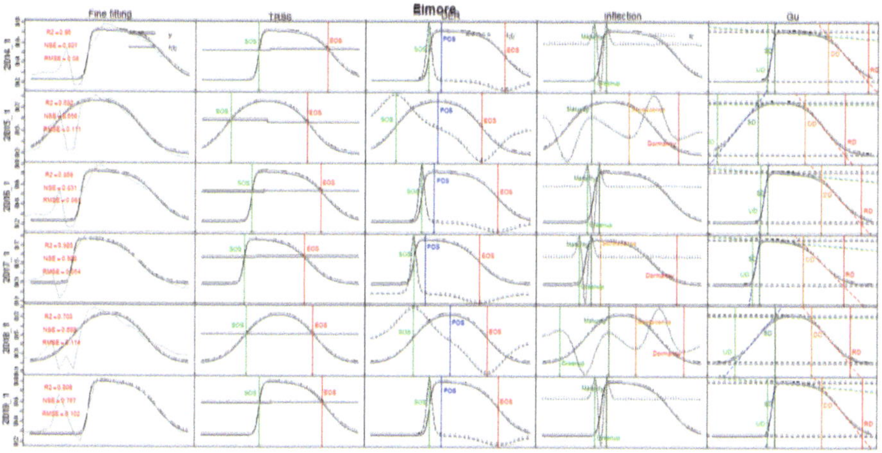

Fig. 10.8 Phenological metrics (2014–2019) for the Tawa River basin derived from the fine-fitting AG model

reaches peak productivity, and begins to senesce or decline. The ability of TIMESAT to fit smoothed curves to the time-series data enables more robust and accurate detection of phenological transition points, particularly in relation to the annual fluctuations and changes in land cover. Over the 19 growing seasons identified in this study, seasonal metrics were computed for each year, allowing for the observation of changes in seasonality patterns over time. The length of the growing period, for instance, fluctuated significantly across the years, with some years experiencing shorter growing periods due to climatic or anthropogenic influences, while others had extended growing seasons, potentially driven by favorable climatic conditions or changes in land management practices. Among all years, the year 2009 was identified as having the longest growing season, suggesting particularly favorable conditions for vegetation growth during that period. This variation highlights the importance of continuous monitoring and analysis of phenological parameters, as such insights can provide early indications of environmental shifts, such as changes in precipitation patterns, temperature trends, or land use alterations.

The provided dataset contains information related to different seasons, including the beginning time (Beg. *t*.), end time (End *t*.), length, base value (Base val.), maximum time (Max *t*.), maximum value (Max val.), amplitude (Ampl.), left derivative (L-der.), right derivative (R-der.), left integral (L-integ.), seasonal integral (S-integ.), starting value of seasonal integral (S-integ. Start val.), and ending value of seasonal integral (S-integ. End val.) (Table 10.3). Analyzing the data, it can be observed that each season has different characteristics in terms of timing, duration, amplitude, and integral values. The length of the seasons varies, ranging from 4.695 to 14.78 units. The base values range from 2195 to 4184, indicating the starting point of the seasonal cycle. The maximum values range from 7274 to 9354, representing the peak value during the respective season. The amplitude values range from 2501 to 6770, indicating the range or difference between the highest and lowest values within a season. The left derivative values range from 512.2 to 1598, representing the progression rate of a particular change of the variable at the start of the season. Similarly, the right derivative values range from 366.5 to 740.4, representing the rate of change of the variable at the end of the season. The left integral values range from 6.31E+04 to 1.23E+05, indicating the cumulative sum of the variable from the beginning of the season. The seasonal integral values range from 4.05E+04 to 7.91E+04, representing the cumulative sum of the variable over the entire season. The starting values of the seasonal integral range from 5450 to 6621, while the ending values range from 5444 to 6665 (Figs. 10.9, 10.10 and 10.11). The features of each season, including their date, duration, peaks, rates of change, and cumulative amounts, can be understood by analyzing the dataset. These characteristics can be utilized to comprehend the dynamics, trends, and interactions between the variable under examination and other environmental variables (Fig. 10.12). To adequately assess the significance of these findings, additional dataset analysis would be required, in addition to additional contextual data and domain expertise. The particular variable and the environment in which it is being investigated will determine how the analyzed data should be interpreted and used.

10.4.3 Identification of Instability and Sensitivity of the Ecosystem

The time series analysis presented here pertains to a model fitted to a dataset covering a specific and stable period, termed the history period, ranging from January 2002 to December 2010. This period spans approximately 8.92 years and was selected

Table 10.3 Phenological parameter by Timesat

Season	Beg. *t*	End *t*	Length	Base val.	Max *t*	Max val.
1	7.173	17.96	10.79	2524	12.39	8370
2	29.81	43.59	13.78	2787	37.74	8597
3	53.37	63.07	9.703	3224	58.82	8749
4	71.85	76.55	4.695	3509	74.26	6009
5	86.84	97.56	10.72	3587	91.46	7274
6	107.1	120.9	13.88	3281	113.2	7301
7	129.7	142.8	13.09	2794	135.9	6931
8	153	167.8	14.78	2578	159.8	8255
9	174.7	189.3	14.69	2846	181	8286
10	199.8	211.6	11.82	2947	204.6	8559
11	223.6	235.7	12.09	2195	228.8	8965
12	248.3	256.9	8.594	2751	250.6	9354
13	269.3	280.3	11.05	4184	273.4	8857
14	290.2	302.9	12.68	3626	295.8	8348
15	316.9	326.2	9.288	2787	320.5	9010
16	338.5	349.4	10.93	2709	342.9	8924
17	363	370.3	7.253	2514	365.7	8706
18	383	395.9	12.99	3282	390.9	9149
19	404.7	418.2	13.52	4006	412.4	9084

Season	Ampl.	L-der.	R-der.	L-integ.	S-integ.	S-integ. Start val.	End val.
1	5845	714.6	477	8.17E+04	5.14E+04	5450	5444
2	5810	509.1	541.3	1.09E+05	6.47E+04	5557	5826
3	5525	527.2	487.7	8.43E +04	4.56E+04	5903	6070
4	2501	827.4	192.6	3.59E+04	1.14E+04	4701	4817
5	3686	1598	520.7	8.96E+04	4.30E+04	5449	5412
6	4020	1195	515.4	1.07E+05	5.79E+04	5426	5157
7	4137	707.6	541.5	9.75E+04	5.56E+04	4972	4754
8	5677	1191	740.4	1.23E+05	7.91E+04	5416	5417
9	5440	1483	517.1	1.20E+05	7.13E+04	5433	5699
10	5612	824	436.2	9.63E+04	5.51E+04	5836	5671
11	6770	566.9	539.2	9.87E+04	6.80E+04	5874	5287
12	6603	2470	366.5	7.48E+04	4.73E+04	5481	6624
13	4674	857.9	373.1	9.84E+04	4.40E+04	6376	6665
14	4723	748.7	484.3	1.03E+05	5.23E+04	6411	5563
15	6223	512.2	510.7	8.36E+04	5.01E+04	5894	5903
16	6215	869.3	537.3	9.39E+04	5.87E+04	5860	5773
17	6191	1305	572.7	6.31E+04	4.05E+04	5664	5556

(continued)

Table 10.3 (continued)

Season	Ampl.	L-der.	R-der.	L-integ.	S-integ.	S-integ. Start val.	End val.
18	5867	580.6	507.8	1.10E+05	6.11E+04	5777	6654
19	5078	380.8	466.7	1.17E+05	5.29E+04	6621	6469

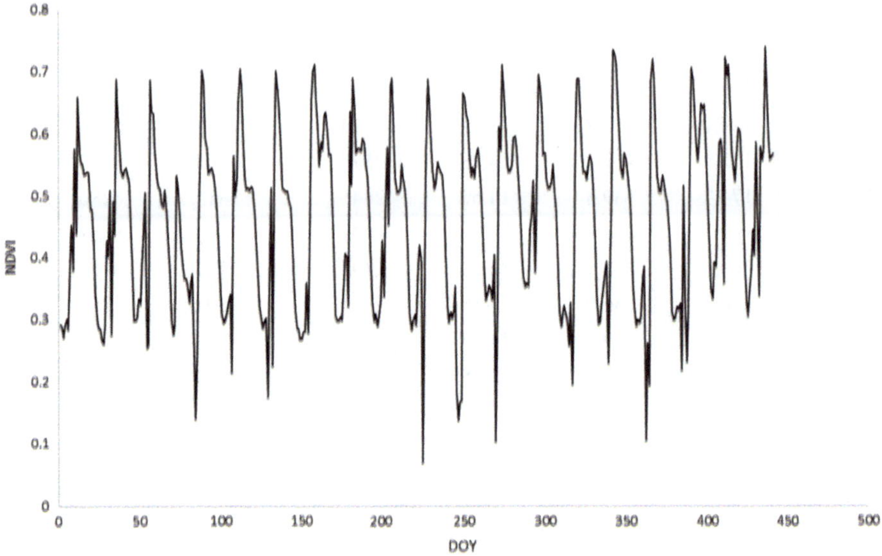

Fig. 10.9 Showing season over the time series data

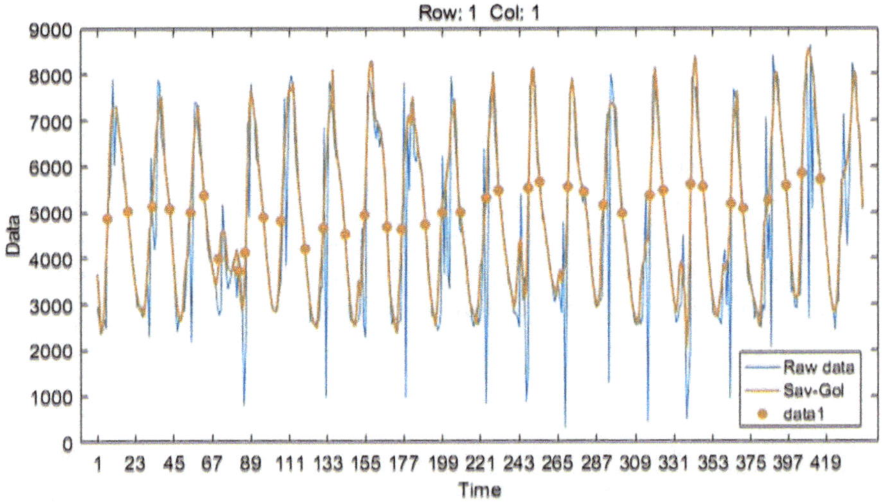

Fig. 10.10 A time-series dataset, consisting of values $y_1, y_2, \ldots, y_N$, spans 20 years. The seasonal parameters are derived from fitted functions based on the 19 complete seasonal cycles

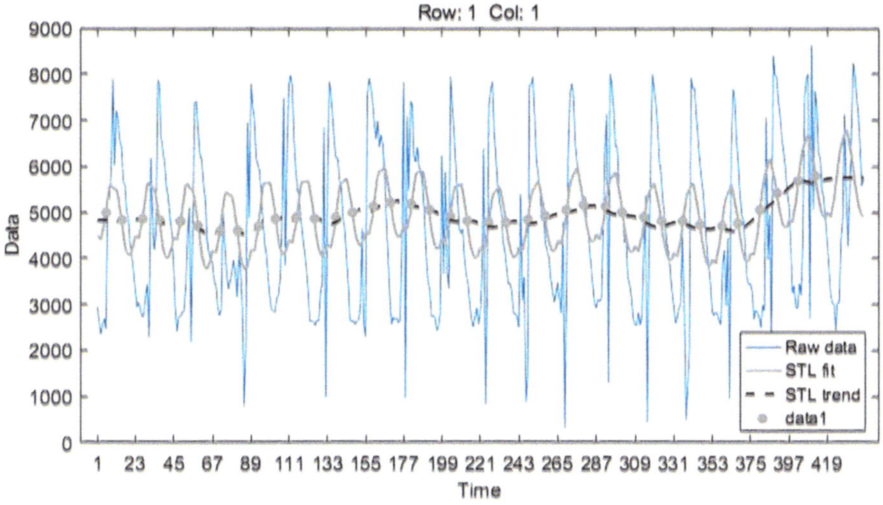

Fig. 10.11 Seasonal trend over the time series data

to ensure stability in the observed trends and seasonal cycles, offering a reliable baseline for assessing long-term changes in vegetation phenology (Table 10.4). The model fit includes several coefficients that represent different aspects of the time series structure. The Intercept represents the baseline or average NDVI (Normalized Difference Vegetation Index) value across the history period. This coefficient gives a reference point from which variations due to seasonal and trend components can be analysis. In the context of forest phenology, the intercept might reflect the average greenness or vegetative productivity in the study region over the stable period. The trend coefficient captures the linear progression of the data over time. A positive trend value indicates an increasing pattern, suggesting that vegetation greenness or productivity has been gradually improving throughout the history period. Conversely, a negative trend would imply a decline in vegetation cover or health over time. The trend component is essential in identifying long-term ecological shifts that could be driven by environmental forces like shifts in climate patterns, land use and land cover changes, or restoration activities. Beyond the linear trend, the model incorporates harmonic components, denoted by coefficients such as harmoncos1, harmonsin1, up to harmoncos3 and harmonsin3. These harmonics model the periodic or cyclical variations in the data, which in this case reflect the seasonality inherent to vegetation growth patterns. Seasonal fluctuations in NDVI are expected due to annual vegetation cycles, influenced by climatic factors like temperature, rainfall, and solar radiation. The first harmonic (e.g., harmoncos1 and harmonsin1) represents the dominant annual cycle, while higher-order harmonics capture more complex sub-annual variations or additional seasonal fluctuations. These components are vital for accurately capturing the temporal dynamics of vegetation phenology and ensuring that the model reflects both long-term trends and recurring seasonal signals (Fig. 10.13). The model demonstrates a strong fit to the data, as evidenced by a high R-squared

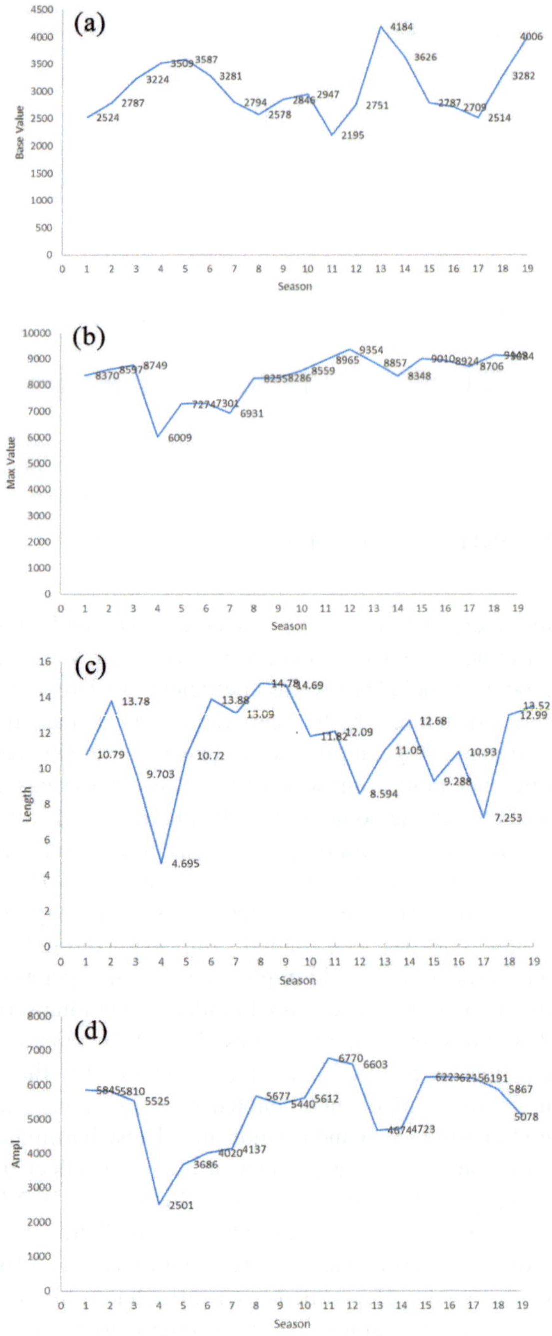

Fig. 10.12 a–d Length of season, base value, amplitude, and maximum value

value of 0.923896. This metric indicates that approximately 92.39% of the variability in the observed NDVI data during the history period is explained by the model. An R^2 value close to 1.0 suggests an excellent model fit, implying that both the linear and harmonic components effectively capture the patterns present in the data. In time series analysis, a high R^2 value is crucial as it demonstrates that the model is robust enough to replicate historical behaviors and can potentially be used for forecasting or detecting anomalies. This level of model performance suggests that the chosen stable period (2002–2010) provides a coherent and consistent dataset with well-defined seasonal behavior and a discernible trend. By fitting a model that explains most of the variation, researchers can confidently interpret deviations or changes occurring in subsequent years as significant departures from the historical baseline. This is particularly important when monitoring ecological health or detecting early signs of ecosystem stress or transformation. Furthermore, the harmonic decomposition provided by the model allows researchers to distinguish between long-term changes and natural seasonal variability. This is particularly useful when identifying anomalies, abrupt declines in vegetation productivity, or disruptions in normal seasonal cycles, which may be indicative of climate impacts, land degradation, or human intervention. In summary, the time series model fitted to the MODIS NDVI dataset over the stable period from 2002 to 2010 demonstrates strong explanatory power and captures both linear and periodic trends effectively. The intercept establishes the baseline vegetation status, the trend component reflects ongoing changes over time, and the harmonic terms model seasonal patterns with high precision. The high R^2 value further reinforces the reliability of the model, indicating a strong alignment between the modeled and observed data. Such analysis provides a solid foundation for Longitudinal Ecological Examining forecasting and the development of responsive land management strategies.

Table 10.4 Liner trend coefficients

1. History period			
Stable period selected: 2002(1)–2010(12)			
Length (in years): 8.916667			
Model fit:			
(Intercept)	trend	harmon cos 1	harmon cos 2
0.4436209847	0.0003680743	0.0057909543	0.0222161142
harmon cos 3	harmon sin 1	harmon sin 2	harmon sin 3
0.0237960330	−0.1682671711	0.0697311454	−0.0123187364
R-squared: 0.923896			

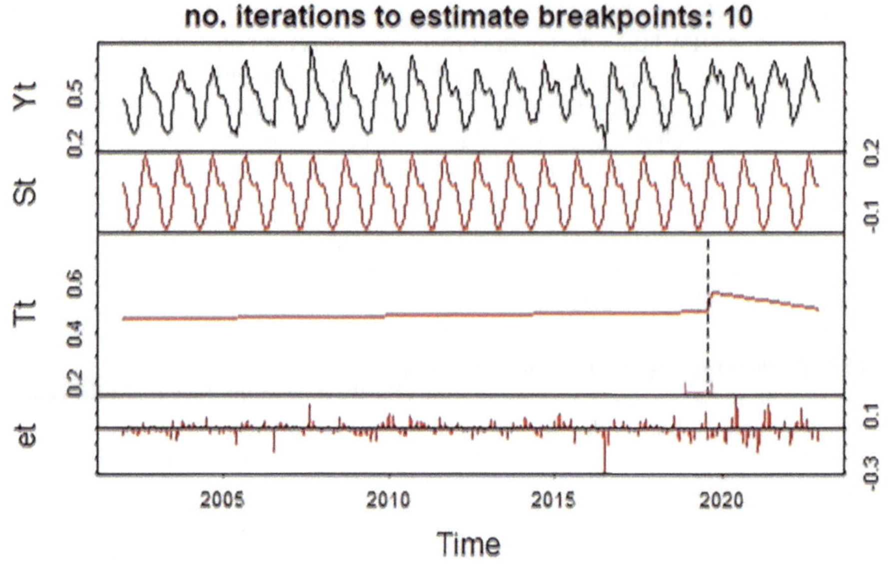

Fig. 10.13 BFAST decomposition of NDVI time, stands. Y_t = original time series, S_t = seasonal component, T_t = trend component, and e_t = remainder component

10.4.4 Break Detection and Trend Identification

In the framework of the BFAST R language product, break detection and trend analysis are essential components used to study land surface phenology over extended time periods. This study utilized the BFAST approach to analyze NDVI time series data spanning from 2002 to 2021 to identify significant structural changes in vegetation patterns and trends. Land surface phenology, which refers to the timing of recurring vegetation developments as observed from satellite data, is influenced by various atmospheric and natural factors, LULC changes, and anthropogenic activities. Detecting anomaly in phenology can provide insights into the health and functioning of ecosystems, mainly in response to climate change variability and land use dynamics. Using BFAST R package, the NDVI time series long term data was first decomposed into its three main components are trend, seasonal, and remainder. This decomposition allowed for a clearer understanding of long-term trends in vegetation, recurring seasonal cycles, and random noise in the data. The BFAST method was employed to examine breakpoints significant changes in the seasonal and trend components which may indicate shifts in phenological behavior due to natural or anthropogenic influences. In this research work, a history period from 2002 January to 2010 December was selected as a stable reference period, which provided a baseline for assessing future changes. This history period, with a length of approximately 8.92 years, was assumed to be relatively stable and free from major disturbances. The monitoring period, defined from 2011 to 2022, covered a span of approximately

11.92 years and was used to detect deviations from the baseline trend established in the history period. During this monitoring period, BFAST was able to identify structural breaks in the NDVI time series, revealing changes in vegetation dynamics, growth cycles, and land surface phenology. The analysis revealed that phenological changes occurred between the stable and monitoring periods, with breakpoints indicating moments when the vegetation patterns significantly deviated from the expected seasonal cycle or long-term trend. The distribution of NDVI values over the 20-year study period, highlighting fluctuations in vegetation greenness and the occurrence of breakpoints in specific years. These breakpoints serve as indicators of phenological shifts, possibly triggered by extreme weather events, land degradation, deforestation, agricultural expansion, or climate change. Table 10.5 summarizes the different factors influencing these phenological changes, such as variations in temperature, rainfall, solar radiation, and human interventions. By integrating trend analysis and breakpoint detection within the BFAST framework, this study provides a robust methodology for monitoring land surface phenology over time. The results highlight the importance of continuous monitoring of vegetation patterns, especially in ecologically sensitive and climate-vulnerable regions. The findings support improved understanding of ecosystem responses to environmental stressors and provide valuable information for land management, conservation planning, and formulating policies to reduce the adverse effects of climate change on terrestrial forest ecosystems.

The analysis provided above pertains to a monitoring period, which spans from January 2011. The analysis presented above focuses on the monitoring period of the research work, which time spans from 2011 January (Month 1) to 2022 December (Month 12), covering a duration of approximately 11.92 years. This period follows a defined history period (2002–2010) and is used to detect significant deviations or structural changes in the Time-based data of NDVI (Normalized Difference Vegetation Index) values. Through the application of the BFAST method within the R programming environment, the study successfully identified a significant breakpoint in July 2016 (Month 7). In time series analysis, a break typically signifies a substantial shift or alteration in the trend, seasonality, or variability of the observed variable, in this case, vegetation greenness represented by NDVI. The detection of this break in July 2016 suggests a notable deviation in vegetation behavior or ecosystem functioning during that time. Such a shift may result from various environmental or anthropogenic factors, including abnormal climatic conditions (e.g., drought, excessive rainfall, heatwaves), land use changes (e.g., deforestation, urban expansion), or agricultural intensification. Identifying this breakpoint is critical for understanding how ecosystems respond to disturbances and how such changes influence vegetation cycles over time. Figure 10.14 visually illustrates this identified change point,

Table 10.5 Monitoring periods

2. Monitoring period
Monitoring period assessed: 2011(1)–2022(12)
Length (in years): 11.916667
Break detected at: 2016(7)

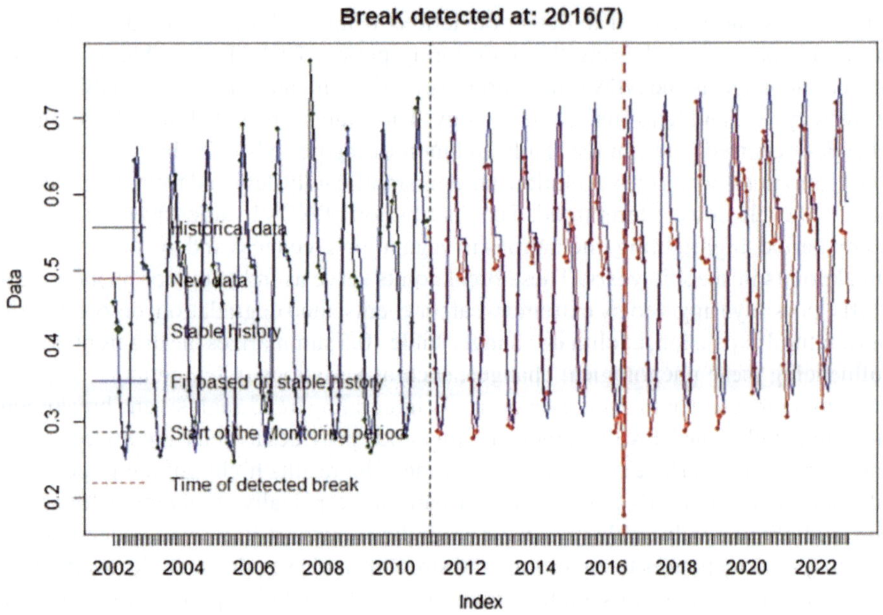

Fig. 10.14 Break detected in 2016 year form the 20-year time series of NDVI data

enabling a clearer understanding of the timing and magnitude of the deviation from the established trend or seasonality. The precise timing of the break can help narrow down the investigation of potential causes, guiding further research into meteorological records, policy changes, or land management practices that occurred around mid-2016. For instance, changes in monsoon patterns, El Niño events, or local development activities might coincide with this observed shift. Moreover, the identification of such a breakpoint underscores the effectiveness of the BFAST method in detecting subtle yet significant changes in long-term time series data. It enables researchers and land managers to track phenological changes, monitor ecosystem health, and implement adaptive management strategies. Understanding when and why such changes occur can also contribute to climate change impact assessments, ecological forecasting, and sustainable planning. The detection of the break in July 2016 adds a critical temporal dimension to the study's findings and opens pathways for in-depth exploration of vegetation dynamics within the broader environmental context.

10.4.5 Climate Analysis

See Figs. 10.15, 10.16 and 10.17.

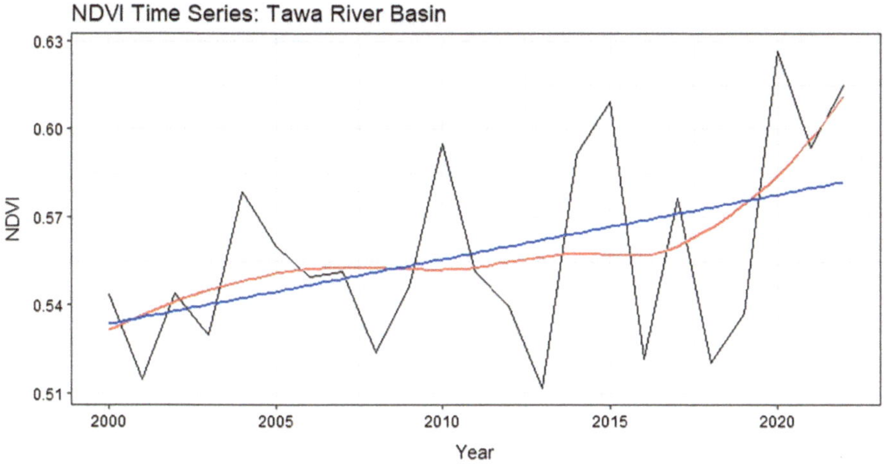

Fig. 10.15 Time series NDVI from 2000 to 2022

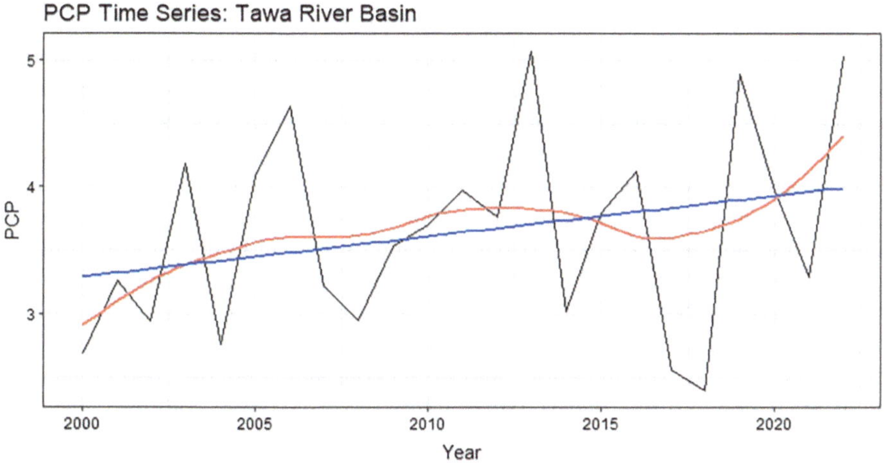

Fig. 10.16 Time series precipitation (PCP) from 2000 to 2022

10.4.6 Mann–Kendall Analysis and Sen's Slope Method for Trend Analysis

The Mann–Kendall trend analysis was used to examine the pattern in vegetation phenology (NDVI), rainfall (PCP), and maximum temperature (TMAX) over 23 observations for the selected region (Figs. 10.15, 10.16 and 10.17). The analysis revealed weak positive trends in all three variables, though none were statistically significant. For NDVI, the test statistic ($z = 1.5318$) indicates a weak positive trend with a Kendall's tau of 0.23320, but the p-value of 0.1256 suggests the trend is not

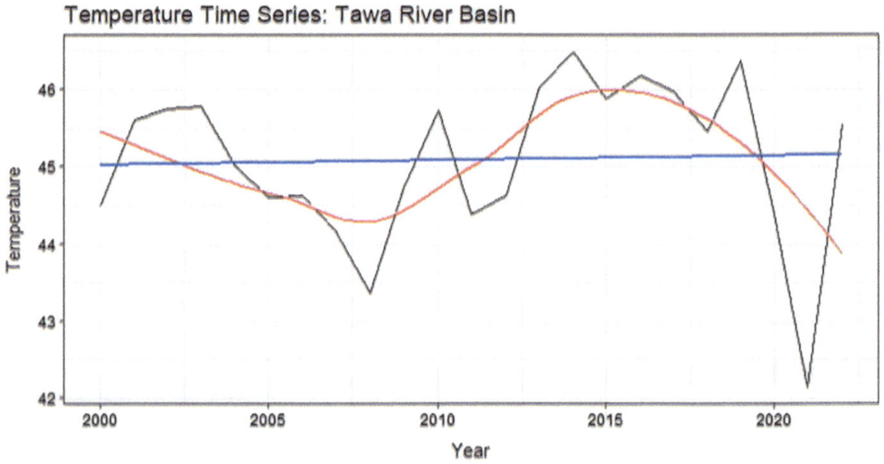

Fig. 10.17 Time series precipitation temperature from 2000 to 2022

significant at the 5% level. The SSS statistic (59) and its variance (1433.67) indicate mild variability in NDVI over the study period. Similarly, for rainfall, the test statistic ($z = 1.3474$) and Kendall's tau (0.20590) suggest a weak increasing pattern, but the p-value of 0.1779 confirms the lack of statistical significance. The SSS statistic (52) and its variance (1432.67) point to moderate variability in rainfall. For maximum temperature, the test statistic ($z = 0.6867$) and Kendall's tau (0.10670) indicate an even weaker positive trend, with a p-value of 0.4923 highlighting the absence of statistical significance. The SSS statistic (27) and variance (1433.67) suggest minimal change in temperature over the study period. In summary, while weak positive trends were observed in NDVI, rainfall, and maximum temperature, none of these trends were statistically significant, indicating no substantial changes occurred during the observation period. This stability could reflect consistent climatic conditions in the study area, though further investigation with a larger dataset or additional variables may be needed to uncover subtle trends or underlying factors. In summary, while all three variables (NDVI, rainfall, and maximum temperature) exhibit weak positive trends, none of these trends are statistically significant. This suggests that over the observation period, no substantial changes occurred in vegetation phenology, rainfall, or temperature, possibly due to stable climatic conditions in the research area. The results of the Sen's slope analysis for NDVI, rainfall (PCP), and maximum temperature (TMAX) from the dataset reveal interesting trends. For NDVI, the estimated Sen's slope is 0.0023, indicating a slight increasing trend in vegetation index over time. However, this trend is not statistically significant ($p = 0.1256$), as the confidence interval (− 0.00036 to 0.00424) includes zero. This suggests that the observed increase in NDVI might be influenced by variability rather than a consistent pattern. Similarly, the rainfall (PCP) data shows a positive Sen's slope of 0.0379, indicating an increasing trend in precipitation. However, like NDVI, this trend is not statistically significant ($p = 0.1779$), with a confidence interval spanning from − 0.0253 to

0.0944. This implies that the observed increase in rainfall is not strong enough to be confidently attributed to a persistent trend. For maximum temperature (TMAX), the Sen's slope is estimated at 0.0238, indicating a marginal increase in temperature over time. However, this trend is also not statistically significant ($p = 0.4923$), with the confidence interval (− 0.0486 to 0.0842) crossing zero. This suggests that temperature fluctuations over time are not consistent enough to define a significant upward or downward trend. While there are slight increasing trends in NDVI, rainfall, and maximum temperature, none of these are statistically significant.

10.4.7 Relationship Between NDVI and Climate Data

Figure 10.18 is a pairs plot, commonly used for exploratory data analysis to visualize pairwise relationships between multiple variables. The variables include NDVI, PCP (representing precipitation), and TMAX (maximum temperature). The diagonal panels display histograms with density curves for each variable, revealing their distributions. NDVI appears slightly left-skewed, PCP shows a moderate spread with a slightly bell-shaped curve, and TMAX also exhibits a bell-shaped distribution. The lower triangular panels contain scatterplots between pairs of variables, overlaid with red smoothing lines (likely LOESS curves) to highlight trends. The relationship between NDVI and PCP shows no significant trend, while NDVI and TMAX exhibit a weak negative correlation (− 0.11), suggesting that higher maximum temperatures may slightly reduce vegetation health. Similarly, PCP and TMAX show a weak positive correlation (0.21), indicating that areas or periods with higher precipitation might experience slightly higher maximum temperatures. The upper triangular panels display correlation coefficients between the variables. NDVI and PCP have a very weak positive correlation (0.03), NDVI and TMAX exhibit a weak negative correlation (− 0.11), and PCP and TMAX show a weak positive correlation (0.21). The correlations between these variables are relatively weak, suggesting no strong linear relationships. However, the weak negative correlation between NDVI and TMAX might hint at a slight adverse effect of maximum temperature on vegetation. Further analysis, such as regression or non-linear modeling, could help clarify these relationships. Additionally, segmenting the data by season or region may uncover stronger patterns or trends. Including other variables could provide deeper insights into factors influencing NDVI.

10.5 Conclusion

Amid global climate change patterns and the escalating impact of Human-induced processes, the ongoing transformation of LULC patterns is influencing various environmental variables. In this scenario, the phenological or life cycle of plants are affected. The research work estimates the phenological change and the start of

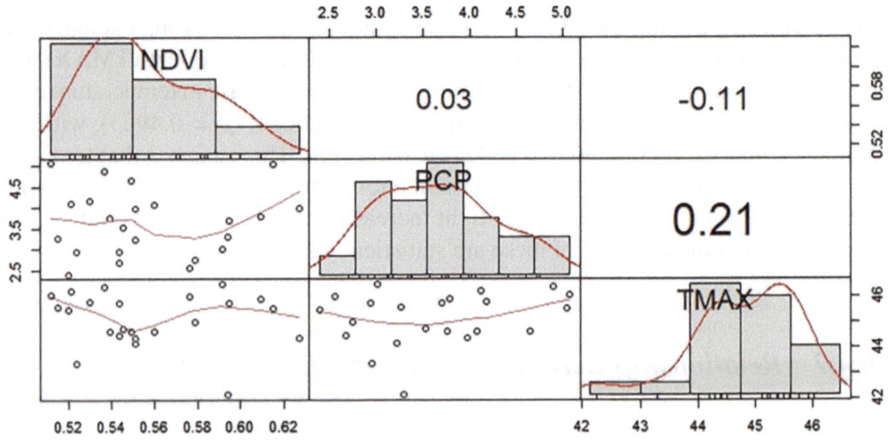

Fig. 10.18 Correlation between NDVI, PCP, and temperature

growing and the end flowering between the 20-year time-based data of NVDI. Over history the plants and animals they developed specific time for response in the system, and this system has various environment attributes, such temperature, Rainfall, precipitation, and humidity which all synchronize. The modis data. Here to study and understand the anthropogenic activities that effects the river basin ecosystem and evolve the problem in that area. To monitor the phenology of the vegetation, cover over a long term period of NDVI values, this study is useful for the making decision support system for development of river basin, based long term and short term planning and provide the nature-based solution. The BFAST method offers a robust approach for detecting various change patterns in NDVI time series data. It decomposes the time ordered data, allowing for the identification of multiple change points within each. Changes in the trend component typically indicate either slow or rapid transitions, while shifts in the seasonal element capture alterations in phenology. One of the key advantages of BFAST is its ability to analyze time ordered data without the need for pre-defining LULC types, reference periods, threshold values, or specific change trajectories. Additionally, simulations have demonstrated BFAST's robustness in handling noisy data and its resilience to fluctuations in seasonal amplitude, as well as its ability to detect changes of varying magnitude and timing. These strengths make BFAST adaptable to a wide range of time ordered, covering diverse ecosystems. In a case study of the Tawa River basin, BFAST was used to a 16-day MODIS NDVI time ordered data from 2002 to 2022 to assess forest changes. The results showed that BFAST effectively identified and characterized changes in forest cover, providing insights into the timing and magnitude of shifts in the landscape.

Authors Contribution All authors have made equal contributions to this research.

Funding This research was conducted without any external funding support.

Data Availability Statement The datasets used and/or analysed in this work can be obtained from the corresponding author upon reasonable request.

Potential Conflicts of Interest The corresponding author affirms, with the full agreement of all co-authors, that there are no conflicts of interest to disclose related to this work.

Informed Consent Permission was acquired from all individual participants who took part in the study.

AI Discloser Artificial Intelligence (AI) tools, were used to support the writing and editing of this manuscript. These tools assisted in language refinement, improving clarity, and generating preliminary drafts or summaries under the close supervision and critical review of the authors.

References

Bendix, J., Jönsson, P., & van der Meijden, R. (1999). Processing of long-term time series of satellite data and the representation of derived biophysical parameters.

Bendix, J., Jönsson, P., & van der Meijden, R. (2002). Timesat: A software programme for crop monitoring and vegetation analysis using remote sensing data.

Cleland, E. E., Chuine, I., Menzel, A., Mooney, H. A., & Schwartz, M. D. (2007). Shifting plant phenology in response to global change. *Trends in Ecology & Evolution, 22*(7), 357–365. https://doi.org/10.1016/j.tree.2007.04.003

Didan, K. (2015). MOD13Q1 MODIS/terra vegetation indices 16-day l3 global 250m SIN grid V006. *NASA EOSDIS Land Processes DAAC*. https://doi.org/10.5067/MODIS/MOD13Q1.006

Eklundh, L., & Jönsson, P. (2016). Timesat for processing time-series data from satellite sensors for land surface monitoring. In Ban, Y. (Ed.), *Remote sensing and digital image processing* (Vol. 20, pp. 177–194). Springer. https://doi.org/10.1007/978-3-319-47037-5_9

Friedl, M. A., Gray, J. M., Sulla-Menashe, D., & Schneider, A. (2006). Land surface phenology in North America from MODIS. *Remote Sensing of Environment, 100*(3), 403–414. https://doi.org/10.1016/j.rse.2005.10.022

Ganguly, S., Samanta, A., Schull, M. A., Shabanov, N. V., Milesi, C., Nemani, R. R., & Myneni, R. B. (2010). Generating vegetation phenology metrics by combining MODIS and ground-based observations in South Asia. *IEEE Transactions on Geoscience and Remote Sensing, 48*(8), 3506–3518. https://doi.org/10.1109/TGRS.2010.2046215

Jeong, S.-J., Ho, C.-H., Gim, H.-J., & Brown, M. E. (2011). Phenology shifts at start vs. end of growing season in temperate vegetation over the Northern Hemisphere for the period 1982–2008. *Global Change Biology, 17*(7), 2385–2399. https://doi.org/10.1111/j.1365-2486.2011.02397.x

Jönsson, P., & Eklundh, L. (2004). TIMESAT—A program for analyzing time-series of satellite sensor data. *Computers & Geosciences, 30*(8), 833–845. https://doi.org/10.1016/j.cageo.2004.05.006

Joshi, P. K., Vyas, A., Roy, P. S., & Singh, R. (2020). Monitoring vegetation phenology using remote sensing data in India: Trends and implications. *Environmental Monitoring and Assessment, 192*, 300. https://doi.org/10.1007/s10661-020-8226-0

Kong, D., McVicar, T. R., Xiao, M., Zhang, Y., Peña-Arancibia, J. L., Filippa, G., … & Gu, X. (2022). Phenofit: An R package for extracting vegetation phenology from time series remote sensing. *Methods in Ecology and Evolution, 13*(7), 1508–1527. https://doi.org/10.1111/2041-210X.13870

Lieth, H. (2013). Phenology and seasonality modeling. *Springer-Verlag*. https://doi.org/10.1007/978-3-642-69400-7

Parmesan, C., & Yohe, G. (2003). A globally coherent fingerprint of climate change impacts across natural systems. *Nature, 421*, 37–42. https://doi.org/10.1038/nature01286

Piao, S., Liu, Q., Chen, A., Janssens, I. A., Fu, Y., Dai, J., et al. (2019). Plant phenology and global climate change: Current progresses and challenges. *Global Change Biology, 25*(6), 1922–1940. https://doi.org/10.1111/gcb.14619

Richardson, A. D., Hufkens, K., Milliman, T., Aubrecht, D. M., Chen, M., Gray, J. M., Johnston, M. R., Keenan, T. F., Klosterman, S., Kosmala, M., Melaas, E. K., Friedl, M. A., & Frolking, S. (2018). Tracking vegetation phenology across diverse North American biomes using PhenoCam imagery. *Scientific Data, 5*, 180028. https://doi.org/10.1038/sdata.2018.28

Richardson, A. D., Keenan, T. F., Migliavacca, M., Ryu, Y., Sonnentag, O., & Toomey, M. (2013). Climate change, phenology, and phenological control of vegetation feedbacks to the climate system. *Agricultural and Forest Meteorology, 169*, 156–173. https://doi.org/10.1016/j.agrformet.2012.09.012

White, M. A., Thornton, P. E., & Running, S. W. (2009). The impact of growing-season length variability on carbon assimilation and evapotranspiration over 88 years in the eastern US deciduous forest. *International Journal of Biometeorology, 53*(5), 429–439. https://doi.org/10.1007/s00484-009-0232-x

Zhang, X., Friedl, M. A., Schaaf, C. B., & Strahler, A. H. (2003). Monitoring the response of vegetation phenology to precipitation in Africa by coupling MODIS and TRMM instruments. *Journal of Geophysical Research: Atmospheres, 108*(D13), 4391. https://doi.org/10.1029/2002JD003275

Part III
Addressing Climate Change, Eco-Resilience and Land Restoration Practices

Chapter 11
Climate Change and Land Suitability Assessment: Global to Regional Perspectives

Neha Awasthi, S. K. Bhardwaj, M. S. Jangra, Chinmaya Sahoo, and Ankit Saini

Abstract Climate change is a critical global challenge that significantly influences land use, crop production, and ecosystem sustainability. The dynamics of climate change and land suitability assessment are examined in this chapter, with a focus on how biophysical, ecological, and socioeconomic aspects are integrated into decision-making processes. Using technologies like Geographic Information Systems (GIS), remote sensing, and machine learning techniques, the chapter explores methodological developments in land suitability modeming. It features case studies from several geographic locations that show how changing precipitation patterns, temperature, and extreme weather events alter the suitability of land for forestry, agriculture, and other land uses. The topic of discussion also includes how participatory methods can be used to involve stakeholders and encourage resilience in land-use planning. By addressing the challenges and opportunities associated with climate change, this chapter provides actionable insights and frameworks for researchers, policy-makers, and practitioners aiming to balance development and sustainability in a rapidly changing environment.

Keywords Agriculture · Temperature · Climate resilience · Suitability · Sustainability · Modelling tools

N. Awasthi
Khalsa College of Charitable Society, Amritsar, India

S. K. Bhardwaj · M. S. Jangra
Dr. Yashwant Singh Parmar University of Horticulture and Forestry, Nauni, Solan, India

C. Sahoo
Department of Agronomy, Dr. Rajendra Prasad Central Agricultural University, Pusa, Samastipur, Bihar, India

A. Saini (✉)
Dr. KSG Akal College of Agriculture, Eternal University, Baru Sahib, Sirmaur, Himachal Pradesh, India
e-mail: ankitsaini970@gmail.com

U. Chatterjee et al. (eds.), *Eco-Resilience: Climate Change, Land Degradation and Sustainable Solutions*, Sustainability Solutions,
https://doi.org/10.1007/978-3-032-00708-7_11

11.1 Introduction

The agricultural land use planning begins with a land suitability evaluation, which is frequently carried out to ascertain whether a specific land use is suitable for a given area (Bodaghabadi et al., 2019). A method for assessing a plot of land, known as land suitability assessment, finds the primary obstacles to cultivating a specific crop (Halder, 2013; Rosa et al., 2004). The assessment of land appropriateness incorporates both qualitative and quantitative considerations. The quantitative assessment yield is estimated and the results are more accurate. In contrast, the qualitative land suitability evaluations take into account soil, vegetation, terrain, climate, and hydrology (Mosleh et al., 2017). The FAO land evaluation framework (FAO, 2007) and physical land evaluation approaches have been used by many to assess a piece of land's suitability.

A major worldwide concern, climate change has a significant influence on both human livelihoods and environmental systems. Its impact on land suitability is a crucial topic of concern, especially when it comes to forestry, urban development, agriculture, and conservation initiatives. The assessment of land suitability is becoming increasingly important for sustainable development and resilience building as global temperatures rise, precipitation patterns shift, and the frequency of extreme weather events increases. In the Paris Agreement, the international community committed to limiting the increase in global temperatures to 1.5 °C and to keeping it well below 2 °C (UNFCCC, 2015). Agriculture is greatly impacted by the type and extent of climate change, including variations in temperature and precipitation patterns, as opposed to the local capacity to tolerate the effects (Li et al., 2011). Significant impacts have been documented in relation to variations in the growing season's duration and the incidence of extreme events at crucial stages of crop development (Hatfield et al., 2011). Globally, climate change is pushing traditional crops out of their comfort zones, especially in climate-sensitive regions. At the same time, new opportunities may emerge in areas becoming more favourable due to changing conditions. At the regional and local levels, these changes are even more complex, influenced by topography, soil, water availability, and human interventions.

Land suitability assessment, when integrated with climate projections and spatial tools like GIS and remote sensing, offers valuable insights into where and how land can be used most effectively under future scenarios. These assessments help identify risks, guide investments, and support decision-makers in creating adaptive strategies tailored to specific contexts.

Future feedstock resource projections are highly variable, with the most uncertainty around biomass estimates from energy-specific crops (Creutzig et al., 2015; Smith et al., 2016). However, there is broad agreement regarding the necessity of biomass. The various number of factors contribute to the uncertainty surrounding the sustainable resource potential of dedicated energy crops, such as the availability of land in light of shifting agricultural practices and food requirements, the appropriateness of the land for agricultural production in a shifting climate, the efficacy of crops not yet extensively cultivated, the prospective yields from neglected or degraded land, the expenses linked to the rehabilitation of degraded land, and emissions associated with land-use alteration and transportation to processing facilities (Anderson & Peters, 2016; Searle & Malins, 2015). It is imperative that these concerns be better understood on a global and regional scale since they may have significant effects on the trade of biomass resources and the possible role of bioenergy.

The energy resources are the main forces behind economic growth; economies rely on them, and development cannot exist without energy (Dai et al., 2022). In actuality, energy and water are two essential and interconnected aspects of political discourse and economic progress. These resources are the most sought-after, yet they have also been the focus of disagreements between economies and governments. These subjects are also at the forefront of contemporary debates concerning the environment, society, and economy. The explanations are straightforward: without energy and water, the term "development" in the context of the modern capitalist world order is meaningless. These resources, which determine international politics and diplomacy, are essential to every nation, economy, and populace.

India has a great deal of potential to address the current climate crisis by implementing agroforestry, considering its high susceptibility to climate change and its position as the third-largest emitter of greenhouse gases globally, with agriculture contributing 18% of all national emissions (Sapkota et al., 2019). According to Nath et al. (2021), the rates of carbon sequestration in various agroforestry systems in India have been noticeably higher. This study found that, due to its capacity for carbon storage, agroforestry could drastically cut emissions from India's energy, agricultural, and industrial sectors with just a 30% increase in area. Moreover, studies indicate that by improving their ability to adapt and providing a variety of ecosystem services, agroforestry may help Indian farmers become more climate resilient (Datta & Behera, 2023a, 2023b; Jana et al., 2015; Pandey et al., 2015, 2017; Tewari et al., 2014) The research findings underscore the importance of adaptive land-use planning, which is informed by dynamic climate models, geospatial analysis, and sustainability indicators. By integrating these components, the study contributes to building long-term eco-resilience, ensuring that land systems can continue to provide essential ecosystem services even in the face of environmental shocks. It also offers a decision-support framework for implementing sustainable land management (SLM) strategies that balance productivity with conservation. In this way, the research advances the discourse on sustainability by offering practical, data-driven solutions that support climate-smart policies and land stewardship practices—critical pillars for achieving eco-resilient landscapes. This chapter integrates dynamic climate forecasts spanning global to regional scales to propose a fresh, climate-informed method to land suitability evaluation. The chapter positions land suitability assessment as a crucial instrument for climate resilience and sustainable development by bridging the gap between scientific methodologies and policy relevance.

11.2 Climate Change: Key Drivers and Impacts

The significant climate drivers during the industrial era include both natural and, to a lesser extent, manmade factors connected to human activity. In the industrial age, the El Niño–Southern Oscillation, volcanic eruptions, and variations in solar irradiance are the only notable natural climate drivers. Although they have fluctuated throughout the industrial era, natural GHG and tropospheric aerosol emissions and sinks have not made a substantial contribution to RF. Global radiative consequences are thought to be negligible, despite the fact that the effects of cosmic rays on cloud formation have been explored (Krissansen-Totton & Davies, 2013). As was mentioned above, the causes of climate change are:

11.2.1 Natural Drivers

Natural drivers of land suitability assessment are the environmental and ecological factors that determine whether a piece of land is suitable for a particular use, such as agriculture, forestry, conservation, or urban development. Some examples of natural drivers were discussed below:

I. **Solar Irradiance**

Earth's primary energy source is solar radiation, and variations in this radiation have a direct impact on the climate system. During the industrial age, the biggest changes in total solar radiation have happened every eleven years (Fröhlich & Lean, 2004). The near-ultraviolet (UV) and shorter wavelengths, where spectral fluctuations in solar irradiance are highest, are the most important wavelengths for generating ozone shifts (Floyd et al., 2003) (Bolduc et al., 2015). Variations in total and spectrum solar irradiance cause noticeable changes in air circulation and heating by altering the quantities of ozone (Gray et al., 2010; Lockwood, 2012; Seppälä et al., 2014).

II. **Volcanoes**

The majority of volcanic eruptions are small-scale occurrences that only have a week or month-long impact on the troposphere. The climate is significantly impacted in the short term by the massive amounts of ash and sulphur dioxide (SO_2) that are released into the stratosphere by strong volcanic eruptions (Myhre et al., 2017). Through its oxidation to sulfuric acid (H_2SO_4), which condenses and either creates new particles or adds mass to existing ones, SO_2 significantly increases the attenuation of sunlight traveling through the stratosphere.

11.2.2 Anthropogenic Drivers

A major environmental concern that impacts people worldwide, human-induced climate change has ramifications for a wide range of industries, including the energy, health, and agricultural sectors (Scott et al., 2023; Tongwane & Moeletsi, 2018). Numerous studies (Chabbi et al., 2017; Outhwaite et al., 2022) have established that global warming is caused by human activities that release greenhouse gases (GHGs), which are gases that warm the globe. Numerous human activities (see in Fig. 11.1), such as deforestation, altering the land's natural use, industrial processes, unsustainable agriculture (such as the overuse of fertilizer and pesticides), and the burning of fossil fuels, release greenhouse gases (GHGs), such as carbon dioxide (CO_2), methane (CH_4), and nitrous oxide (N_2O).

I. **Greenhouse Gas Emissions**: Global warming and climate change are the outcomes of the greenhouse effect, which is brought on by a steady increase in GHG concentrations around the globe (Gopalakrishnan et al., 2019). Globally, GHG emissions have risen by 75% in recent years. About 16% of the present supply of greenhouse gases is made up of methane (CH_4), 6% of which is nitrous oxide (N_2O), and 76% of which is carbon dioxide (CO_2) (Crippa et al., 2021). The relationship between greenhouse gas emissions and human activities, such as burning fossil fuels, using energy, and farming, is important and widely recognized (Lungarska & Chakir, 2018).
II. **Deforestation and Land Use Change**: Deforestation, soil erosion, and machine-intensive agricultural practices are examples of large-scale land use changes that have led to elevated atmospheric carbon concentrations (Ngarava

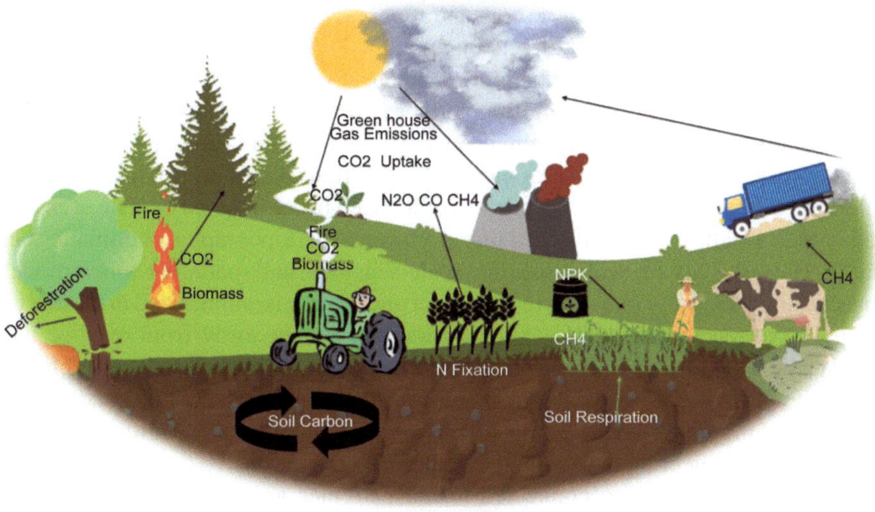

Fig. 11.1 Anthropogenic drivers of environmental change

et al., 2023). The natural environment and agriculture are also impacted by soil erosion caused by wind, water, and excessive tillage practices (Robinson, 2020). According to Ebhota and Tabakov (2021), soil loss and its consequences are among the most significant yet least recognized environmental issues.

III. **Industrialization and Urbanization**: According to Scott et al. (2023), the primary source of CO_2 emissions and the second largest contributor to greenhouse gas emissions was burning of fossil fuels like gasoline and diesel for the transportation of people and products. Air travel, rail, maritime transportation, and highway and passenger vehicles are examples of domestic transportation sources (Kanitkar et al., 2022). Furthermore, a lot of industrial operations use electricity to generate energy, which means that they indirectly supplement CO_2 emissions from the industrial source (Adams & Acheampong, 2019).

11.3 Impacts on Land Systems

According to Wang et al. (2017) and Chaplin-Kramer and George (2013), climate change is the primary driver of changes in the vegetation patterns of large-scale rangelands, the water requirements for producing fodder and meeting animal needs, the efficiency of forage production, and heat stress on cattle. Numerous regional, interregional, and national scales make up LUCC. Interregional scales, or adjacent geomorphological areas, will gradually shift as a result of the primary regional change, which in turn affects the world environment. Recent studies show that changes in soil quality and land productivity were caused by the transition from natural to agricultural land use (Fernandez-Romero et al., 2014). According to Benbi and Brar (2009), prolonged and intense agricultural methods can cause a major degradation of the soil's organic carbon. Due to rising atmospheric concentrations of CO_2 and other greenhouse gases (GHGs) and their consequent effects on agricultural systems are among the factors contributing to warming air temperatures and changing global precipitation patterns in a number of places, including the Northern Hemisphere (Hatfield et al., 2011). Land suitability is impacted by climate change; therefore, it is crucial to incorporate climate-responsive methodologies into land evaluation frameworks. It highlights that land suitability is now a dynamic process influenced by anticipated climate scenarios rather than a static evaluation based just on current soil, topography, or climatic standards. The necessity for sector-specific, scale-sensitive, and spatially explicit models that can direct land-use planning in the face of uncertainty is emphasized in the discussion. The chapter makes the case for a paradigm change away from reactive land management and toward proactive, climate-resilient approaches by fusing climate research with geospatial analysis and decision support systems. In the end, the topic emphasizes the importance of land suitability assessment as a scientific and policy tool for promoting climate change-resilient development.

Deforestation has been connected to hot extremes and has raised the average temperature. Furthermore, it decreased the average precipitation frequency on a global scale as opposed to a local one (Lawrence & Vandecar, 2015). By

Table 11.1 Impact of climate on LULC category

LULC category	Examples	Impact on climate change
Forests	Tropical, temperate, boreal	Carbon sequestration
Agriculture	Croplands, pastures	Emissions of CH_4 (rice paddies) and N_2O (fertilizers)
Grasslands/Savannas	Natural grasslands, meadows	Carbon sequestration potential
Urban Areas	Cities, suburbs	Heat islands increase energy demand for cooling
Wetlands	Marshes, swamps, peatlands	Store vast amounts of carbon
Tundra	Arctic/Alpine Tundra	Permafrost thaw releases CH_4 and CO_2

enhancing carbon sequestration, reforestation effectively reduces high atmospheric CO_2 concentrations and global warming (Wang et al., 2022). Afforestation projects are promoted by the United Nations Framework Convention on Climate Change (UNFCCC) as a workable method of lowering net CO_2 emissions. According to Izaurralde et al. (2011), rising CO_2 has most likely raised rangeland productivity over the past 150 years and will likely do so for the next 30 years in most parts of the world. The impact of LULC of different areas on climate change was established in Table 11.1.

11.4 Land Suitability Assessment: Concepts and Approaches

Definition and Purpose

The capacity of a particular location to sustainably support particular activities, including agriculture, urban growth, or conservation, is assessed by a land suitability assessment. To direct land-use planning, it incorporates biophysical, climatic, and socioeconomic factors.

11.4.1 Key Parameters in Land Suitability

I. **Soil Quality**: In agriculture a promising to boost output without having an adverse effect on the environment since it preserves soil health and boosts agricultural yields because soil testing helps identify how much fertilizer is added to the soil and what crops need. The availability of nutrients influences soil fertility, which in turn influences crop output (Abuzaid & Fadl, 2016; Abdelrahman et al., 2021; Ausari et al., 2020). Growing a crop in a supportive environment that boosts and improves crop output is known as sustainable agriculture.

A thorough land assessment is necessary to determine the best crop placement, balancing the needs of the crops with the condition of the soil. Crop output has increased as a result of the integration of agricultural methods and relevant spatial information.

II. **Climate Variables**: Temperature, sun intensity, rainfall, CO_2, soil salinity, water salinity, and seasonal features have all had a major impact on agricultural activities in Bangladesh (Faisal & Parveen, 2004; Gain et al., 2012). Agricultural productivity is greatly impacted by ongoing temperature increases, which are closely linked to unplanned and fast urbanization. In many regions of the world, changes in land use and land cover (LULC) are directly impacted by the acceleration of population increase (Faisal et al., 2019; Kafy et al. 2021; Pal & Ziaul, 2017; Yuan et al., 2005). Due to significant increase in land surface temperature (LST) and an acceleration of the dangers linked with climate change are two direct effects of changes in LULC on the environment and ecosystems (Al Rakib et al., 2020; Kafy et al., 2020a, 2020b).

III. **Water Resources**: In order to assess the effects of climate change on water resource management, a thorough modeling approach that considered both hydrological responses and reservoir operation was necessary (Gopalan et al., 2020; Lee & Shin, 2021). According to Firoz et al. (2018) and Dong et al. (2023), reservoir operation modelling is now a common method used to address how human activity affects the management of water resources. Firoz et al. (2018) used naturalized and reconstructed streamflow data to analyse hydrological drought using a HEC-ResSim reservoir operation model. The threats of Flood and water supply are clearly important and necessary, but there hasn't been much research done on them simultaneously in the context of climate change and water resource management.

A crucial step in agricultural growth and sustainable land use planning is land suitability assessment, or LSA. In order to assess how well a particular plot of land supports distinct land uses, especially agriculture, it entails analysing a number of biophysical, climatic, and socioeconomic factors. The duration of the growing season, temperature, humidity, rainfall, and evapotranspiration are important climatic elements that affect crop productivity and viability. The risk of soil erosion, water retention, and machinery accessibility are all impacted by topographic elements like slope, elevation, aspect, and landform characteristics. Nutrient availability and root penetration are greatly influenced by the properties of the soil, including texture, depth, pH, organic matter content, cation exchange capacity (CEC), organic matter content, salinity, sodicity, and fertility (particularly nitrogen, phosphorus, and potassium levels). To maintain ecological sustainability, environmental constraints such as flood hazards, drought frequency, erosion susceptibility, and closeness to ecologically sensitive areas must also be taken into account. In order to efficiently integrate and analyse these various parameters and produce more informed and data-driven land suitability classifications, digital tools like Geographic Information Systems (GIS), remote sensing indices (e.g., NDVI, LST), and Multi-Criteria Decision Analysis (MCDA) are frequently used in contemporary assessments.

11.4.2 Methodologies

I. Remote sensing (RS) and geographic information systems (GIS) have been used to assess LULC changes and LST distribution in a specific area (Faisal & Khan, 2018). Many studies have been conducted to understand the LULC transformation and its effect on LST in multitemporal Landsat images (Ramachandran et al., 2020).

II. Several research used Artificial Neural Networks (ANN) (Ahmed et al., 2013; Mas & Flores, 2008) and Logistic Regression (LR) to predict LULC and LST shifts (Arsanjani et al., 2013). The CA model is widely used to simulate possible LULC and estimate the future LULC transition matrix by including the previous trend of change (Al-sharif & Pradhan, 2015; Ozturk, 2015; Yang et al., 2012). Table 11.2 listed the many techniques utilized for the thorough assessment of land adaptability and climate change.

Table 11.2 Techniques for evaluating land suitability and climate change

Assessment of climate change and land suitability		
Methodology	Key tools/models	Purpose
Climate Modelling: Predicts climate scenarios using General Circulation Models (GCMs) or Regional Climate Models (RCMs)	CMIP6, CORDEX	Assess future climate projections and their impacts
Trend Analysis: Statistical analysis of historical climate data (temperature, precipitation)	R, Python, MATLAB	Identifies trends and variability over time
Greenhouse Gas Inventory: Quantifies emissions from sectors like energy, agriculture, and waste	IPCC Guidelines, LEAP, GHG Protocol	Monitors and reports emissions for mitigation strategies
Vulnerability Assessment: Identifies areas and sectors sensitive to climate impacts	GIS, Cristal, AR5 Vulnerability Framework	Guides adaptation strategies by evaluating exposure, sensitivity, and adaptive capacity
Impact Assessment: Analyses sector-specific climate impacts (agriculture, water, biodiversity)	DSSAT, SWAT, SEI WEAP	Supports planning and policymaking for mitigation and adaptation
Remote Sensing: Monitors changes in land cover, forest, or vegetation due to climate impacts	MODIS, Landsat, Sentinel	Tracks large-scale environmental changes over time

11.5 Climate Change Implications for Land Suitability

In accordance with the typical socioeconomic pathways, the amount of land required for urban areas and food crops is believed to change throughout time (Riahi et al., 2017). The effects of climate change itself, such as altered precipitation and temperature patterns, exacerbate future shortages of energy crops. The impact of climate change on the yields of important food crops, including rice, sugarcane, and maize, has been the main focus of regional and international research (Deryng et al., 2016; Porter et al., 2014; Rosenzweig et al., 2014). Process-driven crop models (Haberl et al., 2011), land suitability (LS) modeling (Feng et al., 2017), and field experiments (Dong et al., 2019) have all been used to examine the potential performance of specific energy crops for specific locales. However, as the supply of food and environmental protection should come before the supply of energy crops, it is imperative to assess the climatic consequences in the context of land-use scenarios and to consider the role of marginal quality land.. The potential trade-offs and synergies between energy availability, climate change mitigation, and a number of other SDGs must be carefully considered when planning the expansion of bioenergy pathways (Fuso Nerini et al., 2018). Energy and land-use policy must prioritize safeguarding and enhancing livelihoods, food security, and biodiversity. Due to climatic zone shifts, soil degradation, water resource stress, and ecological disruption, climate change is radically changing the usability of land in different places. Previously arable lands are losing their productivity due to heat stress, drought, and salinization as temperatures rise and rainfall patterns become more unpredictable, especially in susceptible areas like South Asia and Sub-Saharan Africa. New locations, particularly in higher latitudes, may become suitable at the same time, albeit they are frequently constrained by poor soils or brief growth seasons. Coastal areas are experiencing saltwater intrusion due to sea level rise, while irrigation-based agriculture is being undermined by water scarcity brought on by changed hydrological cycles. The loss of biodiversity brought on by climate change and changing habitats makes land use planning even more difficult, which has an impact on forestry and conservation plans. Increased climate migration is also causing urban pressures to worsen, resulting in unplanned settlements in inappropriate or ecologically sensitive locations. Although they provide direction, predictive models are limited by uncertainty, which emphasizes the necessity of adaptive land use strategies that incorporate social, economic, and environmental factors to ensure resilience in a warming world. Nowadays, many implications occur for land suitability in different areas (see Fig. 11.2).

11.6 Policy and Planning Considerations

It is essential to maintain policy and planning for climate change and land suitability assessment. Some of them are listed in Table 11.3.

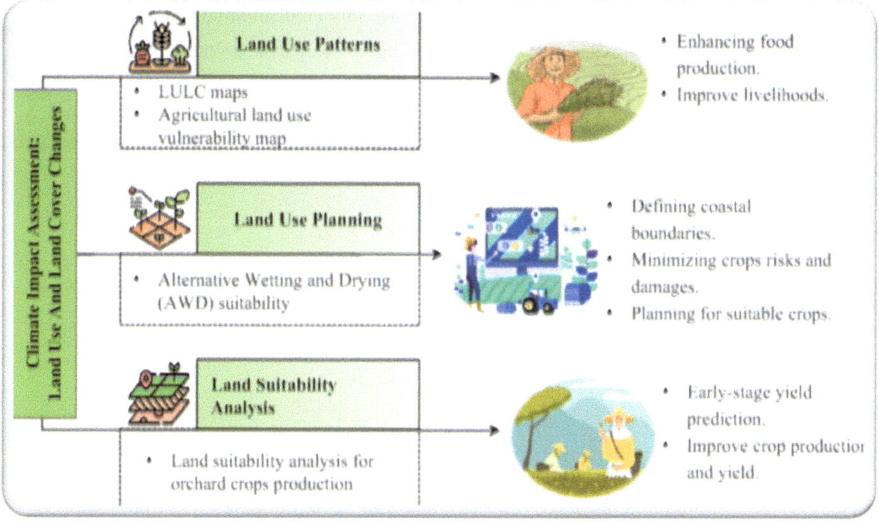

Fig. 11.2 Implications for climate change and land suitability

11.7 Case Study

I. Climate Change and Impact on Agricultural Land Suitability

The UN has stated that lowering hunger and enhancing food security are crucial elements of its sustainable development objectives. Around 735 million people worldwide suffer from hunger and malnutrition as of 2022, and many deaths have been documented. The appropriateness of agricultural land is greatly impacted by climate change, which could result in acute food shortages and the ensuing social and political unrest. We have created a machine learning-based method to forecast the likelihood of significant deterioration in land suitability and modifications to irrigation patterns in order to tackle this problem. Central Eurasia, an area beset by social and economic problems, is the subject of our investigation. The impact of climate change on agricultural land's potential to adapt to various carbon emission scenarios is being assessed for the first time in this study, utilizing interpretable machine learning approaches. Certain topographical and climate traits that could affect land suitability are revealed by the feature importance analysis. In a multi-class land suitability classification assignment, our model's performance measures show its effectiveness, with an accuracy of 86% and a mean average precision of 72%. Addressing the most susceptible areas in Northern Asia and Eastern Europe provides

Table 11.3 Policy/recommendations for climate change and land suitability

Aspect	Description	Policy/planning recommendations	Key stakeholders	Potential barriers
Climate impact	The expected effects of climate change on land (e.g., temperature rise, rainfall changes, etc.)	Promote climate-resilient land-use practices and adaptive agriculture	Government, NGOs, farmers	Limited awareness and funding
Land suitability	Analysis of land suitability for specific uses (e.g., agriculture, forestry, urbanization)	Prioritize land zoning and sustainable practices	Urban planners, policymakers	Resistance to land-use changes
Biodiversity	Impacts of climate change on local flora and fauna	Integrate biodiversity conservation into land-use policies	Conservationists, local communities	Conflicts with development priorities
Water Resources	Effects of climate change on water availability and quality	Implement water conservation and watershed management programs	Irrigation departments, municipalities	Competing demands for water resources
Socio-economic factors	Effects on livelihoods, especially for vulnerable populations	Develop inclusive policies addressing economic resilience	Community leaders, local governments	Inequitable policy implementation
Renewable energy	Land suitability for renewable energy installations (solar, wind, etc.)	Encourage renewable energy projects on degraded lands	Energy departments, private investors	Land acquisition challenges
Urban development	Impacts of urban expansion under changing climate conditions	Plan climate-resilient infrastructure and green spaces	Urban planners, architects	Lack of integrated urban planning frameworks
Disaster risk	Vulnerability to climate-induced disasters (floods, droughts, etc.)	Strengthen early warning systems and disaster management planning	Disaster management agencies	Limited local disaster preparedness
Policy integration	Alignment of climate and land policies with national and international goals	Harmonize policies with SDGs and NDCs	National ministries, international bodies	Policy fragmentation and weak inter-agency coordination

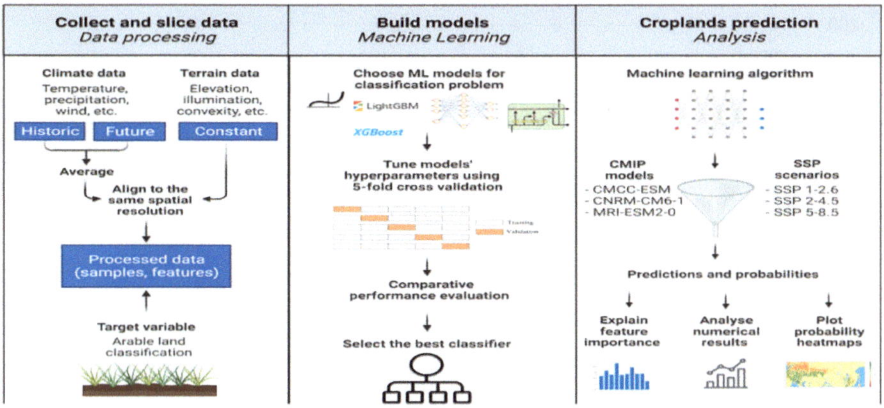

Fig. 11.3 Algorithmic workflow of case study (Shevchenko et al., 2023) on climate change and land assessment

policymakers with important information for decision-making and averting a humanitarian crisis, such as providing more fertilizer and water. This study demonstrates how machine learning can be used to address global issues, especially the reduction of hunger and malnutrition (See Fig. 11.3).

II. **Land Suitability and Crop Vulnerability Assessment Using GIS in the Context of Climate Change in Chtouka Ait Baha, Morocco**

However, among the many problems facing our industry is climate change. This study aims to assess relevant variables that may affect the length of the agricultural growing season, water requirements, and land suitability in the context of climate change. Additionally, in the Chtouka region, which is known for its irrigated agriculture, it might be utilized as a decision support tool. Four factors—slope, land use, temperature, and soil texture—are used in the multiple-criteria decision-making process that underpins the suggested strategy for the appropriate spatial distribution of land. The idea of rising degree days was used to approximate the length of the crop season. The projected growing degree days in 2031–2050 showed a notable drop in comparison to the baseline growth period length of 1985–2006, mainly under the RCP8.5 emission scenario, with reductions varying from 8 to 21% depending on crops. Additionally, agricultural water productivity statistics showed that berries were more than 50% less water productive than other vegetable crops for roughly the same amount of applied irrigation water. These results demonstrate how vulnerable agriculture is to climate change, necessitating significant management and political initiatives to maintain agricultural operations (See Fig. 11.4).

III. **Climate Extremes and Sustainability Issues: A Case of Proposed Hydropower Projects in Lahaul Valley, Himachal Pradesh** (Chandel & Rana, 2023)

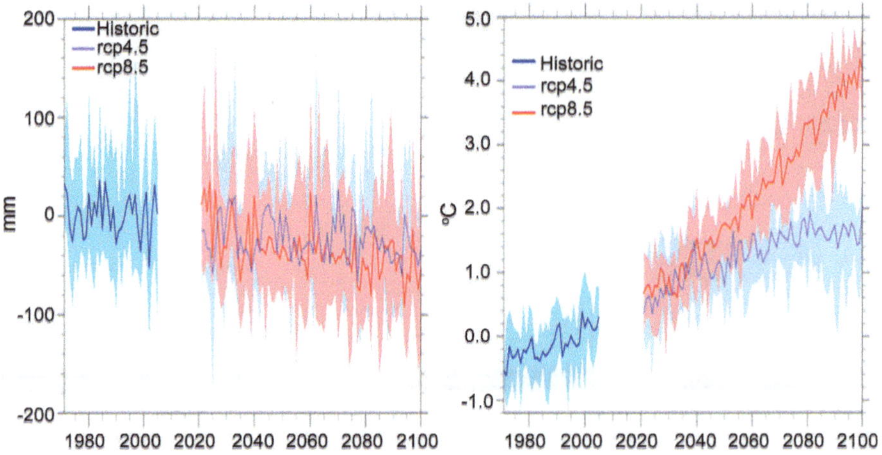

Fig. 11.4 Climate change-related land suitability and crop vulnerability assessment using GIS (Seif-Ennasr et al., 2020)

Energy and water are essential for prosperity, but the argument over hydropower and sustainability in light of climate change and disaster risk is rife with controversy. Hydropower projects have been planned since the 1990s due to the Himalayas' enormous potential for renewable energy. In addition, the social, economic, and environmental issues that come with climate change have made the locals fearful. This study looked at how climatic extremes, disaster risk, and hydropower growth relate to each other in the Lahaul Valley of Himachal Pradesh. While climatic extremes were examined using daily rainfall gridded data, patterns in temperature and precipitation were examined using long-term climate data. The findings indicate a statistically significant rise in winter and post-winter temperatures as well as an increase in precipitation intensity. Additionally, there are increasing tendencies in the intensity of daily rainfall and high and extremely heavy rainfall days, which could have detrimental effects on the local economy, livelihood, and hydropower development. The local population is unhappy with the projected development, according to field observations. To secure a safe future for the region with a solid grasp of the danger of disasters and climate change, it is crucial that this issue be evaluated logically.

11.8 Future Directions

I. **AI & Machine Learning**: Applying AI to analyse large, complicated datasets in order to forecast the suitability of land under different climate scenarios.

II. **GIS and Remote Sensing**: satellite imagery and GIS technologies for monitoring land use changes in real time and determining appropriateness.

III. **Dynamic Crop Models**: These forecast crop performance and land appropriateness by incorporating climate variables.
IV. **Integrated Climate-Land Models**: These models combine climate, socioeconomic, and biophysical factors to provide thorough evaluations.
V. **Green Infrastructure**: Encouraging environmentally friendly infrastructure plans that complement land use that adapts to climate change.
VI. **Water Resource Management**: Identifying lands suitable for water under climatic stress by incorporating hydrological studies.
VII. **Transboundary Assessments**: International cooperation to manage common land resources and lessen the effects of climate change across international borders

11.9 Conclusion

Environmental conditions are being drastically changed by climate change, which also affects the suitability of land for a variety of activities, such as forestry, urbanization, and agriculture. This chapter emphasizes the importance of creating frameworks for evaluating climate-resilient land on a global scale, especially when it comes to food security, ecosystem preservation, and rural livelihoods. Climate-informed land suitability evaluations must be mainstreamed into planning and policy in light of expanding international commitments like the Paris Agreement, the Sustainable Development Goals (SDGs), and the UN Decade on Ecosystem Restoration, a shifting climate. The precision of land suitability evaluations will be improved by ongoing research funding and the creation of extensive databases that integrate biophysical, meteorological, and socioeconomic factors. This research highlights the important role of climate-informed land suitability assessment in addressing the interconnected challenges of climate change and land degradation. The findings emphasize the need for adaptive, climate-smart strategies to ensure long-term environmental sustainability, productive land use, and resilient ecosystems. Ultimately, this work contributes to building a sustainable future through informed decision-making and resilient land management practices.

Authors Contribution All authors have contributed equally to this work.

Funding There was no financing for the research.

Data Availability Statement On reasonable request, the corresponding author will provide the datasets used and/or analyzed in the current study.

Potential Conflicts of Interest The corresponding author declares that there is no conflict of interest on behalf of all authors.

Informed Consent N/A.

References

Abdelrahman, M. A. E., Rehab, H. H., & Yossif, T. M. H. (2021). Soil fertility assessment for optimal agricultural use using remote sensing and GIS technologies. *Applied Geomatics, 13*, 605–618. https://doi.org/10.1007/s12518-021-00376-1

Abuzaid, A. S., & Fadl, M. E. (2016). Land evaluation of Eastern Suez Canal, Egypt using remote sensing and GIS. *Egyptian Journal of Soil Science, 56*(3), 537–548.

Adams, S., & Acheampong, A. O. (2019). Reducing carbon emissions: The role of renewable energy and democracy. *Journal of Cleaner Production, 240*, Article 118245. https://doi.org/10.1016/j.jclepro.2019.118245

Ahmed, B., Kamruzzaman, M., Zhu, X., Rahman, M., & Choi, K. (2013). Simulating land cover changes and their impacts on land surface temperature in Dhaka, Bangladesh. *Remote Sensing, 5*, 5969–5998.

Al Rakib, A., Akter, K. S., Rahman, M. N., Arpi, S., & Kafy, A.-A. (2020). Analyzing the pattern of land use land cover change and its impact on land surface temperature: A remote sensing approach in Mymensingh, Bangladesh. In *1st international student research conferences* (p. 2020).

Al-sharif, A. A. A., & Pradhan, B. (2015). A novel approach for predicting the spatial patterns of urban expansion by combining the chi-squared automatic integration detection decision tree, Markov chain and cellular automata models in GIS. *Geocarto International, 30*, 858–881.

Anderson, K., & Peters, G. (2016). The trouble with negative emissions. *Science, 354*(6309), 182–183. https://doi.org/10.1126/science.aah4567

Arsanjani, J. J., Helbich, M., Kainz, W., & Boloorani, A. D. (2013). Integration of logistic regression, Markov chain and cellular automata models to simulate urban expansion. *International Journal of Applied Earth Observation and Geoinformation, 21*, 265–275.

Ausari, D. K., Singh, B., Aakash, R. K., & Gehlot, Y. (2020). GIS based mapping of soil fertility status of Tehsil Jobat, District Alirajpur, Madhya Pradesh, India. *International Journal of Current Microbiology and Applied Sciences, 9*(10), 60–69. https://doi.org/10.20546/ijcmas.2020.910.009

Benbi, D., & Brar, J. (2009). A 25-year record of carbon sequestration and soil properties in intensive agriculture. *Agronomy for Sustainable Development, 29*, 257–265.

Bodaghabadi, M. B., Faskhodi, A. A., Saleh, M. H., Hosseinifard, S. J., & Heydari, M. (2019). Soil suitability analysis and evaluation of pistachio orchard farming, using canonical multivariate analysis. *Scientia Horticulturae, 246*, 528–534.

Bolduc, C., Bourqui, M. S., & Charbonneau, P. (2015). A comparison of stratospheric photochemical response to different reconstructions of solar ultraviolet radiative variability. *Journal of Atmospheric and Solar-Terrestrial Physics, 132*, 22–32. https://doi.org/10.1016/j.jastp.2015.06.008

Chabbi, A., Lehmann, J., Ciais, P., Loescher, H. W., Cotrufo, M. F., et al. (2017). Aligning agriculture and climate policy. *Nature Clinical Practice Endocrinology & Metabolism, 7*(5), 307–309. https://doi.org/10.1038/nclimate3286

Chandel, V. B. S., & Rana, B. S. (2023). Climate extremes and sustainability issues: A case of proposed hydropower projects in Lahaul valley, Himachal Pradesh. *Current World Environment, 18*(3), 1345–1355. ISSN: 0973-4929.

Chaplin-Kramer, R., & George, M. R. (2013). Effects of climate change on range forage production in the San Francisco Bay Area. *PLoS ONE, 8*, Article e57723.

Creutzig, F., Ravindranath, N. H., Berndes, G., Bolwig, S., Bright, R.,Cherubini, F., et al. (2015). Bioenergy and climate change mitigation: An assessment. *GCB Bioenergy, 7*(5), 916–944. https://doi.org/10.1111/gcbb.12205

Crippa, M., Solazzo, E., Guizzardi, D., Monforti-Ferrario, F., Tubiello, F. N., et al. (2021). Food systems are responsible for a third of global anthropogenic GHG emissions. *Nature Food, 2*(3), 198–209. https://doi.org/10.1038/s43016-021-00225-9

Dai, L., Jia, R., & Wang, X. (2022). Relationship between economic growth and energy consumption from the perspective of sustainable development. *Journal of Environmental and Public Health, 2004*, 10. https://doi.org/10.1155/2022/6884273

Datta, P., & Behera, B. (2023a). Climate change adaptation through agroforestry: Empirical evidence from Indian Eastern Himalayan foothills. *Strategizing agricultural management for climate change mitigation and adaptation* (pp. 167–181). Springer International Publishing.

Datta, P., & Behera, B. (2023b). Assessing the role of agriculture-forestry-livestock nexus in improving farmers' food security in South Asia: A systematic literature review. *Agricultural Systems, 213*, Article 103807.

De la Rosa, D., Mayol, F., Diaz-Pereira, E., & Fernandez, M. (2004). A land evaluation decision support system (MicroLEIS DSS) for agricultural soil protection. *Environmental Modelling & Sofware, 19*, 929–942.

Deryng, D., Elliott, J., Folberth, C., Müller, C., Pugh, T. A., Boote, K. J., Conway, D., Ruane, A. C., Gerten, D., Jones, J. W., & Khabarov, N. (2016). Regional disparities in the beneficial effects of rising CO_2 concentrations on crop water productivity. *Nature Climate Change, 6*(8), 786–790. https://doi.org/10.1038/nclimate2995

Dong, H., Green, S. V., Nishiwaki, A., Yamada, T., Stewart, J. R., Deuter, M., & Sacks, E. J. (2019). Winter hardiness of Miscanthus (I): Overwintering ability and yield of new Miscanthus × giganteus genotypes in Illinois and Arkansas. *GCB Bioenergy, 11*(5), 691–705. https://doi.org/10.1111/gcbb.12588

Dong, N., Guan, W., Cao, J., Zou, Y., Yang, M., Wei, J., Chen, L., & Wang, H. (2023). A hybrid hydrologic modelling framework with data-driven and conceptual reservoir operation schemes for reservoir impact assessment and predictions. *Journal of Hydrology, 129246*, Article 129246. https://doi.org/10.1016/j.jhydrol.2023.129246

Ebhota, W., & Tabakov, P. (2021). Development of domestic technology for sustainable renewable energy in a zero-carbon emission-driven economy. *International Journal of Environmental Science and Technology, 18*, 1253–1268. https://doi.org/10.1007/s13762-020-02920-9

Faisal, A.-A., Hossain, M., Haque, S., Shaunak, M., & Kafy, A.-A. (2019). Remote sensing approach in wetland and land degradation assessment: a scenario of Modhumoti Model Town, Savar, Bangladesh. In *1st International conference on urban & regional planning*

Faisal, A.-A., & Khan, H. (2018). Application of GIS and remote sensing in disaster management: A critical review of flood management. In *International conference on disaster risk mitigation.*

Faisal, I. M., & Parveen, S. (2004). Food security in the face of climate change, population growth, and resource constraints: Implications for Bangladesh. *Environmental Management, 34*, 487–498.

FAO-Food Agricultural Organization. (2007). A framework for LAND evaluation. *FAO Soil Bulletin No. 6.*

Feng, Q., Chaubey, I., Engel, B., Cibin, R., Sudheer, K. P., & Volenec, J. (2017). Marginal land suitability for switchgrass, Miscanthus and hybrid poplar in the Upper Mississippi River Basin (UMRB). *Environmental Modelling & Software, 93*, 356–365. https://doi.org/10.1016/j.envsoft.2017.03.027

Fernandez-Romero, M., Lozano-García, B., & Parras-Alcantara, L. (2014). Topography and land use change effects on the soil organic carbon stock of forest soils in Mediterranean natural areas. *Agricultural Economics and Environment, 195*, 1–9.

Firoz, A. B. M., Nauditt, A., Fink, M., & Ribbe, L. (2018). Quantifying human impacts on hydrological drought using a combined modelling approach in a tropical river basin in central Vietnam. *Hydrology and Earth System Sciences, 22*, 547–565. https://doi.org/10.5194/hess-22-547-2018

Floyd, L. E., Cook, J. W., Herring, L. C., & Crane, P. C. (2003). SUSIM's 11-year observational record of the solar UV irradiance. *Advances in Space Research, 31*, 2111–2120. https://doi.org/10.1016/S0273-1177(03)00148-0

Fröhlich, C., & Lean, J. (2004). Solar radiative output and its variability: Evidence and mechanisms. *The Astronomy and Astrophysics Review, 12*, 273–320. https://doi.org/10.1007/s00159-004-0024-1

Fuso Nerini, F., Tomei, J., To, L. S., Bisaga, I., Parikh, P., Black, M., et al. (2018). Mapping synergies and trade-offs between energy and the Sustainable Development Goals. *Nature Energy, 3*(1), 10–15. https://doi.org/10.1038/s41560-017-0036-5

Gain, A. K., Giupponi, C., & Renaud, F. G. (2012). Climate change adaptation and vulnerability assessment of water resources systems in developing countries: A generalized framework and a feasibility study in Bangladesh. *Water, 4*, 345–366.

Gopalakrishnan, T., Hasan, M. K., Haque, A. S., Jayasinghe, S. L., & Kumar, L. (2019). Sustainability of coastal agriculture under climate change. *Sustainability, 11*(24), 7200. https://doi.org/10.3390/su11247200

Gopalan, S. P., Hanasaki, N., Champathong, A., & Tebakari, T. (2020). Impact assessment of reservoir operation in the context of climate change adaptation in the Chao Phraya River basin. *Hydrological Processes, 35*, Article e14005. https://doi.org/10.1002/hyp.14005

Gray, L. J., Beer, J., Geller, M., Haigh, J. D., Lockwood, M., Matthes, K., Cubasch, U., Fleitmann, D., Harrison, G., Hood, L., Luterbacher, J., Meehl, G. A., Shindell, D., van Geel, B., & White, W. (2010). Solar influences on climate. *Reviews of Geophysics, 48*, RG4001. https://doi.org/10.1029/2009RG000282

Haberl, H., Erb, K.-H., Krausmann, F., Bondeau, A., Lauk, C., Müller, C., Plutzar, C., & Steinberger, J. K. (2011). Global bioenergy potentials from agricultural land in 2050: Sensitivity to climate change, diets and yields. *Biomass and Bioenergy, 35*(12), 4753–4769. https://doi.org/10.1016/j.biombioe.2011.04.035

Halder, J. C. (2013). Land suitability assessment for crop cultivation by using remote sensing and GIS. *Journal of Geography and Geology, 5*, 65–74.

Hatfield, J. L., Boote, K. J., Kimball, B. A., Ziska, L., Izaurralde, R. C., Ort, D., Thomson, A. M., & Wolfe, D. (2011). Climate impacts on agriculture: Implications for crop production. *Agronomy Journal, 103*, 351–370.

Izaurralde, R. C., Thomson, A. M., Morgan, J., Fay, P., Polley, H., & Hatfield, J. L. (2011). Climate impacts on agriculture: Implications for forage and rangeland production. *Agricultural Journal, 103*, 371–381.

Jana, S., Roy, J., Marambe, B., Weerahewa, J., Pushpakumara, G., Silva, P., Miah, M. G., Punyawardena, R., & Premalal, S. (2015). Home gardens in the Paschim Medinipur District of West Bengal in India: A land use system with multiple benefits. *International Journal of Environment and Sustainable Development, 14*(2), 191–206.

Kafy, A.-A., Faisal, A.-A., Sikdar, S., Hasan, M., Rahman, M., Khan, M. H., & Islam, R. (2020a). Impact of LULC changes on LST in Rajshahi District of Bangladesh: A remote sensing approach. *Journal of Geographical Studies, 3*, 11–23. https://doi.org/10.21523/gcj5.19030102

Kafy, A.-A., Naim, M. N. H., Subramanyam, G., Faisal, A.-A., Ahmed, N. U., Al Rakib, A., Kona, M. A., & Sattar, G. S. (2021). Cellular automata approach in dynamic modeling of land cover changes using RapidEye images in Dhaka, Bangladesh. *Environmental Challenges*, 100084.

Kafy, A.-A., Rahman, M. S., Islam, M., Al Rakib, A., Islam, M. A., Khan, M. H. H., Sikdar, M. S., Sarker, M. H. S., Mawa, J., & Sattar, G. S. (2020b). Prediction of seasonal urban thermal field variance index using machine learning algorithms in Cumilla, Bangladesh. *Sustainable Cities and Society, 64*, Article 102542. https://doi.org/10.1016/j.scs.2020.102542

Kanitkar, T., Mythri, A., & Jayaraman, T. (2022). Equity assessment of global mitigation pathways in the IPCC sixth assessment report. https://doi.org/10.31219/osf.io/p46ty

Krissansen-Totton, J., & Davies, R. (2013). Investigation of cosmic ray–cloud connections using MISR. *Geophysical Research Letters, 40*, 5240–5245. https://doi.org/10.1002/grl.50996

Lawrence, D., & Vandecar, K. (2015). Effects of tropical deforestation on climate and agriculture. *Nature Climate Change, 5*, 27.

Lee, J., & Shin, H. (2021). Assessment of future climate change impact on an agricultural reservoir in South Korea. *Water 13*(2125). https://doi.org/10.3390/w13152125

Li, X., Takahashi, T., Suzuki, N., & Kaiser, H. M. (2011). The impact of climate change on maize yields in the United States and China. *Agricultural Systems, 104*(4), 348–353.

Lockwood, M. (2012). Solar influence on global and regional climates. *Surveys in Geophysics, 33*, 503–534. https://doi.org/10.1007/s10712-012-9181-3

Lungarska, A., & Chakir, R. (2018). Climate-induced land use change in France: Impacts of agricultural adaptation and climate change mitigation. *Ecological Economics, 147*, 134–154. https://doi.org/10.1016/j.ecolecon.2017.12.030

Mas, J. F., & Flores, J. J. (2008). The application of artificial neural networks to the analysis of remotely sensed data. *International Journal of Remote Sensing, 29*, 617–663.

Mosleh, Z., et al. (2017). Sustainable allocation of agricultural lands and water resources using suitability analysis and mathematical multiobjective programming. *Geoderma, 303*, 52–59.

Myhre, G., Aas, W., Cherian, R., Collins, W., Faluvegi, G., Flanner, M., Forster, P., Hodnebrog, Ø., Klimont, Z., Lund, M. T., Mülmenstädt, J., Lund Myhre, C., Olivié, D., Prather, M., Quaas, J., Samset, B. H., Schnell, J. L., Schulz, M., Shindell, D., Skeie, R. B., Takemura, T., & Tsyro, S. (2017). Multi-model simulations of aerosol and ozone radiative forcing due to anthropogenic emission changes during the period 1990–2015. *Atmospheric Chemistry and Physics, 17*, 2709–2720. https://doi.org/10.5194/acp-17-2709-2017

Nath, A. J., Sileshi, G. W., Laskar, S. Y., Pathak, K., Reang, D., Nath, A., & Das, A. K. (2021). Quantifying carbon stocks and sequestration potential in agroforestry systems under divergent management scenarios relevant to India's nationally determined contribution. *Journal of Cleaner Production, 281*, Article 124831.

Ngarava, S., Zhou, L., Nyambo, P., Chari, M. M., & Bhungeni, O. (2023). Aquaculture production, GHG emission and economic growth in Sub-Sahara Africa. *Environmental Challenges, 12*(100737), 1–12. https://doi.org/10.1016/j.envc.2023.100737

Outhwaite, C. L., McCann, P., & Newbold, T. (2022). Agriculture and climate change are reshaping insect biodiversity worldwide. *Nature, 605*(7908), 97–102. https://doi.org/10.1038/s41586-022-04644-x

Ozturk, D. (2015). Urban growth simulation of Atakum (Samsun, Turkey) using cellular automata-Markov chain and multi-layer perceptron-Markov chain models. *Remote Sensing, 7*, 5918–5950.

Pal, S., & Ziaul, S. (2017). Detection of land use and land cover change and land surface temperature in English Bazar urban centre. *Egyptian Journal of Remote Sensing and Space Sciences, 20*, 125–145.

Pandey, R., Aretano, R., Gupta, A. K., Meena, D., Kumar, B., & Alatalo, J. M. (2017). Agroecology as a climate change adaptation strategy for smallholders of Tehri Garhwal in the Indian Himalayan region. *Small-Scale for., 16*(1), 53–63.

Porter, J. R., Xie, L., Challinor, A. J., Cochrane, K., Howden, S. M., Iqbal, M. M., Lobell, D. B., & Travasso, M. I. (2014). Food security and food production systems. In C. B. Field, V. R. Barros, D. J. Dokken, K. J. Mach, M. D. Mastrandrea, T. E. Bilir, L. L. White, et al. (Eds.), *Climate change 2014: Impacts, adaptation, and vulnerability. Part A: Global and sectoral aspects. Contribution of working group II to the fifth assessment report of the intergovernmental panel on climate change*. Cambridge University Press.

Ramachandran, R. M., Roy, P. S., Chakravarthi, V., Joshi, P. K., & Sanjay, J. (2020). Land use and climate change impacts on distribution of plant species of conservation value in Eastern Ghats, India: A simulation study. *Environmental Monitoring and Assessment, 192*, 86.

Riahi, K., van Vuuren, D. P., Kriegler, E., Edmonds, J., O'Neill, B. C., Fujimori, S., Tavoni, M., et al. (2017). The shared socioeconomic pathways and their energy, land use, and greenhouse gas emissions implications: An overview. *Global Environmental Change, 42*, 153–168. https://doi.org/10.1016/j.gloenvcha.2016.05.009

Robinson, S. A. (2020). Climate change adaptation in SIDS: a systematic review of the literature pre and post the IPCC Fifth Assessment Report. *Wiley Interdisciplinary Reviews: Climate Change, 11*(4), e653. https://doi.org/10.1002/wcc.653

Rosenzweig, C., Elliott, J., Deryng, D., Ruane, A. C., Müller, C., Arneth, A., Boote, K. J., Folberth, C., Glotter, M., Khabarov, N., Neumann, K., Piontek, F., Pugh, T. A. M., Schmid, E., Stehfest, E., Yang, H., & Jones, J. W. (2014). Assessing agricultural risks of climate change in the 21st century in a global gridded crop model intercomparison. *Proceedings of the National Academy of Sciences of the United States of America, 111*(9), 3268–3273. https://doi.org/10.1073/pnas.1222463110

Sapkota, T. B., Vetter, S. H., Jat, M. L., Sirohi, S., Shirsath, P. B., Singh, R., Jat, H. S., Smith, P., Hillier, J., & Stirling, C. M. (2019). Cost-effective opportunities for climate change mitigation in Indian agriculture. *Science of the Total Environment, 655*, 1342–1354.

Scott, D., Hall, C. M., Rushton, B., & Gössling, S. (2023). A review of the IPCC Sixth Assessment and implications for tourism development and sectoral climate action. *Journal of Sustainable Tourism,* 1–18. https://doi.org/10.1080/09669582.2023.2195597

Searle, S., & Malins, C. (2015). A reassessment of global bioenergy potential in 2050. *GCB Bioenergy, 7*(2), 328–336. https://doi.org/10.1111/gcbb.12141

Seif-Ennasr, M., Bouchaou, L., El Morjani, Z. E. A., Hirich, A., Beraaouz, E. H., & Choukr-Allah, R. (2020). GIS-based land suitability and crop vulnerability assessment under climate change in Chtouka Ait Baha, Morocco. *Atmosphere, 11*(11), 1167; https://doi.org/10.3390/atmos11111167

Seppälä, A., Matthes, K., Randall, C. E., & Mironova, I. A. (2014). What is the solar influence on climate? Overview of activities during CAWSES-II. *Progress in Earth and Planetary Science, 1*, 1–12. https://doi.org/10.1186/s40645-014-0024-3

Shevchenko, V., Lukashevich, A., Taniushkina, D., Bulkin, A., Grinis, R., Kovalev, K., Narozhnaia, V., Sotiriadi, N., Krenke, A., & Maximov, Y. (2023). Climate change impact on agricultural land suitability: An interpretable machine learning-based Eurasia case study. *11*. https://doi.org/10.1109/ACCESS.2024.3358865

Smith, P., Davis, S. J., Creutzig, F., Fuss, S., Minx, J., Gabrielle, B., Yong Sung, C., et al. (2016). Biophysical and economic limits to negative 2 emissions. *Nature Climate Change, 6*(1), 42–50. https://doi.org/10.1038/nclimate2870

Tewari, J. C., Ram, M., Roy, M. M., & Dagar, J. C. (2014). Livelihood improvements and climate change adaptations through agroforestry in hot arid environments. *Agroforestry systems in India: Livelihood security & ecosystem services* (pp. 155–183). Springer.

Tongwane, M. I., & Moeletsi, M. E. (2018). A review of greenhouse gas emissions from the agriculture sector in Africa. *Agricultural Systems, 166*, 124–134. https://doi.org/10.1016/j.agsy.2018.08.011

UNFCCC. (2015). Report of the Conference of the Parties on its twenty-first session, held in Paris from 30 November to 13 December 2015.

Wang, C., Deser, C., Yu, J.-Y., DiNezio, P., & Clement, A. (2017). El Nino and southern oscillation (ENSO): a review. In *Coral reefs of the Eastern Tropical Pacific* (pp. 85–106). Springer.

Wang, P., Tong, X., Qiu, J., Chen, Y., Wu, S., Chan, T. O., & Luo, M. (2022). Amplification effect of urbanization on atmospheric aridity over China under past global warming. *Earth's Future e2021EF002335*

Yang, X., Zheng, X.-Q., & Lv, L.-N. (2012). A spatiotemporal model of land use change based on ant colony optimization, Markov chain and cellular automata. *Ecol. Modell., 233*, 11–19.

Yuan, F., Sawaya, K. E., Loeffelholz, B. C., & Bauer, M. E. (2005). Land cover classification and change analysis of the Twin Cities (Minnesota) Metropolitan Area by multitemporal Landsat remote sensing. *Remote Sensing of Environment, 98*, 317–328.

Chapter 12
Soil Conservation Techniques and Sustainable Land Management

B. Namratha and Santosh L. Gaonkar

Abstract Addressing the phenomenon of soil degradation and ensuring the sustainability of agricultural productivity necessitates carrying out soil conservation methodologies and sustainable land management paradigms. Practices such as crop rotation, contour ploughing, and the application of cover crops represent salient soil conservation techniques that are imperative for the preservation of soil health and easing soil erosion. These soil conservation methodologies are integrated within a broader framework of environmental stewardship through sustainable land management, or SLM. This chapter represents the harmony between agricultural productivity and the preservation of ecosystem services in a holistic manner. There is a focused elucidation on sustainable land management (SLM) techniques that facilitate food security and the enduring viability of natural resources by cultivating a symbiotic relationship between agricultural practices and environmental integrity. Ultimately, to avert soil degradation and uphold agricultural sustainability, it is essential to implement sustainable land management methodologies and soil conservation tactics. These initiatives bolster the resilience and overall health of ecosystems, contributing to the enhancement of soil fertility.

Keywords Soil conservation · Sustainable land management (SLM) · Soil erosion · Soil fertility · Ecosystem

B. Namratha
Department of Chemistry, The Yenepoya Institute of Arts, Science, Commerce and Management, Yenepoya (Deemed to be University), Mangaluru, Karnataka, India

S. L. Gaonkar (✉)
Department of Chemistry, Manipal Institute of Technology, Manipal Academy of Higher Education, Manipal, Karnataka, India
e-mail: sl.gaonkar@manipal.edu

U. Chatterjee et al. (eds.), *Eco-Resilience: Climate Change, Land Degradation and Sustainable Solutions*, Sustainability Solutions,
https://doi.org/10.1007/978-3-032-00708-7_12

12.1 Introduction to Soil Degradation and Agricultural Sustainability

Soil constitutes a natural entity that demonstrates vitality and dynamism, thereby fulfilling critical functions within terrestrial ecosystems (Fig. 12.1). This non-renewable natural resource comprises a complex amalgamation of mineral particles, organic constituents, gaseous elements, nutrients, as well as a diverse array of both microscopic and macroscopic organisms. Soil facilitates the sustenance of life through enhanced fertility, particularly when combined with water (Montgomery, 2007).

The degradation of soil represents a cumulative qualitative and quantitative decline in both soil and water resources. This phenomenon is exacerbated by atmospheric pollution and the erosion of vegetative cover, alongside increasing population density. The situation deteriorates further when peripheral areas are subjected to agricultural practices due to escalating demands for food. Activities such as deforestation, overgrazing, conversion of forested land to agricultural use, and cultivation on steep inclines contribute significantly to the severity of the challenges faced. The adverse conditions are intensified by the depletion of carbon in soil organic matter, which occurs as a result of in-situ burning of post-harvest residues, insufficient application of organic manures, and intensive agricultural practices (Liu et al., 2010).

The introduction of contaminants in the form of toxic substances, as well as the capping and sealing of soil, exacerbates the detrimental effects on agriculture in specific regions. The magnitude and intensity of precipitation are decisive in determining soil health. The management of agricultural practices has a profound influence on soil erosion, which conversely affects agricultural sustainability by undermining farm productivity. Generally, soil erosion is significantly more pronounced in mountainous regions compared to undulating and flat terrains. Land degradation attributable to wind erosion is typically confined to arid and semi-arid regions, where

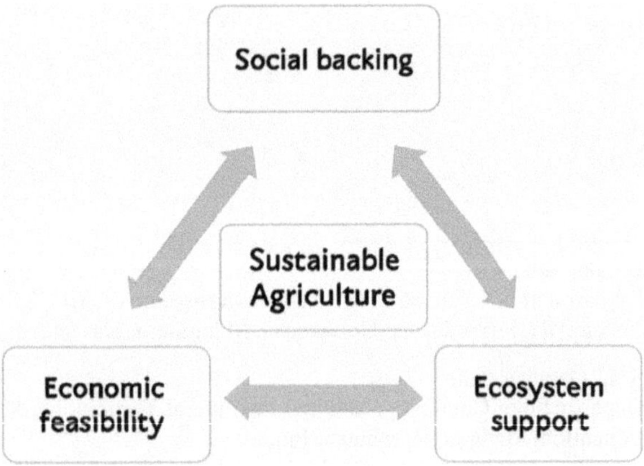

Fig. 12.1 The balance among environmental, social, and economic concerns in sustainable agriculture

the displacement of soil particles is primarily driven by wind action (Singh & Jai, 1995).

The significance of soil organic matter warrants careful consideration. Soils exhibiting organic matter content below 2% organic carbon are generally classified as infertile. The susceptibility to erosion diminishes linearly with an increase in organic content within the range of 0–10%. The patterns of crop growth and the inherent characteristics of soil are profoundly influenced by field tillage systems, which alter the soil profile. Characteristic features of eroded soils include reduced water retention capacity, impaired water infiltration, decreased microbial populations, and enhanced soil firmness. Through a systematic examination of soil characteristics, climatic conditions, and biodiversity, interventions can be implemented to mitigate soil degradation. Such measures yield a restoration of soil productivity and promote an increase in food production rates (Reeves, 1997).

The interplay between poverty and natural resource degradation compels individuals to seek additional land for food production. Social insecurity frequently stems from diminished agricultural productivity, which is intrinsically linked to the acceleration of soil erosion influenced by factors such as salinization, flooding, drought, and waterlogging. Moreover, atmospheric greenhouse gas discharge indirectly contributes to soil degradation. For the attainment of sustainable livelihoods and improved socio-economic equilibrium, it is imperative to conserve natural resources and rejuvenate degraded and wasteland areas (Gambella et al., 2021).

The changing climate has a significant impact on land, causing its degradation or removal. Furthermore, nations that contribute the most significantly to anthropogenic climate change through emissions frequently have more resources to mitigate land-related impacts than countries that may contribute less to temperature escalation but still face land-related consequences. As land degrades and shrinks, competition for freshwater access and land resources increases. Although the international community works to ameliorate the effects of climate change, its effects on land remain persistent and unequally dispersed around the world. Conceptualizing 'land' as a sentient creature deserving of rights has the potential to dramatically improve the protection and flourishing of natural ecosystems, as well as the innumerable organisms that rely on them, paving the path for eco-resilience (Mendoza-Ponce et al., 2021).

This chapter examines the economic incentives, land tenure stability, and technical adoption required for sustainable agricultural practices, with a focus on the crucial challenges of soil degradation and Sustainable Land Management (SLM). Regionally, an analysis of India's Western Ghats is considered, with a focus on the particular issues and adaptive strategies that farmers will implement. The major goal of this research is to provide a model for SLM that can be applied to a variety of worldwide regions with varying socioeconomic circumstances, hence improving eco-resilience, food security, sustainable development, and environmental stress reduction.

12.2 The Importance of Soil Conservation and Sustainable Land Management-Integrated Approaches

Soil serves a crucial function in the processes of biomass production, the provision of ecosystem services, the underpinning of construction foundations, and the facilitation of water purification, storage, and distribution. It constitutes the fundamental basis for habitats and recreational spaces for individuals. The domains of soil and land are critical focal points within the framework of Sustainable Development Goals (SDGs) (Fig. 12.2). Numerous Sustainable Development Goals (SDGs) pertain specifically to soil. Illustrative examples encompass food security (SDG 2), clean water access (SDG 6), resource efficiency (SDG 12), climate action initiatives (SDG 13), and the preservation of terrestrial life (SDG 15). Soil conservation and management strategies are frequently subjected to ongoing enhancements. The components of terrestrial ecosystems exhibit a direct relationship with the generation of goods and services. The greater the advancements in developmental initiatives aimed at serving societal interests, the more severely the essential component of the terrestrial ecosystem, namely soil, becomes degraded (Löbmann et al., 2022).

Sustainable Land Management (SLM) is predicated upon the robust principles of Conservation Agriculture (CA), which advocates for no-till farming, the maintenance of permanent soil cover, and the application of crop rotation techniques. The principles of agroforestry, specialty crop cultivation, and other superior management practices for field crops can enhance market opportunities. SLM may be delineated as the judicious amalgamation of technologies and policies that harmonize socioeconomic principles with environmental considerations, thereby augmenting both soil and water quality and enhancing productivity (Muche et al., 2024).

Despite these advances, soil and land systems face significant societal pressures due to the demand for biomass production for food, livestock feed, textiles, and energy, as well as the imperatives of climate risk prevention and resilience strategy, as well as various other ecosystem services. There is an approximation that nearly one-third of the planet's land is currently experiencing degradation. Climate change represents a significant factor exacerbating this condition through phenomena such as intense precipitation, flooding, droughts, wind erosion, permafrost thawing, salinization resulting from sea-level rise, and the impacts of wave action (Killham, 2011). In this context, coordinated approaches to soil conservation and sustainable land management are critical for ensuring ecosystem resilience and the well-being of future generations.

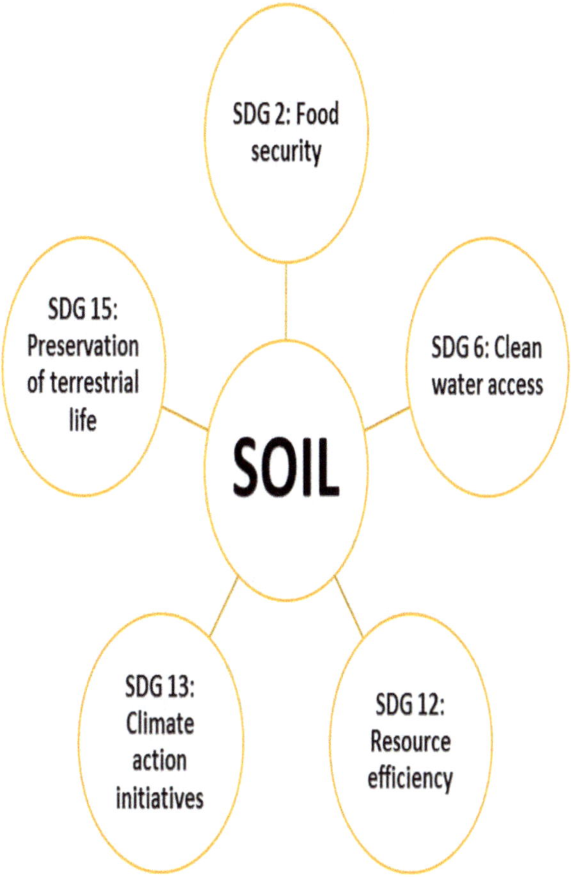

Fig. 12.2 Relationship between soil and the Sustainable Development Goals of the UN

12.3 Fundamentals of Soil Preservation

The primary hazard to soil sustainability and conservation arises from water-induced soil erosion in agricultural fields. The practice of commercial farming significantly obscures this detrimental impact through the extensive use of fertilizers. Concurrently, both traditional and mechanized agricultural systems face persistent threats from soil degradation. Thus, the implementation of effective soil conservation practices remains a formidable challenge in both agricultural paradigms (Blanco & Rattan, 2008). While soil cultivation is indicative of productivity, soil conservation is synonymous with re-productivity. The principal aim of land preservation is to ensure ecological balance. This endeavor, undertaken by individuals or communities, is essential for sustaining soil fertility, thereby securing soil sustainability for future generations (Franzluebbers, 2010).

Soil conservation methodologies are categorized into two distinct levels: the field level and the catchment level (Valentin & Jean, 2018).

12.3.1 Field Level

This level primarily pertains to enhancing the resilience of soil. This enhancement is accomplished through the incorporation of organic carbon into the soil via manure, sewage sludge, compost, and other organic waste products. Such practices lead to improved structural stability of the soil and a concomitant decrease in its susceptibility to erosion. The maintenance of surface cover is instrumental in mitigating the kinetic energy imparted to the soil. For instance, the vegetation that is in closest proximity to the ground serves as the most effective barrier against the impact of raindrops. This factor is critically significant in the context of soil conservation.

The confined dimensions of the agricultural plot also significantly influence the mitigation of erosion. The regulation of runoff velocity is vital in minimizing the formation of rills and gullies. The construction of terraces is a time-honored technique employed for this purpose. Nonetheless, the challenges associated with terrace farming include the considerable effort required for both construction and ongoing maintenance. Wind erosion can only commence when wind velocity exceeds a certain threshold, at which point particles with a diameter of approximately 100 μm begin to be dislodged. Therefore, augmenting surface roughness can substantially curtail wind speed. The establishment of windbreak hedges represents a traditional solution in this context. Soil fragmentation resulting from cattle trampling or tillage on steep terrains exacerbates soil erosion. Consequently, the foundation of sustainable agriculture relies on practices such as direct seeding, minimal tillage, continuous soil cover through harvest leftovers, and the implementation of crop rotation strategies. The integration of these methodologies collectively contributes to the reduction of soil loss.

12.3.2 Catchment Level

At this level, all conservation strategies delineated under the field level are taken into account. A diverse landscape mosaic can be established in which forested or grassed areas effectively reduce runoff velocity. This configuration enables the trapping of sediments along the slopes. To mitigate soil erosion linked to heavy rainfall, the implementation of hedges and contour drainage ditches is imperative. The planting of trees with deep taproots serves to stabilize slopes effectively. The construction of terraces can also be a viable strategy, as well as the aquifer drawdown, which leads to a decrease in soil moisture and consequently soil mass. Generally, the catchment scale emphasizes the deceleration of runoff and the facilitation of sediment deposition.

12.4 Key Soil Conservation Practices

The discourse herein pertains to three fundamental conservation methodologies—crop rotation, contour ploughing, and cover cropping—that are particularly well-suited for the humid and sub-humid hilly regions characterized by seasonal precipitation within the Western Ghats. The pronounced monsoonal precipitation in this area precipitates surface runoff and soil erosion, with notable manifestations in the sloping agricultural terrains. The selection of the Western Ghat region for this examination is predicated upon its distinctive amalgamation of rich biodiversity, substantial rainfall, precipitous topography, and intensive agricultural practices. As a recognized biodiversity hotspot and a UNESCO-designated World Heritage Site, the Western Ghats are susceptible to phenomena such as soil erosion, deforestation, and unsustainable land management practices. The conservation techniques delineated herein are instrumental in augmenting agricultural productivity while maintaining ecological integrity. Regions such as the East African Highlands, the Andean Highlands in South America, and various locales in Southeast Asia exhibit mountainous terrains and significant rainfall patterns akin to those found in the Western Ghats. Consequently, this study bears substantial global relevance (Sowjanya et al., 2021; Surendran et al., 2024).

12.4.1 Crop Rotation

Crop rotation represents a time-honored agronomic practice wherein various plant species are cultivated sequentially on the same parcel of land (Fig. 12.3). A notable characteristic of crop rotation is the proficiency of leguminous plants to bio-fix nitrogen (N) from atmospheric sources. Contemporary farmers have acquired specialization in augmenting productivity and enhancing marketing strategies through the adoption of short cereal or soybean rotations (Shah et al., 2021).

Crop sequencing assumes a critical role within organic farming systems, wherein the use of synthetic fertilizers and pesticides is categorically prohibited. To sustain nutrient richness, humus content, and soil texture, the implementation of crop rotation is indispensable. Rotational cropping further serves to mitigate pest proliferation, weed growth, and the incidence of diseases. The practice of crop rotation disrupts the reproductive cycles of insects and pathogens, thereby interrupting their life cycles (Ball et al., 2005).

An extensive investigation into the rotational cropping technique is essential for the effective management of soil structure. Comprehensive and sustainable plans are essential for developing an in-depth understanding of its application. Crop rotation necessitates ongoing monitoring, the regular assessment of progress, the evaluation of crop performance, and the adaptation of management strategies as required. The phenomenon of population growth has escalated the demand for food. Furthermore, the planet is experiencing diminished crop yields, culminating in a productivity

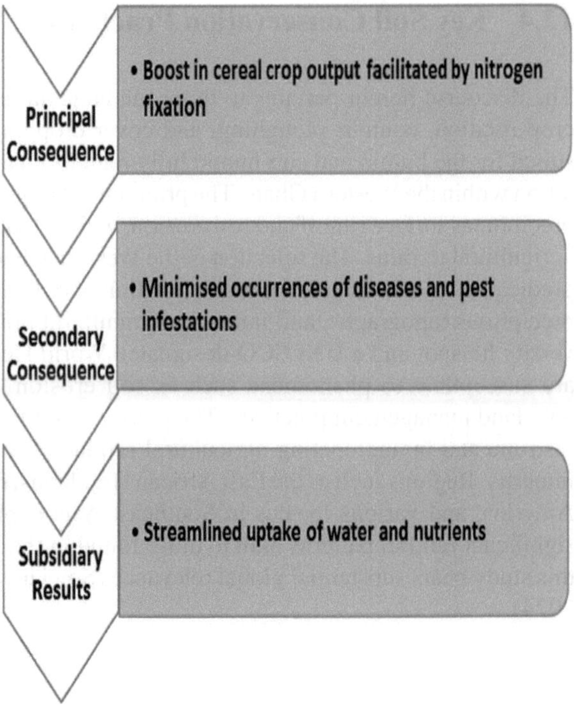

Fig. 12.3 Farming outcomes of crop rotation

crisis. Given the increasing scarcity of arable land alongside an ever-expanding food demand, the enhancement of crop yield and soil productivity is imperative (He et al., 2021).

12.4.2 Contour Ploughing

Contour ploughing is an agricultural technique that entails the establishment of furrows in alignment with the land's slope. This method effectively impedes the flow of running water, thereby safeguarding the soil from erosion and preventing sedimentation. Furthermore, this practice fosters natural water retention within the soil, eliminating the necessity for artificial water retention ponds and terraces. Ultimately, this results in the enhancement of both the quality and quantity of crops cultivated on sloped terrains. Contour farming, when integrated with strip cropping, constitutes a robust strategy for soil conservation that yields significant benefits for society, transcending the dichotomy between economic and environmental considerations. These practices facilitate the realization of sustainable agricultural systems through advanced water management techniques, optimized soil composition, and a marked minimization of soil loss. Empirical evidence suggests that by harmonizing agricultural practices with the natural contours of the terrain, farmers can attain reductions in soil erosion by as much as 50–60% and experience considerable enhancements in crop yields (Stevens et al., 2009).

Although terrace farming and contour farming are occasionally used interchangeably, terrace farming involves the construction of walls and the alteration of the slope's configuration to create level areas that function as water catchments and mitigate erosion, whereas contour farming preserves the natural contour of the slope. In geographical terms, a contour refers to an abstract line that connects two points of equal elevation. Contour bunding represents a soil conservation technique specifically designed to address soil erosion in accordance with the contours of the land (Gebreegziabher et al., 2009).

The implementation of bunding and the establishment of vegetative barriers along the slope are imperative to mitigate erosion on exceedingly steep slopes when contour farming alone may prove insufficient. In addition to these measures, optimal results are achieved when integrated with practices such as mulching, residue management, and the utilization of alley and cover crops, among others. In smaller parcels of land with uniform gradients, a single contour guideline is deemed adequate; however, in areas characterized by irregular slopes, two or three guidelines are requisite. Contour farming serves to preserve nutrient-rich top soils from excessive erosion. By fostering biological activity that facilitates the liberation of supplementary nutrients into the soil, the augmented soil moisture diminishes the necessity for input expenditures (Heckrath et al., 2006).

Similar to other agricultural methodologies, the profitability of contour farming is contingent upon various factors, including geography, landholding size, crop selection, and seed quality. Preliminary research indicates that contour farming tends to be more profitable than allowing land to remain in its natural state, particularly because steep slopes are prone to soil degradation and necessitate continuous nutrient replenishment. However, there remains a pressing need for additional long-term systematic research to fully understand and document the income-enhancing potential of this agricultural practice (Gebreegziabher et al., 2009).

The advent of contemporary technology has rendered contour trenching more precise and adaptable. Scientific methodologies such as GPS-based contour mapping, drone surveillance, and precision agriculture have significantly advanced the practice of scientific farming. Addressing existing challenges through focused research, the formulation of improved guidelines for tailoring tillage practices to specific agro-ecological conditions, the development of cost-efficient machinery for smallholder farmers, and the implementation of supportive policies can facilitate broader adoption and optimize the advantages of contour tillage. This approach not only promotes sustainable land management but also enhances resilience against the impacts of climate change (Pásztor & Izolda, 2019).

12.4.3 Cover Crops

Cover cropping represents an ancient agricultural strategy in which leguminous broadleaves, non-leguminous broadleaves, or grasses are cultivated in the interstitial spaces between cash crops. The principal purpose of implementing cover crops is to safeguard and enhance the soil during periods of non-production between regular annual crop cycles or alongside tree growth. The augmented root development associated with cover crops confers substantial advantages to the soil, including enhancements in organic carbon content, nutrient availability, and soil aggregation. Additional benefits derived from cover crops encompass pest management, weed suppression, and the facilitation of nutrient cycling. The retention of soil moisture is significantly improved as cover crops contribute to the mitigation of water evap oration from the soil matrix. In scenarios characterized by limited water availability and the absence of irrigation, cover crops have frequently demonstrated a capacity to preserve soil hydration status for successive plantings (Komlan et al., 2022).

Cover crop maintenance yields an extensive array of advantages concerning the physical, chemical, and biological attributes of the soil. Microbial dynamics in soil is assumed to experience an increase, alongside their abundance and diversity. A diminution in soil compactness and an enhancement of soil organic carbon are anticipated outcomes. In this context, cover crops possess the capability to ameliorate both soil structure and hydraulic properties. However, the degree to which cover crops can effectuate such alterations may be contingent upon the selection of cover crop species and mixtures, the duration of the growth phase, the cumulative biomass produced, and prevailing natural surroundings (Adewole et al., 2020).

According to studies, the treatment of covering mulch on the soil surface, in conjunction with amendments, can lead to a significant reduction in salinity phenomena. Similarly, the decomposition of leguminous crops exerts a solubilizing effect on essential soil micronutrients, including nitrogen, phosphorus, and potassium. In this context, nutrient recycling may result in deficiencies of various soil nutrient components. Moreover, cover crops contribute to a reduction in the reliance on synthetic fertilizers and diminish agricultural operations, thereby decreasing fossil fuel consumption (Komlan et al., 2022).

The realization of desired agricultural outcomes is contingent upon the meticulous selection of species (Fig. 12.4), cultivation cycles, and harvesting techniques, tailored to the specific type of farm and its intended purpose. The integration of cover crops into conventional farming systems not only facilitates crop diversification but also contributes to food security. Cover crops enhance the carbon and nitrogen content of the soil, provide biomass support, engage in photosynthetic processes, and promote rhizosphere secretions and microbial population density. They offer protection during seeding and contribute to soil enhancement during intervals of standard crop production (Sanyal et al., 2023).

Cover crop species constitute the cornerstone of a comprehensive approach that facilitates the management of biotic constraints on crops, including pathogens, herbivorous insects, and competitive weeds. By sequestering excess nutrients prior

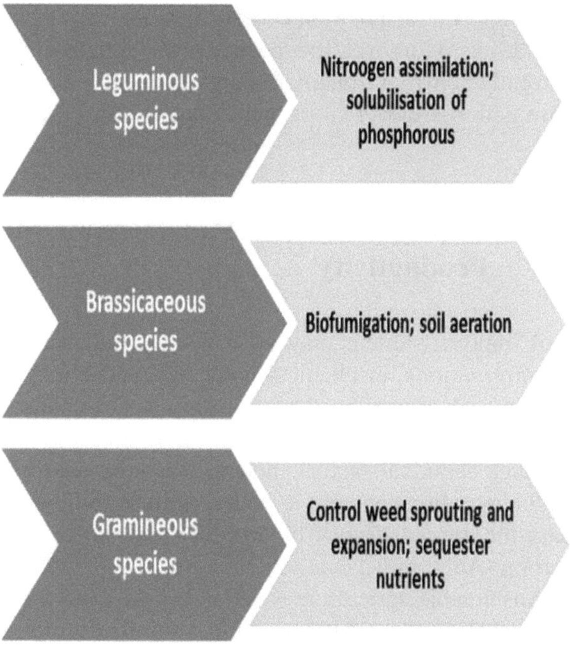

Fig. 12.4 Ecological functions provided by various categories of cover crops

to their leaching from the soil matrix, cover crops—particularly those within the Poaceae family—can enhance soil and water quality indirectly (Scavo et al., 2022). Certain brassica cover crop varieties release glucosinolates, a category of sulfur-containing compounds, upon decomposition, which possess nematode-suppressing properties. Furthermore, these cover crops act as a soil safeguard, thereby promoting the penetration of subsequent crop root systems. Empirical studies indicate that various leguminous cover crop species have exhibited the potential to augment the yields of the following cash crops (Abdalla et al., 2019). Upon being terminated, select cruciferous cover crop species undergo hydrolysis of glucosinolates to yield isothiocyanates, which are bioactive compounds. This transformation transpires during the decomposition of their tissues. Isothiocyanates represent potent biofumigants that can be utilized in the management of various soil-borne pathogens and pests. Consequently, employing cruciferous plants as cover crops may facilitate the natural regulation of potential diseases. In perennial cropping systems, such as those prevalent in vineyards and orchards, cover crops contribute to improved soil health and a reduction in mineral fertilization requirements. These systems, often afflicted by excessive tillage practices and depletion of organic matter, derive substantial benefits from cover mulching (Sharma et al., 2018).

However, it is imperative to acknowledge that the implementation of cover crops may encounter practical constraints that necessitate additional diligence. While cover crops enhance soil quality, organic matter content, soil water infiltration, and overall soil water holding capacity in the long term, they may induce a temporary water deficit for cash crops. The financial implications linked to the establishment of cover crops and the acquisition of new agricultural machinery represent two significant limitations. Moreover, irrespective of the agricultural system employed, practitioners currently confront a distinctive and formidable management challenge regarding the termination of cover crops, which necessitates meticulous attention.

12.5 Sustainable Land Management and Agricultural Productivity

Soil impoverishment may arise from a multitude of factors, including chemical transformations, mechanical processes, and biological activities. Manipulations of matter, such as erosion, desertification, environmental contamination, and the degradation of soil structure, are particularly noteworthy among these phenomena. The primary chemical factors include leaching, acidification, diminished fertility and cation retention capacity, as well as salinization, while the biological factors encompass the degradation of biodiversity and the reduction of both total and biomass carbon (Adetomiwa et al., 2022).

Sustainable agriculture is defined as the capacity of agricultural systems to maintain continuous production while minimizing detrimental environmental impacts. SLM constitutes the basis for sustainable agricultural practices and contributes significantly to mitigating the economic burdens faced by society. SLM can be articulated as the judicious utilization of land resources—including soils, water, fauna, and flora—for the generation of goods that address the evolving needs of humanity. This approach not only preserves the environmental functions of these resources but also guarantees their long-term productivity potential. It facilitates the integration of land, water, biodiversity, and environmental management to satisfy the escalating requirements for food and fiber. The implementation of SLM represents a strategy that fosters income generation while safeguarding the sustainability of land resources as a foundational element for production. There exist four fundamental criteria that underpin SLM: the maintenance of production levels; the minimization of risk; the preservation of soil and water quality; and the assurance that systems remain economically viable and socially acceptable (Kassie et al., 2010).

The practice of SLM necessitates moderation in light of the pervasive resource degradation stemming from suboptimal land use practices. Moreover, there is a burgeoning global demand for nutrition, coupled with the erratic patterns associated with climate change. The establishment of an effective SLM framework leads to a more integrated, productive, and robust ecosystem, which forecasts significant benefits for human health (Bindraban et al., 2000). An examination of historical trends indicates that the escalation in global food demand has been met through the expansion of arable land, the augmentation of agricultural territories, and the heightened utilization of soil ameliorants and pest management chemicals. However, an excessive amplification of these practices ultimately results in exacerbated environmental degradation on a global scale. Consequently, it is essential that sustainable agricultural methodologies be universally embraced (Motavalli et al., 2013).

12.6 Sustainable Land Management and Food Security

With respect to the policies associated with rural development and food security, a multitude of actions and amendments have been implemented. In evaluating the correlations between agricultural policy and the supply responses of farmers, it is imperative to consider alterations at the macro level of agricultural households. The combined effects of land overexploitation and the detrimental implications of climate change are responsible for the decline in agricultural productivity, which in turn results in food scarcity and exacerbates poverty levels (Branca et al., 2013).

Sustainable Land Management (SLM) appears to be the most effective strategy for enhancing food security, as it promotes diversification of production, increases average productivity, and aids in mitigating seasonal fluctuations in yield. Given that soil quality exerts a direct influence on global food security, it is advantageous to possess a comprehensive understanding of SLM in the context of the ever-evolving climate. SLM bolsters the global economy by minimizing seasonal yield variability while preserving soil and water resources (Swaminathan, 2001). Populations in different regions will possess distinct demographic and economic attributes. As a result, the evolving dietary patterns and overall demand for food will further influence the pressures exerted on land resources. Nevertheless, disparities in nutritional access are likely to persist on a global scale, thereby intensifying the stress on the limited availability of arable land in certain regions (Holden et al., 2006).

Food shortages can materialize swiftly in areas characterized by significant productivity variability due to seasonal fluctuations and extreme weather conditions that directly and indirectly affect crops and pastures. Nutritional access is also influenced by additional factors such as market dynamics, political conditions, conflict, as well as the prevalence of pests and diseases affecting plants and livestock. By increasing the likelihood of land degradation, climate variability and extremes, water scarcity, steep terrain, and soils that are shallow, fragile, and nutrient-deficient may all contribute to undernutrition. Addressing the reduction of individuals experiencing food insecurity constitutes one of the most pressing humanitarian challenges faced globally, and this issue is intricately linked to the development and implementation of SLM practices. The second goal of the Sustainable Development Agenda acknowledges this imperative: to eliminate hunger, achieve food security and improved nutrition, and foster sustainable agricultural practices. In regions where undernourishment is prevalent, the necessity for action is most acute; however, to meet the Sustainable Development Goals (SDGs), a concerted international response through SLM is essential. Nonetheless, the sustainability of land management practices for food production is influenced by hunger in various capacities, including the phenomenon of undernutrition (Ojo et al., 2024).

Urbanization is anticipated to exert additional strain on water resources. Consequently, the enhancement of water-use efficiency within the agricultural sector will be paramount, particularly in regions where access to safe drinking water falls short of fulfilling human requirements. In the absence of SLM practices, the degradation of land and water resources will increasingly challenge food security across all nations, including those currently possessing sufficient nutritional resources (Viana et al., 2022).

12.7 The Role of SLM in Soil Health and Ecosystem Resilience

The utilization of land for agricultural production invariably transforms the resources from their natural condition and modifies the inherent biodiversity to a certain degree. Owing to competition with alternative land uses and the potential adverse effects on climate and biodiversity, opportunities to expand cropland area to satisfy the escalating demand for food and feed are markedly constrained (Reed et al., 2015).

Biological systems inherently possess a degree of resilience to mild and moderate disturbances; thus, recovery from certain levels of degradation is achievable without significant disruption to livelihoods, provided that the climatic conditions are favorable and SLM practices are employed. However, degradation that surpasses a critical threshold will lead to enduring productivity losses, and the restoration of ecological functionality may prove to be protracted, impractical, or prohibitively expensive (Mallick et al, 2025).

A methodology within SLM known as "wildlife-friendly farming" integrates agricultural practices with conservation efforts on farms in a manner that enhances ecosystem functionality. The objective is to augment biological diversity and mitigate the risk of species extinction through "land sharing," which involves the adoption of low-intensity, low-input agricultural practices that allow for the coexistence of agricultural and biodiversity conservation efforts within the same geographical area. To promote more wildlife-friendly farming, it is essential to diminish chemical inputs such as pesticides and fertilizers, while also maintaining extensive areas of low-impact farming or grazing within the agricultural landscape (Xie et al., 2020).

Land sparing presents an alternative strategy to wildlife-friendly farming, concentrating high-input, high-yield agricultural practices on a reduced land area, thereby preserving other lands for biodiversity conservation. The prospective role of land sparing in addressing future food security demands while simultaneously conserving biodiversity is contingent upon the degree to which crop yields can be augmented on existing agricultural land, thereby alleviating the impetus for the conversion of additional land into agricultural use (Mallick et al, 2025).

The concept of soil health encompasses the potential of soil to operate effectively within the constraints of land use and ecosystem boundaries, thereby encouraging biological yield, preserving environmental integrity, and fostering the health of plants, animals, and humans alike. Soil health represents one of the myriad factors that affect the land's potential to offer ecosystem services in a manner that is environmentally sustainable. A comprehensive understanding of the soil properties that define soil capability and condition, along with an analysis of how land management practices influence these properties, facilitates the identification of soil limitations and risks, the selection of SLM practices, and the establishment of indicators to evaluate the effectiveness of SLM interventions. The monitoring of soil conditions is imperative for the identification of land degradation, the evaluation of unsustainable practices, and the assessment of the efficacy of SLM strategies (Jeffrey, 2000).

Practices such as forest farming, bio-organic farming, forest regeneration, and ecological rehabilitation exemplify SLM strategies that are instrumental in conserving biodiversity while enhancing the quality of ecosystem services. These approaches contribute significantly to ecological equilibrium, soil health, water availability, and carbon sequestration. SLM ensures the provision of ecosystem services, including clean water, pollination, and climate regulation. Moreover, the implementation of these SLM strategies can yield long-term economic advantages for local communities and bolster the regional economy (Löbmann et al., 2022).

12.8 Towards Soil Conservation and SLM: Land Tenure Security, Incentives and Policy Synergy

The economic determinants affecting soil conservation are significantly influenced by the technological characteristics employed. Specifically, within the context of the Western Ghats region in India, the successful implementation of Sustainable Land Management (SLM) is contingent upon the availability, accessibility, and compatibility of technology with pre-existing farming processes and technologies. Economic limitations are a direct consequence of the technology employed, particularly if it entails substantial financial outlay, such as the acquisition of new machinery or investments in agricultural infrastructure. The accessibility to financial capital serves as a prerequisite for such investments to materialize. The assessment of a technology is typically based on its profitability or financial yield, encompassing either the costs incurred or the costs mitigated (for instance, those associated with fuel, fertilizer, labor, and time), or alternatively, by the profits realized through enhanced yields. Furthermore, expenses associated with land leasing and externalities may also have a significant influence. The technology must effectively address the issues of degradation while simultaneously exhibiting both environmental and economic advantages, thereby demonstrating its overall effectiveness. In a similar vein, production factors and agricultural practices exert considerable influence. Adequate labor must be available to execute the necessary operations, and reliance on family labor may facilitate the acceptance of new methods. The agricultural system must exhibit sufficient flexibility to incorporate the new technology or practice, or at a minimum, permit adaptation without incurring substantial costs. The lesser the modifications required by a conservation practice, the greater the likelihood of its adoption (Boardman et al., 2003).

The accessibility to markets and the potential profitability derived from the selection of specific crops and their cultivation during particular seasons are critical determinants of profitability. Incentive programs are designed to offset the costs incurred in the application of soil conservation measures and may encompass incentive payments aimed at covering transitional expenses. Indirect costs, such as those associated with learning and application costs, are influenced by individual characteristics and institutional contexts. Costs associated with compliance to regulations or penalties for non-compliance (fines) are universally applicable yet may be perceived in varying ways (Kuhlman et al., 2010).

Land tenure security pertains to the safeguarding of an individual's land rights, encompassing access, utilization, control, and transfer rights to land, as well as the modalities through which these rights are acknowledged, secured, and enacted, whether through legal or customary means. Land tenure study constitutes a substantial and intricate domain owing to the considerable heterogeneity and complexity inherent in land tenure systems, which has consequently led to the emergence of specialized land tenure scholars. The escalation of land scarcity globally, particularly in economically disadvantaged nations confronting significant climate vulnerabilities, amplifies the policy significance of the interconnections between land access and tenure security. Urbanization processes generate ambiguity surrounding tenure rights on agricultural lands adjacent to urban centers, where competition for land resources is notably intense (Holden & Ghebru, 2016).

Land tenure underpins the socio-economic relations surrounding natural resource use and, especially, decisions around whether to increase agricultural production, decrease environmental degradation, support urban development, empower women, and support property rights for indigenous people and local communities. Tenure security is inextricably linked to current global sustainability agendas that merge the aspirations of humanity with dignity and the respect for human rights, especially those of persons involved in regeneration and restorative actions, seeking to reboot the health and functioning of our environmental systems, biodiversity, and ecological richness (Dagdeviren et al., 2023).

All manifestations of land tenure, encompassing private, communal, state, and common forms, are vulnerable to insecurity. The degree of this vulnerability is contingent upon the configuration of the associated rights within a specific context: the manner in which rights are delineated or the mechanisms through which they are upheld. Each context presents a distinctive amalgamation of rights, definitions, and demarcations—resulting in a diversity that can occasionally engender confusion or lead to disparate developmental trajectories. Secure tenure has a triple dimension of legality, legitimacy, and certainty, which is essential to realizing the benefits within a bundle of rights, mediating investment, safeguarding incentives, securing inclusion in governance, and guaranteeing the safety of property over land. The cascading effects from insecure tenure have implications for climate change, biodiversity, and sustainable economic development at all scales (Abdulai et al., 2011).

In essence, land tenure security is a landholder's perception that rights will be upheld by society. Whether planned or autonomous, adaptation in the rural developing world is intrinsically connected to land tenure security, which can enable or impede people's capacity to adapt to shifting resource availability and climate conditions. As land tenure realities and the effects of climate change vary drastically by location, studying specific cases in the rural, developing world can help policymakers better understand the relationships between these variables. Policies that are flexible, attuned to environmental rights, and is coupled with effective citizen participation models can help ensure sustainable implementation of land tenure programs and heightened climate adaptation planning for some of the most vulnerable communities (Deininger & Jin, 2006).

Global sea levels have been rising for a century; however, the velocity of this rise has intensified in recent decades and is anticipated to accelerate substantially in the forthcoming years. The submergence of land into the sea engenders complex land tenure challenges, including inquiries regarding whether individuals who previously owned the submerged land will receive compensation, the recourse available to those lacking secure tenure who lose their residence, and the strategies for resettling communities onto land projected to become uninhabitable due to rising sea levels. Thus, the necessity for a more efficacious and enduring global sustainable development policy that concurrently addresses land resources, land tenure, and climate change is paramount in the context of contemporary climate change dynamics. Enhancements in land tenure policy could be achieved through two principal avenues—by adaptively integrating environmental rights and fostering active citizen participation. The fortification of land tenure security is pivotal for facilitating responsive, localized autonomous climate adaptation and ecological resilience (Murken & Gornott, 2022).

The Sustainable Development Goals serve as a paradigmatic framework that integrates environmental rights and protections within the ambit of broader developmental objectives. There exists a pressing requirement for adaptable policy mechanisms that can respond effectively to diverse locally specific adaptation strategies in varying contexts and climate-induced stressors. The possession of legal title to land can function as a motivator for the adoption of SLM practices, involving structural alterations to the agricultural landscape. Given that SLM practices are inherently long-term endeavors and require substantial investment, the absence of land tenure security renders their adoption increasingly precarious. Empirical studies indicate that legal land tenure security incentivizes farmers to commit resources to SLM practices. A multitude of policy instruments exist that empower governmental entities to intervene in the aftermath of natural disasters, particularly through various forms of subsidized insurance schemes. Crop insurance emerges as a potent financial mechanism to adapt to climate change, as it provides indemnification to farmers during adverse conditions (Eekhout & de Vente, 2019).

12.9 Discussion

The adoption of diversified crop rotation must navigate several challenges, including land fragmentation, insufficient resource allocation, and inefficient use of space across boundaries, as well as a general reluctance to employ specific types of machinery. Crop diversification, however, is among the most economical ways to lessen farmers' income unpredictability, particularly for smallholders in poverty. Contour cultivation, which encompasses practices such as terracing and bunding, is a well-established agronomic strategy for water conservation in sloping terrains and hilly agro-ecosystems. The barriers formed by the ridges and furrows traversing the hillside effectively reduce water flow. By enhancing infiltration rates in arid regions and mitigating soil erosion in areas experiencing higher precipitation, it bolsters soil moisture levels in both circumstances and elevates agricultural productivity. As water gradually traverses down the slope, it concurrently prevents fields from becoming waterlogged, thereby diminishing the incidence of waterborne or root diseases. Field trials involving cover crops, whether of a short-term or long-term nature, should be elucidated to agricultural producers. It is essential for them to possess comprehensive knowledge regarding the types of cover mulches that are suitable for varying climatic conditions. The growth of sustainable farming methods and the reduction of climate change depend on the effective integration of all these aspects.

The implementation of Sustainable Land Management (SLM) and conservation techniques within the humid and sub-humid hilly regions experiencing heavy rainfall is feasible when an effective monitoring framework is adeptly managed at the local level. The successful application of resource utilization hinges on the collaborative efforts of forest departments, non-governmental organizations, local governance, and research institutions. Furthermore, public investment in research and extension services is imperative to promote soil conservation initiatives and the adoption of SLM technologies. SLM approaches are essential to biodiversity conservation, especially in light of the continuous extinction of species and ecosystems and the growing environmental problems. In addition to human well-being, biodiversity—which includes the variety of life forms, ecosystems, and genetic diversity found on Earth—is crucial for the resilience, functionality, and general well-being of ecosystems.

This chapter closely fits with the book's theme, "Eco-resilience: Climate Change, Land Degradation, and Sustainable Solutions," by highlighting the critical role of resilient ecosystems in minimizing the impact of climate change. The chapter focuses on the technoeconomic and regulatory aspects of SLM in the regional and global contexts. A thorough examination of conservation strategies and land tenure security adds to a complete view of the overarching issue of resilience-focused and sustainable environmental solutions.

12.10 Conclusion and Future Directions

Assessing the extent and ecological ramifications of soil degradation represents a formidable challenge entrusted to soil scientists and environmental specialists. The prospective enhancement of land utilization is contingent upon a methodical examination of the distribution, attributes, and issues pertinent to the soil. This endeavor is feasible solely through the adoption of advancements in both scientific inquiry and agricultural methodologies. In this technologically advanced epoch, it is crucial to employ remote-sensing satellites and cutting-edge database analytical systems to acquire supplementary information regarding natural resources such as soil.

It is posited that the global food system presently generates sufficient food energy to meet the needs of the entire population. Nevertheless, access to adequate and economically feasible nutrition remains inequitable, and addressing the environmental challenges linked to both current and prospective food production—particularly the degradation of agricultural land and susceptibility to climate change—will necessitate the engagement of the comprehensive food system, encompassing consumption and waste management. Enhancing the capacity of farmers to invest in technologies that promote efficiency and advocating for sustainable land management practices will fortify their resilience against the cyclical issues of land degradation, poverty, and compromized food security.

The future trajectory of SLM is anticipated to be characterized by precision agriculture systems. These systems can be defined as technologically and information-driven agricultural management frameworks designed to identify, evaluate, and address variability within fields to optimize profitability, sustainability, and the conservation of land resources. They are adept at recognizing spatial heterogeneity among soil types and plant species that arises from differences across agricultural landscapes, providing additional in-season data to facilitate improved decision-making for farmers. Often referred to as prescription farming or site-specific farming, these methodologies employ technologies grounded in the Global Positioning System (GPS), encompassing a spectrum of land management techniques ranging from rudimentary to advanced. This contemporary practice incorporates field mapping and monitoring, grid sampling, sensor-based mapping, remote sensing, and an array of variable rate application techniques, such as variable rate fertilization (Shafi et al., 2019).

This chapter is important on a worldwide scale because it covers the difficulties of land degradation and climate unpredictability, both of which have a significant impact on food security, biodiversity, and individual wellbeing. This chapter provides vital knowledge for agricultural stakeholders, policymakers, and environmental campaigners around the world by investigating SLM approaches. The measures described here can be applied to a variety of regional contexts to promote land resource restoration.

AI Discloser AI not used in this book chapter.

References

Abdalla, M., Astley, H., Kaixiang, C., Qi, Y., David, C., Mari-Liis, E., Jaak, T., Robert, M. R., & Pete, S. (2019). A critical review of the impacts of cover crops on nitrogen leaching, net greenhouse gas balance, and crop productivity. *Global Change Biology, 25*, 2530–2543. https://doi.org/10.1111/gcb.14644

Abdulai, A., Owusu, V., & Goetz, R. (2011). Land tenure differences and investment in land improvement measures: Theoretical and empirical analyses. *Journal of Development Economics, 96*(1), 66–78. https://doi.org/10.1016/j.jdeveco.2010.08.002

Adetomiwa, K., Adekunle, J. D., Oluwatoba, J. A., & Oluwatosin, E. O. (2022). Adoption of multiple sustainable land management practices and its effects on productivity of smallholder maize farmers in Nigeria. *Resources, Environment and Sustainability, 10*, Article 100084. https://doi.org/10.1016/j.resenv.2022.100084

Adewole, T. A., Bongani, N., Reckson, M., & Francis, B. L. (2020). Management impact and benefit of cover crops on soil quality: A review. *Soil and Tillage Research, 204*, Article 104717. https://doi.org/10.1016/j.still.2020.104717

Ball, B. C., Bingham, I., Rees, R. M., Watson, C. A., & Litterick, A. (2005). The role of crop rotations in determining soil structure and crop growth conditions. *Canadian Journal of Soil Science, 85*, 557–577. https://doi.org/10.4141/S04-078

Bindraban, P. S., Stoorvogel, J. J., Jansen, D. M., Vlaming, J., & Groot, J. J. R. (2000). Land quality indicators for sustainable land management: Proposed method for yield gap and soil nutrient balance. *Agriculture, Ecosystems & Environment, 81*, 103–112. https://doi.org/10.1016/S0167-8809(00)00184-5

Blanco, H., & Rattan, L. (2008). *Principles of soil conservation and management* (Vol. 167169). Springer. https://doi.org/10.1007/978-1-4020-8709-7

Boardman, J., Poesen, J., & Evans, R. (2003). Socio-economic factors in soil erosion and conservation. *Environmental Science and Policy, 6*(1), 1–6. https://doi.org/10.1016/S1462-9011(02)00120-X

Branca, G., Leslie, L., Nancy, M., & Maria, C. J. (2013). Food security, climate change, and sustainable land management. A review. *Agronomy for Sustainable Development, 33*, 635–650. https://doi.org/10.1007/s13593-013-0133-1

Dagdeviren, H., Elangovan, A., & Parimalavelli, R. (2023). Land tenure and food security in South India. *Land Use Policy, 132*, Article 106837. https://doi.org/10.1016/j.landusepol.2023.106837

Deininger, K., & Jin, S. (2006). Tenure security and land-related investment: Evidence from Ethiopia. *European Economic Review, 50*(5), 1245–1277. https://doi.org/10.1016/j.euroecorev.2005.02.001

Eekhout, J. P. C., & de Vente, J. (2019). Assessing the effectiveness of sustainable land management for large-scale climate change adaptation. *Science of the Total Environment, 654*, 85–93. https://doi.org/10.1016/j.scitotenv.2018.10.350

Franzluebbers, A. J. (2010). Principles of soil conservation and management. *Vadose Zone Journal, 9*, 199–200. https://doi.org/10.2136/vzj2009.0110br

Gambella, F., Quaranta, G., Morrow, N., Vcelakova, R., Salvati, L., Gimenez, M. A., & Rodrigo-Comino, J. (2021). Soil degradation and socioeconomic systems' complexity: Uncovering the latent Nexus. *Land, 10*, 30. https://doi.org/10.3390/land10010030

Gebreegziabher, T., Jan, N., Bram, G., Fekadu, G., Mintesinot, B., Mitiku, H., & Jozef, D. (2009). Contour furrows for in situ soil and water conservation, Tigray, Northern Ethiopia. *Soil and Tillage Research, 103*, 257–264. https://doi.org/10.1016/j.still.2008.05.021

He, D. C., Ma, Y. L., Li, Z. Z., Zhong, C. S., Cheng, Z. B., & Zhan, J. (2021). Crop rotation enhances agricultural sustainability: From an empirical evaluation of eco-economic benefits in rice production. *Agriculture, 11*, 91. https://doi.org/10.3390/agriculture11020091

Heckrath, G., Ulrich, H., Jørgen, D., & Govers, G. (2006). The effect of tillage direction on soil redistribution by mouldboard ploughing on complex slopes. *Soil and Tillage Research, 88*, 225–241. https://doi.org/10.1016/j.still.2005.06.001

Holden, S. Shiferaw, B., & Pender, J. (2006). Policies for poverty reduction, sustainable land management, and food security: A bioeconomic model with market imperfections. In J. Pender, F. Place, & S. K. Ehui (Eds.), *Strategies for sustainable land management in the East African highlands* (Chap. 14, pp. 333–356). International Food Policy Research Institute (IFPRI). https://hdl.handle.net/10568/160568

Holden, S. T., & Ghebru, H. (2016). Land tenure reforms, tenure security and food security in poor agrarian economies: Causal linkages and research gaps. *Global Food Security, 10*, 21–28. https://doi.org/10.1016/j.gfs.2016.07.002

Jeffrey, E. H. (2000). Soil quality: An indicator of sustainable land management? *Applied Soil Ecology, 15*, 75–83. https://doi.org/10.1016/S0929-1393(00)00073-1

Kassie, M., Precious, Z., John, P., & Gunnar, K. (2010). The economics of sustainable land management practices in the Ethiopian highlands. *Journal of Agricultural Economics, 61*, 605–627. https://doi.org/10.1111/j.1477-9552.2010.00263.x

Killham, K. (2011). Integrated soil management–moving towards globally sustainable agriculture. *The Journal of Agricultural Science, 149*(S1), 29–36. https://doi.org/10.1017/S0021859610000845

Komlan, K., Samuel, C. A., & Koffi, D. (2022). Critical review of the impact of cover crops on soil properties. *International Soil and Water Conservation Research, 10*, 343–354. https://doi.org/10.1016/j.iswcr.2022.03.003

Kuhlman, T., Reinhard, S., & Gaaff, A. (2010). Estimating the costs and benefits of soil conservation in Europe. *Land Use Policy, 27*(1), 22–32. https://doi.org/10.1016/j.landusepol.2008.08.002

Liu, X. B., Zhang, X. Y., Wang, Y. X., Sui, Y. Y., Zhang, S. L., Herbert, S. J., & Ding, G. (2010). Soil degradation: A problem threatening the sustainable development of agriculture in Northeast China. *Plant, Soil and Environment, 56*, 87–97. https://doi.org/10.17221/155/2009-PSE

Löbmann, M. T., Linda, M., Gundula, P., Jos, B., Johannes, B., Antonio, B., & Katharina, H. (2022). Systems knowledge for sustainable soil and land management. *Science of the Total Environment, 822*, Article 153389. https://doi.org/10.1016/j.scitotenv.2022.153389

Mallick, R., Banerjee, A., & Poddar, S. (2025). Sustainable ecosystem management for strengthening climate resilience. In M. Choudhury, G. Dixit, & S. Majumdar (Eds.), *Sustainable synergy: Harnessing ecosystems for climate resilience. Climate change management*. Springer. https://doi.org/10.1007/978-3-031-77957-2_15

Mendoza-Ponce, A., Corona-Núñez, R. O., Nava, L. F., Estrada, F., Calderón-Bustamante, O., Martínez-Meyer, E., Carabias, J., Larralde-Corona, A. H., Barrios, M., & Pardo-Villegas, P. D. (2021). Impacts of land management and climate change in a developing and socioenvironmental challenging transboundary region. *Journal of Environmental Management, 300*, Article 113748. https://doi.org/10.1016/j.jenvman.2021.113748

Montgomery, D. R. (2007). Soil erosion and agricultural sustainability. *Proceedings of the National Academy of Sciences, 104*, 13268–13272. https://doi.org/10.1073/pnas.0611508104

Motavalli, P., Kelly, N., Ranjith, U., Shibu, J., & Sougata, B. (2013). Global achievements in sustainable land management. *International Soil and Water Conservation Research, 1*, 1–10. https://doi.org/10.1016/S2095-6339(15)30044-7

Muche, A. T., Yohannes, S. K., & Belete, M. A. (2024). Assessing the effectiveness of integrated watershed management practices and suggesting innovative strategies in southern Ethiopia. *Heliyon, 10*, Article e38619. https://doi.org/10.1016/j.heliyon.2024.e38619

Murken, L., & Gornott, C. (2022). The importance of different land tenure systems for farmers' response to climate change: A systematic review. *Climate Risk Management, 35*, Article 100419. https://doi.org/10.1016/j.crm.2022.100419

Ojo, T. O. O., Adesiyan, O. F., Ige, A. O., Emenike, C. U., Kassem, H. S., Elhindi, K. M., Kitole, F. A., & Sesabo, J. K. (2024). The role of sustainable land management practices in alleviating household food insecurity in Nigeria. *Frontiers in Sustainable Food Systems*. https://doi.org/10.3389/fsufs.2024.1414243

Pásztor, J., & Izolda, P. (2019). The dynamic particularities of contour ploughing. *Papers on Technical Science, 10*, 67–72. https://doi.org/10.33894/mtk-2019.10.09.

Reed, M. S., Stringer, L. C., Dougill, A. J., Perkins, J. S., Atlhopheng, J. R., Mulale, K., & Favretto, N. (2015). Reorienting land degradation towards sustainable land management: Linking sustainable livelihoods with ecosystem services in rangeland systems. *Journal of Environmental Management, 151*, 472–485. https://doi.org/10.1016/j.jenvman.2014.11.010

Reeves, D. W. (1997). The role of soil organic matter in maintaining soil quality in continuous cropping systems. *Soil and Tillage Research, 43*, 131–167. https://doi.org/10.1016/S0167-1987(97)00038-X

Sanyal, D., Avik, M., Amin, R., Johnathon, W., David, K., Jason, D. C., & Anthony, B. (2023). Cover crops did not improve soil health but hydroclimatology may guide decisions preventing cash crop yield loss. *Frontiers in Soil Science.* https://doi.org/10.3389/fsoil.2023.1111821

Scavo, A., Stefania, F., Alessia, R., Gaetano, R. P., Cristina, A., & Giovanni, M. (2022). The role of cover crops in improving soil fertility and plant nutritional status in temperate climates: A review. *Agronomy for Sustainable Development, 42*, 93. https://doi.org/10.1007/s13593-022-00825-0

Shafi, U., Rafia, M., José, G., Syed, A. H., Syed, Ali, R. Z., & Naveed, I. (2019). Precision agriculture techniques and practices: From considerations to applications. *Sensors, 19*, 3796. https://doi.org/10.3390/s19173796

Shah, K. K., Bindu, M., Hari, P. P., Arjun, S., Geeta, A., Meena, P., & Jiban, S. (2021). Diversified crop rotation: An approach for sustainable agriculture production. *Advances in Agriculture, 2021*, 8924087. https://doi.org/10.1155/2021/8924087

Sharma, P., Atinderpal, S., Charanjit, S. K., Amandeep, S. B., Kulbhushan, K. G., Mahendra, D., & Robert, L. S. (2018). The Role of Cover Crops towards Sustainable Soil Health and Agriculture—A Review Paper. *American Journal of Plant Sciences, 9*, 1935–1951. https://doi.org/10.4236/ajps.2018.99140

Singh, J., & Jai, P. S. (1995). Land degradation and economic sustainability. *Ecological Economics, 15*, 77–86. https://doi.org/10.1016/0921-8009(95)00037-A

Sowjanya, A., Sekhar, D., Sagal, B. V., Sravanthi, S., Sandeep Naik, B. N., & Rama Rao, G. (2021). Soil and water conservation problems and measures in hilly area. *JustAgriculture, 1*(11), 1–11.

Stevens, C. J., Quinton, J. N., Bailey, A. P., Deasy, C., Silgram, M., & Jackson, D. R. (2009). The effects of minimal tillage, contour cultivation and in-field vegetative barriers on soil erosion and phosphorus loss. *Soil and Tillage Research, 106*, 145–151. https://doi.org/10.1016/j.still.2009.04.009

Surendran, U., Raja, P., Liu, K., Bilotto, F., & Sridevi, G. (2024). Comparative analysis of soil organic carbon and soil properties in landscapes of Kerala: Insights from the Western Ghats of India. *Environmental Monitoring and Assessment, 196*(9), 838. https://doi.org/10.1007/s10661-024-12984-6

Swaminathan, M. S. (2001). Food security and sustainable development. *Current Science, 81*, 948–954. https://www.jstor.org/stable/24106519.

Valentin, C., & Jean, L. R. (2018). Erosion and principles of soil conservation. In *Soils as a key component of the critical zone 5: Degradation and rehabilitation*(vol. 5, pp. 39–82). https://doi.org/10.1002/9781119438298.ch3

Viana, C. M., Dulce, F., Patrícia, A., Jorge, R., & Paulo, P. (2022). Agricultural land systems importance for supporting food security and sustainable development goals: A systematic review. *Science of the Total Environment, 806*, Article 150718. https://doi.org/10.1016/j.scitotenv.2021.150718

Xie, H., Zhang, Y., Zeng, X., Li, X., Li, S., Li, Y., & Wang, L. (2020). Sustainable land use and management research: A scientometric review. *Landscape Ecology, 35*, 2381–2411. https://doi.org/10.1007/s10980-020-01002-y

Chapter 13
Ecosystem-based Disaster Risk Reduction (Eco-DRR) and Mangrove Restoration: Present Knowledge and Bibliometric Analysis

Raghav Prakash and Abhiroop Chowdhury

Abstract Mangrove ecosystems are vital for co-existence of humans and nature in a coastal region. It is known to have several ecological services, not limited to increased fishing productivity, cleaning the water and soil of pollutants and protecting the coastlines from erosion. But the major ecological service of the mangroves is the protection of the coast from natural disasters. So, Ecosystem-based disaster risk reduction (Eco-DRR) is an effective process to reforest degraded tidal flats with mangroves, which can also sequester blue carbon and help reduce the global warming burden. This research sheds light on existing peer-reviewed literature in the domain of Eco-DRR and mangroves. Preferred Reporting Items for Systematic reviews and Meta-Analyses (PRISMA) have been used to narrow down research in the cross-section of mangroves and Eco-DRR, till March 2024. This chapter also focuses on conservation policies revolving around mangrove ecosystems and case studies related to mangrove conservation with community participation.

Keywords Ecosystem-based disaster risk reduction · Mangrove restoration · Community participation · Natural disasters · Disaster resilience · Coast · Nature based solution

---

Electronic Supplementary Material The online version of this chapter (https://doi.org/10.1007/978-3-032-00708-7_13) contains supplementary material, which is available to authorized users.

R. Prakash · A. Chowdhury (✉)
Jindal School of Environment and Sustainability, O.P. Jindal Global University, Sonipat, Haryana, India
e-mail: abhiroop.chowdhury@gmail.com

U. Chatterjee et al. (eds.), *Eco-Resilience: Climate Change, Land Degradation and Sustainable Solutions*, Sustainability Solutions,
https://doi.org/10.1007/978-3-032-00708-7_13

13.1 Introduction

The world is battling the challenges posed by climate change and global warming. None of the ecosystems is safe from climate change and environmental degradation. Recent research indicates that tropical forests can only withstand around 4 °C rise in air temperature before impacting their physiological function (Doughty et al., 2023). The Amazon rainforest has been losing resilience and biodiversity since 2000 (Boulton et al., 2022), and central African rainforests are facing the challenges due to the loss of habitat (Abernethy et al., 2016); around 59 Mha of forest land was degraded in Indonesia between 1950 and 1997 (Tsujino et al., 2016), while mangrove forests endured a 35% loss of habitat between 1980 and 2000 (Valiela et al., 2001). With these irreversible changes, ecological solutions need to be undertaken across the globe to build resilience against climate change and global warming. Man-made changes to the natural state of the coastal area can cause disasters in coastal regions. The coastal region is observed to have different physical characteristics, problems and conditions. The International Union for Conservation of Nature (IUCN) indicated that around fifty per cent of mangrove forests worldwide are under serious extinction risk (Chowdhury et al., 2024; IUCN, 2024). Therefore, various strategies for reducing disaster risk in coastal regions are considered to impart eco-resilience along the mangrove forested coastline against climate change vagaries. In today's times, disaster resilience is executed with ECO-DRR (Renaud et al., 2013). In the 1960s, the primary reason for mangrove depletion was aquaculture development, specifically the stress on the creation of fish ponds (mostly milkfish and other carp culture). However, most fishponds were abandoned due to low fish fertility rates. In Thailand, most mangroves were destroyed to create prawn ponds. The major issue of converting mangroves to fish ponds are the acid sulphate conditions and degradation of mangrove habitat (Jin-Eong, 1995).

Abrasion happens due to oceanic waves and strong ocean currents that cause coastal erosion. Abrasion is simply coastal erosion (Indarsih & Masruri, 2019). Abrasion can happen due to many factors; it can be man-made or even natural. For example, strong winds blowing over the ocean make currents and waves that erode the coastal area (Rahman, 2015). Changes in land use affect mangroves in the arid regions, an also affect groundwater flows, aquifer recharge rates. This may negatively impact mangroves and coastal habitats (Ferguson & Gleeson, 2012). This is a major issue as reducing freshwater supply will reduce mangrove growth and productivity, especially for arid-region where mangrove forests are under high water stress. Another stressor for mangroves is getting excessive nutrient flow from terrestrial systems. Additionally, hypersaline environments can exacerbate mortality rates of mangrove saplings and seedlings in arid environmental settings. Reduction in sediment supply is also seen as a threat to mangrove forests (Krauss and Osland, 2020). This is a major issue, as a crucial service offered by mangrove forests is being compromised—their ability to adapt to a rise in the level of the sea (Lovelock et al., 2015a, 2015b). In some regions, it was observed that the foremost reason for low sediment supply is the low amount of precipitation and diversion of river for irregation purposes. However, in semi-arid regions, installations such as Dams alter the sediment and surface water flow along the coast (Ward et al., 2016). This was noticed in Senegal, along the region where the Somone River meets the Atlantic Ocean, due to the construction of upstream dams and high droughts, leading to a staggering water reduction in mangrove forests. The result was mangroves being replaced by greater salinity-tolerant taxa (Sakho et al., 2017).

Major climatic events, hurricanes and cyclones, and severe droughts in subtropical arid regions are major factors for mangrove mortality (Dittmann et al., 2022). More than 5000 ha of mangrove coverage being impacted is seen as high mortality (Duke et al., 2017). For instance, these events have been observed in Australia; they are also linked with El Nino events, resulting in the soil's hyper-salinity, as well as changing mangrove biodiversity as observed from other parts of the world (Lovelock et al., 2017; Thivakaran et al., 2020; Chowdhury et al. 2023). Reduced precipitation lowers canopy height and biomass, causes a drop in carbon sequestration ability (Adame et al., 2021; Feher et al., 2017).

Human interference with mangrove forests presents a significant threat. For instance, to mitigate the effects of rising sea levels and protect communities, seawalls and other man-made barriers have been constructed (Chowdhury et al., 2021). This poses challenges for mangroves, as the establishment of wall-like structures along the sea and other erosion-resistant architecture leads to scouring of mangroves along these structures (Fletcher et al., 1997; Tait & Griggs, 1990). Regions experiencing elevated groundwater extraction can pose significant challenges for mangroves, leading to an increase in sea level in relation to mangrove (Krauss et al., 2003). Rising precipitation sometimes requires the stormwater canals to mitigate water rise or inundation in elevated zones. This threatens mangroves due to the reduced availability of water for their productivity (Gilman et al., 2008). The growing population significantly stresses mangrove forests. Like all ecosystems, coastal systems have a fixed carrying capacity. Governments and industries have begun reclaiming coastal land, creating tremendous pressure on these communities. This has been viewed as a means of generating quick profits, driven by high demand due to population growth. The mangrove wood-chip industry has endangered large areas of mangrove forests, primarily in Japan. Significant regions of the Asia–Pacific have been depleted of mangroves to obtain timber. The activities in Sabah (involving an annual felling of approximately 4000 ha) have been halted, and those in Sarawak (with annual felling of about 1600 ha) have reached a dead end due to the minimal amount of timber remaining in the Rajang River system within 25 years (Jin-Eong, 1995).

This chapter assesses the extent of research at the intersection of Eco-DRR and mangroves and finds future research avenues in the domain. Climate-positive actions are on the rise worldwide to find processes of removing greenhouse gases from our world and comply with SDG 13 (Climate action). Mangroves are known to have a significant ecological service in terms of carbon sequestration. The interconnection between Eco-DRR and mangroves is also important to plan restorations to address land degradation and address the issues through sustainable solutions. Though there has been an increased focus on study in this domain, there is a need for a bibliometric study to understand the major research to date, which can pave the way for the expansion of research across the globe.

13.2 Material and Methods

The data derived was from Web of Science and the Scopus website. The words searched were "Mangrove" or "Mangrove Forests". The research focuses on the relevance and benefits of these forests. The authors also searched "Eco-system Based Disaster Risk Reduction" or "Natural Barriers against climate change". The study discusses the various ecosystem-based disaster risk reduction methods, focusing specifically on the disaster risk associated with mangrove reduction. Google Scholar was utilised to find the most relevant and up-to-date data to present in the paper. Google Scholar is a highly efficient mode of finding relevant data; it has numerous articles from credible journals. Other relevent publications have been used to validate the information and data presented in this review paper. The bibliometric package in R was used to analyse SCOPUS database results using the keywords "Ecosystem-based Disaster Risk Reduction" AND "mangrove" till March 2024. For bibliometric analysis, only SCOPUS database results has been taken into consideration ($n = 10$), while for the general review other articles were also been considered (Fig. 13.1).

13.3 Mangroves and Eco-DRR

Mitigation can take two primary forms: structural and non-structural. Structural mitigation uses physical structures to lessen the impact of disasters. This includes building walls to prevent flooding. Non-structural mitigation involves knowledge development, policy recommendations, and, among other things, community capacity building (Amri et al., 2016; Indarsih & Masruri, 2019). Moos et al. describe Eco-DRR as the conservation, restoration, and management of ecosystems that aim to reduce disaster risk and promote sustainable, resilient community development (Moos et al., 2018).

Green Infrastructure is a common term for ECO-DRR. It has gained importance for protection against coastal flooding following the 'Intergovernmental Panel on Climate Change Fifth Assessment Report' and the 'Sendai Framework for Disaster Risk Reduction 2015–2030' (Pearson & Pelling, 2015). It is an economically viable option, and adjusting to different climates has enhanced green infrastructure (Guannel et al., 2016; Reguero et al., 2018; Sutton-Grier et al., 2015). In line with Green Infrastructure, natural barriers to coastal hazards include mangroves, coastal dunes, coral reefs, forests, wetlands, and sandy beaches. Coastal forests and mangroves are known for saving coastal communities against storms and reducing wave/hydrodynamic energy during floods. Numerous reports have highlighted how coastal pines aided in reducing the energy of small and medium-sized tsunamis during the Tohoku

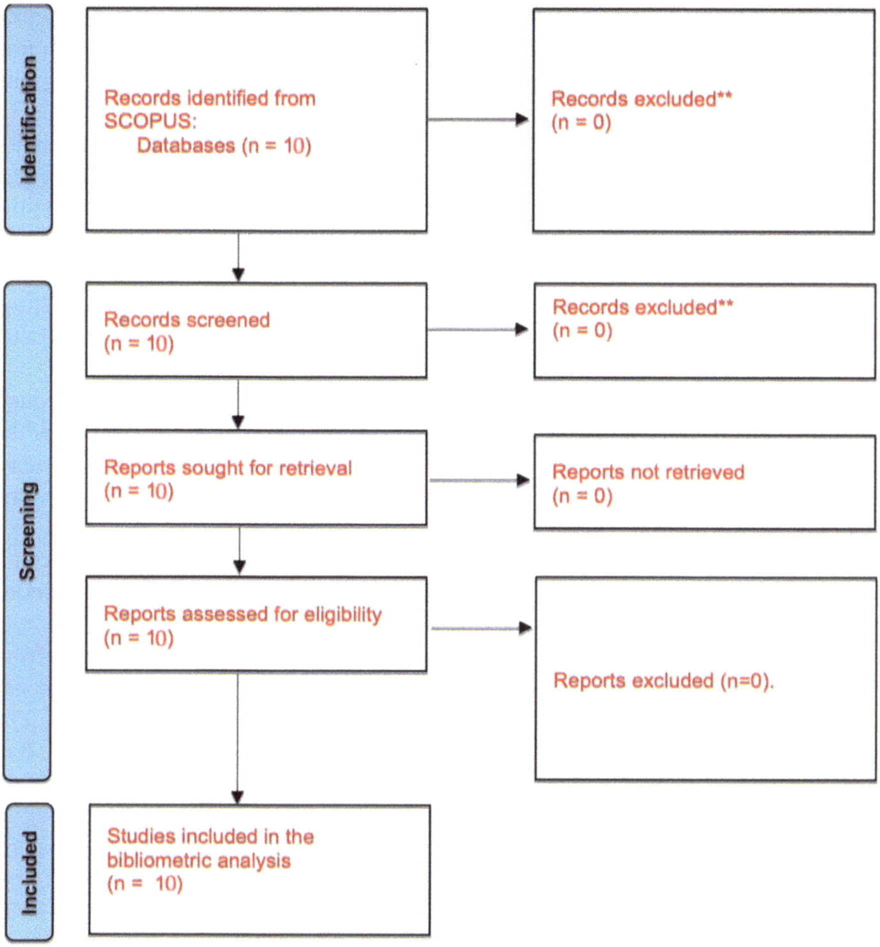

Fig. 13.1 Flow chart depicting the process of selecting studies for a bibliometric analysis. The chart is divided into three main phases: Identification, Screening, and Included. – **Identification**: Records identified from SCOPUS databases (= 10). No records were excluded at this stage. – **Screening**: All 10 records were screened, with none excluded. Reports sought for retrieval were 10, with none not retrieved. – **Eligibility**: Reports assessed for eligibility were 10, with none excluded. – **Included**: All 10 studies were included in the bibliometric analysis. The flow chart uses arrows to indicate progression through the stages, with annotations in red for excluded records

Earthquake Tsunami of 2011 (Tanaka, 2012). In the 2004 Indian Ocean Earthquake Tsunami, mangroves acted as an excellent buffer in the tropics and sub-tropics (Danielsen et al., 2005; Goda et al., 2019). Additionally, in Southeast Asia and the Pacific islands, efforts have started to enhance carbon storage and environmental health through afforestation and the restoration of mangroves, which also provide protection against natural disasters (Mori et al., 2022).

Eco-DRR has been widely used in numerous countries. In Indonesia, the central government is adopting regulations that align with the DRR of the ecosystem. Action is being taken to tackle the adversities of climate change, such as planting and restoring mangroves along the coast to mitigate the effects of rising sea levels. Another project was the "one-billion trees", which was initiated to tackle the major problem of deforestation. Indonesia can utilise mangrove forests, which are expected to grow to approximately 9,204,840.32 ha. This is further categorised based on the quality of the mangrove forests: those in great form (27%), poor state (48%), and destroyed/ ruined condition (23%) (Dalimunthe, 2018). Mangroves are often seen as the most productive ecosystem in the world. They inhabit brackish waters along the tropical and sub-tropical coasts. Mangroves are beneficial for the local coastal communities; they help in stabilising coastal waters, which aids in shielding the coastline from flooding, filtering out toxic and waste material that comes in through the water, aids in controlling the erosion rate, and providing shelter and food for the aquatic life. Furthermore, mangrove forests provide the communities with tannins, wax, honey, timber, etc. Lastly, Mangrove forests have an aesthetic and cultural value that must be considered (Dalimunthe, 2018).

Various coastal ecosystems, such as mangroves, coral reefs, seagrass, and coconut palms, work like natural barricades against disasters like tsunamis and floods in Indonesia. Coastal mitigation strategies utilise nature-based approaches. For instance, in Indonesia's coastal regions, coastal forests shield communities from flooding, strong waves, erosion, and tsunamis. Moreover, the economic advantages of these forests provide an additional reason for employing natural barriers. The numerical analysis highlighted that incorporating green belts with a width of 50 m leads to a 13.57% decrease in the peak flooding depth and an 18.4% reduction in the maximum tsunami speed (Rudianto, 2016).

Additionally, 50% of mangroves with a trunk diameter of 20–25 cm can withstand a tsunami of about 6–7 m in height. A computer simulation depicted a coastal forest of approximately 500 m and highlighted how it can reduce the damage by 38%. Moreover, the hydrodynamic force on the mangrove forests was reduced by about 70% when the tsunami reached almost 3 m. An important topic the paper shed light on was the time and effectiveness of the forests. For example, a ten-year-old forest can reduce the impact of a 3-m tsunami, while one that is aged twenty to thirty years can safeguard against a 4-m tsunami. Ecosystems require time to evolve; this will give improved adaptation opportunities as they quickly adapt to climatic changes (Cheong et al., 2013; Purwanto et al., 2024).

Ihinegbu et al. highlighted that a rise in sea level would threaten communities across the globe. To protect humanity, solutions are being made other than engineered and man-made solutions to climate change. Nature-based solutions can play a pivotal part in saving the global coastal population (Mclvor et al., 2015). Thus, the study focuses on mangroves that offer coastal protection, income generation, and livelihood creation. Mangrove forests have numerous benefits for the local communities. They supply the community with timber and other resources for construction, fuel, and aid in fisheries.

Moreover, regarding regulatory services, mangroves aid in controlling erosion rates, attenuating waves, and sediment accretion. Mangroves are also beneficial in spiritual enrichment, recreational experiences, and cognitive development. Several studies shed light on mangrove forests' role as natural barriers against disasters. For example, Akbar et al. compared two villages hit by storms. One of the villages had a cover of mangrove forests to protect the coastline, and the other had no protection. The damage to the village without any protection was devastating. In contrast, the village containing the mangrove forest suffered minimal damage to material goods and people's lives (Akbar et al., 2018). Ataur Rahman and Rahman, after conducting studies to understand the protection mangroves offer against cyclones, attested to the fact that mangrove forests aid in protecting coastal communities (Ataur Rahman & Rahman, 2015). Auerbach et al. studied the potential of mangroves compared to engineered solutions to work against flooding and sea level rise. It was clearly stated that mangroves play a major role in fighting rising sea levels and protecting communities against flooding (Auerbach et al., 2015). Sudmeier-Rieux et al. also conducted extensive work on mangroves and their great potential in disaster risk reduction (Ihinegbu et al., 2023; Sudmeier-Rieux et al., 2021).

Table 13.1 has been presented to help you understand different ecosystem-based DRRs. The table will highlight how they are working to support the communities and the environment.

Mangrove forests are among the best natural barriers against hydrodynamic storm surges, ocean tides, high waves, and tsunamis. Constructing a mangrove plantation zone will reduce coastal flooding, reduce flow velocity, and facilitate smoother flow effects (Takagi et al., 2016a, 2016b; Bayas et al., 2011; Chowdhury et al., 2021; Krauss et al., 2014). Moreover, the vegetation further reduces the effects of the waves with greater plant density and diameter. For example, salt marshes aid in lessening wave loads and coastal dikes (Vuik et al., 2016). They also help enhance shoreline resilience through ecosystem services like trapping sediments, providing habitat to local aquatic creatures, aiding in flood mitigation and acting as nurseries and nutrient recyclers (Alongi, 2008; Primavera & Esteban, 2008). The benefits derived from mangroves are called ecological resilience (Gunderson et al., 2002; Walker et al., 2004).

A case study of Jakarta was conducted, yielding the following insights. The city has experienced rising subsidence rates, leading to more frequent coastal flooding. This situation poses a significant challenge for the urban area (Takagi et al., 2016a, 2016b). While coastal dikes can mitigate flooding in the short term, they will become ineffective if land subsidence worsens (Takagi et al., 2016a, 2016b). In the Philippines, major coastal flooding occurred following the Bohol earthquake in 2013 (Jamero et al., 2017). Excessive groundwater extraction, combined with natural consolidation, has led to increased subsidence rates in the Chao Phraya Delta, Thailand, and the Mekong Delta, Vietnam, reaching up to 5 cm per year and 2 cm per year, respectively (Giosan et al., 2014; Takagi et al., 2016a, 2016b). In contrast, Sediment buildup in mangrove forests occurs at a slow pace, typically averaging only a few millimetres annually (Krauss et al., 2013; Lovelock et al., 2015a, 2015b). Takagi

Table 13.1 Ecosystem-based DRRs projects across the globe in the coastal areas

S.No.	Name of ecosystem-based DRR	Description	Citation
1	Mangrove forests	In Vietnam, specifically in Tam Hai commune and Binh Trung village, the local communities vouch for mangroves as a livelihood source. For instance, the fisheries business is thriving due to the mangroves in the village area	Nehren et al. (2017)
2	Protective forests	Forests are seen as the best example of ecosystem-based DRR in mountainous regions. This is simply because forests protect the population, infrastructure, and settlements against avalanches, floods, or landslides. Thus, these forests have also come to be known as protective forests	Moos et al. (2018)
3	Coastal dunes	Coastal dunes are observed as key in shoreline prevention, diminishing the effects of coastal erosion and sea processes. Coastal dunes provide many services, such as coastal protection, diverse flora and fauna habitat, and groundwater retention and purification	Alessandrini et al. (2024)
4	Protective forests	The primary aim of protection forests is to protect the population and environment from disaster. They are also seen as a great example of nature-based solutions. This is primarily because they directly prevent disasters from happening and mitigate their impact in the runout zone	Poratelli et al. (2020)

(continued)

Table 13.1 (continued)

S.No.	Name of ecosystem-based DRR	Description	Citation
5	Mangrove forests	A study conducted in Mauritius highlighted the importance of mangroves in protecting and providing for the local communities. For instance, the Rhizophora mangrove species is seen to lower the wave height by 70 cm and cut wave energy by 99% over a 500 m wide mangrove area forest. Moreover, these forests also protect the local communities against storm surges as they curb the amount of water and even store it if the forest is of proper size. Lastly, they aid in stabilising the coast and preventing erosion. This is due to the complex root system as the incoming silt load gets captured in the mangrove's twigs, stems and fallen leaves	Sunkur et al. (2023)
6	Coastal dunes	Coastal dunes and their vegetative systems are excellent barriers against storm surges and wind waves, and reduce coastal erosion. These vegetative communities also crucially support a favourable or nearly balanced coastal budget	Fernández-Montblanc et al. (2020)

illustrated how the mangrove landfill adjacent to the port works like an active barrier against waves (Takagi, 2018).

Developed countries can afford expensive technological solutions for protection against storms, coastal flooding and damage caused by waves. Though developing countries/ underdeveloped nations still need to have this option and rely on ecological solutions for protection against these issues (Esteban et al., 2017; Takagi et al., 2014). For example, a few countries can afford a 10 m high coastal dike to protect the coastal community against a 5 m high tsunami. It highlighted that one of Japan's ports spent 4000 USD/m^2 to enhance the coastal dike by 1 m (Esteban et al., 2020). However, developing countries need help to afford these expensive technological solutions. Thus, making affordable, adaptive and sustainable solutions is of high importance. Moreover, such developing countries have great potential for ecological solutions because they are most likely situated in tropical and sub-tropical regions (Takagi, 2019).

The west coast of Taiwan has soft sediments that lead to coastal erosion. This is a major issue for the coastal communities in Taiwan's central and southwest areas (Hsu et al., 2007). The coastal ecosystems present in those areas are mangroves, coastal dunes, tidal flats, and windbreak forests. Due to the significant development in the west coast region of Taiwan, these ecosystems are under pressure from human activities (Huang et al., 2016). Healthier ecosystems will be able to provide more protection against climate change and also offer more services for the population (Satake et al. 2018). The study captured how mangrove plantations have aided in reducing coastal erosion. These are considered a highly dependable natural barrier against coastal threats. Mangroves, with the support of their unique root systems, help reduce wave energy and increase coastal accretion (Duarte et al., 2013; Gracia et al., 2018). These forests significantly contribute to carbon sequestration and provide a habitat for the wildlife (Gracia et al., 2018). In Taiwan, mangroves are found naturally along the coastline. Typhoons often hit the coastal regions of Taiwan. Coastal windbreak forests are recognised for shielding communities from intense winds and salt spray originating from the ocean (Lee et al., 2017). These forests also aid in saving agricultural land near the coastline. The article emphasised how coastal dunes function as natural barriers against coastal erosion and rising sea levels. Moreover, like mangroves, they are also a unique ecological habitat (Onselen et al., 2023; Triyanti & Chu, 2018).

Mangroves thrive in tropical and subtropical regions and are recognised as efficient systems globally. In Sri Lanka, these ecosystems stretch along a coastline of 170 km. Biogeographically, they are part of the Indo-West Pacific region, which showcases greater plant diversity than the Caribbean, West Pacific, and Atlantic areas (Amarasinghe & Perera, 2017). They are scattered in the seaside areas of Sri Lanka and found in a few dense forests. Mangroves are viewed as a financially viable ECO-DRR capability in Sri Lanka, especially in socio-economically weak regions. A noteworthy point is that in 2015, Sri Lanka became the inaugural country to safeguard its mangrove forests. This is also seen as an exemplar for other countries in conserving and restoring mangrove forests. Mangroves have a crucial role in mitigating the severe effects of climate change. For example, mangroves play a vital role in stabilising shorelines, controlling floods, managing groundwater, maintaining water quality, protecting biodiversity, and supporting fishing, among other benefits (Giri et al., 2015). These forests are included in carbon pricing policies in numerous countries because of their significant impact on carbon sequestration (Jakovac et al., 2020; Veettil et al., 2023). Unless disturbed or uprooted due to human-related activities or strong natural phenomena, mangroves are durable and adaptable to diverse environmental conditions (rising sea levels and high soil salinity) (Veettil et al., 2023).

Mangrove forests in South Asia are located in India, Bangladesh, Sri Lanka, and Pakistan. The Sundarbans, spanning the coasts of India and Bangladesh, is the biggest mangrove forest in the world (Samanta et al., 2021). India hosts the most incredible variety of mangrove species, with 30–35 species, while Sri Lanka has 18–20 species (Giri et al., 2015). In the coastal regions of Sri Lanka, mangrove vegetation is found in irregular patches, covering an area of about 6000–7000 ha (Giri et al., 2015; Ranawana, 2017; Veettil et al., 2023). These forests also contribute to the region's economic security by providing activities such as fisheries and various direct and indirect benefits (Karunathilake, 2010). The sale of fish, shrimp, prawns, and lobsters provides substantial profits for the local community, resulting in the transformation of these forests into aquaculture ponds. Furthermore, mangroves in Sri Lanka are regarded as a natural habitat for approximately 440 distinct species across five vertebrate groups (Fernando et al., 2022; Veettil et al., 2023).

13.4 Mangroves-Restoration, Conservation Projects, and Policy Planning

Mangroves have come under tremendous threat and are near extinction in some regions. Deforestation has increased in most areas inhabited by mangroves. However, strides are being taken to ensure that mangrove conservation and management are given importance. For instance, more than six international agreements and regional agreements have been formed that are linked to mangrove conservation and restoration. The article quoted two main frameworks that aid in the conservation of mangroves: These include (1) international agreements for regulating pressures on marine resources; (2) bioregional management approaches; and (3) protected area frameworks, such as the RAMSAR Convention and Man and the Biosphere (MAB) Programme" (Macintosh & Ashton, 2002). Regarding coastal zone ecosystems and mangrove forests, a key advantage of these international policies is their potential to assist in transboundary situations. The major policies regulating Eco-DRR are elucidated in Table 13.2.

The number of Transboundary Protected Areas (TBPAs) has also increased. In 1988, approximately 59 areas existed, primarily in Europe and North America. By 1997, this figure had grown to 136, spreading worldwide. This growth was primarily driven by considerable support from donors and international assistance focused on preserving biodiversity and encouraging the sustainable utilisation of natural resources.

Mangrove conservation can be understood as forests with good natural coverage and the ability to provide essential ecosystem services (Sidik et al., 2018).

13.5 Case Studies on Mangrove Restoration and Community Impacts

A study conducted in Mauritius explains the benefits of using mangroves as a form of ECO-DRR technique. Mauritius is an eight-million-year-old island situated in the southwest Indian Ocean and has an area of 1865 km^2. Mangroves are located in the coastal wetlands near river mouths and cover an area of approximately 14 km2. These forests appear in tiny patches, typically only tens of square meters, and hardly reach sizes of two thousand square meters. The primary species of mangroves present on the island are- *Rhizophora mucronata* Lam. and *Bruguiera gymnorrhiza* (L.) Lam.

Table 13.2 Distinct policies, frameworks, and agreements for the conservation and management of mangrove forests

S.No.	Name of the policy	Description of the policy	Year of implementation
1	Convention on International Trade in Endangered Species of wild flora and fauna (CITES)	The policy was implemented in Washington, D.C., and aimed to regulate the wildlife trade, which includes both living or deceased animals and plants. Furthermore, parts and derivatives were also meant to be seized	Adopted in March 1973, it took effect in July 1975
2	The Convention on Biological Diversity (CBD)	Adopted in Rio de Janeiro, Brazil and signed by 150 countries. This legally binding agreement has three primary goals: "(1) Biodiversity Conservation, (2) Sustainable use of biodiversity, and (3) Fair and equitable sharing of the resulting benefits"	Adopted in June 1992, and came into force in December 1973
3	Agenda 21	Adopted by Rio, these are a set of actions that are promoting sustainable development in the twenty-first century	Adopted in June 1992
4	United Nations Framework Convention on Climate Change (UNFCCC) and The Kyoto Protocol	This convention's ultimate aim is to lower greenhouse gas emissions in the atmosphere and limit man-made stress on the climate system. Biodiversity conservation is the primary pillar built into the UNFCCC objectives	UNFCCC was signed in 1992 and came into force in 1997 The Kyoto Protocol was adopted in 1997
5	United Nations Convention on the Law of the Sea (UNCLOS)	This agreement limits the exploitation and use of oceans. As of 2000, 132 countries had ratified the UNCLOS	Signed in 1982 and took effect in 1994

Mangroves provide many benefits to the coastal communities of Mauritius. The study detailed the major benefits of mangroves: reducing storm surges, large waves and swells, retaining sediment and pollutants, minimising erosion, absorbing carbon dioxide from the atmosphere, serving as nursery habitats for marine species, functioning as eco-tourism destinations, providing traditional medicine for wound healing and ailments, and offering livelihoods and food resources (Sunkur et al., 2023).

This case study is of the Sayung Sub-district, Demak District, Central Java, Indonesia. The emphasis of the study is showcasing the many benefits mangroves provide to the four villages in the area. The results highlighted mangroves playing a significant role in the fisheries sector in these villages. For instance, mud crabs and mud skippers were being commercialised and used for self-consumption in a few cases. The coastal communities have utilised the 'Avicennia marina' seeds, commonly called 'Brayo', as food. Despite the high seed demand, it has been commercialised and sold in the markets for profit. Another unique use of mangroves was seen in one of the villages (Damastuti & Groot, 2017). Work of Chowdhury et al.

(2021) indicated a case study where the Eco-DRR model has been implemented at Satjelia Island in the Indian Sundarbans. Later it was observed that the mangrove bioshield has been successful in ameliorating impact of storm surges and cyclones.

13.6 Bibliometric Analysis

This section revisits the major recent publications with the major keywords. Figure 13.2 mentions the major universities/institutions publishing in the domain.

Concerning the most relevant affiliations, Kyoto University tops the list with ten published articles. Kyoto University has published more than the mentioned societies, departments, or universities. The least number of affiliations was by the 'School of Environment and Society', 'Marine Sciences Sector', and the Environmental Engineering Department, which only managed to publish one article. The most prominent reason for this could be a lack of research in the specific domain. 'Tokyo University of Marine Science and Technology', 'Tohoku University', and 'National Institute for Environmental Studies' have published two articles. The low number of published articles is attributed to the relatively new and technical research field. The National Taiwan Normal University (NTNU) and Nagoya University affiliation have published three articles. This is a fairly decent number of publications, given that

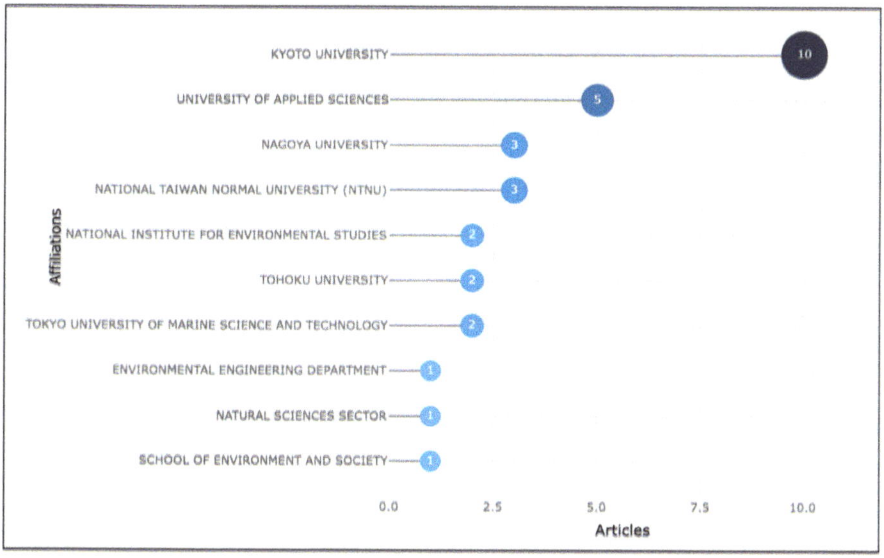

Fig. 13.2 Bubble chart displaying affiliations on the y-axis and a metric labeled "Articles" on the x-axis. Each bubble represents a different institution, with size indicating the volume of articles. Notable institutions include Kyoto University and University of Natural Sciences. The largest bubble corresponds to Kyoto University, indicating the highest number of articles

Fig. 13.3 Word cloud with prominent terms related to environmental and coastal management. Key words include "mangrove," "risk assessment," "decision making," "Philippines," "ecosystem management," "policy implementation," "carbon," "sea level," "coastal zone," "disasters," and "anthropogenic effect." The words vary in size and color, indicating their frequency and importance

the highest number of publications by Kyoto University is 10. The 'University of Applied Sciences' has published five articles. Kyoto University is the best affiliation in this list, with a total of 10 articles published. The high number of published articles can be attributed to the increase in knowledge base, good know-how of technical aspects, and appropriate use of resources to conduct the research. Kyoto University has been a remarkable and the most relevant affiliation in understanding the concept of ECO-DRR management. The word cloud highlights the 50 most crucial and relevant words used in the space of ECO-DRR (Fig. 13.3). The size of these keywords depicts the frequency of the words used in studies.

The most relevant and frequently used word is mangrove. The reason for this is the ability of these forests to protect against disasters such as tsunamis, high and strong waves, flooding, etc. Mangroves represent a highly effective choice for ECO-DRR. Firstly, these forests are categorised as natural barriers against disasters. A few reasons for their importance are that they are a source of livelihood for coastal communities, protection against natural disasters, and shelter for small aquatic species. Forestry, ecosystem and ecology are also relevant when it comes to mangroves. Keywords such as decision-making, surveys, management, risk assessment, and perception are also given high weightage when discussing the topic of ECO-DRR. These words are associated with the policy aspect. Surveys gather insights into coastal communities' perceptions and responses to the mangrove area. These aid in clarifying how the coastal communities manage mangroves and what major support these forests provide for the community. Words of the same colour primarily represent the same idea. For instance, policy implementation and ECO-DRR point to the same idea—protecting the ecosystem. Keywords such as ecological approach, habitat structure, design, and green infrastructure fall under the same umbrella of protecting the environment through natural barriers. The lower the anthropogenic interference, the better it is for the environment. The words environmental management, management, and decision-making highlight the urgency of proper ecological protection. Moreover, household surveys and democratisation depict the participation of the communities. The keyword local participation is also used quite a bit and points to the fact that local involvement is highly valued.

The major journals publishing in the field of Eco-DRR and mangroves are elucidated in Figs. 13.4 and 13.5.

The Fig. 13.5, showcases the top 10 journals (Sources) and the number of documents published. Sustainability (Switzerland) is the only publication house with two documents. This is mainly because of the increased knowledge and access to resources that the publications are more than compared to the other centres/journals. 'The Water Resources Development and Management', 'Science and Technology in Disaster Risk Reduction', 'Results in Engineering', 'Journal of Forest Research', 'Journal of Disaster Research', 'IOP Conference Series: Earth and Environmental Science', 'Geological Society Special Publication', and 'Frontiers in Built Environment' all published one article. This can be attributed to the need for more research in the field of ECO-DRR.

Frontiers in Built Environment has had a cumulative occurrence of 1 from 2017 to 2023. This consistency is due to knowledge of ECO-DRR and access to resources.

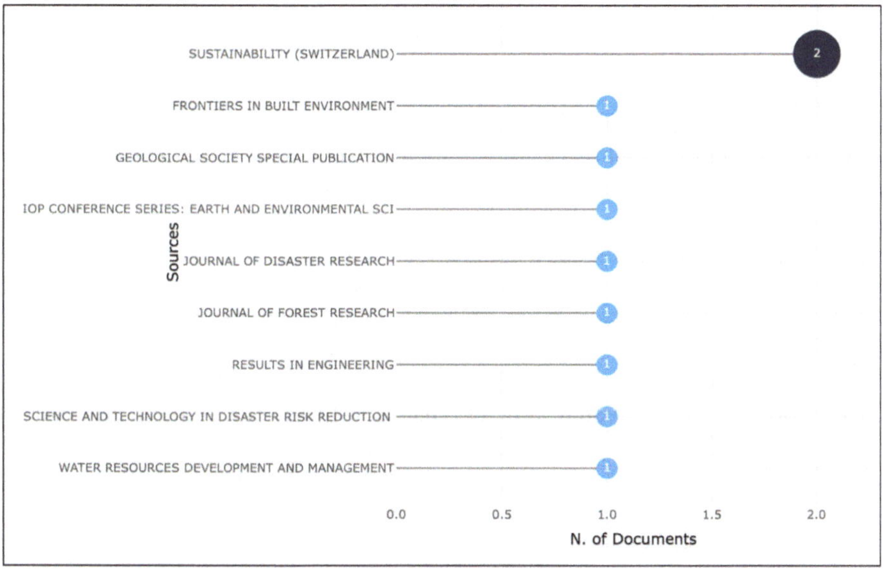

Fig. 13.4 Horizontal bar chart displaying the number of documents across various journals. Journals listed include "Sustainability (Switzerland)," "Frontiers in Built Environment," and others. Each bar is labeled with the journal name and extends horizontally to indicate the document count, with numerical values on the x-axis ranging from 0 to 20. The chart highlights the distribution of documents per journal, with "Sustainability (Switzerland)" having the highest count

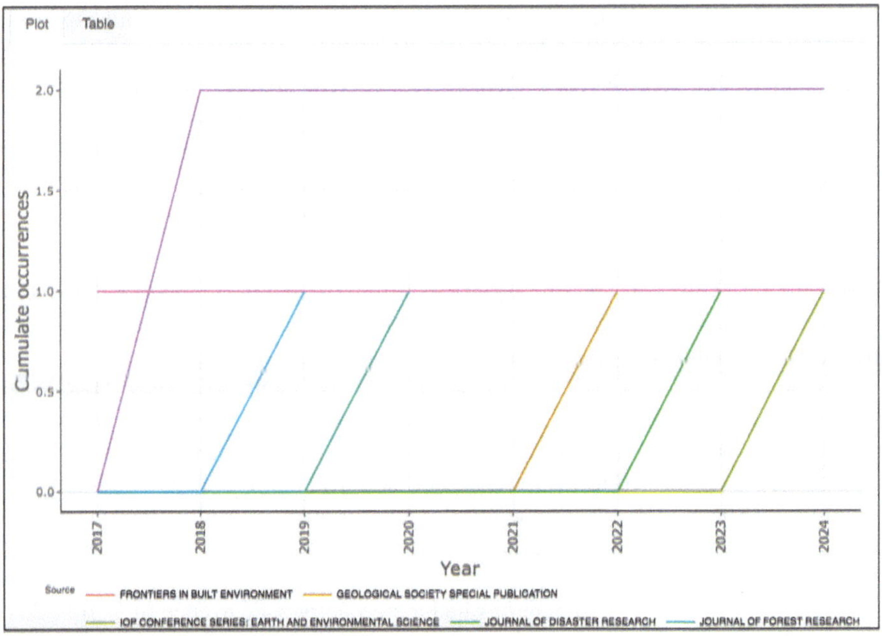

Fig. 13.5 A line chart showing cumulative occurrences over time from 2019 to 2024. The x-axis represents the year, and the y-axis represents cumulative occurrences. Multiple colored lines indicate different categories, with each line starting at different years and increasing at various rates. The legend at the bottom identifies the categories corresponding to each line color

From 2017 to 2018, the Journal of Forest Research had a cumulative occurrence of 1. In 2022, the Journal of Disaster Research had a cumulative occurrence of 1. The low occurrences in 2020 and 2021 can be attributed to the onset of the Coronavirus Pandemic and the less possibility for field based research. The Geological Society Special Publication has a cumulative occurrence of 1. The IOP Conference Series: Earth and Environmental Series was published in 2021 and then again in 2023. Average citations acquired by the publications in the journal are presented in Fig. 13.6.

The graph represents the number of documents by different authors being cited globally. The research articles that constitute our bibliometric survey are, Quevedo et al. (2020), Mori et al. (2022), Nehren et al. (2017), Oki et al. (2022), Dalimunthe (2018), DasGupta and Shaw (2018), Takagi (2019), Purwanto et al. (2024), van Onselen et al. (2023), and Takagi (2018). The graph has nine documents, and the global citations range from 0 to 40. The lowest is 0, and the highest number of global citations is 45. The highest citation was noted for the research by Quevedo et al. (2020), where the community perception was studied in response to Eco-DRR project at Eastern Samar, Philippines. The work of Takagi (2019), focuses on an applied approach to integrate mangroves with ecological engineering to improve the resilience of the vulnerable coastlines. Mori et al. (2022) studied the

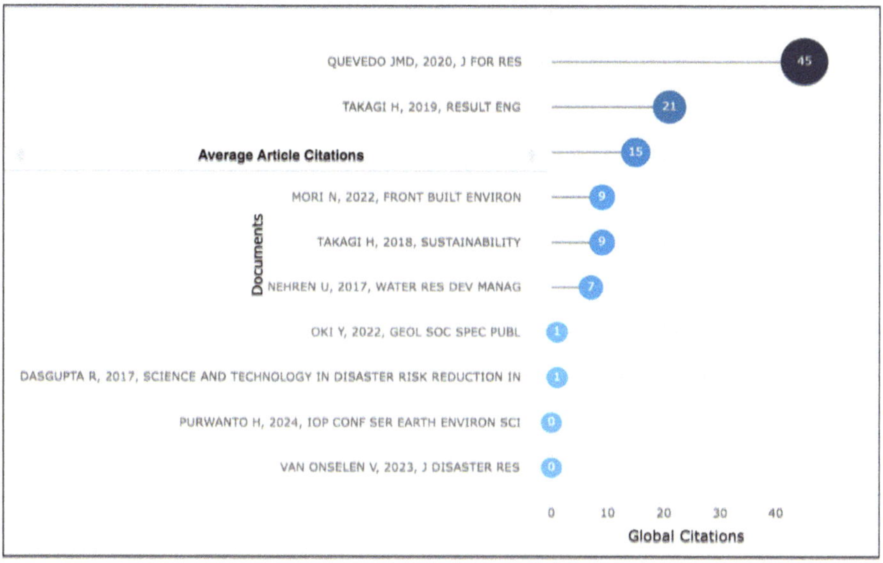

Fig. 13.6 Bubble chart displaying global citations versus average article citations for various authors. Each bubble represents an author, with size indicating the number of global citations. Authors are listed on the vertical axis, and citation numbers are on the horizontal axis. Larger bubbles indicate higher global citations, with one notably larger bubble at the top right

hydrological model to integrate the root dynamics of a true mangrove-- Rhizophora stylosa. Takagi (2018), focused on mangroves to attenuate the impact of sea level rise. Nehren (2017), focuses on the integration of sand dunes and mangroves to build coastal resilience in the Quang Nam Province, Vietnam. DasGupta and Shaw (2018), focused on the ecosystem servce benifits that are associated with the Eco-DRR models and their context. A fairly large number of global citations, i.e., 21, was achieved by Takagi (2019). Citation(s) indicate a focus of research towards applicability and case studies of Eco-DRR project which also focus on the trend of research in this domain. A total of 45 global citations is an excellent number and is also triple the average number of average article citations. The co-word analysis of the keywords is elucidated in Fig. 13.7.

Mangroves have different uses, and a few of them are captured in the above image. Mangroves are often seen as natural barriers against sea-level rise and coastal hazards. They are a protective front for keeping the local communities safe from these disasters. These forests also aid in reducing the speed and flow of waves.

Another significant use of mangroves is viewed as actively engaging the local population by providing livelihood. Many studies prove that mangroves are of great economic importance. For instance, mangroves offer the chance for pond-based farming. The local population uses shrimp, fish, and prawns to sustain their families.

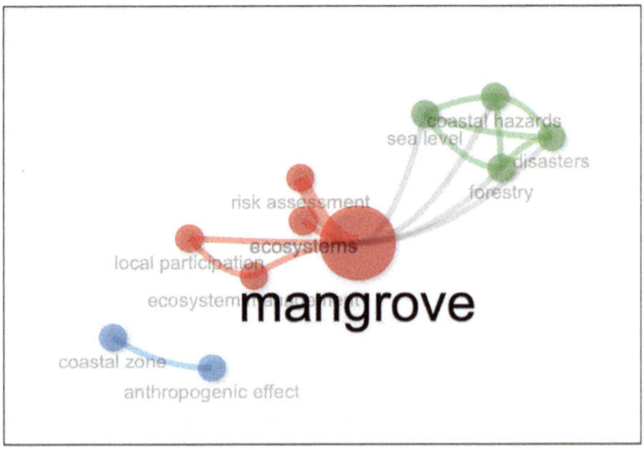

Fig. 13.7 Flow chart depicting a mind map centered around the keyword "mangrove." The map includes interconnected nodes with terms such as "ecosystem," "risk assessment," "biodiversity," "local authorities," and "coastal zone." Nodes are color-coded, with green, pink, and blue clusters representing different thematic areas. Lines connect related concepts, illustrating relationships and associations.

13.7 Conclusion and Recommendations

Mangroves are an ecosystem home to many small fish species and other small creatures. Aquatic life depends on mangrove forests for food, shelter, and protection. Natural disasters and the rapid rise in sea levels also threaten mangrove forests. Ecosystem management by the local population will eventually benefit the forests and the local population. If these forests are protected and utilised judiciously for their resources, they can aid the socio-economic conditions of marginalised communities in coastal areas.

Mangroves inhabit coastal zones and are frequently threatened by natural disasters. The maximum human population inhabits coastal habitats and deltas. Development leads to the decline and disappearance of mangrove ecosystems. The interconnected challenges of a changing climate and global warming also impact the carbon sequestration solutions required. Due to global warming, the number of cyclones is increasing across tropical and subtropical habitats. Eco-DRR with mangroves can address these issues while building resilience in the communities' disaster preparedness. Mangroves also have ecological services of sequestering blue carbon in soil and biomass. The bibliometric analysis highlights the need for focused research in the field of Eco-DRR and mangroves, as this is a key ecological service that can positively impact the lives of millions residing globally along mangrove sheltered coastlines.

References

Abernethy, K., Maisels, F., & White, L. J. (2016). Environmental issues in central Africa. *Annual Review of Environment and Resources, 41*(1), 1–33.

Adame, M. F., Reef, R., Santini, N. S., Najera, E., Turschwell, M. P., Hayes, M. A., Masque, P., & Lovelock, C. E. (2021). Mangroves in arid regions: Ecology, threats, and opportunities. *Estuarine, Coastal and Shelf Science, 248*, Article 106796. https://doi.org/10.1016/j.ecss.2020.106796

Akber, Md. A., Patwary, M. M., Islam, Md. A., & Rahman, M. R. (2018). Storm protection service of the Sundarbans mangrove forest, Bangladesh. *Natural Hazards, 94*(1), 405–418. https://doi.org/10.1007/s11069-018-3395-8

Alessandrini, V., Bertoni, D., Rangel-Buitrago, N., & Ciccarelli, D. (2024). Implementing a vegetation-based risk index to support management actions in Mediterranean coastal dunes. *Ocean & Coastal Management, 252*, Article 107105. https://doi.org/10.1016/j.ocecoaman.2024.107105

Alongi, D. M. (2008). Mangrove forests: Resilience, protection from tsunamis, and responses to global climate change. *Estuarine, Coastal and Shelf Science, 76*(1), 1–13. https://doi.org/10.1016/j.ecss.2007.08.024

Amarasinghe, M., & Perera, R. (2017). Historical biogeography of Sri Lankan mangroves. *Ceylon Journal of Science, 46*, 111. https://doi.org/10.4038/cjs.v46i5.7458

Amri, Mohd. R., Yulianti, G., Yunus, R., Wiguna, S., Adi, A. W., Ichwana, A. N., Randongkir, R. E., & Septian, R. T. (2016). Babi Gambaran Umum Indonesia. Risiko Bencana Indonesia.

Ataur Rahman, M., & Rahman, S. (2015). Natural and traditional defense mechanisms to reduce climate risks in coastal zones of Bangladesh. *Weather and Climate Extremes, 7*, 84–95. https://doi.org/10.1016/j.wace.2014.12.004

Auerbach, L. W., Goodbred Jr, S. L., Mondal, D. R., Wilson, C. A., Ahmed, K. R., Roy, K., Steckler, M. S., Small, C., Gilligan, J. M., & Ackerly, B. A. (2015, January 5). Flood risk of natural and embanked landscapes on the Ganges–Brahmaputra tidal delta plain. *Nature Climate Change*. https://www.nature.com/articles/nclimate2472

Bayas, J. C. L., Marohn, C., Dercon, G., & Cadisch, G. (2011, November 7). Influence of coastal vegetation on the 2004 tsunami wave impact in west Aceh. *PNAS*. https://www.pnas.org/doi/full/10.1073/pnas.1013516108

Boulton, C. A., Lenton, T. M., & Boers, N. (2022). Pronounced loss of Amazon rainforest resilience since the early 2000s. *Nature Climate Change, 12*(3), 271–278.

Cheong, S.-M., Silliman, B., Wong, P. P., van Wesenbeeck, B., Kim, C.-K., & Guannel, G. (2013, August 28). Coastal adaptation with ecological engineering. *Nature Climate Change*. https://www.nature.com/articles/nclimate1854

Chowdhury, A., Naz, A., Iyer, A. S., & Bhattacharyya, S. (2021). Ecosystem based disaster risk reduction at Indian Sundarbans: A lesson learned from AMPHAN supercyclone. *IOP Conference Series. Earth and Environmental Science, 796*(1). https://doi.org/10.1088/1755-1315/796/1/012042

Chowdhury, A., Naz, A., & Maiti, S. K. (2023). Variations in soil blue carbon sequestration between natural mangrove metapopulations and a mixed mangrove plantation: A case study from the world's largest contiguous mangrove forest. *Life, 13*(2), 271.

Chowdhury, A., Naz, A., & Maiti, S. K. (2024). Community-based, cost-effective multispecies mangrove restoration innovation to maximize soil blue carbon pool and humic acid and fulvic acid concentrations at Indian Sundarbans. *Environmental Science and Pollution Research*, 1–14.

Dalimunthe, S. A. (2018). Who manages space? Eco-DRR and the local community. *Sustainability (Switzerland), 10*(6). Scopus. https://doi.org/10.3390/su10061705

Danielsen, F., Sørensen, M., Olwig, M., Selvam, V., Parish, F., Burgess, N., Hiraishi, T., Karunagaran, V., Rasmussen, M., Hansen, L., Quarto, A., & Suryadiputra, N. (2005). The Asian Tsunami: A protective role for Coastal Vegetation. *Science (New York, N.Y.), 310*, 643. https://doi.org/10.1126/science.1118387

DasGupta, R., & Shaw, R. (2018). Utilizing ecosystem services for disaster risk reduction: The role of "scale" and "context". In *Science and technology in disaster risk reduction in Asia* (pp. 197–206). Academic Press.

Damastuti, E., & de Groot, R. (2017). Effectiveness of community-based mangrove management for sustainable resource use and livelihood support: A case study of four villages in Central Java, Indonesia. *Journal of Environmental Management, 203*, 510–521. https://doi.org/10.1016/j.jenvman.2017.07.025

Dittmann, S., Mosley, L., Stangoulis, J., Nguyen, V. L., Beaumont, K., Dang, T., & McGrath, A. (2022). Effects of extreme salinity stress on a temperate mangrove ecosystem. *Frontiers in Forests and Global Change, 5*, Article 859283.

Doughty, C. E., Keany, J. M., Wiebe, B. C., Rey-Sanchez, C., Carter, K. R., Middleby, K. B., et al. (2023). Tropical forests are approaching critical temperature thresholds. *Nature, 621*(7977), 105–111.

Duarte, C. M., Losada, I. J., Hendriks, I. E., Mazarrasa, I., & Marbà, N. (2013, October 29). The role of coastal plant communities for climate change mitigation and adaptation. *Nature Climate Change*. https://www.nature.com/articles/nclimate1970

Duke, N. C., Kovacs, J. M., Griffiths, A. D., Preece, L., Hill, D. J. E., Van Oosterzee, P., Mackenzie, J., Morning, H. S., & Burrows, D. (2017). Large-scale dieback of mangroves in Australia's Gulf of Carpentaria: A severe ecosystem response, coincidental with an unusually extreme weather event. *Marine and Freshwater Research, 68*, 1816–1829. https://doi.org/10.1071/MF16322

Esteban, M., Takagi, H., Mikami, T., Aprilia, A., Fujii, D., Kurobe, S., & Utama, N. A. (2017). Awareness of coastal floods in impoverished subsiding coastal communities in Jakarta: Tsunamis, typhoon storm surges and dyke-induced tsunamis. *International Journal of Disaster Risk Reduction, 23*, 70–79. https://doi.org/10.1016/j.ijdrr.2017.04.007

Esteban, M., Takagi, H., Nicholls, R. J., Fatma, D., Pratama, M. B., Kurobe, S., & Avelino, E. (2020). Adapting ports to sea-level rise: empirical lessons based on land subsidence in Indonesia and Japan. *Maritime Policy & Management, 47*(7), 937–952.

Feher, L. C., Osland, M. J., Griffith, K. T., & Grace, J. B. (2017, October). *(PDF) Linear and nonlinear effects of temperature and precipitation on ecosystem properties in tidal saline wetlands.* https://www.researchgate.net/publication/320191522_Linear_and_nonlinear_effects_of_temperature_and_precipitation_on_ecosystem_properties_in_tidal_saline_wetlands

Ferguson, G., & Gleeson, T. (2012, February 19). Vulnerability of coastal aquifers to groundwater use and climate change. *Nature Climate Change*. https://www.nature.com/articles/nclimate1413

Fernández-Montblanc, T., Duo, E., & Ciavola, P. (2020). Dune reconstruction and revegetation as a potential measure to decrease coastal erosion and flooding under extreme storm conditions. *Ocean & Coastal Management, 188*, Article 105075. https://doi.org/10.1016/j.ocecoaman.2019.105075

Fernando, E., Wickramasinghe, D., & Kulasena Fernando, V. (2022). An overview of vertebrate faunal diversity in Sri Lankan Mangroves. *Journal of Coastal Research, 38*. https://doi.org/10.2112/JCOASTRES-D-21-00041.1

Fletcher, C., Mullane, R., & Richmond, B. (1997). Beach Loss along Armored Shorelines on Oahu, Hawaiian Islands. *Journal of Coastal Research, 13*, 209–215.

Gilman, E. L., Ellison, J., Duke, N. C., & Field, C. (2008). Threats to mangroves from climate change and adaptation options: A review. *Aquatic Botany, 89*(2), 237–250. https://doi.org/10.1016/j.aquabot.2007.12.009

Giosan, L., Syvitski, J., Constantinescu, S., & Day, J. (2014, March 12). Climate change: Protect the world's deltas. *Nature*. https://www.nature.com/articles/516031a

Giri, C., Long, J., Abbas, S., Murali, R. M., Qamer, F. M., Pengra, B., & Thau, D. (2015). Distribution and dynamics of mangrove forests of South Asia. *Journal of Environmental Management, 148*, 101–111. https://doi.org/10.1016/j.jenvman.2014.01.020

Goda, K., Mori, N., Yasuda, T., Prasetyo, A., Muhammad, A., & Tsujio, D. (2019, October 4). *Frontiers | Cascading Geological Hazards and Risks of the 2018 Sulawesi Indonesia Earthquake and Sensitivity Analysis of Tsunami Inundation Simulations*. https://www.frontiersin.org/articles/10.3389/feart.2019.00261/full

Gracia, A., Rangel-Buitrago, N., Oakley, J. A., & Williams, A. T. (2018, April 15). *Use of ecosystems in coastal erosion management—ScienceDirect*. https://www.sciencedirect.com/science/article/abs/pii/S0964569117301588

Guannel, G., Arkema, K., Ruggiero, P., & Verutes, G. (2016, July 13). The power of three: Coral reefs, seagrasses and mangroves protect coastal regions and increase their resilience. *PLOS ONE*. https://journals.plos.org/plosone/article?id=10.1371/journal.pone.0158094

Gunderson, L. H., Holling, C. S., Pritchard, L., & Peterson, G. D. (2002). *Resilience of large-scale resource systems | Health & Environmental Research Online (HERO) | US EPA*. https://hero.epa.gov/hero/index.cfm/reference/details/reference_id/3286406

Hsu, T.-W., Lin, T.-Y., & Tseng, I.-F. (2007). Human Impact on Coastal Erosion in Taiwan. *Journal of Coastal Research, 23*(4 (234)), 961–973. https://doi.org/10.2112/04-0353R.1

Huang, W.-P., Chien, L.-K., & Lee, L.-H. (2016). Risk maps and coastal defense criteria in Taiwan. *Journal of Marine Science and Technology, 24*(6). https://doi.org/10.6119/JMST-016-1128-1

Ihinegbu, C., Mönnich, S., & Akukwe, T. (2023). Scientific evidence for the effectiveness of Mangrove forests in reducing floods and associated hazards in Coastal Areas. *Climate, 11*(4). Scopus. https://doi.org/10.3390/cli11040079

Indarsih, R., & Masruri, M. S. (2019). Mangrove conservation as an abration strategy risk reduction based on ecosystem in the coastal area of the Rembang Regency. *IOP Conference Series: Earth and Environmental Science, 271*(1), Article 012021. https://doi.org/10.1088/1755-1315/271/1/012021. https://www.researchgate.net/publication/334712462_Mangrove_conservation_as_an_abration_strategy_risk_reduction_based_on_ecosystem_in_the_coastal_area_of_the_Rembang_Regency

IUCN. (2024). More than half of all mangrove ecosystems at risk of collapse by 2050, first global assessment finds. https://www.iucn.org/press-release/202405/more-half-all-mangrove-ecosystems-risk-collapse-2050-first-global-assessment. Accessed December 22, 2024.

Jakovac, C. C., Latawiec, A. E., Lacerda, E., Leite Lucas, I., Korys, K. A., Iribarrem, A., Malaguti, G. A., Turner, R. K., Luisetti, T., & Baeta Neves Strassburg, B. (2020). Costs and carbon benefits of Mangrove conservation and restoration: A global analysis. *Ecological Economics, 176*, 106758. https://doi.org/10.1016/j.ecolecon.2020.106758

Jamero, M. L., Onuki, M., Esteban, M., Billones-Sensano, X. K., Tan, N., Nellas, A., Takagi, H., Thao, N. D., & Valenzuela, V. P. (2017). Small-island communities in the Philippines prefer local measures to relocation in response to sea-level rise. *Nature Climate Change, 7*(8), 581–586. https://doi.org/10.1038/NCLIMATE3344

Jin-Eong, O. (1995). The ecology of mangrove conservation & management. *Hydrobiologia, 295*(1), 343–351. https://doi.org/10.1007/BF00029141

Karunathilake, K. (2010). Status of mangroves in Sri Lanka. *Journal of Coastal Development, 7*.

Krauss, K. W., Allen, J. A., & Cahoon, D. R. (2003). Differential rates of vertical accretion and elevation change among aerial root types in Micronesian mangrove forests. *Estuarine, Coastal and Shelf Science, 56*(2), 251–259. https://doi.org/10.1016/S0272-7714(02)00184-1

Krauss, K. W., McKee, K. L., Lovelock, C. E., Cahoon, D. R., Saintilan, N., Reef, R., & Chen, L. (2014). How mangrove forests adjust to rising sea level. *New phytologist, 202*(1), 19–34.

Krauss, K. W., & Osland, M. J. (2020). Tropical cyclones and the organization of mangrove forests: a review. *Annals of Botany, 125*(2), 213–234.

Lee, J.-T., Tsai, S.-M., & Lee, M.-J. (2017). Uprooting resistance of two tropical tree species for sand dune stabilization. *African Journal of Agricultural Research, 12*(45), 3214–3220. https://doi.org/10.5897/AJAR2017.12715

Lovelock, C. E., Ball, M. C., Martin, K. C., & Feller, I. C. (2009, May 19). Nutrient enrichment increases mortality of mangroves. *PLOS ONE*. https://journals.plos.org/plosone/article?id=10.1371/journal.pone.0005600

Lovelock, C., Cahoon, D., Friess, D., Guntenspergen, G., Krauss, K., Reef, R., Rogers, K., Saunders, M., Sidik, F., Swales, A., Saintilan, N., Thuyen, L., & Tran, T. (2015). The vulnerability of Indo-Pacific mangrove forests to sea-level rise. *Nature, 526*. https://doi.org/10.1038/nature15538. https://www.nature.com/articles/nature15538

Lovelock, C. E., Feller, I. C., Reef, R., Hickey, S., & Ball, M. C. (2017, May 10). Mangrove dieback during fluctuating sea levels. *PubMed*. https://pubmed.ncbi.nlm.nih.gov/28490782/

McIvor, A., Spencer, T., Spalding, M., Lacambra, C., & Möller, I. (2015). Mangroves, tropical cyclones, and coastal hazard risk reduction. *ScienceDirect*. https://www.sciencedirect.com/science/article/abs/pii/B9780123964830000145

Moos, C., Bebi, P., Schwarz, M., Stoffel, M., Sudmeier-Rieux, K., & Dorren, L. (2018). Ecosystem-based disaster risk reduction in mountains. *Earth-Science Reviews, 177*, 497–513. https://doi.org/10.1016/j.earscirev.2017.12.011

Mori, N., Chang, C. W., Inoue, T., Akaji, Y., Hinokidani, K., Baba, S., & Shimura, T. (2022). Parameterization of mangrove root structure of Rhizophora stylosa in coastal hydrodynamic model. *Frontiers in Built Environment, 7*, 782219.

Nehren, U., Ho Dac Thai, H., Trung, N. D., Raedig, C., & Alfonso, S. (2017). Sand dunes and mangroves for disaster risk reduction and climate change adaptation in the coastal zone of Quang Nam Province, Vietnam. In *Land use and climate change interactions in Central Vietnam: LUCCi* (pp. 201–222). Singapore: Springer Singapore.

Thivakaran, G. A., Sharma, S. B., Chowdhury, A., & Murugan, A. (2020). Status, structure and environmental variations in semi-arid mangroves of India. *Journal of forestry research, 31*(1), 163–173.

Oki, Y., Kitazato, H., Fujii, T., & Yasukawa, S. (2022). Habitat mapping for human well-being: a tool for reducing risk in disaster-prone coastal environments and human communities. https://doi.org/10.1144/SP505-2021-26

Oni, O. (2019). Success story Rehabilitasi Ekosistem Mangrove di Pantai Karangsong Kabupaten Indramayu. *Jurnal Pengelolaan Sumberdaya Alam Dan Lingkungan (Journal of Natural Resources and Environmental Management), 9*, 787–796. https://doi.org/10.29244/jpsl.9.3.787-796

Pearson, L., & Pelling, M. (2015, September 9). The UN Sendai framework for disaster risk reduction 2015–2030: Negotiation process and prospects for science and practice. *Journal of Extreme Events*. https://www.worldscientific.com/doi/abs/10.1142/S2345737615710013

Poratelli, F., Cocuccioni, S., Accastello, C., Steger, S., Schneiderbauer, S., & Brun, F. (2020). State-of-the-art on ecosystem-based solutions for disaster risk reduction: The case of gravity-driven natural hazards in the Alpine region. *International Journal of Disaster Risk Reduction, 51*, Article 101929. https://doi.org/10.1016/j.ijdrr.2020.101929

Primavera, J. H., & Esteban, J. M. A. (2008, August 6). A review of mangrove rehabilitation in the Philippines: Successes, failures and future prospects. *Wetlands Ecology and Management*. https://link.springer.com/article/10.1007/s11273-008-9101-y

Purwanto, H., Paripurno, E. T., & Prasetya, J. D. (2024, March). Valuing hybrid engineering approach: ecosystem and structural based DRR using InVEST model of coastal zone Sikka Flores-Literature Review. In *IOP Conference Series: Earth and Environmental Science* (Vol. 1314, No. 1, p. 012029). IOP Publishing.

Quevedo, J. M. D., Uchiyama, Y., & Kohsaka, R. (2020). Perceptions of local communities on mangrove forests, their services and management: Implications for Eco-DRR and blue carbon management for Eastern Samar, Philippines. *Journal of Forest Research, 25*(1), 1–11.

Rahman, A. Z. (2015). Kajian Mitigasi Bencana Tanah Longsor Di Kabupaten Banjarnegara. *Gema Publica, 1*(1), 1–14. https://doi.org/10.14710/gp.1.1.2015.1-14

Ranawana, K. (2017). *Mangroves of Sri Lanka. 1*.

Reguero, B. G., Beck, M. W., Bresch, D. N., Calil, J., & Meliane, I. (2018). Comparing the cost effectiveness of nature-based and coastal adaptation: A case study from the Gulf Coast of the United States. *PLoS ONE, 13*(4), Article e0192132. https://doi.org/10.1371/journal.pone.0192132

Renaud, F., Sudmeier-Rieux, K., & Estrella, M. (2013). *The Role of Ecosystems in Disaster Risk Reduction.*

Rudianto, E. (2016). *Tsunamis and Earthquakes in Coastal Environments.* Springerprofessional.De. https://www.springerprofessional.de/en/tsunamis-and-earthquakes-in-coastal-environments/10031752

Sakho, I., Dupont, J.-P., Cisse, M., El Janyani, S., & Loum, S. (2017). Hydrological responses to rainfall variability and dam construction: A case study of the upper Senegal River basin. *Environmental Earth Sciences, 76*, 253. https://doi.org/10.1007/s12665-017-6570-4

Samanta, S., Hazra, S., Mondal, P. P., Chanda, A., Giri, S., French, J. R., & Nicholls, R. J. (2021). Assessment and attribution of Mangrove Forest changes in the Indian Sundarbans from 2000 to 2020. *Remote Sensing, 13*(24), Article 24. https://doi.org/10.3390/rs13244957

Satake, K., McLean, C., & Alcántara-Ayala, I. (2018). Understanding disaster risk: The role of science and technology. *Journal of Disaster Research, 13*(7), 1168–1176. https://doi.org/10.20965/jdr.2018.p1168

Sidik, F., Supriyanto, B., Krisnawati, H., & Muttaqin, M. Z. (2018). Mangrove conservation for climate change mitigation in Indonesia. *Wires Climate Change, 9*(5), Article e529. https://doi.org/10.1002/wcc.529

Silver, J. M., Arkema, K. K., Griffin, R. M., Lashley, B., & Lemay, M. (2019, September 24). *Frontiers | Advancing Coastal Risk Reduction Science and Implementation by Accounting for Climate, Ecosystems, and People.* https://www.frontiersin.org/articles/10.3389/fmars.2019.00556/full

Sippo, J. Z., Lovelock, C. E., Santos, I. R., Sanders, C. J., & Maher, D. T. (2018). Mangrove mortality in a changing climate: An overview. *Estuarine Coastal and Shelf Science, 215*, 241–249. https://doi.org/10.1016/j.ecss.2018.10.011

Sudmeier-Rieux, K., Arce-Mojica, T., Boehmer, H. J., Doswald, N., Emerton, L., Friess, D. A., Galvin, S., & Hagenlocher, M. (2021, June 28). Scientific evidence for ecosystem-based disaster risk reduction. *Nature Sustainability.* https://www.nature.com/articles/s41893-021-00732-4

Sunkur, R., Kantamaneni, K., Bokhoree, C., & Ravan, S. (2023). Mangroves' role in supporting ecosystem-based techniques to reduce disaster risk and adapt to climate change: A review. *Journal of Sea Research, 196*, Article 102449. https://doi.org/10.1016/j.seares.2023.102449

Sutton-Grier, A. E., Wowk, K., & Bamford, H. (2015, September). Future of our coasts: The potential for natural and hybrid infrastructure to enhance the resilience of our coastal communities, economies and ecosystems. *ScienceDirect.* https://www.sciencedirect.com/science/article/pii/S1462901115000799

Tait, J. F., & Griggs, G. (1990, January). *(PDF) Beach response to the presence of a seawall: A comparison of field observations.* https://www.researchgate.net/publication/279689500_Beach_response_to_the_presence_of_a_seawall_a_comparison_of_field_observations

Takagi, H. (2018). Long-term design of mangrove landfills as an effective tide attenuator under relative sea-level rise. *Sustainability, 10*(4), 1045.

Takagi, H. (2019). "Adapted mangrove on hybrid platform"—coupling of ecological and engineering principles against coastal hazards. *Results in Engineering, 4*, 100067

Takagi, H., Esteban, M., Mikami, T., & Fujii, D. (2016, September). Projection of coastal floods in 2050 Jakarta. *ScienceDirect.* https://www.sciencedirect.com/science/article/abs/pii/S2212095516300281

Takagi, H., Esteban, M., & Tam, T. T. (2014). 7—Coastal Vulnerabilities in a Fast-Growing Vietnamese City. In N. D. Thao, H. Takagi, & M. Esteban (Eds.), *Coastal Disasters and Climate Change in Vietnam* (pp. 157–171). Elsevier. https://doi.org/10.1016/B978-0-12-800007-6.00007-1

Takagi, H., Fujii, D., Esteban, M., & Yi, X. (2017, April 16). *Sustainability | free full-text | effectiveness and limitation of Coastal Dykes in Jakarta: The need for prioritizing actions against land subsidence.* https://www.mdpi.com/2071-1050/9/4/619

Takagi, H., Thao, N. D., & Anh, L. T. (2016, September 21). *Sustainability | free full-text | sea-level rise and land subsidence: Impacts on flood projections for the Mekong Delta's Largest City.* https://www.mdpi.com/2071-1050/8/9/959

Takagi, H., Mikami, T., Fujii, D., Esteban, M., & Kurobe, S. (2016). Mangrove forest against dyke-break-induced tsunami on rapidly subsiding coasts. *Natural Hazards and Earth System Sciences, 16*(7), 1629–1638. https://doi.org/10.5194/nhess-16-1629-2016

Tanaka, N. (2012). Effectiveness and limitations of coastal forest in large tsunami: conditions of Japanese pine trees on coastal sand dunes in tsunami caused by great east japan earthquake. *Journal of Japan Society of Civil Engineers, Ser. B1 (Hydraulic Engineering), 68*, II_7-II_15. https://doi.org/10.2208/jscejhe.68.II_7

Triyanti, A., & Chu, E. (2018). A survey of governance approaches to ecosystem-based disaster risk reduction: Current gaps and future directions. *International Journal of Disaster Risk Reduction, 32*, 11–21. https://doi.org/10.1016/j.ijdrr.2017.11.005

Tsujino, R., Yumoto, T., Kitamura, S., Djamaluddin, I., & Darnaedi, D. (2016). History of forest loss and degradation in Indonesia. *Land Use Policy, 57*, 335–347.

Valiela, I., Bowen, J. L., & York, J. K. (2001). Mangrove Forests: One of the World's threatened major tropical environments: At least 35% of the area of mangrove forests has been lost in the past two decades, losses that exceed those for tropical rain forests and coral reefs, two other well-known threatened environments. *BioScience, 51*(10), 807–815.

van Onselen, V., Bayrak, M. M., & Lin, T. Y. (2023). Assessment of ecosystem-based disaster risk reduction strategies in coastal environments of Taiwan. *Journal of Disaster Research, 18*(7), 700–707.

Veettil, B. K., Wickramasinghe, D., & Amarakoon, V. (2023). Mangrove forests in Sri Lanka: An updated review on distribution, diversity, current state of research and future perspectives. *Regional Studies in Marine Science, 62*. Scopus. https://doi.org/10.1016/j.rsma.2023.102932

Vuik, V., Jonkman, S. N., Borsje, B. W., & Suzuki, T. (2016, October). Nature-based flood protection: The efficiency of vegetated foreshores for reducing wave loads on coastal dikes. *ScienceDirect*. https://www.sciencedirect.com/science/article/abs/pii/S0378383916301004

Walker, B., Holling, C. S., Carpenter, S. R., & Kinzig, A. (2004, December). *Resilience, adaptability and transformability in social–ecological systems on JSTOR*. https://www.jstor.org/stable/26267673

Walters, B. B., Rönnbäck, P., Kovacs, J. M., Crona, B., Hussain, S. A., Badola, R., Primavera, J. H., Barbier, E., & Dahdouh-Guebas, F. (2008, July). Ethnobiology, socio-economics and management of mangrove forests: A review. *ScienceDirect*. https://www.sciencedirect.com/science/article/abs/pii/S0304377008000417

Ward, R., Friess, D., Day, R., & Mackenzie, R. (2016). Impacts of climate change on mangrove ecosystems: A region by region overview. *Ecosystem Health and Sustainability, 2*, 1–25. https://doi.org/10.1002/ehs2.1211

Wijayanti, H., Marfai, M. A., & Mei, E. T. W. (2019). Collaboration practice on ecosystem-based disaster risk reduction in the coastal area of Semarang City, Indonesia. In Y. Setiawan, L. B. Prasetyo, Y. Murayama, K. D. Kanniah, G. J. Perez & P. T. Dat (Eds.), *Proceedings of SPIE—The International Society for Optical Engineering* (Vol. 11372). SPIE; Scopus. https://doi.org/10.1117/12.2539059

Wolf, S., My Pham, Matthews, N., & Bubeck, P. (2020, February 23). *Understanding the implementation gap: Policy-makers' perceptions of ecosystem-based adaptation in Central Vietnam: Climate and Development: Vol 13, No 1.* https://www.tandfonline.com/doi/abs/10.1080/17565529.2020.1724068

Chapter 14
Climate Change and Community-Based Initiatives in Land Restoration in Coastal Areas

Dimpi Dutta, Debasmita Nath, Pallabi Dutta, Mahima Begum, Chinmaya Sahoo, and Ankit Saini

Abstract Climate change reflects a significant threat to coastal ecosystems in the global scenario. The Coastal ecosystems are the most vulnerable areas facing the adverse effects of the changing climatic patterns. In India, the rising sea levels, increased storm intensity, and extreme weather events heightened risks to habitat degradation, threatening the biodiversity and the livelihoods of millions of vulnerable coastal communities. The community-based initiatives in land restoration have emerged as effective strategies for enhancing resilience in these regions, with significant developments observed in India. To exacerbate the impacts of climatic variation, worldwide coastal communities are grappling with several initiatives, including the UN's "Blue Carbon" project that aims to protect and restore coastal ecosystems, particularly mangroves, salt marshes, and seagrasses, which are crucial for carbon sequestration and biodiversity. In the Philippines and Indonesia, community-driven mangrove restoration projects revived degraded lands but also improved fisheries and

D. Dutta · M. Begum
Department of Agronomy, Assam Agricultural University, Jorhat, India
e-mail: dimpi.dutta.adj23@aau.ac.in

M. Begum
e-mail: mahima.begum@aau.ac.in

D. Nath
Department of Plant Breeding and Genetics, Assam Agricultural University, Jorhat, India

P. Dutta
Department of Education, Gauhati University, Guwahati, Assam, India

C. Sahoo (✉)
Department of Agronomy, Dr. Rajendra Prasad Central Agricultural University, Pusa, Samastipur, Bihar, India
e-mail: chinmaya97sahoo@gmail.com

A. Saini (✉)
Department of Agronomy, Dr. KSG Akal College of Agriculture, Eternal University, Baru Sahib, Sirmaur, Himachal Pradesh, India
e-mail: ankitsaini9777@gmail.com

U. Chatterjee et al. (eds.), *Eco-Resilience: Climate Change, Land Degradation and Sustainable Solutions*, Sustainability Solutions,
https://doi.org/10.1007/978-3-032-00708-7_14

provided natural barriers against storm surges. Similarly, in Indian coastline extends over 7500 km encompassing diverse ecosystems and millions of people dependent on marine resources. Such land restoration initiatives are also gaining traction within Indian boundaries. Recent developments in India highlight the effectiveness of community-based approaches in coastal land restoration. In West Bengal, the Sundarbans, a renowned UNESCO World Heritage site, glorifies the potential of community involvement in restoration efforts under such adversities. The involvement of local coastal communities, NGOs, along with the peer support from government agencies, enhances the engagement in mangrove planting and sustainable fisheries management, significantly contributing to both ecological perspective and economic stability. Another approach "Mangrove Restoration Program" in Orissa, where the local villagers actively participate in the planting and maintenance of mangrove saplings, contributed significantly to the restoration of the coastal ecosystem. This initiative not only combats coastal erosion but also enhances fish populations, benefiting local fishermen's financial sufficiency. The program has fostered a true sense of ownership among the coastal communities, leading to increased vigilance and patriotism against illegal logging and poaching in the protected surroundings. Community-based initiatives in land restoration are vital for mitigating the challenges posed by climate change in coastal areas and upon coastal communities. The intensive learning as well as collaboration between communities, governments, and other private and cooperative sectors will be essential for creating adaptable and effective strategies in coastal land restoration. By acquiring knowledge from successful examples and overcoming existing challenges, we can safeguard our coastal ecosystems for future generations.

Keywords Ecosystem · Climate · Mangroves · Coastal-community · Land-restoration

14.1 Introduction

The coastal ecosystem is one of the most productive ecosystems of the biome, serving several functions, including sustaining biodiversity, protection against storms, and carbon sequestration to establish ecological equilibrium within and among other systems. Climate change has severe negative implications for the coastal ecosystem by exacerbating changes such as a rise in sea level, increasing frequency and intensity of storms, and destruction of natural habitats for both micro and macro flora and fauna. These impacts pose severe threats to biodiversity and the livelihoods of millions, particularly in countries with extensive coastlines, such as India. In countries such as the Maldives, Bahamas, and Bahrain are nations with having significantly higher proportion of people living nearby to coastal areas (Table 14.1).

Table 14.1 Population percentage in coastal regions of a few nations

S.No.	Nation	Population percentage in coastal regions (%)
1	Maldives	100
2	Bahamas	88
3	Bahrain	78
4	Suriname	78
5	Netherlands	60
6	Macao	59
7	Guyana	55
8	Djibouti	40
9	Vietnam	53
10	Bangladesh	39
11	Egypt	36
12	Thailand	25
13	Japan	24
14	Indonesia	20
15	Philippines	20
16	China	10
17	USA	8
18	India	6

Source Maribus (2010)

Global climate change and its impacts have established itself as a burning topic in the twenty-first century, simulating severe adversities in every aspect. The Intergovernmental Panel for Climate Change (IPCC) emerged as the prime body encountering complex scenarios from ecological, social, and environmental perspectives. The impact of climate change on societal, economic, and socio-political issues was thoroughly evaluated by numerous scientists, including Feliciano et al. (2022), Filho et al. (2021), Adger et al. (2005). Global climatic variability has resulted in shifts in climate zones, mainly expanding the dry zones and shrinking the polar zones. Further, it was predicted that the warming of the Earth's atmosphere would shift tropical zones towards poles in mid-high latitudes and upward regions. Climatic extremities created vulnerable ecosystems, disrupting the food chain, water availability, morbidity and mortality, and health hazards for human beings (Burkett et al., 2014). These climatic changes affect millions of people, particularly those residing in vulnerable coastal regions. Globally, community-based initiatives have emerged as effective strategies for mitigating these risks and restoring degraded coastal lands. The coastal area of India, extending 7500-km comprises plentiful ecosystems consisting of estuaries, mangroves, coral reefs, etc., which support the livelihoods of numerous people who are directly or indirectly interconnected with these ecosystems (Table 14.2). The extreme weather variations, including the rise of sea level, habitat land degradation, etc., threaten the biodiversity and the coastal communities (Fig. 14.1). The community-based approaches on the Indian coastline have made a noteworthy impact on the vindication of climate change adversities.

India, being the largest democracy of the world with diverse ecology, geography, and socio-economic is one of the most or highly vulnerable countries to the variable weather or climatic status. Environmental apprehensions associated with climate change have become a prominent concern of the nation. The capital of the country is suffering from unadorned air pollution creating hazardous situations caused by vehicles, industrial emissions, fossil fuel burning, etc. (Hussain & Hoque, 2022b; Sahu & Saxena, 2015; Sonwani, 2016; Sonwani et al., 2021a, 2021b, 2021c, 2022a). The World Air Quality Report, State of Global Air (SoGA), indexed Delhi as the most polluted city in the world. It can be noted that climate change has induced rapid air pollution and vice versa, both of these are intertwined closely (Singh & Yadav, 2021; Sonwani et al., 2022b). The emission of greenhouse gases, viz., CO_2 (carbon dioxide), H_2O (water vapor), HFCs (hydrofluorocarbons), CH_4 (methane), N_2O (nitrous oxide), HCFCs (hydrochlorofluorocarbons), and O_3 (Tropospheric

Table 14.2 Coastal databases on India

Coastal data		
1.	Length of coastline area	7516.6 km
2.	Coastal states	Nine states (a) Gujarat (b) Maharashtra (c) Goa (d) Karnataka (e) Kerala (f) Tamil Nadu (g) Andhra Pradesh (h) Odisha (i) West Bengal
3.	Coastal union territories	Two union territories 1. Daman & Diu 2. Puducherry
4.	Coastal districts	66
5.	Total population of coastal districts	171 million
6.	Coastal wetlands	43,230 km^2
7.	Major estuaries	97
8.	Major lagoons	34
9.	Mangrove Areas	31
10.	Area under mangroves	6740 km^2 (57% East coast, 23% west coast, 20% Andaman & Nicobar Islands)
11.	Coral reef areas	5

Source Venkatraman and Wafar (2005), Singh (2003)

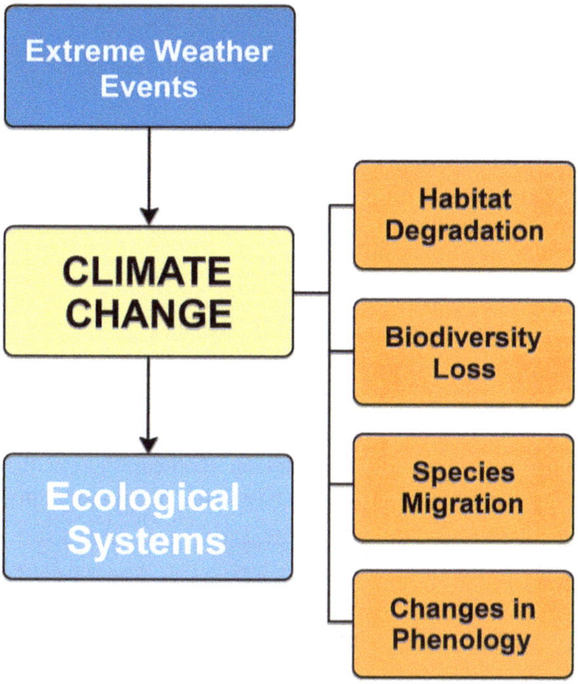

Fig. 14.1 Impact of extreme weather and climate change events on ecological systems. *Source* Ummenhofer and Meehl (2017)

ozone) along with pollutants including aerosol, brown carbon (BrC) and black carbon (BC) act as primary agents in warming the globe creating global climate change (Ramanathan & Carmichael, 2008; Sonwani et al., 2021a, 2021b, 2021c, 2022a). The Earth's temperature enhancement or increase resulted in melting glaciers and ecological damage (Saxena & Sonwani, 2019). These consequences are not limited to one country but have far reached around or all over the earth (Haibach & Schneider, 2013; Klingenfeld & Schellnhuber, 2012). The impact of changing climate resulted in rising earth surface temperatures, resulting global warming (Krishnan et al., 2020) along with heavy/low i.e., erratic precipitation behavior (Behera & Roy, 2019; Sonwani & Kulshrestha, 2019), altered weather phenomenon (Hussain & Hoque, 2022a; Saxena et al., 2022), rise in sea level resulting submergence of low-lying areas (Krishnan et al., 2020), the occurrence of forest fire and other negative impacts of forest and wildlife (Jain et al., 2021), incidence of dust/ dirt or sand storm in dry/ desert areas (Maji & Sonwani, 2022). The Indian coast has risen by an average of 3.3 mm per year, creating a high risk to the low-lying areas such as Sundarbans. The coastal areas of India, affected by cyclones such as Amphan (2020) and Tauktae (2021), have caused extensive damage to the coastline and its biodiversity, which requires proper attention and community-based approaches for mitigation and adaptation (Chatterjee & Roy, 2021).

The key objectives of this chapter are:

1. To scrutinize the impacts of climate change on coastal ecosystems with a global-to-local perspective.
2. To evaluate initiatives in terms of community-based land restoration in Indian coastal regions and compare them with international models.
3. To identify the challenges, opportunities, and policy implications of such initiatives for long-term climate resilience.

14.2 Materials and Methods

This chapter adopts a qualitative, integrative review approach to analyze the impact of climate change on coastal ecosystems and evaluate community-based land restoration initiatives. The study focuses on synthesizing existing knowledge from both global and Indian perspectives to draw comparative insights and policy recommendations.

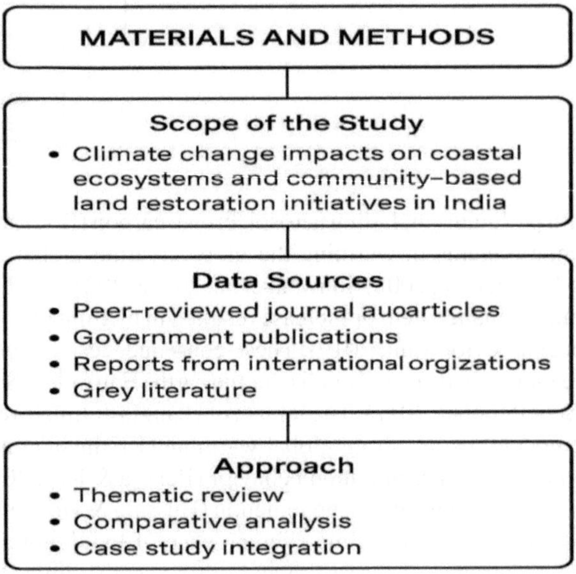

14.2.1 Scope of the Study

The analysis covers coastal regions vulnerable to climate change impacts, with a particular emphasis on India's eastern and western coastlines, including critical ecosystems such as the Sundarbans (West Bengal) and Odisha's coastal mangroves.

International examples from the Philippines, Indonesia, and global initiatives like the UN's "Blue Carbon" project were included for comparative evaluation.

14.2.2 Data Sources

The study is based on a comprehensive review of secondary data sources, including:

Peer-reviewed journal articles (e.g., IPCC Reports, scientific journals like *Global Environmental Change, Climate Policy*).

Government publications (e.g., India's Ministry of Environment, Forest and Climate Change reports, National Coastal Mission reports).

Reports from international organizations (e.g., UNEP, UNDRR, FAO, World Bank).

Grey literature from NGOs and research institutions focusing on climate change, coastal restoration, and community engagement.

14.2.3 Approach

Thematic Review: A thematic analysis was conducted to identify major climate-induced challenges in coastal zones and categorize restoration initiatives based on their strategies (e.g., mangrove reforestation, wetland management, integrated blue-green infrastructure).

Comparative Analysis: Indian community-based initiatives were compared against selected international models to highlight best practices, contextual challenges, and adaptation strategies.

Case Study Integration: Selected case studies from Odisha, West Bengal, and international examples were used to illustrate community participation models and their outcomes.

14.2.4 Analytical Framework

The study applies a resilience-based lens to assess the effectiveness of community initiatives. Key evaluation criteria include:

Ecological outcomes (e.g., biodiversity conservation, carbon sequestration).

Socio-economic benefits (e.g., livelihood improvement, community ownership).

Institutional engagement (e.g., involvement of local governance, NGOs, government support).

By triangulating information across diverse sources and comparing case studies at global and regional levels, the chapter aims to generate insights relevant to both scientific research and policy formulation in coastal land restoration.

14.3 Results and Discussions

14.3.1 Climate Change and Land Degradation in Coastal Areas

The alteration in physical, chemical as well and biological properties of the soil is referred to as land degradation. It hampers the quality and quantity of plant growth by altering soil properties (Douglas, 1994). Climatic alteration affects land cover and land surface processes, typically raising the mean surface temperature of the Earth from the prehistoric period. Soil, as it defines the soul of infinite life, is the key source of food security also affected by climatic alteration due to its effect on soil's physical properties. Therefore, it is of prior importance to understand and evaluate the interaction of weather/climatic variables and physico-chemical and biological characteristics of soil degrading yield and environmental quality. The long-term impact of climatic aberration at the continental, national, and regional levels should be considered for tackling various changes in global aspects, including heavy precipitation, ocean salinity, incidence of heat waves. Wind pattern variation, intense tropical cyclones, and rapid and uncontrolled generation of carbon dioxide (CO_2) resulted in climate change, profoundly exacerbating land degradation and desertification. Another important aspect related to climate change is the increase in river runoff load, which is projected to increase by up to 10–40% at higher or upper latitudes while decreasing by 10–30% in mid-latitudes, as shown. The improper management of land resulted in land degradation, resulting in infertile soil. The degraded land includes eroded land, removal of soil carbon, cat soils, soil acidification, etc. The aberrant climatic changes, including heavy precipitation, are another important agent of water erosion. Water erosion is a series of processes, viz., splash erosion, removal of sheet of soil, i.e., sheet erosion, formation of rills and gullies (Douglas, 1994). Changes in cyclonic behavior also resulted in erosion in coastal areas.

Soil erosion is the process of the removal of soil mainly due to the agents, including wind and water. Bare soil without vegetation or due to overgrazing is most prone to severe erosion. The initial step of water erosion is the splash erosion, where soil particles are splashed out from their coherent mass and disbursed to a wider horizon, followed by sheet erosion, removing thin layers without being noticed, following the formation of rills and wider gullies termed as rill and gully erosion. Formation of gullies removed a huge amount of soil, which cannot be easily removed by tillage. Soil organic matter is the prime constituent characterizing its quality and health.

It infers soil structure, stability, water holding capacity and oxygen diffusion rate, nutrient storage, and microbial biomes. The rate and quantity of addition of organic residues and organic matter are highly sensitive to the climatic parameters, mainly temperature fluctuations. The decomposition and incorporation rate of the organic matter into the soil is directly related to the increase in temperature regime, i.e., the higher the temperature, more is the rate of decomposition. While the agricultural intensification leads to a decrease in organic matter level in soil, as a major portion of it will be lost as carbon dioxide to the atmosphere. Along with carbon dioxide, other greenhouse gases are also released from agricultural land (Fig. 14.2). The increase in carbon dioxide concentrations enhances the plant growth characteristics utilizing a higher amount of organic matter from the soils. The climatic aberration highly affects the organic matter content of the soil.

Climate change also creates disturbance to soil structure (Lal, 1994). Soil structure is the fundamental property of soil affecting nutrient transformations and availability, water regime in soil, and its movement, nutrient chemistry, etc. (Nordstrom, 2009). The higher clay content in the soil resulted in swelling and shrinkage properties of the soil led to the formation of large cracks and fissures, particularly in temperate areas (Boden & Driscoll, 1987). The climate change increases the intensity of accumulation of sesquioxides inside the cracks, resulting in heavy metal contamination in the ecosystem (Wild, 1993). The increase in global climatic temperature and rainfall resulted in nitrate leaching, causing acidification of soil (Fenton & Helyar, 2007). This acidification process is the consequence of nitrate leaching in high rainfall areas. The acidification of the soil affects crop production, highly degrades the soil, making it unproductive for other uses. An increase in rainfall amount and intensity resulted in severe acidification in the humid tropics. Along with humid and sub-humid areas, arid zones are also highly influenced by evaporative and precipitation alterations. Another type of acidified soil, i.e., cat soils or acid sulfate soils, is the degraded soil that exists on the earth surface due to the aberrant climatic interference. Acid sulfate soils are formed when pyrite-rich parent materials are drained, causing the oxidation of pyrite. Mainly, two kinds of acid-sulfate soils are present in nature. One of which is the result of the production of acid sulphate soil by oxidation of pyrites,

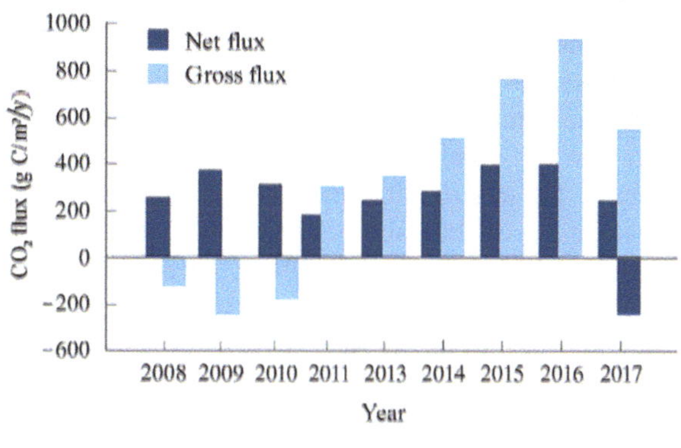

Fig. 14.2 Carbon dioxide fluxes (net and gross) from soil (2008–2017). *Source* Jia et al. (2019)

releasing sulphuric acid. This process also liberates huge quantities of metals to the surroundings, causing environmental hazard in areas where sulfides are present, mainly in coastal regions (De Nicola & Stapleton, 2002; Dinelli et al., 2001; Lee et al., 2005; Lin & Lin, 2003; Lu et al., 2011; Pereira, 1993). The severity of climate change on the formation of acid sulfate soils. The climate change events, including heavy rains and droughts, have accelerated the cat soil formation process (Levings et al., 2005). The inclusion of acid sulphate to aquatic bodies creates a severe threat to the aquatic life system, disbalancing terrestrial and marine ecosystems altogether (Mirza, 2003).

Climatic factors highly influence soil nutrient status. The change in climatic parameters, including temperature and rainfall, influences the nutrient status in several ways. The increase in temperature regimes increases the evaporative demands, extracting of nutrients from the soil along with increasing the ease of leaching (Dent, 1986). The movement of water towards gravity transports a significant amount of nutrient loss, making it unavailable to the life forms staying in the biosphere, causing the inclusion of more synthetic inputs, which subsequently resulted in precipitating global climate change. Contrary to this, the reduction of downward movement could result in other problems, viz., salinization, and surface crusting, etc., which is another major problem of arid, tropical, semiarid and subtropical ecosystems (Bullock et al., 1995). A better understanding of the climatic variabilities and land degradation is needed to cope and adapt to the climatic incongruities.

14.3.2 Community-Based Initiatives in Land Restoration in Coastal Areas

Various community-based initiatives must be undertaken to mitigate the climatic aberrations for ecosystem well-being, along with proper planning and measures. One of the important approaches that can be undertaken includes green and blue infrastructure. The temporal changes of land surface and mean surface temperature highly influence the living status of the coastal population via influencing agriculture, environment and climatic scenario of the people (Fig. 14.3).

For such long-term changes, community-based initiatives are the need of the hour. The establishment of such infrastructure can act as a storehouse for carbon storage when combined with traditional forms of infrastructure, i.e., gray infrastructure. These structures can also mitigate the weather anomalies, viz., flooding, heatwaves, droughts, and heavy rainfall, which furthermore enhances health benefits, enhancing quality of life (IPCC, 2013). Adoption of climate-resilient agriculture approaches also reduces the risk of weather-related disasters in coastal ecosystems. These approaches tackle such adversities, which will further help in achieving wider approaches of Sustainable Development Goals (Jha et al., 2013). The government should incorporate climatic risk, its impact, and mitigation as serious issues and

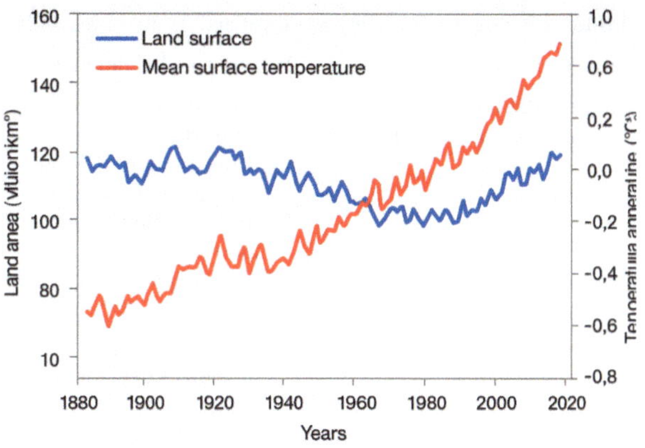

Fig. 14.3 Temporal change in land surface and mean surface temperature from 1860 to 2020. *Source* Allen et al. (2010)

comprehend them in their decision-making advisory to upscale the strategies towards mitigation of risk (IPCC, 2013). Along with proper decision-making, appropriate prediction of the climatic events will enhance the risk-taking abilities for effective action against climatic variation (IPCC, 2013). The transformation of climatic risk into advancing the use of renewable energy sources not only upholds economic sustainability provides safeguards to resource conservation (IPCC, 2013). The transition of fossil fuels to green energy is another better approach towards mitigating climate change from a broader perspective. REDD + (Reducing Emissions from Deforestation and Forest Degradation), along with deforestation and degradation, are the major constraints contributing a significant impact on the mitigation of climate change (Jenkins & Schaap, 2018). Therefore, a better understanding of the climatic changes and addressing them in a resilient and sustainable way should be of the utmost need.

In India also the rising sea level also causes numerous adverse impact including inundation of low-lying seashore or coastline areas, erosion and morphological anomalies, alteration of coastal wetland areas and intrusion of sea water (salty water) into surface or into ground water limited the ecological values of the coastal ecosystem which should be properly managed to reduced its impacts and complications (Table 14.3).

14.3.3 Global and Indian Context of Coastal Ecosystem Restoration

An international perspective, the "Blue Carbon" Project has gained significant importance in combating climate change with main initiatives in the mangrove ecosystem, seagrasses, and salt marshes. These ecosystems act as carbon stores, storing huge amounts of carbon stock in the form of CO_2 and reducing the emission of other greenhouse gases from the study system. Countries like the Philippines and Indonesia

Table 14.3 Impacts of rising sea level and its measures and constraints

Effects of rising sea level	Measures to be undertaken	Constraints
(a) Inundation of low-lying seashore or coastline areas	(a) Identifying and marking high-risk zones (b) Construction of flood barriers (c) Building floodproof homes (d) Erecting artificial dwelling structures	(a) Required very high maintenance (b) Costly technologies
(b) Alteration of coastal wetland areas	(a) Adoption of land resources development program (b) Reclamation of salty marshes (c) Sediment stabilization, beach cleanliness drive, etc. to be undertaken	(a) Construction cost is high (b) Participation of all the stakeholders is important
(c) Erosion and morphological anomalies	(a) Creation of naturally possible sea walls (b) Construction of erosion-preventing dams (c) Reforestation	(a) High maintenance cost (b) Recurrent appraisal required
(d) Intrusion of seawater (salty water) into surface or groundwater	(a) Establishing dams and preventive gates (b) Withdrawal or pumping of fresh marine	(a) Involvement of high cost (b) Advanced technology
(e) Drainage limitations	(a) Establishment of a pumping station (b) Alternate land use planning	(a) Higher cost requirement (b) Scientific interventions

Source Maribus (2010)

establish community-led approaches for improving or restoring of ecosystem that have significant influence on the mitigation of global climate change.

- Philippines: Community-driven mangrove restoration projects have rejuvenated degraded lands, improved fisheries, and created natural storm barriers.
- Indonesia: Initiatives like Mangroves for the Future (MFF) have successfully engaged local populations in planting mangroves, showcasing ecological and socioeconomic benefits.

In the Indian scenario, the collaborative approach of local communities along with NGO's (non-governmental organisations) in coastal mangrove restoration in Sundarbans, West Bengal, critically addresses the biodiversity loss and reduced soil erosion in such areas. Along with these remunerations, the local communities also sustained a stable livelihood in terms of fisheries and other ecological benefits. This model integrates equilibrium stability and economic benefits and mitigates climate change adversities. In the Odisha mangrove restoration model, the villagers' interest plays the crucial role in coastal ecosystem restoration, providing financial security along with a reduction in illegal activities such as logging and poaching, ensuring long-term ecosystem protection. In Delhi, an initiative named "Blue-Green Master Plan" has emerged as a suitable approach for mitigating and building resilience against the adverse effects of adversities of changes in climatic regimes. This strategy accelerates the establishment of blue-green infrastructure, which will focus on the improvement of both the water and vegetation complex. This aspect has shown positive attributes in combating environmental pollution, such as air pollution, water contamination, etc. The integration of both water and land systems allows the establishment of a symbiotic relationship among the elements, offering both social and environmental benefits. While in Gujarat, the developmental plans such as "Climate Change and Environment Action Plan" mainly focus on combating rising temperature and climate change. The two-way efforts of both the Gujarat government and the Department of Climate Change resulted in a reduction of the adverse effects of climate change. The main objectives of such action are to establish a climate resilient city promoting renewable energy use, water storage initiatives, improving air quality, etc. While in Bangalore, the indiscriminate establishment of industrial and infrastructural development elevated pollution in the environment. This approach establishes long-term resilience against climate change. Cities like Nagpur face resource scarcity problems, including water, problems related to flash floods and pollution of environmental properties, exacerbated by the ongoing urban exploitation and expansion. In such situations, plantation drives and approaches like vertical gardening on buildings and the establishment of green roofs can mitigate the adversities of climate change on coastal restoration (Dhyani et al., 2021). From these examples, the conclusion can be drawn that community engagement is of utmost importance for sustainability and effectiveness (Govindarajulu, 2014). The conservation and restoration of green spaces and degraded ecosystems will improve biodiversity health and also nurture the ecosystem, establishing ecological balance. The adoption of all possible pathways towards advancements of a sustainable future and mitigation of climate change will establish a futuristic paradigm for the existence of a harmonious ecological balance.

14.4 Constraints or Problems in Climate Change Mitigation in the Indian Scenario

The role of institutions in harnessing climate-related benefits in Indian cities should be properly acknowledged (Boyd & Ghosh, 2013; Chu, 2018; De Oliveira et al., 2015; Revi, 2008; Revi et al., 2016; Sethi & Mohapatra, 2013). The studies on constraints related to climate change mitigation identified various climate-related adversities faced due to global climatic variation phenomena, including temperature or thermal deviation, precipitation variation, rainfall departure, heatwaves, etc. These changes not only impacted human health but also agriculture and the economy.

Gupta et al. (2020) also described the challenges related to climate change on the vulnerability of agriculture, susceptibility of the environment, economic downfall, socio-economic discrimination, socioenvironmental hazards, health issues, etc. In the Indian context, there's a lack of scientific interventions in understanding the interaction between climate change and government policies. Various short-term actions undertaken without proper evaluation don't result in the desired objectives. The holistic approach of considering long-term benefits during the planning stage at the government level to the individual level is of utmost importance. The governmental initiatives, viz., Smart Cities Mission (SCM), Transit Oriented Development Policy (TODP), Unified Metropolitan Transport Authorities (UMTA), Green Urban Mobility Scheme (GUMS) etc., imposed their significant impact to withstand climate change aspects mainly in urban planning. The hierarchy of governance and central schemes must be in the proper direction to mitigate climate change action on different perspectives. India's diverse climatic conditions and socio-economic status pose difficulties in uniform policy intervention across the boundaries of the nation.

The involvement of various societal and environmental policies improves accountability of the policies toward the achievement of various interventions (Fig. 14.4). Recently, various nations have announced various schemes regarding reducing greenhouse gas emissions (GHG), i.e., net zero emissions. While doing so, numerous obstacles are present during the adoption of such measures, viz., remote sensing expertise, innovative sensor mechanisms, expertise in artificial intelligence, machine learning tools and technology, etc. Therefore, enhanced policymaking with partnerships at different levels should be given as utmost importance to overcome challenges related to climate change.

14.5 Conclusion

Community-based restoration initiatives have proven themselves to be an influential tool for the mitigation of climate change adversities and degradation of habitat. This initiative not only combats coastal erosion but also enhances fish populations, benefiting local fishermen's financial sufficiency. The program has fostered a true sense of ownership among the coastal communities, leading to increased vigilance and patriotism against illegal logging and poaching in protected surroundings. Community-based initiatives in land restoration are vital for mitigating the challenges posed by climatic variability in coastal areas as well as in coastal communities. The intensive learning as well as collaboration between communities, governments, and other private and cooperative sectors will be essential for creating adaptable and effective strategies in coastal land restoration. By acquiring knowledge from successful examples and overcoming existing challenges, we can safeguard our coastal ecosystems for future generations. Community engagement, government support, and long-term scientific research on climatic variabilities will enhance the development of successful models and protect these critical ecosystems for future

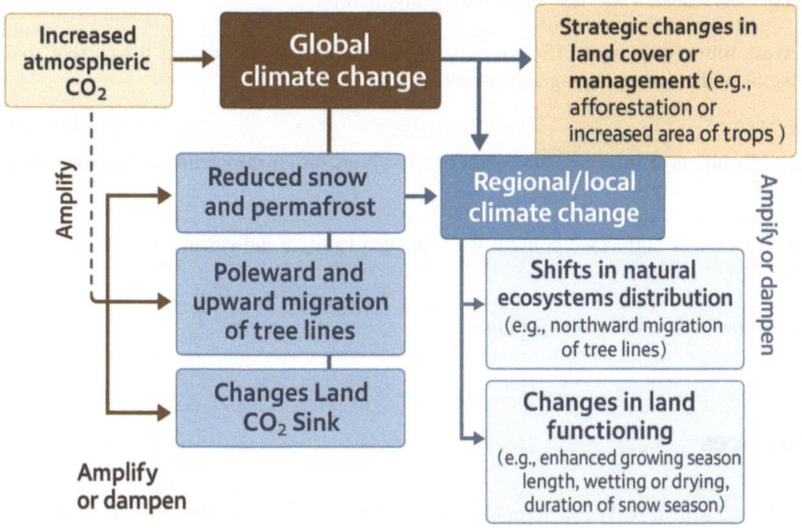

Fig. 14.4 Innovative ways for mitigation of climate change impact. *Source* Jia et al. (2019)

generations. Enhanced collaboration, capacity building, and proper monitoring and evaluation will help to combat the climate change adversities.

Other persistent constraints include:

- Policy fragmentation and short-term project cycles
- Weak coordination between institutions
- Limited access to funding and modern restoration techniques
- Rising pressures from tourism, aquaculture, and urbanization.

These findings highlight the importance of enabling frameworks and locally grounded strategies for scalable and sustainable outcomes. This chapter aligns with the broader theme of the book by offering evidence of how local communities can act as *frontline agents of land restoration* under the growing threat of climate change. It provides a lens to understand *adaptive capacity*, *resilience-building*, and *climate justice* through ground-level practices in India's vulnerable coastal regions issues that are central to the book's core narrative. The analysis illustrates that community-led restoration, though context-specific, holds valuable lessons for climate adaptation globally. The Indian case demonstrates how decentralized, inclusive approaches can restore ecological balance while improving socio-economic conditions. If supported by cohesive policy and long-term investment, these models can contribute significantly to national and global restoration targets, including the UN Decade on Ecosystem Restoration (2021–2030).

Authors' Contribution All authors have contributed equally to this work.

Funding The research was not supported by any funding.

Data Availability Statement The datasets during and/or analyzed during the current study are available from the corresponding author on reasonable request.

Potential Conflicts of Interest On behalf of all authors, the corresponding author states that there is no conflict of interest.

Informed Consent Informed consent was obtained from all individual participants included in the study.

AI Disclosure AI was not used for this chapter.

References

Adger, W. N., Arnell, N. W., & Tompkins, E. L. (2005). Successful adaptation to climate change across scales. *Global Environmental Change, 15*(2), 77–86.

Allen, C. D., Macalady, A. K., Chenchouni, H., Bachelet, D., McDowell, N., Vennetier, M., & Hogg, E. T. (2010). A global overview of drought and heat-induced tree mortality reveals emerging climate change risks for forests. *Ecological Management, 259*(4), 660–684.

Behera, M. D., & Roy, P. S. (2019). Pattern of distribution of angiosperm plant richness along latitudinal and longitudinal gradients of India. *Biodiversity and Conservation.* https://doi.org/10.1007/s10531-019-01772-1

Boden, J. B., & Driscoll, R. M. C. (1987). House foundations—A review of the effect of clay soil volume change on design and performance. *MunEngr, 4*, 181–213.

Boyd, E., & Ghosh, A. (2013). Innovations for enabling urban climate governance: Evidence from Mumbai. *Environment and Planning c: Government and Policy, 31*(5), 926–945. https://doi.org/10.1068/c12172

Bullock, P., Le Houerou, H., Hoffman, M. T., Rounsevell, M., Sehgal, J., Varallyay, G., Aidoud, A., Balling, R., Long-Jun, C., & Goulding, K. W. T. (1995). *Impacts adaptations and mitigation of climate change: Scientific and technical analyses* (pp. 173–189). Cambridge University Press (CUP).

Burkett, V. R., Suarez, M., Bindi, C., Conde, R., Mukerji, M. J., Prather, A. L., Clair, S., & Yohe, G. W. (2014). *Point of departure. Impacts, Adaptation and Vulnerability. Part A: Global and Sectoral Aspects, Contribution of Working Group II to the Fifth Assessment Report of the Intergovernmental Panel on Climate Change.* Cambridge University Press, Cambridge, United Kingdom and New York, NY, USA, pp. 169–194.

Chatterjee, S., & Roy, S. (2021). A complete study on the costliest super Cyclone Amphan (May 2020) with its devastating impact on West Bengal, India. *Remote Sensing in Earth Systems Sciences, 4*, 249–263. https://doi.org/10.1007/s41976-022-00066-5

Chu, E. K. (2018). Urban climate adaptation and the reshaping of state–society relations: The politics of community knowledge and mobilisation in Indore, India. *Urban Studies, 55*(8), 1766–1782. https://doi.org/10.1177/0042098016686509

DeNicola, D. M., & Stapleton, M. G. (2002). Impact of acid mine drainage on benthic communities in streams: The relative roles of substratum vs. aqueous effects. *Environmental Pollution, 119*, 303–315.

De Oliveira, R., Lal, R., Aguiar, D. A., De Figueiredo, E. B., Perillo, L. I., Adami, M., Rudorff, B. E. T., & La Scale, N. (2015). Greenhouse gas balance from cultivation and direct land use change of recently established sugarcane (*Saccharum officinarum*) plantation in south-central

Brazil. *Renewable and Sustainable Energy Review, 52*, 547–556. https://doi.org/10.1016/j.rser.2015.07.137

Dent, D. (1986). *Acid sulphate soils: A baseline for research and development* (pp. 196–200). ILRI Publication Netherlands.

Dhyani, S., Singh, A., Gujre, N., & Joshi, R. K. (2021). Quantifying tree carbon stock in historically conserved Seminary Hills urban forest of Nagpur, India. *Acta Ecologica Sinica, 41*(3), 193–203. https://doi.org/10.1016/j.chnaes.2021.01.006

Dinelli, E., Lucchini, F., Fabbri, M., & Cortecci, G. (2001). Metal distribution and environmental problems related to sulfide oxidation in the Libiola copper mine area (Ligurian Apennines, Italy). *Journal of Geochemical Exploration, 74*, 141–152.

Douglas, A. E. (1994). Symbiotic interactions. *Oxford University Press, London, 74*, 743–745.

Feliciano, D., Recha, J., Ambaw, G., MacSween, K., Solomon, D. & Wollenberg, E (2022) Assessment of agricultural emissions, climate change mitigation and adaptation practices in Ethiopia. *Climate Policy*, 1–18.

Fenton, G., & Helyar, K. (2007). Soil acidification. In P. E. V. Charman & B. W. Murphy (Eds.), *Soils: Their properties and management* (3rd ed.). Oxford University Press.

Filho, W. L., Azeiteiro, U. M., Balogun, A. L., Setti, A., Mucova, S. A. R., Desalegn, Y. A., Totin, E., Lydia, A. K., Felix, K. K., & Oguge, N. (2021). The influence of ecosystems services depletion to climate change adaptation efforts in Africa. *Science of the Total Environment, 779*(5), Article 146414. https://doi.org/10.1016/j.scitotenv.2021.146414

Govindarajulu, D. (2014). Urban green space planning for climate adaptation in Indian cities. *Urban Climate, 10*(1), 35–41. https://doi.org/10.1016/j.uclim.2014.09.006

Gupta, A. K., Negi, M., Nandy, S., Kumar, M., Singh, V., Valente, D., Irene, P., & Pandey, R. (2020). Mapping socio-environmental vulnerability to climate change in different altitude zones in the Indian Himalayas. *Ecological Indicators, 109*, Article 105787. https://doi.org/10.1016/j.ecolind.2019.105787

Haibach, H., & Schneider, K. (2013). The politics of climate change: Review and future challenges. In O. C. Ruppel, Roschmann Ch., & K. Ruppel-Schlichting (Eds.), *Climate change: International law and global governance. Volume II: Policy, diplomacy and governance in a changing environment*. Nomos Verlagsgesellschaft mbH.

Hussain, S., & Hoque, R. R. (2022a). Ecological and natural-based solutions as green growth strategies for disaster and emergency management of air pollution extremes. In *Extremes in atmospheric processes and phenomenon: Assessment, impacts and mitigation* (pp. 369–395). Springer Nature Singapore.

Hussain, S., & Hoque, R. R. (2022). Biomonitoring of metallic air pollutants in unique habitations of the Brahmaputra Valley using moss species—Atrichumangustatum: Spatiotemporal deposition patterns and sources. *Environmental Science and Pollution Research International, 29*, 10617–10634.

IPCC. (2013). Summary for policymakers. *Clim Chang Phys Sci Basis Contrib Work Gr I Fifth Assess Rep.*

Jain, M., Saxena, P., Sharma, S., & Sonwani, S. (2021). Investigation of forest fire activity changes over the central India domain using satellite observations during 2001–2020. *GeoHealth, 5*, Article e2021.

Jenkins, M., & Schaap, B. (2018). *Forest Ecosystem Services. Background Analytical Study, 1. United Nations Form on Forests.*

Jha, A. K., Miner, T. W., & Stanton-Geddes, Z. (2013). *Building urban resilience: Principles, tools, and practice.* World Bank Publications.

Jia, G. E., Shevliakova, P., Artaxo, N., De Noblet-Ducoudré, R., Houghton, J., House, K., Kitajima, C., Lennard, A., Popp, A., Sirin, R., Sukumar, L.& Verchot, K. (2019). Land–climate interactions. In: *Climate Change and Land: An IPCC special report on climate change, desertification, land degradation, sustainable land management, food security, and greenhouse gas fluxes in terrestrial ecosystems.*

Krishnan, R., Sanjay, J., Gnanaseelan, C., Mujumdar, M., Kulkarni, A., & Chakraborty, S. (2020). *Assessment of Climate Change Over the Indian Region: A Report of the Ministry of earth Sciences (MOES)*, Government of India, Springer Nature, Cham, p. 226.

Klingenfeld, D. & Schellnhuber, H. J. (2012). *Climate change as a global challenge–And its implications for knowledge generation and dissemination.*

Lal, R. (1994). Sustainable land use systems and soil resilience. In D. J. Greenland & I. Szabolcs (Eds.), *Soil resilience and sustainable land use* (pp. 41–67). CAB International, Wallingford, UK.

Lee, P. K., Kang, M. J., Choi, S. H., & Touray, J. C. (2005). Sulfide oxidation and the natural attenuation of arsenic and trace metals in the waste rocks of the abandoned Seobo tungsten mine, Korea. *Applied Geochemistry, 20*, 1687–1703.

Levings, C. D., Varela, D. E., Mehlenbacher, N. M., Barry, K. L., Piercey, G. E., Guo, M., & Harrison, P. J. (2005). Effect of an acid mine drainage effluent on phytoplankton biomass and primary production at Britannia Beach, Howe Sound, British Columbia. *Marine Pollution Bulletin, 50*(12), 1585–1594.

Lin, C., & Lin, J. (2003). Heavy metals in a sulfidic mine spoil: Fractions and column leaching. *Pedosphere, 13*, 75–80.

Lu, W., Ma, Y., & Lin, C. (2011). Status of aluminium in environmental compartments contaminated by acidic mine water. *Journal of Hazardous Materials, 189*, 700–709.

Maji, S., & Sonwani, S. (2022). *Nature of Sand and Dust Storm in South Asian Region: Extremities and Environmental Impacts in Extremes in Atmospheric Processes and Phenomenon: Assessment, Impacts and Mitigation.* Springer Nature Singapore, Singapore, pp. 113–139.

Maribus. (2010). *World ocean review: Living with the oceans.* Retrieved from https://worldoceanreview.com

Mirza, M. M. Q. (2003). Climate change and extreme weather events: Can developing countries adapt? *Climate Policy, 3*, 233–248.

Nordstrom, D. K. (2009). Acid rock drainage and climate change. *Journal of Geochemical Exploration, 100*, 97–104.

Pereira, H. C. (1993). Food production and population growth. *Land Use Policy, 10*, 187–190.

Ramanathan, V., & Carmichael, G. (2008). Global and regional climate changes due to black carbon. *Natural Geoscience, 1*, 221–227. https://doi.org/10.1038/ngeo156

Revi, A. (2008). Climate change risk: an adaptation and mitigation agenda for Indian cities. *Environment and Urbanization, 20*, 207–229. https://doi.org/10.1177/0956247808089157

Revi, A., Gajjar, S. P., Basu, R., Jain, G., & Bazaz, A. B. (2016). *Cities on a finite planet*. Routledge.

Sahu, L. K. & Saxena, P. (2015). High time and mass resolved PTR-TOF-MS measurements of VOCs at an urban site of India during winter: Role of anthropogenic, biomass burning, biogenic and photochemical sources. *Atmospheric Research, 164*, 84–94. https://doi.org/10.1016/j.atmosres.2015.04.021

Saxena, P., Shukla, A., & Gupta, A. K. (2022). *Extremes in atmospheric processes and phenomenon: Assessment, impacts and mitigation.* Springer.

Saxena, P., & Sonwani, S. (2019). Secondary criteria air pollutants: environmental health effects. In *Criteria Air pollutants and their impact*.

Sethi, M., & Mohapatra, S. (2013) Governance framework to mitigate climate change: challenges in urbanizing India. In *Governance approaches to mitigation of and adaptation to climate change in Asia* (pp. 200–230). Palgrave Macmillan UK.

Singh, H. S. (2003). Marine protected areas in India. *Indian Journal of Marine Sciences, 32*(3), 226–233.

Singh, P., & Yadav, D. (2021). Link between air pollution and global climate change. In *Global Climate Change (Elsevier)*, 79–108. https://doi.org/10.1016/B978-0-12-822928-6.00009-5

Sonwani, S. (2016). Source apportionment of polycyclic aromatic hydrocarbons in the urban atmosphere of South Delhi, India. In *2nd International Conference on Atmospheric Dust-DUST2016, Pro Science Conference Proceedings of Scientific Events (Bari)*, pp. 111–116.

Sonwani, S., Hussain, S., & Saxena, P. (2022). Air pollution and climate change impact on forest ecosystems in Asian region–a review. *Ecosystem Health and Sustainability, 8*, 2090448.

Sonwani, S., & Kulshrestha, U. C. (2019). PM 10 carbonaceous aerosols and their real-time wet scavenging during monsoon and non-monsoon seasons at Delhi, India. *Journal of Atmospheric Chemistry, 76*, 171–200.

Sonwani, S., Madaan, S., Arora, J., Suryanarayan, S., Rangra, D., Mongia, N., Vats, T., & Saxena, P. (2021). Inhalation exposure to atmospheric nanoparticles and its associated impacts on human health: A review. *Frontiers in Sustainable Cities, 3*, Article 690444.

Sonwani, S., Saxena, P., & Khillare, P. S. (2022). Profile of atmospheric particulate PAHs near busy roadway in tropical megacity, India. *Inhalation Toxicology, 34*, 39–50.

Sonwani, S., Saxena, P., & Shukla, A. (2021b). Carbonaceous aerosol characterization and their relationship with meteorological parameters during summer monsoon and winter monsoon at an industrial region in Delhi, India. *Earth Space Science 8*, e2020EA001303.

Sonwani, S., Yadav, A., & Saxena, P. (2021c). Atmospheric brown carbon: a global emerging concern for climate and environmental health. *Managing Contaminants of Emerging Concern (CEC) Environment, 1*, 225–247.

Ummenhofer, C. C., & Meehl, G. A. (2017). Extreme weather and climate events with ecological relevance: A review. *Philosophical Transactions of the Royal Society B: Biological Sciences, 372*, 20160135.

Venkatraman, K., & Wafar, M. (2005). Coastal and marine biodiversity of India. *Indian Journal of Marine Science, 34*(1), 57–75.

Wild, A. (1993). Soils and environment: An introduction. *Cambridge University Press, Cambridge.* https://doi.org/10.1017/CBO9780511623530

Chapter 15
Soil Conservation Techniques and Sustainable Land Management

Santanu Kundu, Suthari Vijay Kumar, Vijay Pooniya, Boppidi Manasa Reddy, Nisani Hussain Basha, Kajal Das, Anamika Barman, Suman Dutta, Maksud Hasan Shah, and Pooja Erapuram

Abstract Land is one of the most valuable resources and the foundation of human civilization. Although the terms "soil" and "land" are often used synonymously, "soil" is typically used to describe the medium, while "land" indicates the resource. Nowadays, owing to the population pressure, deforestation, intensive farming with injudicious use of agrochemicals, climate change, and improper land use patterns, productive land is getting degraded. According to the UNEP report, more than 25% of global land has been degraded, and around 40% of global agricultural land has been seriously degraded, posing a threat to the well-being of 3.2 billion people globally. If current trends continue, 95% of the land area could become degraded within the next 30 years. In this context, soil conservation techniques play a pivotal role in sustainable land management by mitigating soil degradation and enhancing environmental resilience. Therefore, soil conservation through cultural (*e.g.*, carbon

S. Kundu (✉) · V. Pooniya · A. Barman
Division of Agronomy, ICAR-Indian Agricultural Research Institute, New Delhi, India
e-mail: santanukundu120@gmail.com

S. V. Kumar
Department of Agronomy, Uttarbanga Krishi Viswavidyalaya, Coochbihar, West Bengal, India

B. M. Reddy · N. H. Basha · P. Erapuram
Department of Agronomy, Professor Jayshankar Telengana Agricultural University, Hyderabad, Telangana, India

K. Das
Division of Germplasm Evaluation, ICAR-National Bureau of Plant Genetic Resources, New Delhi, India

S. Dutta
Department of Genetics and Plant Breeding, School of Agriculture and Rural Development, Ramakrishna Mission Vivekananda Educational and Research Institute, Belur Math, Howrah, India

M. H. Shah
Department of Agronomy, Bidhan Chandra Krishi Viswavidyalaya, Mohanpur, Nadia, West Bengal, India

U. Chatterjee et al. (eds.), *Eco-Resilience: Climate Change, Land Degradation and Sustainable Solutions*, Sustainability Solutions,
https://doi.org/10.1007/978-3-032-00708-7_15

farming, conservation agriculture, organic farming, integrated nutrient management), mechanical (*e.g.*, construction of artificial structure), and chemical (*e.g.*, amelioration of chemically degraded soil) methods holds paramount importance in this contemporary situation to protect our soil health, biodiversity, and sustainable production system. The main aim of soil conservation is to protect the fertile soil from erosion, and to cease the depletion of soil organic carbon, desertification, soil acidification, salinization, etc. In this chapter, different advanced soil conservation techniques based on the multidisciplinary approach and ongoing research have been identified in relation to the anthropogenic and natural causes of soil/land degradation in the climate change scenario.

Keywords Land degradation · Conservation agriculture · Carbon farming · Chemical intervention · Alternate land use system

15.1 Introduction

The world's cultivable land area is a vital resource for food production and indispensable to serve food for the growing population. In 2021, 4.79 billion ha of global land was under agriculture, and cropland (1.58 billion ha) occupied one-third of the total agricultural land (FAO, 2023). The global population is growing at a rapid pace of nearly 1 billion every 12 years, and is expected to reach around 9.8 billion by 2050 (UNEP, 2007). To address the challenge of the ever-increasing population, sustainability in the agricultural production system and in the total environment should be maintained. Land is one of the important components of agriculture. Overexploitation of land resources in an injudicious manner exacerbates land degradation, and threatens agricultural sustainability and productivity (Mahadevan et al., 2024).

The term "Land degradation" refers to the soil quality caused by various factors, including soil erosion, fertility depletion, salinization, desertification, etc. As per the estimation, almost 1/4th of the global land area has been degraded and causing direct and indirect problems to the farmers and common people. If the current trend of land degradation is not checked properly, ~ 95% of the total world land area will be degraded in the next 30 years (UNEP, 2024). The ecology and society are significantly impacted by land degradation. The most direct effect is the inferior agricultural output due to decreased soil fertility, which jeopardizes food security. It indicates an ensuing disaster in farmers' livelihoods, especially in rural areas.

In response to the multifaceted challenges posed by land degradation in the context of climate change, the concept of Land Degradation Neutrality (LDN) came into the picture. For instance, India has committed to restoring 26 million ha of degraded land by 2030 to create an additional carbon sink (~ 3 million tonnes of CO_2 equivalent) (PIB, 2021). To restore the degraded land and to restrict the process of land degradation, alteration of the existing system with an improved package and practices is highly intended. The major cause of land degradation in the climate change scenario is poor soil fertility and structure due to the loss of soil organic carbon and essential plant nutrients by the process of soil erosion and intensive cropping with conventional methods. Along with these reasons, soils with some improper chemical properties are also considered degraded soil. To overcome the process of degradation, there are some promising cultural, mechanical, and chemical methods that are to be implemented by replacing the conventional methods.

This chapter aims to explore the different causes of land degradation in the climate change scenario. On the other hand, in this chapter, effective cultural, mechanical, and chemical interventions have been identified to restore degraded land by enhancing soil health and building up eco-resilience under changing climatic conditions. The rationale lies in supporting global efforts to achieve LDN by advancing sustainable land management strategies that ensure long-term food security while maintaining environmental sustainability.

15.2 Material and Methods

The methodological approach (Fig. 15.1) of this chapter involved an initial phase of extensive literature review, encompassing research articles and reviews focused on land degradation and soil conservation practices. This foundational step facilitated the identification of key factors contributing to land degradation, which in turn guided the selection of the most relevant and ecologically sound soil conservation techniques. These identified techniques were subsequently classified into three distinct categories: cultural methods/non-chemical interventions, mechanical methods, and chemical interventions. The culmination of this review process involved the practical implementation and subsequent evaluation of these categorized soil conservation techniques to assess their effectiveness and implications in mitigating environmental degradation.

15.3 Result and Discussion

15.3.1 Ways and Means of Land Degradation

Land degradation is caused by a number of man-made and natural forces that work together to reduce the productivity of the soil. From a physical, chemical, and biological point of view, there are different ways/types of land degradation (Fig. 15.2). The following are the main causes of land degradation.

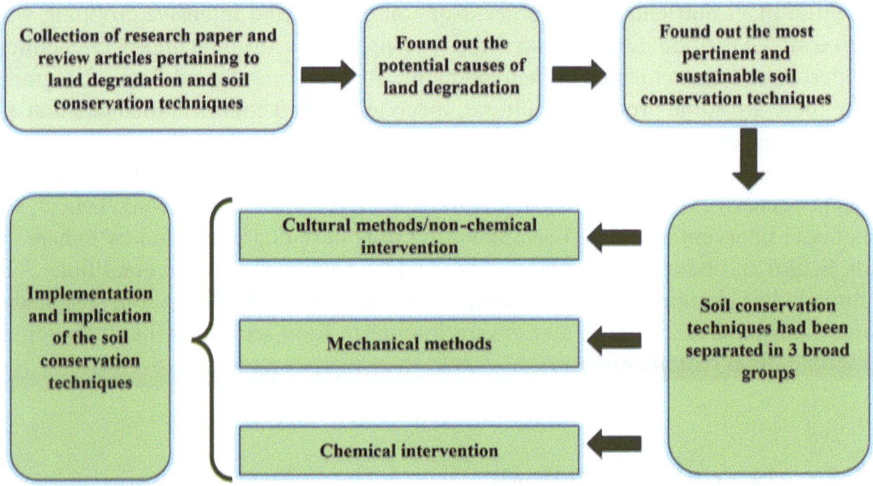

Fig. 15.1 Methodological flowchart

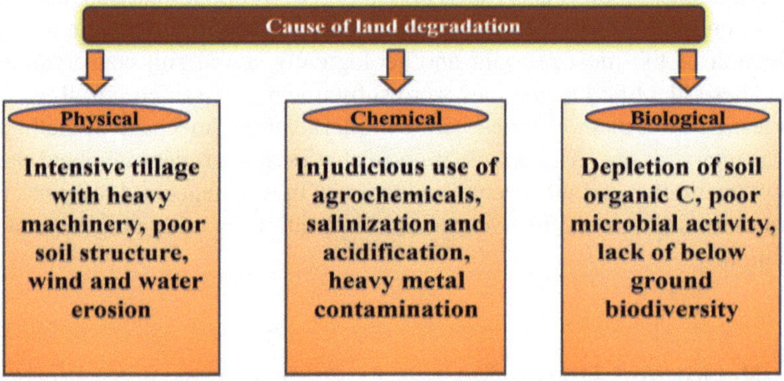

Fig. 15.2 Different causes of land degradation

15.3.1.1 Water Erosion

One of the most common types of land degradation is water erosion, which impacts agricultural land in a variety of agroclimatic zones, especially in areas with sloping terrain. It starts when rainfall hits exposed soil surfaces, dislodging soil particles into the air and dissolving soil aggregates. Runoff water further detaches and transfers soil as it moves down slopes, resulting in the loss of fertility and topsoil. This type of degradation is especially troublesome in places with a lot of rainfall and little vegetation. Numerous factors, including both natural (such as rainfall, temperature, and topography) and man-made (like deforestation, cultivation, and grazing), influence water erosion (Bhat et al., 2019; Massa et al., 2012).

15.3.1.2 Wind Erosion

Wind erosion, a natural phenomenon that typically happens in dry situations with high wind velocity or on bare soils when plant biomass is sparse, has affected almost one-third of the earth's land area over the previous centuries (Weinan & Fryrear, 1996). Soil loss is accelerated by anthropogenic pressures, including monoculture systems, overharvesting of vegetation, deforestation, overgrazing rangelands, or leaving farmed areas fallow for an extended period of time (Chen et al., 2014; He et al., 2006). Wind erosion primarily affects arid and semi-arid regions, where the soil is loose and exposed.

15.3.1.3 Soil Compaction

It is a process of reduction of soil pore space, and consequently, the soil aeration is negatively affected, and soil bulk density becomes higher. In compacted soil, plant growth is hampered due to poor nutrient availability and root growth. Soil compaction is mainly caused by using heavy machinery. Puddling and excessive ploughing at improper soil moisture conditions also cause subsoil compaction (Srinivasarao et al., 2021).

15.3.1.4 Acidic Soils

Acidic soils develop in humid regions where excessive rainfall facilitates the leaching of basic cations, resulting in a lower pH and the loss of essential nutrients necessary for plant growth. Intensive agriculture enhances soil acidification through rapid and unconscious fertilization without proper soil testing. Such soils are typically less fertile and less suitable for agriculture without chemical amendments. In India, ~ 48 M ha of arable land is affected by acid soil (Mishra et al., 2020).

15.3.1.5 Salt-Affected Soils

Salt-affected soils are characterized by an excess of soluble salts or exchangeable sodium, which adversely affects plant growth and agricultural productivity. These soils are typically found in arid, semi-arid, and coastal areas, where high evaporation rates lead to the accumulation of salts. Depending on the type and concentration of salts, these soils can be classified as saline, sodic, or saline-sodic. Being a second highest arable land holder in the world, India is also facing the problem of salt-affected soil in ~ 6.74 million ha of lands. In India, it is estimated that almost 50% of the arable land will be degraded by salt by 2050, with an annual increment of 10% (Kumar & Sharma, 2020). Excess use of fertilizers and irrigation water also causes the secondary salinization (Srinivasarao et al., 2021).

15.3.1.6 Soil Pollution with Chemical Fertilizers and Pesticides

In order to get higher productivity, excessive use of chemical fertilizers and pesticides jeopardizes soil's natural ecological cycles and biodiversity. As a result, partial factor productivity and resilience of the system are deteriorated. From 2000 to 2021, the use of pesticides in the world increased by ~ 62% and in 2021, chemical fertilizer use was recorded as 195 million tonnes of nutrients, and N alone accounted for 56% of the total nutrients (FAO, 2023).

15.3.1.7 Deterioration of Soil C and Biodiversity

Soil carbon and biodiversity are determinants of the soil's biological health. Continuous use of chemical fertilizers without using organic amendments or crop residues results in deterioration of soil organic C. Insufficient levels of soil organic C deteriorate soil biological activity and nutrient cycling processes. Excessive ploughing is one of the main causes of soil C depletion due to the high rate of C mineralization through microbial intervention. A depleted level of soil C is an impediment for desirable soil structure (Siedt et al., 2021).

Together, these factors contribute to the widespread degradation of land resources and threaten agricultural sustainability, food security, and environmental health in a changing climate scenario. To address these problems, different soil conservation techniques could be adopted to restore the degraded land and to cease the further process of degradation.

15.3.2 Sustainable Land Management with Different Soil Conservation Techniques and Its Implications

There are different approaches to soil conservation, and it can be broadly classified into cultural, physical/mechanical, and chemical methods. Based on the location and type of degradation, the appropriate method of soil conservation is followed. However, different types of soil conservation techniques could be followed at the same time to hasten the process of rejuvenation of degraded land.

15.3.2.1 Cultural Methods/Non-chemical Intervention

- **Carbon Farming**

It refers to a range of land use and land management practices designed to reduce emissions from farming activities, or sequester carbon in natural sinks such as soil and vegetation (Fig. 15.3) (Saikanth et al., 2023). Through C sequestration, and by protecting the soil organic C, soil physical, chemical, and biological health can be improved. Soil C could be stabilized by three main mechanisms, viz., physical inaccessibility, chemical interaction, and biological intervention.

Physical inaccessibility refers to the protection of soil organic carbon (SOC) through stable macro- and micro-aggregate formation and its physical location at depth within the soil profile. This physical barrier limits the accessibility of organic matter to microbial decomposers, thereby reducing the rate of C mineralization. *Chemical interaction* involves the stabilization of SOC through interactions with the mineral components of the soil, specifically silt, clay particles, and various cations. These interactions create organo-mineral complexes that are less susceptible to microbial breakdown. *Biological intervention* encompasses molecular transformations and the incorporation of carbon into the structural components of living cells within the soil ecosystem. This pathway highlights the role of soil biodiversity and microbial activity in both sequestering and transforming carbon (Fig. 15.4), and these mechanisms are facilitated by different farming practices such as conservation agriculture, organic farming, crop diversification, integrated nutrient management, etc. Improvement of soil C not only ascertains the food security but also reduces the impact of climate change due to GHG emissions. In the 21st season of the United Nations Framework Convention on Climate Change, it was unanimously proposed

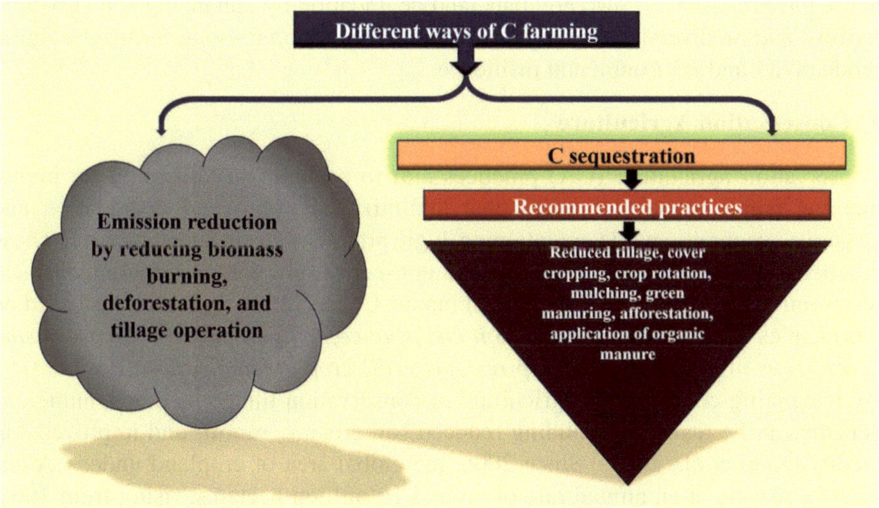

Fig. 15.3 Two major aspects of C farming

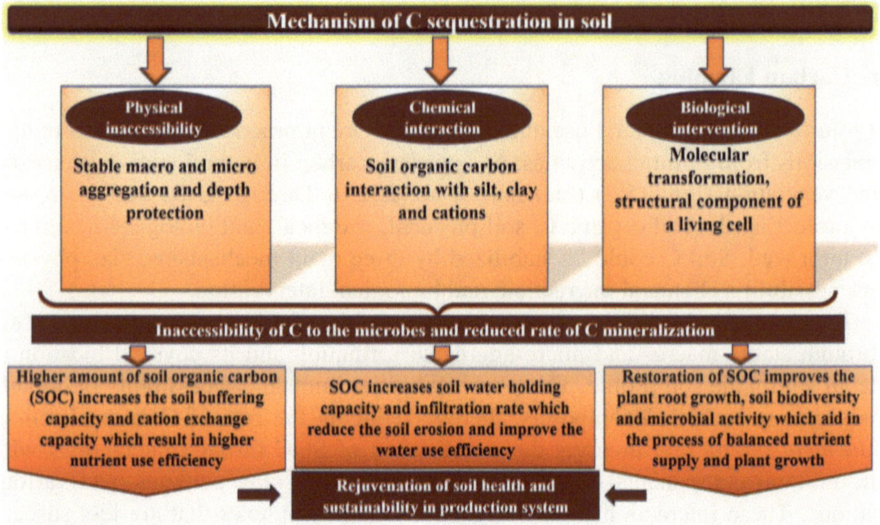

Fig. 15.4 Mechanism of C sequestration and its implication

that each year 0.4% of the existing topsoil C stock (~ 3.4 GT C year^{-1}) could be enhanced by capturing atmospheric C, which will nearly compensate for the effect of anthropogenic CO_2 emission. Improved technology in agriculture can fulfil the intended target even more than that; in addition, soil fertility enhancement and yield increment are the concomitant effects of soil C buildup (Fig. 15.4) (Soussana et al., 2019). Carbon farming directly addresses climate change by capturing atmospheric carbon dioxide and storing it in soils and vegetation, thereby mitigating greenhouse gas emissions. It also combats land degradation by enhancing soil structure, fertility, and biodiversity, making it a sustainable solution for long-term agricultural productivity and environmental resilience.

- **Conservation Agriculture**

Conservation agriculture (CA) practices aim to promote sustainability by implementing management techniques that minimize environmental degradation and conserve resources, while maintaining high productivity and profitability. These practices build up an ecologically sound agro-ecosystem with minimal mechanical intervention and careful use of external inputs. Conservation agriculture is based on 3 principles: *(i) no-till or minimal soil disturbance; (ii) permanent soil cover using cover crops or mulches with crop residues; (iii) crop diversification* (FAO, 2015). By integrating conservation agriculture or conservation tillage methods, numerous benefits can be realized, including reduced soil erosion, runoff, and improved soil quality (Xiao et al., 2021). Since 2008, the global area of cropland under CA has been increasing at an annual rate of around 10 million hectares, rising from 106.5 million hectares in 2008 to 205.4 million hectares in 2018 (Kassam et al., 2022).

The major advantage of following conservation agriculture is the reduction of soil erosion due to permanent soil cover, reduced runoff, and greater infiltration (Erenstein, 2002; Roose & Barthes, 2001). Almost 97% of the erosion was reduced in the US after the adoption of conservation agriculture (USDA, 2013). No other technique is as effective as no-tillage for reducing soil erosion and making food production highly sustainable (Baker et al., 1996). Tillage often accelerates soil erosion, which leads to organic matter depletion. By reducing tillage, soil erosion rates will be reduced, aggregate stability of soil will be improved, and a healthier soil profile can be maintained. Cover crops such as green manures have the potential to improve soil strength, aggregation, and infiltration by reducing soil bulk density (Singh et al., 2007). Retention of crop residues on the soil surface and minimal soil disturbance can improve soil physical conditions and can reduce soil erosion by ~ 80% (Du et al., 2022). The soil cover acts as a barrier against water and wind erosion and improves the infiltration. In degraded landscapes, maintaining the soil cover is an immediate solution to halt erosion and simultaneously foster long-term soil health. The crop rotation with different rooting patterns, along with minimal disturbance of soil in no-till systems, encourages an elaborate network of macropores and root channels in the soil. This helps in the process of infiltration and ultimately reduces soil erosion. Conservation agriculture (CA) notably influences the soil organic carbon (SOC) storage (Table 15.1). CA improves SOC through reduced tillage/no-till (NT) systems. By reducing soil disturbance, NT helps to conserve soil macroaggregates, which protect organic matter from microbial degradation. As a result, organic matter is more securely retained in the soil, which minimizes soil loss and increases organic material turnover time (Bayer et al., 2006; Beare et al., 1994). In addition to tillage reduction, residue management is critical for improving SOC in CA systems. Crop residues on the soil surface provide a consistent source of organic matter, which contributes to the accumulation of SOC (Duiker & Lal, 2002). However, in places with low fertility or limits on plant development, the benefits of residue retention may be reduced, necessitating the use of additional nutrients to maintain SOC accumulation (Lal, 2015).

Table 15.1 Relation between conservation agriculture and change in soil organic carbon (SOC)

Cropping system	Soil depth	Rate of increment in SOC (kg ha^{-1} year^{-1})	References
Cereal-based system	Up to 30 cm	~ 450	Franzluebbers (2010)
Cereal-based system	Up to 50 cm	~ 510	González-Sánchez et al., (2012)
Cereal-based system	Average 34 cm	~ 440	Aguilera et al. (2013)
Cereal and legume-based rotations	Up to 40 cm	~ 300	Francaviglia et al. (2017)
Cereal and pulse-based system	Up to 100 cm	~ 250	Du et al. (2017)

Furthermore, crop rotation within CA systems, particularly the inclusion of legumes, improves SOC levels. Legumes supply nitrogen to the soil, increasing fertility and biomass production both above and below ground. These rotations, together with the root contributions, ensure the enhancement of SOC stocks. Finally, conservation agriculture boosts soil fertility, and supports sustainable agricultural systems.

Conservation agriculture (CA) has a notable influence on soil biological activity (Table 15.2). The increment of soil organic carbon (SOC) in CA systems provides a key energy source for soil microorganisms, resulting in higher microbial biomass than in conventional agricultural systems (González-Chávez et al., 2010). Crop residue retention and the absence of tillage increase microbial abundance at the soil surface, especially fungal communities (Helgason et al., 2010). These modifications promote a more diversified microbial population, which is necessary for a variety of soil functions, including nutrient cycling, plant growth promotion, and disease suppression (Peters et al., 2003). Higher microbial biomass and diversity in CA systems indicate improved soil health, which in turn promotes soil aggregation, moisture retention, and temperature regulation (Zhang et al., 2018). These improvements collectively contribute to the restoration process of degraded land.

Hence, Conservation agriculture plays a crucial role in addressing climate change by enhancing soil organic carbon sequestration, thereby reducing atmospheric CO_2 levels. It also mitigates land degradation through improved soil structure, fertility, and erosion control, offering a sustainable and resilient approach to long-term agricultural productivity.

- **Integrated Nutrient Management**

Integrated nutrient management (INM) indicates the conjoint application of chemical fertilizers and organic manure/organic source of nutrients to increase crop productivity and soil fertility *vis-à-vis* SOC and essential nutrients. It influences the soil physiochemical and biological properties and helps in soil conservation (Table 15.3).

Table 15.2 Soil MBC under conservation (CA) and conventional (CT) agricultural practices

Cropping system	Location	Difference in MBC between CA and CT (CA − CT) (ppm)	% change in MBC	References
Wheat-based system	USA	274	~ 94 (25 years)	González-Chávez et al., (2010)
	UK	225	~ 53 (7 years)	Mangalassery et al. (2015)
Soybean-based system	India	27	~ 8.3 (3 years)	Somasundaram et al. (2019)
Maize-legume rotation	India	86	~ 27 (3 years)	
Maize-based system	China	34	~ 20 (7 years)	Zhang et al. (2012)

INM improves soil physical parameters through the combination of organic and inorganic inputs. According to research, combining NPK fertilizers with organic manures, biofertilizers, and lime increases water-stable aggregates (WSA), moisture retention, and infiltration rate while decreasing bulk density. Organic supplements such as farmyard manure (FYM) and green manure enhance soil aggregation, structure, and moisture retention. The buildup of organic carbon in macroaggregates improves soil health (Bhattacharyya et al., 2007). The application of organic manure in combination with the recommended fertilizer dosage significantly enhances crop yield by improving nutrient use efficiency, synchronizing nitrogen availability with crop demand, and reducing nitrogen losses through denitrification and leaching. The balanced use of mineral fertilizers, organic manures, biofertilizers, and micronutrients maximizes available NPK and promotes long-term soil health and crop yields (Bamboriya et al., 2023; Jat et al., 2013). Furthermore, the combined application of NPK + FYM resulted in significant increases in organic carbon, mineralizable nitrogen, and ammonium-N, with notable improvements in nutrient uptake in various cropping systems (Kusro et al., 2014). INM practices, particularly in wheat–maize systems, enhanced soil enzymatic activity, thereby improving soil health and soil biological processes (Kannan et al., 2013; Liang et al., 2014).

Table 15.3 Effect of INM on soil physiochemical and biological properties

<table>
<tr><th>Crop/cropping system</th><th colspan="2">Impact of INM on soil health</th><th>References</th></tr>
<tr><td colspan="4">Soil physical properties</td></tr>
<tr><td rowspan="2">Wheat-based cropping system</td><td colspan="2">RDF + FYM @ 5 t ha^{-1} significantly reduced bulk density</td><td>Kusro et al. (2014)</td></tr>
<tr><td colspan="2">Increased WSA, total porosity, and water holding capacity</td><td>Bhattacharyya et al. (2007)</td></tr>
<tr><td colspan="4">Soil chemical properties</td></tr>
<tr><td>Wheat-based system</td><td colspan="2">Significantly increased SOC, total N, and enzymatic activities</td><td>Liang et al. (2014)</td></tr>
<tr><td>Soybean-based system</td><td colspan="2">RDF with biofertilizers resulted in improved productivity, soil fertility, and nutrient balance</td><td>Gharpinde et al. (2014)</td></tr>
<tr><td colspan="4">Soil biological properties</td></tr>
<tr><td>Maize-based system</td><td colspan="2">NPK + Biofertilizers (Azotobacter + PSB) + 0.5% Zn foliar spray resulted in higher microbial biomass C, N, P, and higher soil enzymatic activity</td><td>Bamboriya et al. (2023)</td></tr>
<tr><td colspan="4">C sequestration</td></tr>
<tr><td rowspan="2">Rice–wheat system (15 years)</td><td>C sequestration</td><td>NPK + FYM: 5.31 Mg C ha^{-1}
NPK: 3.57 Mg C ha^{-1}</td><td rowspan="2">Choudhary et al. (2018)</td></tr>
<tr><td>C sequestration rate</td><td>NPK + FYM: 0.46 Mg C ha^{-1} yr^{-1}
NPK: 0.37 Mg C ha^{-1} yr^{-1}</td></tr>
</table>

Thus, INM mitigates climate change by enhancing carbon sequestration and nutrient efficiency, while simultaneously addressing environmental degradation and agricultural productivity, and ecosystem resilience.

- **Organic Farming**

Organic farming is the agriculture where the use of chemical fertilizers and pesticides is strictly prohibited. According to IFOAM, organic farming should be based on the principle of the ecological cycle and system. As a result, above and below-ground biodiversity in organic farming is very prominent. Organic farming also enhances the health of the soil. A healthy soil can produce healthy plants and healthy plants can produce healthy food, which will protect the health and well-being of the present and future generations. There is a notion that organic farming yields are less than those of conventional farming; it is not true in all cases, the yield gap in organic farming is highly contextual and system specific (Seufert et al., 2012). Balanced supply of nutrients through organic fertilizer is the biggest challenge. It has been substantiated that a long-term organic farming system is highly comparable with a conventional system (Das et al., 2024), additionally, organic farming renders other significant ecosystem services. Thus, environmental sustainability, food security could both be achieved by organic farming. Implications of long-term organic farming systems on soil fertility improvement, C sequestration, and soil biodiversity have been pointed out in Table 15.4.

Hence, organic farming contributes to climate change mitigation by enhancing soil carbon sequestration and by reducing greenhouse gas (N_2O) emissions through the elimination of synthetic N-fertilizers. It also combats land degradation by improving soil structure and biodiversity, offering a sustainable solution that supports long-term environmental health and food security.

- **Agroforestry**

To create diversified and productive land-use systems, agroforestry combines forestry and agricultural technology. It is extensively employed to reduce land degradation via a variety of ecological processes. Crops by themselves cannot slow down runoff, but trees planted alongside crops help to strengthen the soil, which helps to preserve it. Under the different tree species, water-stable aggregates grow dramatically, and the erodibility of the soil is decreased (Saha et al., 2007). This works best in hilly and sloping places where soil erosion is a big issue and there is a high likelihood of heavy rains. Additionally, agroforestry systems boost soil carbon sequestration, which improves soil nutrient status and infiltration rate by improving soil organic matter content inputs through root biomass and leaf litter (Schoeneberger et al., 2012). Agroforestry systems produce habitat heterogeneity, which fosters a variety of soil biological habitats that enhance soil fertility through nutrient recycling and mineralization. Furthermore, tree-based agroecosystems can protect against negative climate changes, especially drought (Jose, 2009). Drought enhances soil loss by degrading soil structure and by wind erosion. Key parameters like pH and nitrogen content improvement indicate a positive impact on soil fertility in the agroforestry system.

Table 15.4 Implication of long-term organic farming systems on soil health

Organic inputs	Crop	Duration of experiment	Remarks	References
Continuous application of FYM @ 10 t/ha + green manure (5.48 t/ha—dry weight) in rice followed by FYM @ 10 t/ha + leaf manure (5 t/ha—fresh weight) in wheat	Rice–Wheat	21 years	128.2%, 145.2%, 127.5% increment of soil available N, P, K After 21 years, total organic carbon (TOC)—1.48% and MBC—619.3 ppm	Das et al. (2024)
Continuous application of vermicompost @ 80 kg N ha^{-1} + Crop residues	Rice–wheat–Mungbean	13 years	Total N—451.8 ppm, TOC—1.21%, microbial biomass N (MBN)—55.5 ppm	Meena et al. (2022)
	Rice–Wheat	13 years	Total N—445.2 ppm, TOC—1.13%, MBC—51.5 ppm	
Reduced tillage + organic farming (cattle slurry-based fertilizers)	Wheat, Maize	4 years	30% reduction of mean sediment delivery by organic farming, organic farming + reduced tillage minimized sediment delivery by 61% compared to intensive tillage	Seitz et al. (2019)

Hence, agroforestry is one of the most effective ways to stop overall environmental deterioration (Schwab et al., 2015).

Apart from these practices, there are some other agronomic interventions in sloping land such as contour farming (ploughing and growing of crops across the slope and along the contour line when slope is ~ 2–10%), strip cropping (growing of erosion resistant and erosion permitting crop in alternate rows), bunding, and vegetative barrier with grass (e.g., *Vetriveria zizanoides*, *Saccharum spontaneum*) across the slope. These practices reduce soil erosion and increase infiltration rate. When the land slope is very steep and rainfall is very high, mechanical intervention for controlling land degradation/ soil erosion is indispensable.

15.3.2.2 Mechanical Interventions

To support the cultural intervention and to strengthen the soil conservation methods against the water erosion in sloping land, there are some methods such as contour bunding, graded bunding, bench terracing, contour trenching, etc. To reduce the impact of wind erosion, windbreaks and shelterbelts are constructed (Salehi et al., 2022). Details of mechanical soil conservation techniques have been highlighted below.

- **Contour and Graded Bunding**

This method is useful where the slope is about 6% and the annual rainfall is < 600 mm. A trapezoidal bund is formed along the contour line and rainwater is impounded behind this structure. Permeable soil is required for better infiltration of water and proper durability of this structure. On the other hand, a Graded bund is meant for channelizing runoff. A graded bund can withstand annual rainfall > 600 mm.

- **Bench Terracing**

It is constructed when the slope is ranged between 16 and 33%, to reduce the intensity of scouring action of runoff, different types of bench terraces are constructed, viz., table-top bench terrace, inward sloping bench terrace, outwardly sloping bench terrace. Conservation bench terraces (CBT) are constructed to protect the land with a mild slope. In the CBT system, two parts will be there, one portion is the doner and another portion is the level bench (recipient). From doner portion, runoff is accumulated into the level bench due to the presence of the terrace ridge. In sloping land, upland crops are grown and in level bench water-intensive crop like rice are grown. In a semi-arid region, this CBT is very useful to protect the erosion and to conserve soil moisture. The ratio between doner and recipient area is determined based on the situation and requirement.

- **Contour Trenching**

This method is helpful when the slope is up to 20%. Soil and water conservation both objectives are both fulfilled with this structure. In this structure, trenches are excavated across the slope to reduce the runoff speed and its carrying capacity.

- **Shelterbelts and Windbreaks**

Rows of trees or shrubs are planted across the strong wind direction. These structures reduce the velocity of the wind to an extent of 60–80% at the leeward side, which leads to prevention of the soil from being lifted and carried away. This method is particularly effective in arid and semi-arid regions where wind erosion is prevalent. Shelterbelts and windbreaks can be used synonymously (Ramanan et al., 2023); however, in some cases, it is said that shelterbelts are a living structure, whereas windbreak can be a living or non-living structure. The efficiency of these structures depends on height (H) and the type of vegetation. However, width or length (L) is an important factor for a shelterbelt. The optimum L/H is considered as 10–13. Maximum wind speed reduction (~36%) by a shelterbelt was observed up to $2H$ distance from the shelterbelt base at the leeward side and beyond $20H$ distance, the effect of windbreak/shelterbelt was nullified (Prasad et al., 2013). Morphology of shelterbelt determines the efficacy of shelterbelt, and *Acacia tortilis, Eucalyptus camaldulensis, Dalbergia sissoo, Tecomella undulata* tree species could be used for shelterbelt.

Thus, these mechanical interventions play a vital role in mitigating the adverse effects of climate change by enhancing the resilience of landscapes to erratic rainfall and extreme weather events. Moreover, they contribute to combating land degradation and serve as sustainable solutions for maintaining soil health, agricultural productivity, and ecological balance in vulnerable regions.

15.3.2.3 Chemical Intervention

- **Reclamation of Salt-Affected Soils**

Sodic/alkali soils have higher amounts of exchangeable sodium (Na^+) that displace other essential nutrients, leading to poor soil structure with high pH (Ziskin et al., 2024). The low infiltration rate of these soils causes waterlogging, which further intensifies the problem. Saline soils contain higher amounts of soluble salts, primarily calcium sulphate ($CaSO_4$), sodium chloride (NaCl), and magnesium sulphate ($MgSO_4$). These soils have high electrical conductivity (EC), which retards plant growth by affecting the uptake of water (Hail & Mehari, 2021). The reclamation of these soils focuses on minimizing the salt concentration to a level where crops are unaffected. The first step in reclaiming sodic soils is the gypsum application. Ca^{2+} ions in gypsum replace the Na^+ ions present in soil particles, which improves the soil structure, and ultimately Na_2SO_4 is leached out. Application of sulfur or sulfur-containing compounds is also regarded as a promising approach for increasing the pH of sodic soil. Organic matter, such as compost or farmyard manure/green manure, helps in the process of reclamation of salt-affected soils by releasing organic acids, and by improving the soil structure, and ultimately, soil physiochemical properties get improved (Shaygan & Baumgartl, 2022).

- **Reclamation of Acid Soil**

Acidic soils have a low pH (< 6.5), and are characterized by higher amounts of hydrogen ions (H^+). These soils are formed due to excessive use of acid-forming fertilizers, high rainfall, or decomposition of organic matter. Reduced microbial activity, nutrient imbalances, and poor plant growth are caused by acidic soils (Agegnehu et al., 2021). Lime (calcium carbonate) application is the most prevalent method for reclaiming acidic soils. It neutralizes the excess H^+, and increases the soil pH (Warner et al., 2023). It also supplies calcium, an important nutrient for the growth of plants. The quantity of lime depends on the soil's buffering capacity and present pH. In some cases, fertilizers like dolomitic lime (which has both magnesium and calcium) or basic slag are applied to increase the pH, ultimately, soil productivity can be enhanced for a wide range of crops.

- **Advanced Chemical Intervention**

The use of biodegradable polymers such as hydrogels is an innovative approach to improve water retention in degraded and salt-affected soils (Tefera et al., 2022), and it also retards the process of desertification. Polymers, such as polyacrylamide, also aid in the soil aggregation process and reduce erosion. Hydrogels can store a significant amount of water and release it slowly to the soil. This process reduces evaporation losses and ultimately mitigates drought stress (Ghobashy et al., 2024). Other advanced chemical interventions have been highlighted in Table 15.5.

Therefore, these chemical interventions directly address the challenges posed by climate-induced soil degradation, such as salinization and acidification, which are intensifying under changing environmental conditions. By restoring soil health and enhancing resilience, these strategies offer sustainable, long-term solutions for maintaining agricultural productivity and ecosystem stability under a changing climate scenario.

Table 15.5 Advanced chemical interventions and their reliability in soil conservation

Type of soil degradation	Strategies	Remarks	References
Salt-affected soil	Application of Si nano particle @ 500 mg L^{-1} + Plant growth promoting rhizo-bacteria (*Azotobacter chroococcum* + *Pseudomonas koreensis*) @ 950 g ha^{-1}	Barley plant growth and chlorophyll content were improved, stress enzymes and proline content were significantly reduced, soil enzymatic activity was improved	Alharbi et al. (2022)
	Application of nano-gypsum @ 960 kg ha^{-1}	Improvement of stable soil structure (46.2%), and hydraulic conductivity (~ 2.4 times), Na removal efficiency was increased by 68.9% compared to control	Abd El-Halim et al. (2023)
	Application of liquefied biomass hydrogel (liquefied sugarcane bagasse + polyvinyl alcohol + sodium tetraborate decahydrate) @ 3% w/w	27% reduction of sodium absorption ratio and 58% reduction of Cl^{-} ion concentration in salt-affected soil	Tefera et al. (2022)
Heavy metal-contaminated soil	Application of 3 mmol kg^{-1} soil N, N-bis glutamic acid	Phytoremediation with a higher absorption rate of Cd by *Solanum nigrum* (28.65–68.74% higher extraction rate of Cd compared to control)	Teng et al. (2022)

15.4 Limitations and Challenges Pertaining to the Implementation of Soil Conservation Techniques

Soil conservation techniques are crucial in maintaining long-term agricultural productivity and ecological sustainability by mitigating land degradation. Despite the well-documented benefits, their widespread implementation is affected by a complex array of challenges. A holistic approach is necessary to overcome these challenges and to ensure the successful adoption of soil conservation strategies.

15.4.1 Educational and Awareness Challenges

A major non-monetary limitation arises from insufficient knowledge and awareness (Shinogi et al., 2021). Many farmers are either unaware of the existence of advanced conservation techniques or do not fully understand their long-term benefits. Traditional systems frequently emphasize short-term productivity, and in the absence of robust empirical evidence or demonstrative practices, soil conservation measures are often regarded by farmers as non-essential or ancillary rather than integral to sustainable production (Kagoya et al., 2018). Moreover, literacy and education levels influence a farmer's ability to comprehend and adopt conservation practices (Sharna et al., 2022). Extension services play a pivotal role in bridging this gap, but in many regions, these services are either poorly resourced or inconsistently delivered. Training programs and knowledge dissemination efforts are often urban-centred or limited in scale, failing to reach the communities most in need.

15.4.2 Technical Limitations

The implementation of conservation practices is frequently hindered by a lack of technical proficiency and appropriate infrastructure. Many practices, such as carbon farming or agroforestry, demand precise knowledge of ecological interactions, plant physiology, and soil science. However, such technical knowledge is often inaccessible to grassroots practitioners due to limited extension services or a lack of formal training opportunities (Kagoya et al., 2018; Shinogi et al., 2021).

Practices like organic farming, agroforestry, and contour trenching are often labour-intensive and require a long-term commitment. Mechanical interventions like bench terracing are feasible only on slopes with specific gradients (16–33%). Similarly, contour and graded bunding require precise land assessment and adequate rainfall regimes. If applied in unsuitable conditions, their effectiveness diminishes, and maintenance becomes burdensome.

15.4.3 Economic Risk and Delayed Returns

The adoption of soil conservation practices is often hindered by delayed economic returns and market-related uncertainties. Techniques such as Conservation Agriculture (CA), Integrated Nutrient Management (INM), and organic farming typically yield long-term benefits, such as enhanced soil fertility, improved moisture retention, and reduced erosion, rather than immediate productivity gains. For example, the build-up of soil organic carbon and restoration of microbial activity in CA systems may take 3 to 5 years to significantly influence yields (Mbaga et al., 2024). This delay presents a major disincentive for resource-poor farmers focused on short-term food and income security.

Moreover, economic analysis indicated that during the transition period (within 1–3 years), CA-based practices resulted in 11.9% lower profit. However, in the long-term (> 20 years), 13% higher profit was witnessed in CA-based practices compared to the conventional system (Madarász et al., 2025). The result of a meta-analysis of organic farming (covered 55 crops and 5 continents) revealed that when the premium price of organic produce was applied, organic agriculture was 22–35% more profitable than conventional farming, with a significantly higher (20–24%) benefit–cost ratio. It implied that long-term organic farming was highly sustainable and profitable with multiple ecosystem services (Crowder & Reganold, 2015). However, many smallholders lack the financial resilience for early-stage losses without institutional support or subsidies. In many developing regions, organically or sustainably grown produce lacks reliable access to certification systems and specialized markets (Sharna et al., 2022).

15.4.4 Other Associated Challenges

Beyond technical and economic barriers, a range of institutional, resource-based, and socio-cultural factors further constrain the adoption of soil conservation practices. Weak institutional frameworks and government policy often hinder coordinated action among government agencies, research institutions, and local communities. The absence or poor implementation of policies, subsidies, and crop insurance for smallholders and landless farmers who frequently face bureaucratic hurdles is one of the major problems because in a developing country like India, more than 85% farmers are small and marginal (Paramesh et al., 2022; Sharna et al., 2022).

Resource constraints and climate variability also hinder the adoption of conservation practices. Techniques like integrated nutrient management and no-till farming often require inputs and equipment that are scarce or unaffordable in rural areas (Mbaga et al., 2024). Agro-ecological differences such as soil type, topography, and rainfall limit the applicability of standardized practices, and which hinders long-term adoption. Socio-cultural dynamics and insecure land tenure further complicate implementation. Traditional uses of crop residues for fuel and fodder also contradict the principles of CA (Sharna et al., 2022). Farmers without secure land rights are less likely to invest in practices with delayed benefits (Nahayo et al., 2016). A coordinated approach integrating financial incentives, policy support, technical training, and culturally sensitive community engagement is essential for scaling up sustainable soil conservation.

15.5 Conclusion

The ensuing disaster caused by land degradation can be mitigated with the collaborative effort towards the goal of "Land Degradation Neutrality" (LDN), conceptualized by the United Nations. To achieve the LDN target, the existing process of land degradation should be reduced, further land degradation should be prevented, and restoration of already degraded land could be carried out. To bring about these processes, there are some promising sustainable methods that not only facilitate present food security but also take care of the demand of future generations while enhancing eco-resilience in the context of climate change. Among different cultural methods, conservation agriculture, organic farming, and agroforestry systems cover all necessary cultural interventions to protect soil health. As per the constraints and existing situation, and to harness the full benefits of sustainable land management, integration of cultural, mechanical, and chemical interventions could be followed. The key indication of soil fertility or degraded land restoration is the improvement of Soil Organic Carbon (SOC), and the SOC rejuvenates the soil's physical, chemical, and biological properties. Through crop breeding, new varieties could be developed to grow in degraded land. Along with considerable yield, those varieties may aid in the process of soil fertility restoration. Therefore, a multidisciplinary and integrated approach to soil conservation may bring sustainability to food production system and the environment.

References

Abd El-Halim, A. E., Salama, A. M., Ibrahim, M. M., Aiad, M. A., & Shokr, M. (2023). Nano-gypsum in low dose improves the physicochemical properties of saline-sodic soil. *Archives of Agronomy and Soil Science, 69*(12), 2286–2299.

Agegnehu, G., Amede, T., Erkossa, T., Yirga, C., Henry, C., Tyler, R., Nosworthy, M. G., Beyene, S., & Sileshi, G. W. (2021). Extent and management of acid soils for sustainable crop production system in the tropical agroecosystems: A review. *Acta Agriculturae Scandinavica, Section B—Soil Plant and Science, 71*(9), 852–869.

Aguilera, E., Lassaletta, L., Gattinger, A., & Gimeno, B. S. (2013). Managing soil carbon for climate change mitigation and adaptation in Mediterranean cropping systems: A meta-analysis. *Agriculture, Ecosystems & Environment, 168*, 25–36.

Alharbi, K., Rashwan, E., Mohamed, H. H., Awadalla, A., Omara, A. E. D., Hafez, E. M., & Alshaal, T. (2022). Application of silica nanoparticles in combination with two bacterial strains

improves the growth, antioxidant capacity and production of barley irrigated with saline water in salt-affected soil. *Plants, 11*(15), 2026.
Baker, C. J., Saxton, K. E., & Ritchie, W. R. (1996). *No-tillage seeding: Science and practice.* CAB International.
Bamboriya, J. S., Purohit, H., Naik, B. S. S. S., Pramanick, B. S. D., Bamboriya, S. D., Doodhawal, K., Sunda, S. L., Medida, S. K., Gaber, A., & Alsuhaibani, A. M. (2023). Monitoring the effect of integrated nutrient management practices on soil health in maize-based cropping system. *Frontiers in Sustainable Food Systems, 7*, 1242806.
Bayer, C., Lovato, T., Dieckow, J., Zanatta, J. A., & Mielniczuk, J. (2006). A method for estimating coefficients of soil organic matter dynamics based on long-term experiments. *Soil Tillage Research, 91*, 217–226.
Beare, M. H., Hendrix, P. F., Cabrera, M. L., & Coleman, D. C. (1994). Aggregate-protected and unprotected organic matter pools in conventional and no-tillage soils. *Soil Science Society of America Journal, 58*, 787–795.
Bhat, S. A., Dar, M. U. D., & Meena, R. S. (2019). Soil erosion and management strategies. In R. Meena, S. Kumar, J. Bohra, & M. Jat (Eds.), *Sustainable management of soil and environment* (pp. 73–122). Springer.
Bhattacharyya, R., Chandra, S., Singh, R. D., Kundu, S., Srivastava, A. K., & Gupta, H. S. (2007). Long-term farmyard manure application effects on properties of a silty clay loam soil under irrigated wheat–soybean rotation. *Soil Tillage Research, 94*, 386–396.
Chen, L., Gao, J., Ji, Y., Bai, Z., Shi, S., & Liu, H. (2014). Effects of particulate matter of various sizes derived from suburban farmland, woodland and grassland on air quality of the central district in Tianjin. *Aerosol and Air Quality Research, 14*, 829–839.
Choudhary, M., Jat, H. S., Datta, A., Yadav, A. K., Sapkota, T. B., Mondal, S., Meena, R. P., Sharma, P. C., & Jat, M. L. (2018). Sustainable intensification influences soil quality, biota, and productivity in cereal-based agroecosystems. *Applied Soil Ecology, 126*, 189–198.
Crowder, D. W., & Reganold, J. P. (2015). Financial competitiveness of organic agriculture on a global scale. *Proceedings of the National Academy of Sciences, 112*(24), 7611–7616.
Das, B., Pooniya, V., Shivay, Y. S., Zhiipao, R. R., Biswakarma, N., Kumar, D., Saikia, N., Kumar, A., Govindasamy, P., Barman, A., & Critykar, J. (2024). Twenty-one years' impact of using organic amendments on the productivity of rice-wheat rotation and soil properties. *Field Crops Research, 309*, Article 109311.
Du, X., Jian, J., Du, C., & Stewart, R. D. (2022). Conservation management decreases surface runoff and soil erosion. *International Soil and Water Conservation Research, 10*(2), 188–196.
Du, Z., Angers, D. A., Ren, T., Zhang, Q., & Li, G. (2017). The effect of no-till on organic C storage in Chinese soils should not be overemphasized: A metanalysis. *Agriculture, Ecosystems & Environment, 236*, 1–11.
Duiker, S. W., & Lal, R. (2002). Mulch rate and tillage effects on carbon sequestration and CO_2 flux in an Alfisol in central Ohio. In *Agricultural practices and policies for carbon sequestration in soil* (pp. 77–86). CRC Press.
Erenstein, O. (2002). Crop residue mulching in tropical and semi-tropical countries: An evaluation of residue availability and other technological implications. *Soil Tillage Research, 67*(1), 115–133.
FAO. (2015). Conservation agriculture. https://www.fao.org/conservation-agriculture/en/. Accessed on January 10, 2025.
FAO. (2023). World Food and Agriculture—Statistical yearbook 2023. https://openknowledge.fao.org/server/api/core/bitstreams/6e04f2b4-82fc-4740-8cd5-9b66f5335239/content. Accessed on January 10, 2025.
Francaviglia, R., Di Bene, C., Farina, R., & Salvati, L. (2017). Soil organic carbon sequestration and tillage systems in the Mediterranean Basin: A data mining approach. *Nutrient Cycling in Agroecosystems, 107*, 125–137.
Franzluebbers, A. J. (2010). Achieving soil organic carbon sequestration with conservation agricultural systems in the southeastern United States. *Soil Science Society of America Journal, 74*(2), 347–357.

Gharpinde, B. G. B., Gabhane, V. V., Nagdeve, M. B., Sonune, B. A., & Ganvir, M. M. (2014). Effect of integrated nutrient management on soil fertility, nutrient balance, productivity and economics of soybean in an Inceptisol of semiarid region of Maharashtra. *Karnataka Journal of Agricultural Science, 27*(3), 303–307.

Ghobashy, M. M., Amin, M. A., Mustafa, A. E., El-Diehy, M. A., El-Damhougy, B. K., & Nady, N. (2024). Synthesis and application of a multifunctional poly (vinyl pyrrolidone)-based superabsorbent hydrogel for controlled fertilizer release and enhanced water retention in drought-stressed Pisum sativum plants. *Science and Reports, 14*(1), 27734.

González-Chávez, M. D. C. A., Aitkenhead-Peterson, J. A., Gentry, T. J., Zuberer, D., Hons, F., & Loeppert, R. (2010). Soil microbial community, C, N and P responses to long-term tillage and crop rotation. *Soil Tillage Research, 106*, 285–293.

González-Sánchez, E. J., Ordóñez-Fernández, R., Carbonell-Bojollo, R., Veroz González, O., & Gil-Ribes, J. A. (2012). Meta-analysis on atmospheric carbon capture in Spain through the use of conservation agriculture. *Soil Tillage Research, 122*, 52–60.

Hail, B., & Mehari, H. (2021). Impacts of soil salinity/sodicity on soil-water relations and plant growth in dry land areas: A review. *Journal of Natural Science Research, 12*, 1–10.

He, X., Zhou, J., Zhang, X., & Tang, K. (2006). Soil erosion response to climatic change and human activity during the Quaternary on the Loess Plateau, China. *Regional Environmental Change, 6*, 62–70.

Helgason, B. L., Walley, F. L., & Germida, J. J. (2010). No-till soil management increases microbial biomass and alters community profiles in soil aggregates. *Applied Soil Ecology, 46*, 390–397.

Jat, L. K., Singh, S. K., Latare, A. M., Singh, R. S., & Patel, C. B. (2013). Effect of dates of sowing and fertilizer on growth and yield of wheat (*Triticum aestivum*) in an Inceptisol of Varanasi. *Indian Journal of Agronomy, 58*(4), 611–614.

Jose, S. (2009). Agroforestry for ecosystem services and environmental benefits: An overview. *Agroforestry Systems, 76*(1), 1–10.

Kagoya, S., Paudel, K. P., & Daniel, N. L. (2018). Awareness and adoption of soil and water conservation technologies in a developing country: A case of Nabajuzi watershed in central Uganda. *Environmental Management, 61*(2), 188–196.

Kannan, R., Lalith, M., Dhivya, D., Abinaya, R., Krishna, L., & Krishnakumar, S. (2013). Effect of integrated nutrient management on soil fertility and productivity in maize. *Bulletin of Environment, Pharmacology and Life Sciences, 8*, 61–67.

Kassam, A., Friedrich, T., & Derpsch, R. (2022). Successful experiences and lessons from conservation agriculture worldwide. *Agronomy, 12*(4), 769.

Kumar, P., & Sharma, P. K. (2020). Soil salinity and food security in India. *Frontiers in Sustainable Food Systems, 4*, Article 533781.

Kusro, P. S., Singh, D. P., Paikra, M. S., & Kumar, D. (2014). Effect of organic and inorganic additions on physicochemical properties in Vertisol. *American International Journal of Research in Formal Applied and Natural Sciences, 5*(1), 51–53.

Lal, R. (2015). A system approach to conservation agriculture. *Journal of Soil and Water Conservation, 70*(4), 82–88.

Liang, Q., Chen, H., Gong, Y., Yang, H., Fan, M., & Kuzyakov, Y. (2014). Effects of 15 years of manure and mineral fertilizers on enzyme activities in particle-size fractions in a North China Plain soil. *European Journal of Soil Biology, 60*, 112–119.

Madarász, B., Járási, E. Z., Jakab, G., Szalai, Z., & Ladányi, M. (2025). Economic comparison of conventional and conservation tillage in a long-term experiment: Is it worth shifting? *International Soil and Water Conservation Research*. https://doi.org/10.1016/j.iswcr.2025.02.012

Mahadevan, M., Noel, J. K., Umesh, M., Santhosh, A. S., & Suresh, S. (2024) Climate change impact on water resources, food production and agricultural practices. In: The climate-health-sustainability Nexus: Understanding the interconnected impact on populations and the environment (pp. 207–229). Springer Nature.

Mangalassery, S., Mooney, S. J., Sparkes, D. L., Fraser, W. T., & Sjögersten, S. (2015). Impacts of zero tillage on soil enzyme activities, microbial characteristics and organic matter functional chemistry in temperate soils. *European Journal of Soil Biology, 68*, 9–17.

Massa, C., Bichet, V., Gauthier, É., Perren, B. B., Mathieu, O., Petit, C., Monna, F., Giraudeau, J., Losno, R., & Richard, H. (2012). A 2500-year record of natural and anthropogenic soil erosion in South Greenland. *Quaternary Science Reviews, 32*, 119–130.

Mbaga, S. G., Ngaga, Y. M., & Nyamoga, G. Z. (2024). Comparison of the profitability of conservation and conventional agriculture in Bahi District, Dodoma, Tanzania. *Journal of Agriculture and Environmental Sciences, 9*(2), 111–123.

Meena, A. L., Pandey, R. N., Kumar, D., Sharma, V. K., Meena, M. D., Karwal, M., Dutta, D., Meena, L. K., Narwal, E., Mishra, R. P., & Panwar, A. S. (2022). Impacts of long-term rice-based organic farming on fractions and forms of soil organic carbon and nitrogen in the Indo-Gangetic Plain. *Soil Res, 61*(2), 159–175.

Mishra, A., Nayak, R. K., Chander, G., Reddy, M., & Choudhari, P. (2020). Management of acidic soils. *Mapping the nutrient status of Odisha's soils* (pp. 1–55). ICRISAT.

Nahayo, A., Pan, G., & Joseph, S. (2016). Factors influencing the adoption of soil conservation techniques in Northern Rwanda. *Journal of Plant Nutrition and Soil Science, 179*(3), 367–375.

Paramesh, V., Ravisankar, N., Behera, U., Arunachalam, V., Kumar, P., Solomon, R. R., Dhar, M. S., Mohan, K. R., Prusty, A. K., Jacob, D., & Panwar, A. S. (2022). Integrated farming system approaches to achieve food and nutritional security for enhancing profitability, employment, and climate resilience in India. *Food Energy Security, 11*(2), 321.

Peters, R. D., Sturz, A. V., Carter, M. R., & Sanderson, J. B. (2003). Developing disease-suppressive soils through crop rotation and tillage management practices. *Soil Tillage Research, 72*, 181–192.

PIB. (2021). https://pib.gov.in/PressReleasePage.aspx?PRID=1727045. Accessed on January 10, 2025.

Prasad, R., Mertia, R. S., Gajja, B. L., Singh, D., & Prakash, C. (2013). Morphological characteristics of tree shelterbelts and their impact on wind regimes, air temperature and soil quality. *Indian Journal of Agroforestry, 15*(1), 74–81.

Ramanan, S., Manojkumar, S., Arunachalam, A., Handa, A. K., & Priyanka, V. (2023). Conceptual analysis of 'windbreaks' and 'shelterbelts' in agroforestry. *Indian Journal of Agroforestry, 24*(2), 39–45.

Roose, E., & Barthes, B. (2001). Organic matter management for soil conservation and productivity restoration in Africa: A contribution from Francophone research. *Nutrient Cycling in Agroecosystems, 61*(1), 159–170.

Saha, R., Tomar, J. M. S., & Ghosh, P. K. (2007). Evaluation and selection of multipurpose tree for improving soil hydro-physical behaviour under hilly eco-system of north east India. *Agroforestry Systems, 69*(3), 239–247.

Saikanth, D. R. K., Kishore, A. J., Sadineni, T., Singh, V., Upadhyay, L., Kumar, S., & Panigrahi, C. K. (2023). A review on exploring carbon farming as a strategy to mitigate greenhouse gas emissions. *International Journal of Plant Soil and Science, 35*(23), 380–388.

Salehi, Y., Moghadam, N. K., Lajayer, B. A., & Astatkie, T. (2022). Soil conservation using mechanical and non-mechanical methods. In A. Rakshit, S. Ghosh, V. Vasenev, H. Pathak, & V. D. Rajput (Eds.), *Soils in urban ecosystem* (pp. 219–241). Springer.

Schoeneberger, M., Bentrup, G., de Gooijer, H., & De Rosier, J. (2012). Branching out: Agroforestry as a climate change mitigation and adaptation tool for agriculture. *Journal of Soil and Water Conservation, 67*(5), 128–136.

Schwab, N., Schickhoff, U., & Fischer, E. (2015). Transition to agroforestry significantly improves soil quality: A case study in the central mid-hills of Nepal. *Agriculture, Ecosystems and Environment, 205*, 57–69.

Seitz, S., Goebes, P., Puerta, V. L., Pereira, E. I. P., Wittwer, R., Six, J., van Der Heijden, M. G., & Scholten, T. (2019). Conservation tillage and organic farming reduce soil erosion. *Agronomy for Sustainable Development, 39*, 1–10.

Seufert, V., Ramankutty, N., & Foley, J. (2012). Comparing the yields of organic and conventional agriculture. *Nature, 485*, 229–232.

Sharna, S. C., Anik, A. R., Rahman, S., & Salam, Md. A. (2022). Impact of social, institutional and environmental factors on the adoption of sustainable soil management practices: An empirical analysis from Bangladesh. *Land, 11*(12), 2206.

Shaygan, M., & Baumgartl, T. (2022). Reclamation of salt-affected land: A review. *Soil Systems, 6*(3), 61.

Shinogi, K. C., Srivastava, S., Rashmi, I., Jayaraman, S., Sinha, N. K., Gurav, P. P., & Kamble, A. L. (2021). Socioeconomic challenges and prospects in the adoption of conservation agriculture practices in India. In S. Jayaraman, R. C. Dalal, A. K. Patra, & S. K. Chaudhari (Eds.), *Conservation agriculture: A sustainable approach for soil health and food security: Conservation agriculture for sustainable agriculture* (pp. 611–621). Springer.

Siedt, M., Schäffer, A., Smith, K. E., Nabel, M., Roß-Nickoll, M., & Van Dongen, J. T. (2021). Comparing straw, compost, and biochar regarding their suitability as agricultural soil amendments to affect soil structure, nutrient leaching, microbial communities, and the fate of pesticides. *Science of the Total Environment, 751*, Article 141607.

Singh, G., Jalota, S. K., & Singh, Y. (2007). Manuring and residue management effects on physical properties of a soil under the rice-wheat system in Punjab, India. *Soil Tillage Research, 94*(1), 229–238.

Somasundaram, J., Salikram, M., Sinha, N. K., Mohanty, M., Chaudhary, R. S., & Dalal, R. C. (2019). Conservation agriculture effects on soil properties and crop productivity in a semiarid region of India. *Soil Research, 57*, 187–199.

Soussana, J. F., Lutfalla, S., Ehrhardt, F., Rosenstock, T., Lamanna, C., Havlík, P., Richards, M., Chotte, J. L., Torquebiau, E., Ciais, P., & Smith, P. (2019). Matching policy and science: Rationale for the '4 per 1000-soils for food security and climate' initiative. *Soil Tillage Research, 188*, 3–15.

Srinivasarao, C., Rakesh, S., Kumar, G. R., Manasa, R., Somashekar, G., Lakshmi, C. S., & Kundu, S. (2021). Soil degradation challenges for sustainable agriculture in tropical India. *Current Science, 120*(3), 492.

Tefera, B. B., Bayabil, H. K., Tong, Z., Teshome, F. T., Wenbo, P., Li, Y. C., Hailegnaw, N. S., & Gao, B. (2022). Using liquefied biomass hydrogel to mitigate salinity in salt-affected soils. *Chemosphere, 309*, Article 136480.

Teng, Y., Li, Z., Yu, A., Guan, W., Wang, Z., Yu, H., & Zou, L. (2022). Phytoremediation of cadmium-contaminated soils by Solanum nigrum L. enhanced with biodegradable chelating agents. *Environmental Science and Pollution Research, 29*(37), 56750–56759.

UNEP. (2007). Global environment outlook 4. https://www.unep.org/geo/geo-resources/geo-4. Accessed on January 10, 2025.

UNEP. (2024). Science, policy and innovation for land restoration. https://www.unep.org/gef/focal-areas/land-degradation. Accessed on January 10, 2025.

USDA. (2013). Summary report: 2010 National resources inventory, natural resources conservation service, Washington, DC, and Center for Survey Statistics and Methodology, Iowa State University, Ames, Iowa. http://www.nrcs.usda.gov/Internet/FSE_DOCUMENTS/stelprdb1167354.pdf. Accessed on January 10, 2025.

Warner, J. M., Mann, M. L., Chamberlin, J., & Tizale, C. Y. (2023). Estimating acid soil effects on selected cereal crop productivities in Ethiopia: Comparing economic cost-effectiveness of lime and fertilizer applications. *PLoS ONE, 18*(1), 0280230.

Weinan, C., & Fryrear, D. W. (1996). Grain-size distributions of wind-eroded material above a flat bare soil. *Physical Geography, 17*, 554–584.

Xiao, L., Kuhn, N. J., Zhao, R., & Cao, L. (2021). Net effects of conservation agriculture principles on sustainable land use: A synthesis. *Global Change Biology, 27*(24), 6321–6330.

Zhang, B., He, H., Ding, X., Zhang, X., Zhang, X., & Yang, X. (2012). Soil microbial community dynamics over a maize (*Zea mays* L.) growing season under conventional and no-tillage practices in a rainfed agroecosystem. *Soil Tillage Research, 124*, 153.

Zhang, X., Xin, X., Zhu, A., Yang, W., Zhang, J., Ding, S., Mu, L., & Shao, L. (2018). Linking macro aggregation to soil microbial community and organic carbon accumulation under different tillage and residue managements. *Soil Tillage Research, 178*, 99–107.

Ziskin, R., Dag, A., Yermiyahu, U., & Levy, G. J. (2024). Different amendments for combating soil sodicity in an olive orchard. *Agricultural Water Management, 299*, Article 108837.

Chapter 16
Climate Change and Community-Based Initiatives in Land Restoration in Hilly Areas

Bolagam Ravikumar, Ankita Shinde, Chinmaya Sahoo, and Ankit Saini

Abstract Hill ecosystems and the communities that depend on them are significantly impacted by climate change. The increasing effects of climate change, also referred to as tipping points, pose an uncertain future for hill ecosystems, raising the possibility of extensive, non-linear, catastrophic alterations. Hill communities are more exposed to and vulnerable to climate-related dangers due to their rough terrain, steep slopes, and drastic climatic fluctuations throughout the year. Even while many people in the Indian Himalayan region have adapted to living and existing in a changing environment for decades, if not centuries, the current rate of climate change is believed to be too rapid for them to do so. Furthermore, remoteness, economic marginalization, and political marginalization further hinder stakeholders' capacity to adapt to and manage the different impacts of climate change in mountain communities. Furthermore, climate change leads to changes in water availability, deforestation, biodiversity loss, and soil erosion. The inherent functions of these fragile ecosystems, as well as the means of subsistence for dependent populations, are at risk due to their degradation, which is crucial to the stability of the global environment. The current chapter explores the connection between climate change and land degradation in hilly areas, emphasizing the importance of community-based initiatives in restoration efforts. By leveraging local knowledge, collaborative efforts, and integrated strategies, these grassroots initiatives tackle problems through practices including afforestation, soil conservation, and rainwater collection. The potential for

B. Ravikumar
Department of Horticulture, Lovely Professional University, Phagwara, Jalandhar, India

A. Shinde
Division of Agricultural Engineering, ICAR-Indian Agricultural Research Institute, New Delhi, India

C. Sahoo
Department of Agronomy, Dr. Rajendra Prasad Central Agricultural University, Pusa, Bihar, India

A. Saini (✉)
Department of Agronomy, Dr. KSG Akal College of Agriculture, Eternal University, Baru Sahib, Sirmaur, Himachal Pradesh, India
e-mail: ankitsagr@eternaluniversity.edu.in

U. Chatterjee et al. (eds.), *Eco-Resilience: Climate Change, Land Degradation and Sustainable Solutions*, Sustainability Solutions,
https://doi.org/10.1007/978-3-032-00708-7_16

inclusive, sustainable solutions to the twin issues of environmental degradation and climate change is highlighted in the present chapter.

Keywords Climate change · Land restoration · Community-based initiatives · Hilly areas · Biodiversity · Soil conservation

16.1 Introduction

Ecosystems, communities, and economies around the world are all greatly impacted by climate change, making it one of the most pressing concerns of the twenty-first century. Hilly and mountainous regions, which are already prone to environmental deterioration because of their delicate ecosystems and distinctive geographical features, belong to the groups most at risk from climate change. Additionally, hilly areas are distinguished by the unique biological services they provide. They preserve essential genetic resources, control local temperature patterns, and function as repositories of spiritual and cultural heritage. In addition to reducing these landscapes' ability to deliver these services, deterioration has an impact on the health of the ecosystem worldwide. For example, increased soil erosion and deforestation in steep regions accelerate climate change by releasing stored carbon. Therefore, it is essential for global climate adaptation and mitigation efforts to meet the difficulties these regions face.

Hilly areas are vulnerable to a variety of natural disasters, including those caused by climate change. Some risks are unique to hilly regions, or they strike harder and more frequently than in lowlands. These dangers included geological (rock falls), climatological (glacial lake outburst floods), and hydrological (landslides and avalanches). Floods and other meteorological (storms, high temperatures) and climatological (droughts, wildfires) dangers are common to both hilly and lowland environments (Wymann von Dach et al., 2017). The hazards associated with climate change in hilly regions are summarized in the current book chapter. It highlights the characteristics of dangers associated with climate change, the vulnerability and exposure of human and ecological systems in hills to these risks, and the ways in which they interact. This chapter examines the complex connection between hilly region degradation and climate change, as well as the crucial role that community-based initiatives play in maintaining and restoring these important landscapes. With an emphasis on recognizing the particular environmental issues that hilly regions experience, like soil erosion and deforestation, this chapter explores the intricate connection between these places' deterioration and climate change. It also looks at the critical role that community-based projects play in restoring these settings, as well as how grassroots efforts could mitigate the consequences of climate change and advance long-term environmental sustainability.

16.2 Consequences of Climate Change in the Hills

Mountainous and hilly terrain is crucial to the stability of the environment worldwide. By affecting precipitation patterns and maintaining freshwater sources, they serve as essential carbon sinks, absorbing and storing large amounts of atmospheric carbon dioxide. They also play a crucial role in controlling hydrological cycles. In addition, these areas sustain biodiversity, serve as buffers against severe weather, and offer a variety of ecosystem services that are vital to both ecological and human well-being. They support abundant biodiversity, house communities whose livelihoods rely on resources from the earth, and act as vital water towers, supplying freshwater to billions of people downstream. Despite their significance, the effects of climate change pose a serious threat to these regions. Because of their steep slopes, shallow soils, and diminished capacity to withstand disturbances, hilly regions are especially susceptible to the effects of climate change. The deterioration of hilly ecosystems is exacerbated by extreme weather events, erratic rainfall patterns, and elevated temperatures, which lead to a domino effect on the environment and the economy. Extreme weather occurrences, higher temperatures, and changed rainfall patterns are the results of the ongoing changes in the weather and recurrent drought-like conditions (Fig. 16.1). Both sudden-onset catastrophes (flash floods) and slow-onset changes (temperature and precipitation fluctuations and glacier retreat) are occurring in hilly regions. Major hydrometeorological dangers have already had a significant socioeconomic impact on hilly places. Table 16.1 represents the hydrometeorological dangers caused in various regions between the years 1985 and 2014. The changing climate is expected to have a more detrimental impact on the hilly places than on the world average. Table 16.2 shows a comparison between observed and expected consequences of climate change on hilly areas with world average estimates. Risks of flooding disasters can arise from the size, severity, and interconnection of hazards as well as the terrain of hills. Complications can happen over a short or lengthy period of time when one hazard sets off one or more other hazards. A landslide dam, for instance, is created when a river is naturally blocked by material due to landslides. As seen in Nepal in 2014, Pakistan in 2010, and China in 2008, this may be a contributing factor to later outburst floods (Vaidya et al., 2019). In January 2010, a massive slope collapse in northern Pakistan caused a landslide that obstructed two kilometers of the Hunza River basin. In the months that followed, the artificial lake that was created as a result of this started to bury the nearby settlements, including Ainabad, completely. This scenario demonstrates the catastrophic socioeconomic implications of highland-lowland, cascading multifaceted effects.

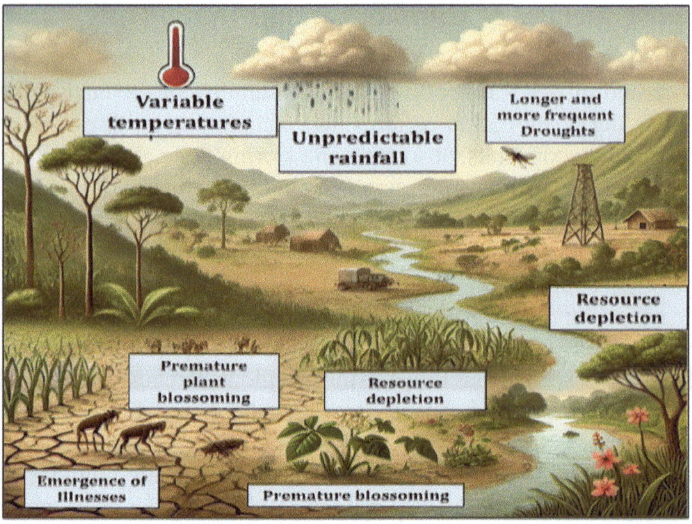

Fig. 16.1 Key effects of warming planet

Table 16.1 Hydro-meteorological hazards and impacts in the hills over the past few years

Hilly area	Disasters (No.)	Losses (in million USD)	Affected people
Central Asia	39	257.40	3,518,763
European Alps	38	7245.00	33,011
Eastern and North Africa	163	1246.80	76,127,779
Hindu Kush Himalaya	323	44,690.40	165,694,879

16.3 Exposure and Vulnerability of Hilly Areas

Over the past few decades, hilly areas have become more vulnerable to climate-related risks for people, property, and ecosystems, and a similar trend will probably persist (IPCC, 2019). At the local, regional, and worldwide levels, biodiversity may suffer significantly due to the effects of climate change detrimental effects on mountain communities. The hills, which are also repositories of biodiversity, water, and ecosystem services, may be home to over 85% of the world's bird, animal, and amphibian species. Many of these species are exclusively found in hilly regions and are particularly vulnerable to climate-related risks. Hills' topography also increases the exposure and vulnerability of people and their means of subsistence to climate-related risks. Compared to plains, hills have less exposed land that can be used by humans, and this land is frequently concentrated in the valleys. This increases the rivalry between different land uses, including transportation routes, tourism, townships, vital infrastructure, and productive agriculture (Patru-Stupariu et al., 2020).

Table 16.2 Global and regional impacts of climate change on hilly areas

Factors	Regional level effects	Global level effects
Temperature fluctuations	The mean ambient air temperature has risen by 0.3 °C every decade over the past five decades, with an expected variance of ± 0.2 °C. Although there is some geographic evidence for a higher frequency of unusually warm days and fewer instances of very cold temperatures, there is no conclusive evidence for mountains globally	The pace of global warming over the past few decades is 0.2 ± 0.1 °C per decade. While the frequency and importance of extraordinarily warm days and nights will almost surely rise, the incidence and intensity of severely cold days and nights will almost certainly decline
Precipitation	Although there has been a decrease in snowfall, especially at lower altitudes, hilly regions do not show clear yearly precipitation dynamics during the past few decades. Extreme precipitation events vary in frequency and intensity depending on the season and geographic location. There is a very strong probability that snowfall at lower elevations will decline in the near future (2031–50). Snowfall at greater heights may increase in response to increases in total winter precipitation	There has been a rise in precipitation since 1901 in the Northern Hemisphere's mid-latitude terrestrial regions. Area-averaged long-term either positive or adverse trends are not very confident for other latitudes. In many regions of the world, extreme precipitation events are predicted to become more frequent and more severe
Droughts	There is a strong degree of certainty that several places, including Africa, the Caucasus, and the Eastern Himalaya, will see warmer summers with more frequent droughts	In the Mediterranean, Asia, South America, and Africa areas, the extent and severity of droughts are getting worse. It is anticipated that droughts will increase in frequency, duration, and severity, particularly in the Mediterranean and Africa
Floods	In glaciered mountains, floods caused by glaciers are a frequent threat. Outburst flood dangers are increased by new glacial lakes close to unstable slopes. According to projections, the frequency of rain-on-snow events, which combine heavy rain and snowmelt, has grown in recent decades and is expected to shift to earlier springs and later autumns	Due to a lack of testimony, global estimates of increases in flood intensity and frequency are not very reliable. In certain streams or regions, anticipated increases in heavy precipitation will exacerbate local flooding caused by precipitation
Landslides	Landslides, rockfalls, and avalanches have been made worse by permafrost thawing, and more thawing is predicted to exacerbate slope instability and events in the Himalaya, Alps, Pyrenees, and Caucasus	The observed worldwide patterns of significant landslides in certain areas are not very reliable. However, there is a high degree of confidence that landslides in some areas will be impacted by future changes in heavy precipitation

In recent decades, the land in hilly regions has been under more strain due to socioeconomic development (such as agriculture, tourism, and energy) and population growth. As a result, vulnerable individuals and vital infrastructure have been forced into dangerous locations, increasing the risk of natural disasters for them and their possessions (Fig. 16.2). In the Himalayas, for example, almost two-thirds of the hydroelectric installations are situated in the route of possible glacier floods. Due to a variety of socioeconomic circumstances and capacity limitations, mountain communities in poor nations are also more susceptible to climate-related dangers than those in plains (Alfthan et al., 2018). These mountain communities frequently experience economic, political, and geographic marginalization, which isolates them from services, markets, and decision-making processes. Additionally, communities lack the resources necessary to cope with or adapt to the adverse effects of climate change due to their inadequate institutional, financial, or technological capabilities. Such capacity shortfalls are further exacerbated by hill communities' restricted access to health and education systems. In the Hindu Kush Himalayan region, two-thirds of the residents live in poverty, but poverty rates are higher in the highlands than in the lowlands. Both short-term forecasting for impending disasters and long-range strategies for lowering climate and disaster risks are hampered by limited access and affordability of meteorological and climate data and information unique to distant places. Remoteness and accessibility issues (such as inadequate road infrastructure) make it difficult for disaster relief and rehabilitation efforts to reach communities in need when catastrophes hit (Wymann von Dach et al., 2017). In mountainous areas, women are often among the groups most at risk from the consequences of climate change, for a number of causes, including financial prospects and social institutions, and norms (Bhadwal et al., 2019). Disasters, especially those that occur in hilly regions, are likely to claim the lives of more women than men. This happens for a number of reasons, including the fact that women have less access to financial resources, training, accessibility, and decision-making authority.

Women's ability to adapt can also be adversely affected by gender-based social and cultural conventions, such as those pertaining to the division of labor. Working-age men from the highlands are increasingly leaving the area, which frequently means that women and children have more work and obligations. Consequently, girls are more likely to drop out of formal schooling. In the long term, this may further limit their ability to enhance their adaptive capacity (e.g., increased access to technology solutions, improved comprehension of exposure and hazards) (Vaidya et al., 2019). Additionally, out-migration puts more strain on senior citizens, who must tend to the land in the difficult mountainous terrain. For example, those in the workforce in southern Kyrgyzstan's Batken and Osh oblasts leave the hilly villages, while pensioners and children remain in the disaster-prone areas. Migrants' remittances also help their family members improve living circumstances and secure their livelihoods, making the families more financially resilient to climate hazards. Furthermore, out-migration can reduce the stress on natural resources and biodiversity in high areas, potentially reducing the susceptibility of mountain ecosystems. Migration has been utilized to react to environmental changes in the past and the possibility of changes in the future. Natural disasters and climate change are making small-scale mountain

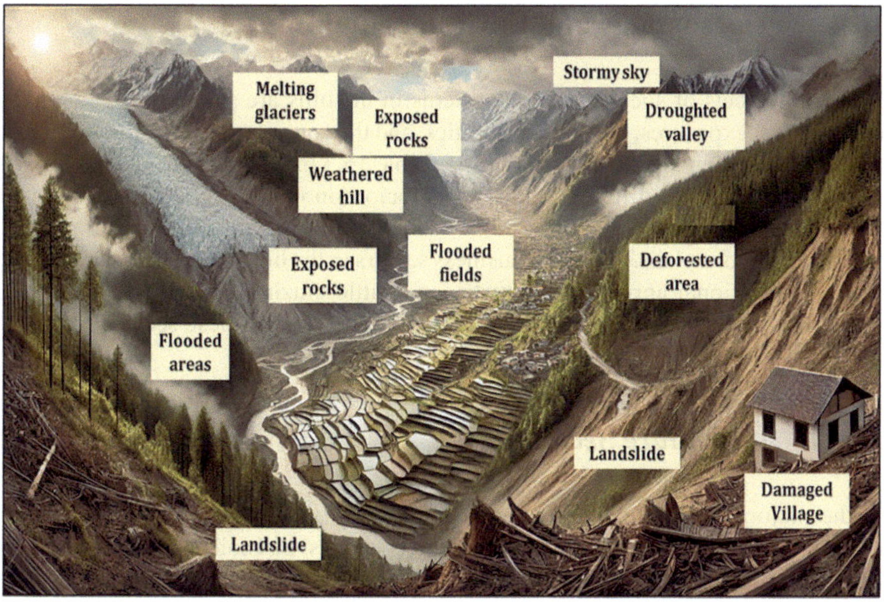

Fig. 16.2 Impact of climate change on hill ecosystem

farmers' already perilous condition even dire. However, the effects of climate change are frequently not the main or only factors driving migration in mountainous regions when contrasted to economic resources outside of the region (Dinshaw et al., 2019).

Ecosystems in the mountains are seriously threatened by rising temperatures, especially those that have effectively adapted to prolonged periods of extremely cold temperatures. For example, permafrost degradation can change the moisture content and nutritional availability of the soil, affecting species composition, and a decrease in snow cover can directly result in the loss of physical habitat. Another significant carbon storage resource is permafrost. Additionally, there is already evidence of species migration upslope, which results in a decrease in the abundance of endemic species that are adapted to cold climates, as well as an increase in species richness at higher elevations. The latter increases the risk of local extinctions, particularly for species like salmon and trout that live in cold freshwater. Desertification is another risk associated with permafrost degradation. Animal food chains and plant growth variables are changing as a result of variations in the length and thickness of snow cover (Rasul et al., 2019). By changing the water cycle, climate change can make water scarcer, especially when coupled with increased demand. Because populations in mountains around glaciers depend mostly on snow, ice, and glacier meltwater, they are especially vulnerable during dry months and droughts. Mountains are sometimes referred to be "water towers" due to the fact that they store and provide water to meet downstream human and environmental demands.

16.4 Socioeconomic and Environmental Impact

Numerous facets of society and ecosystems in the hills are susceptible to climate-related risks, exposures, and susceptibilities, and the consequences of climate change on these are anticipated to shift with time. Table 16.3 lists some detrimental impacts of climate change on different natural and socioeconomic systems in mountainous areas that need to be taken into account when developing and putting into practice climate resilience plans. Different business sectors in the hills face particular climatic risks. Conventional agriculture, forestry and cattle grazing, mining and extractive industries, hydropower generation, and tourism are some of these sectors.

16.4.1 Agriculture Sector

In the majority of mountainous areas, agriculture is the primary revenue source; nevertheless, it is frequently hampered by steep slopes, bad soil, and a lack of arable land. The quantity and seasonality of water flow are already changing in river basins that are supplied by glaciers and snow. It is anticipated that spring peaks will take place earlier in the year and that the average winter snowmelt runoff will rise. Changes in runoff are anticipated to have an impact on ecosystems, associated risks, and water management. Overall, it is anticipated that decreasing runoff in river basins supplied by glaciers and snow will lower irrigated agricultural productivity (Fig. 16.3). In order to stop soil erosion and preserve water, farmers use unusual techniques like terracing, which entails cutting flat steps into mountain slopes. Depending on temperature and height, the crops grown are typically climate-resilient, such as millet, barley, and maize. Cash crops like tea, coffee, and spices also support local economies in some areas.

16.4.2 Energy Sector

The production of hydropower, particularly from small hydro, depends heavily on the steady supply of water from rivers and glaciers, both of which are becoming more and more endangered due to climate change. Water flow is initially increased by glacier melting brought on by warming temperatures, but as glaciers recede over time, this flow decreases and causes long-term water shortages. Furthermore, seasonal precipitation patterns and variations in water temperature impair hydropower efficiency and lower energy production reliability. Given that hydropower is a vital renewable energy source in many hilly areas, these obstacles impede attempts to lower greenhouse gas emissions. Furthermore, communities find it more difficult to switch to low-carbon energy systems due to the unpredictability in hydropower output, which jeopardizes energy security.

Table 16.3 Climate variability and its impact on hill ecosystems

System	Relevant area	Impacts in hilly areas	Example of risk
Natural	Ecosystems	The species composition has changed; some species have found new habitats, some have lost their ability to reproduce, and terrestrial species have migrated upslope	Livelihoods
	Water quantity	Degradation of permafrost, decreased snow cover, glacier retreat, snowmelt, and seasonal and peak changes in precipitation and river runoff	Livelihoods, energy and water security
	Water quality	Heavy metals and downstream nutrients are released and move as a result of glacial loss	Livelihoods, people, assets at risk, downstream ecosystems
Assets	Mineral exploration/energy production	Glacier shrinking and retreat; shift in water supply	Changes to energy production (e.g., hydropower generation), mining operations and operational dependability (e.g., cooling for power plants)
	Roads and buildings	Decline of permafrost	Infrastructure deformation and instability
People	Cultural	Retreating glaciers and melting permafrost	Cultural behaviors and beliefs are disrupted, and human history is better understood
	Health	Decreased water quality due to glacier retreat; altered disease vectors	Migration of labor and displacement brought on by extreme occurrences
Livelihoods	Economic activity (demand)	Future demand uncertainty, shifting consumption and trade trends, the possibility of residential property flooding, and interruptions to import/export operations	Finance, consumption, and commerce
	Economic activity (supply)	Decreased agricultural production, resource diversion, decreased productivity (natural catastrophes), food and other vital input shortages, and infrastructure damage	Technology, labor migration, capital stock, and energy inputs

(continued)

Table 16.3 (continued)

System	Relevant area	Impacts in hilly areas	Example of risk
	Food	Growing season change; productivity shift as a result of altered irrigation water availability; soil drying from less snow cover; increased crop evapotranspiration as a result of warming temperatures; and upslope cropping zone movement	Income loss as a result of lower yields and the potential to grow new crops
	Tourism	Due to landslides and floods, access to the tourist destination is restricted	Income loss

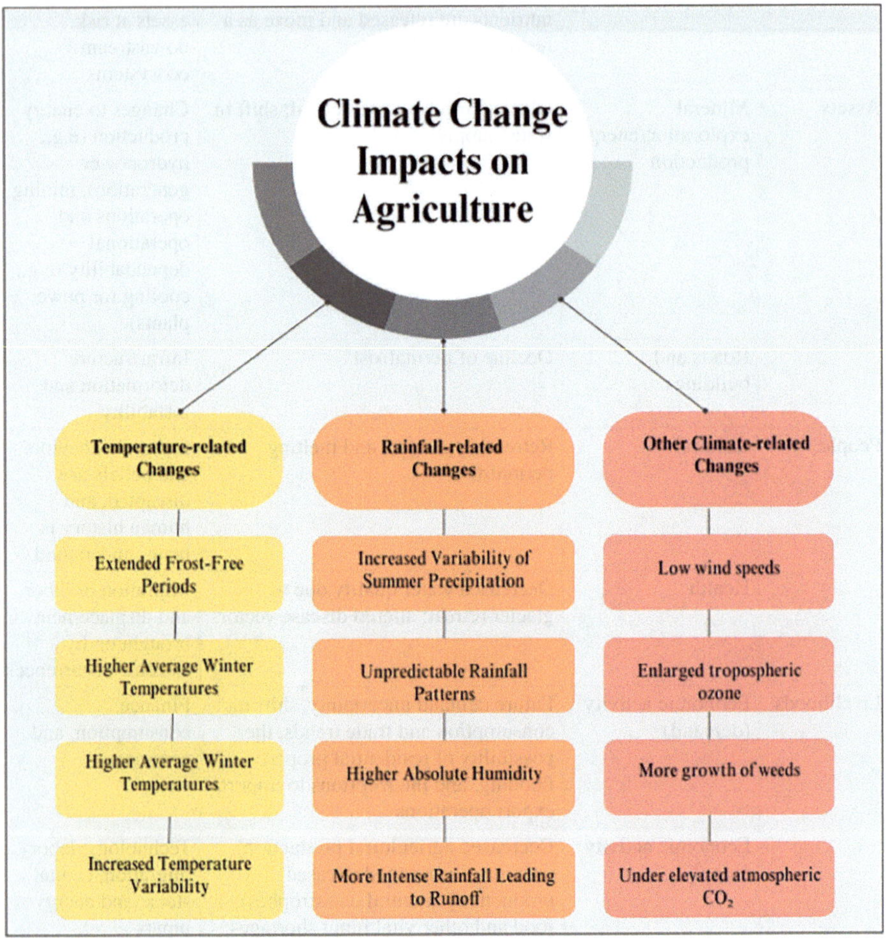

Fig. 16.3 Impact of climate variability on farming system

16.4.3 Livestock Rearing

Raising livestock, such as sheep, goats, and yaks, is another crucial aspect of livelihoods, particularly in regions with high elevations where cultivation is less practical. Because of their adaptability to the challenging alpine climate, these animals offer people a steady source of food and revenue. Local inhabitants depend on the production of meat, milk, and wool from animal husbandry to meet their subsistence needs. In addition to being used for subsistence, animals play a major role in economic endeavors like the creation of textiles made from wool, which are frequently highly prized in both domestic and foreign markets. Furthermore, livestock dung is applied as a natural fertilizer on marginal lands to increase soil fertility for crop development. Climate change, however, presents difficulties for raising cattle, such as decreased grazing availability as a result of plant loss, and heightened vulnerability of animals to illnesses and heat stress. In mountainous areas, addressing these problems is crucial to maintaining livelihoods and guaranteeing food security.

16.4.4 Tourism Sector

The natural beauty, cultural richness, and adventure options (such as trekking, mountaineering, and skiing) of mountainous regions draw tourists. Jobs in businesses like hospitality, transportation, and regional handicrafts are made possible by tourism. The Himalayan region, for instance, benefits from ecotourism, which also encourages environmental and cultural preservation. Decreased availability and snowfall, particularly at lower elevations, would likely result in shorter seasons for snow-based winter activities. Seasonal variations in hiking have already been noted in the Himalayas and, to a lesser degree, in the Andes and African highlands.

16.4.5 Welfare of People Living in Hilly Regions

The cultural services of the alpine landscape, such as cultural identification, spiritual, inherent, and beautiful attributes, in addition to the contributions made by glacial archaeology, may be impacted by a decrease in glaciers and snow cover. For example, lofty mountain peaks and glaciers are considered by indigenous people in the Andes to be the home of deities and life-giving spirits. Thus, these populations have experienced anxiety and emotional anguish as a result of retreating glaciers. Deteriorated water quality and quantity can also have an impact on people's health. The choice to uproot populations during glacial lake outburst floods is one example of how certain solutions to climate-related disasters in alpine regions can cause trauma and psychological issues.

Risks associated with climate change will continue to evolve for ecosystems, people, and economic, social, and environmental situations. Although this climate risk dynamic is valid for many different types of landscapes, the progression of climate hazards may be much more pronounced in specific mountainous regions. For example, up until a certain point, cryospheric changes (such as snowmelt and glacier retreat) lead to an increase in water availability; after that, the discharge of water begins to decrease. It is anticipated that different hills would experience this pivotal moment, known as the "water peak," at varying periods. It is estimated that 82–95% of the glacier regions in the Tropical Andes and over fifty percent of the glacier areas in Central Europe and the Caucasus are currently above a water peak. Several of the large glaciers in the Hindu Kush Himalayan region are expected to reach their water peak by the mid-twentieth century (Rasul et al., 2019).

16.5 Community-Based Approaches to Land Restoration in Hilly Areas

In order to address environmental concerns, community-based projects place a high priority on inclusive involvement and utilize local knowledge. With an emphasis on inclusive involvement, local knowledge, and long-term sustainability, community-based efforts have become a key component of sustainable land restoration in mountainous and degraded areas. These strategies address socioeconomic and environmental issues while utilizing the distinctive assets of nearby communities. Key principles include:

16.5.1 Planning

Coordinated landscape strategies often aim to enhance landscape governance through the establishment and maintenance of multi-stakeholder services. These institutional coordination structures allow stakeholders from different sectors in a specific environment to have conversations, negotiate, and plan together. Multi-stakeholder platforms frequently entail intricate procedures with a wide range of participants, and their goals and priorities may be periodically reassessed, revised, or reformed (Kusters et al., 2018). Engaging local stakeholders in identifying problems, setting goals, and designing restoration activities ensures that interventions are context-specific and widely accepted.

16.5.2 Traditional Ecological Knowledge

The sustainability of livelihoods and ecosystems in highlands is greatly dependent on mountain farming practices, and the progress of these systems is greatly influenced by ancestral wisdom. Native farmers, who are the main actors, fully utilize a variety of conventional knowledge, which results in a series of action situations at the farm and landscape levels. In addition to helping the inhabitants satisfy their fundamental requirements, this improved the ability to save land and water (Jiao et al., 2024). Integrating indigenous knowledge, such as terrace farming, agroforestry, and water conservation practices, enhances the effectiveness of restoration efforts.

16.5.3 Capacity Building and Education

Training programs and awareness campaigns equip communities with the skills and knowledge needed to implement sustainable land management practices. To successfully rehabilitate damaged hilly terrain, teach communities' sustainable practices like eco-friendly farming, water management, and soil conservation.

16.5.4 Policy and Institutional Support

By offering steady funding, scalability, and institutional support, coordinating community-based land restoration projects with regional and federal policy guarantees long-term success. While neighbourhood efforts are strengthened by legal and technical backing, integration with national environmental goals aids in securing resources through government and international programs. Additionally, this connection makes it possible to create restoration models that are replicable in different areas, guaranteeing sustainability and a wider impact. Aligning community initiatives with regional and national policies ensures long-term support, funding, and scalability.

16.5.5 Women-Led Initiatives

Women are essential to land restoration, especially in hilly regions where community involvement and traditional knowledge are crucial. Communities can guarantee the survival of local plant species, increase agricultural resilience, and improve food security by giving women leadership roles in seed banks and biodiversity conservation initiatives. They have extensive traditional expertise in this area. In addition to strengthening local ecosystems, women's participation in the selection, preservation, and propagation of indigenous and drought-resistant seeds also meets nutritional needs and promotes long-term sustainability. Adapting to environmental changes and climate variability requires this information. Eighty to ninety percent of the seed needs of crops in the Himalayan Mountain region are satisfied by traditional seed management and exchange networks, in which women play a crucial role. Through in-situ conservation, women preserve a varied genetic pool of this priceless commodity and are the keepers of traditional knowledge pertaining to seed conservation (Dhakal & Leduc, 2010).

16.5.6 Eco-tourism

Eco-tourism provides a sustainable means of funding restoration initiatives while also giving local communities access to alternative sources of income. Communities may preserve their environment and reap financial rewards by encouraging tourism focused on biodiversity, restored landscapes, and cultural heritage. By providing financial incentives to protect ecosystems, drawing tourists, and increasing knowledge of the value of sustainable land management, eco-tourism promotes conservation.

16.5.7 Incentives

One effective strategy for encouraging communities to participate in restoration work is Payment for Ecosystem Services (PES). Under PES programs, people or organizations receive financial compensation for ecologically beneficial endeavours like soil protection, reforestation, or water catchment management. By rewarding communities for their hard work and guaranteeing that vital ecosystem services, like carbon sequestration and biodiversity restoration, are maintained for future generations, these incentives produce a win–win scenario (Fig. 16.4).

16.6 Climate Change Adaptation and Mitigation Measures in Hilly Areas

Adaptation describes actions that improve one's ability to deal with shifting circumstances. Through learning by doing, trying out novel concepts, and knowledge-making processes, adaptation techniques are continuously updated to enable hill communities to adapt and change their behavior in response to shifting climatic conditions. In hilly areas, community-led land restoration usually takes a multi-stage, planned approach. Assessment and planning are the first steps in the process, during which communities determine the degree of land degradation and create restoration

Fig. 16.4 Sustainable land restoration through community engagement

plans specific to their area. The implementation phase comes next, which entails gathering resources, carrying out rehabilitation projects, and fusing old wisdom with contemporary methods. In order to improve long-term success, the last stage concentrates on sustainability and monitoring, making sure that the restoration efforts are sustained and modified to changing environmental circumstances while maintaining community involvement (Fig. 16.5).

People who live in hilly and rural regions are keenly aware of the changes brought about by climate change in their immediate surroundings. Variable temperatures, longer and more frequent droughts, unpredictable rainfall, premature blossoming in certain plant species, the gradual natural resource depletion, and the rise of emerging illnesses are the most prevalent effects of climate change. Using the resources at their disposal, Communities on hills are adjusting to the alterations in some way. However, they now lack mobility, the ability to make decisions, knowledge, services, and technologies (Chulu, 2015; Joshi & Bhardwaj, 2015). Farmers in certain hilly regions have already begun using adaptation strategies, such as switching to vegetable crops from grain crops (Gum et al., 2009). Even in smaller regions with fewer resources, these crops produce more than grain crops and work. They also get much-needed nutrition and money from this. Producing commercial crops and off-season veggies together can be a way to guarantee a sustainable standard of living. Farmers in hilly regions might begin planting crops earlier due to rising temperatures and shorter, milder winters. To guarantee the sustainability of ecosystems and livelihoods, the

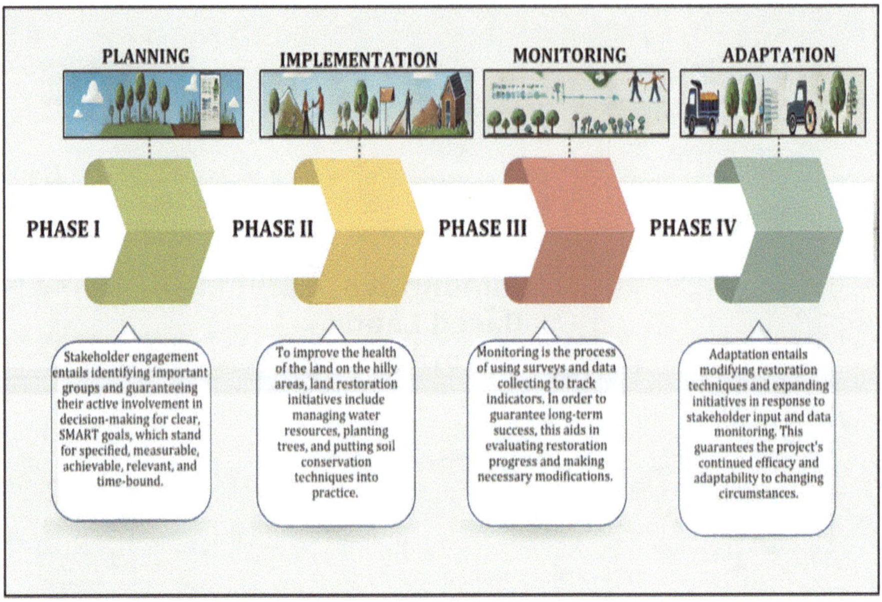

Fig. 16.5 Stages of community-led land restoration in hilly regions

following methods include efforts intended to adapt and respond to the changing environmental conditions.

16.6.1 Integrated Landscape Management

A thorough restoration strategy that recognizes the connections between forestry, agriculture, water resources, and local livelihoods is known as integrated landscape management. This approach seeks to balance practices like farming, forestry, and water conservation by viewing the entire landscape as a system in order to provide sustainable results in mountainous regions. For instance, agroforestry—the practice of growing trees next to farmlands can improve water penetration, stabilize soil, and decrease erosion while giving farmers access to other resources like fruits or lumber. Likewise, coordinating irrigation techniques with crop and forest management guarantees efficient water use and lowers runoff in hilly regions. This strategy creates a scenario where both humans and the environment benefit by addressing community economic demands while simultaneously enhancing ecological balance.

16.6.2 Financial Incentives and Livelihood Support

In environments with limited resources, like in hilly areas, financial incentives are essential for encouraging communities to participate in restoration efforts. With the well-proven Payment for Ecosystem Services (PES) concept, communities or landowners receive incentives for actions that improve the environment, such as reforestation, tree planting, or water conservation. China's Grain for Green Program, for example, pays farmers to turn their land into forests, resulting in extensive restoration (Deng et al., 2022). Furthermore, offering substitute sources of income, such as encouraging ecotourism, agroforestry, or non-timber forest products, lessens reliance on damaging methods like slash-and-burn farming. Financial incentives foster long-term community involvement and ownership by guaranteeing that restoration initiatives produce observable economic rewards. Providing economic incentives, such as payments for ecosystem services (PES), and supporting alternative livelihoods can encourage community participation.

16.6.3 Observation and Assessment

For restoration efforts to be successful, strong monitoring and assessment procedures are necessary. These tools monitor development, assess results, and assist in locating implementation flaws or inefficiencies. For instance, community surveys can evaluate social and economic effects, and satellite images can track changes in vegetation cover. Stakeholders can quantify gains in biodiversity, soil quality, water retention, and community well-being using data-driven techniques. Furthermore, adaptive management guarantees that projects maintain their flexibility; tactics can be changed in light of new information if a specific intervention is not producing the expected results. Frequent reviews increase responsibility, draw in new money, and offer insightful information for expanding successful models.

16.6.4 Climate-Resilient Practices

The effects of climate change, including increased rainfall variability, landslides, and droughts, are especially likely to affect hilly environments. The goal of climate-resilient practices is to reduce risks while increasing these landscapes capacity for adaptation. Long-term agricultural productivity is ensured by contour farming, which plants crops along the lands natural contours to minimize soil erosion and water runoff. Landslides are avoided and water retention is supported by bioengineering methods like employing live check dams or planting vegetation to stabilize slopes. Communities in mountainous areas are further protected in terms of food security by the introduction of climate-resilient crop varieties, such as drought-tolerant or flood-resistant crops. Communities can ensure the longevity of their restoration efforts while adjusting to shifting environmental conditions by incorporating these methods.

16.7 Conclusions and Future Prospects

Hilly areas are highly vulnerable to changes in climate due to their fragile ecosystems and reliance on environmental resources. Receding glaciers, erratic rainfall, rising temperatures, and drying streams are among the effects that are upsetting ecosystems and livelihoods, especially in the agricultural sector. These alterations alter cropping patterns and pose a hazard to plant and animal species. Because women have fewer resources and less decision-making authority, gender inequality makes the effects much worse. Policy interventions must diversify incomes and encourage gender-sensitive practices, such as acknowledging women's involvement in resource management and policy creation, in order to decrease dependence on rainfall and subsistence farming and increase resilience. Effective climate adaptation and mitigation require multidisciplinary, cooperative methods.

References

Alfthan, B., et al. (2018). Mountain Adaptation Outlook Series—Synthesis report. United Nations Environment Programme, GRID-Arendal. Nairobi and Arendal. https://www.grida.no/publications/426

Bhadwal, S., et al. (2019). Livelihoods, gender, and climate change in the Eastern Himalayas. *Environmental Development, 31*, 68–77. https://doi.org/10.1016/j.envdev.2019.04.008

Chulu, J. (2015). A feminist perspective that poverty is gendered: Do women have lesser access to resources in comparison with men? SSRN. Retrieved from http://ssrn.com/abstract=266338

Deng, Y., Jia, L., Guo, Y., Li, H., Yao, S., Chu, L., Lu, W., Hou, M., Mo, B., Wang, Y., Yang, H., & Zhang, T. (2022). Evaluation of the ecological effects of ecological restoration programs: A case study of the Sloping Land Conversion Program on the Loess Plateau, China. *International Journal of Environmental Research and Public Health, 19*. https://doi.org/10.3390/ijerph19137841

Dhakal, T. D., & Leduc, B. (2010). Women's role in biodiversity management in the Himalayas. *Sustainable Mountain Development, 57*, 16–17. ICIMOD.

Dinshaw, A., et al. (2019). *Mainstreaming adaptation in action: Case studies from two states in India.* World Resources Institute. Washington, DC. Retrieved from https://www.wri.org/research/mainstreaming-adaptation-action-case-studies-two-states-india

Gum, W., Singh, P. M., & Emmett, B. (2009). *Even the Himalayas have stopped smiling: Climate change, poverty, and adaptation in Nepal.* Oxfam International, Lalitpur.

IPCC. (2019). High mountain areas. In V. Pörtner, et al. (Eds.), *Intergovernmental panel on climate change* (pp. xxx–xxx). Geneva. Retrieved from https://www.ipcc.ch/srocc/chapter/chapter-2/

Jiao, W., Yang, X., & Li, Y. (2024). Traditional knowledge's impact on soil and water conservation in mountain agricultural systems: A case study of Shexian Dryland stone terraced system. *China. Ecological Indicators, 159*, Article 111742. https://doi.org/10.1016/j.ecolind.2024.111742

Joshi, K., & Bhardwaj, N. (2015). Women and natural resource management: A study of 'communities of practice' prevailing in women farmers' community for management of water and forests of the lesser Himalayan region in India. *International Journal of Advanced Research, 3*, 363–374.

Kusters, K., Buck, L., de Graaf, M., Minang, P., van Oosten, C., & Zagt, R. (2018). Participatory planning, monitoring, and evaluation of multi-stakeholder platforms in integrated landscape initiatives. *Environmental Management, 62*(1), 170–181. https://doi.org/10.1007/s00267-017-0847-y

Patru-Stupariu, I., et al. (2020). A review of changes in mountain land use and ecosystem services: From theory to practice. *Land, 9*(9), 336. https://doi.org/10.3390/LAND9090336

Rasul, G., et al. (2019). Adaptation to mountain cryosphere change: Issues and challenges. *Climate and Development, 12*(4). https://doi.org/10.1080/17565529.2019.1617099

Vaidya, R., et al. (2019). Disaster risk reduction and building resilience in the Hindu Kush Himalaya. In P. Wester, et al. (Eds.), *The Hindu Kush Himalaya assessment* (pp. xxx–xxx). Springer.

Wymann von Dach, S., et al. (2017). *Safer lives and livelihoods in mountains: Making the Sendai framework for disaster risk reduction work for sustainable mountain development*. Centre for Development and Environment, University of Bern, Bern Open Publishing. https://doi.org/10.7892/boris.99068

Chapter 17
Resilient Landscapes: Community Approaches to Land Restoration in Climate-Vulnerable Hilly Regions: A Case of Nilgiris, India

Apurv Bhogibhai Patel and Dharmesh Shantilal Solanki

Abstract Climate change poses significant environmental and socio-economic challenges, particularly in hilly regions characterized by fragile ecosystems and complex topographies. These areas are increasingly vulnerable to climate-induced phenomena such as soil erosion, landslides, and biodiversity loss, which disrupt local livelihoods and degrade natural resources. Rising temperatures, altered precipitation patterns, and increased frequency of extreme weather events have accelerated land degradation, making hilly areas one of the most climate-sensitive zones. Thus, the urgency for adaptive land restoration practices in these regions cannot be overstated. This chapter focuses on the Nilgiris district in India, a key biodiversity hotspot within the Western Ghats, which exemplifies the broader challenges hilly areas face under the influence of climate change. The region's delicate ecosystems, including Shola forests and watershed areas, are under immense strain, with climate-induced disasters such as landslides becoming more frequent. Simultaneously, the socio-economic fabric of the region is being tested, as local communities struggle to adapt to the changing environmental conditions. Research evaluates the effectiveness of community-based initiatives in addressing climate change and restoring degraded lands. Strategies such as afforestation, agroforestry, and watershed management have been implemented with varying degrees of success, drawing on traditional knowledge and integrating modern technological tools to enhance land restoration efforts. The study concludes that scaling up these initiatives, supported by robust policy frameworks and technological innovations, offers a sustainable pathway for hilly regions to adapt to climate change and ensure long-term ecological and economic stability.

Keywords Climate change adaptation · Land degradation · Community-based initiatives · Hilly regions · Landslide mitigation · Watershed

A. B. Patel (✉)
Urban Planner—Jamnagar Area Development Authority (JADA), Jamnagar, Gujarat, India
e-mail: apurvab1112@gmail.com

D. S. Solanki
Urban Planner—Junagadh Municipal Corporation (JuMC), Junagadh, Gujarat, India

U. Chatterjee et al. (eds.), *Eco-Resilience: Climate Change, Land Degradation and Sustainable Solutions*, Sustainability Solutions,
https://doi.org/10.1007/978-3-032-00708-7_17

17.1 Introduction

Hilly terrain is one of the most climate-sensitive landscapes in the world and is ecologically fragile because of the steepness of slopes, complex hydrology, and typical biological diversity (Jaiswal et al., 2011; Sankar et al., 2018). They are sensitive to climate variability, uncontrolled land use dynamics, and increasing socio-economic pressures promoting processes of land degradation, soil erosion, and ecosystem instability (Fayas et al., 2019; Jima et al., 2021). The changing climatic conditions at a global scale have far-reaching consequences on the sensitive ecosystems, which require adaptive and sustainable land management that can mitigate environmental degradation and promote socio-economic resilience (Borrelli et al., 2017a; Hobbs et al., 2008).

Being an international biodiversity hotspot, Nilgiris is recognized by UNESCO as a biosphere reserve (Saravanan et al., 2021; Senthilkumar & Selvaraj, 2025). That makes it carry an important set of ecosystems of native communities with agricultural landscapes inside it (Lakshumanan et al., 2012; Mamtha et al., 2016; Senthilkumar & Selvaraj, 2025. Regarding climate, Nilgiris witnessed growing challenges all over the previous decades in its increased temperatures; rainfall patterns remain erratic; it shows an overall rise in extremes of weather situations (Sukhdev, 2005). These climatic changes have fueled the rate of deforestation, soil degradation, and water resource loss, among direct impacts on local livelihoods, agricultural productivity, and biodiversity conservation (Jaiswal et al., 2011). The annual average temperature rose, the downfall in rain reached up to 15–20% of previous figures, along with extreme frequencies of events in cloudbursts and landslides in all regions (Senthilkumar & Selvaraj, 2025). On account of the rise in landslides and soil erosion, major parts appear to take high altitude places for plantation, and coffee plantations in High and Middle-economical loss from both angles with stability for environments, as well (Chandrasekaran et al., 2013b). Moreover, the forest landscape fragmentation and human settlement have further disturbed the natural drainage system and biodiversity corridors of the region (Chandrasekaran et al., 2013a; Heidi Hendersson, 2019; Ravichandran et al., 2020). Globally, analogous mountain ecosystems—such as the Andes in South America and the Ethiopian Highlands —have been identified with high vulnerability to land degradation and climate extremes, which emphasizes the necessity of prompt implementation of adaptive land restoration approaches combining community engagement and technological advancements. But there is an important knowledge deficit in evaluating the performance of such community-driven restoration models in ecologically sensitive Indian environments, especially regarding long-term sustainability, climate resilience, and socio-economic sustainability.

This chapter critically discusses the nature of change in land degradation for the Nilgiris in relation to climate change impacts while analyzing the effectiveness of community-led restoration projects. Using GIS-based spatial analysis, questionnaires, and climate modeling, this research explores how local communities are responding to climate challenges with innovative land management strategies. The

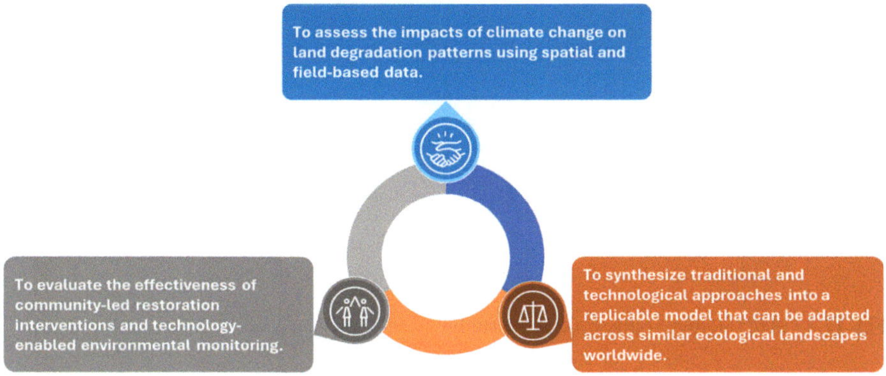

Fig. 17.1 Key objectives of the study

methodology follows a mixed-method paradigm that integrates spatial mapping via GIS applications, trend analysis of climate through secondary data, and participatory research via surveys and focus group interviews with indigenous stakeholders. These procedures enable both quantitative verification and community-level interpretation of environmental change. The study aims to provide an inclusive framework of land restoration in hill regions through a synthesis of traditional ecological knowledge with modern technological inputs, including climate modeling driven by AI and environmental monitoring using the Internet of Things. The key objectives are (Fig. 17.1):

17.2 Methodology

Study uses a mixed-method, interdisciplinary methodology to examine the effectiveness of community-led land restoration approaches to address climate-induced degradation in the Nilgiris. The research design includes a combination of spatial analysis, qualitative field surveys, and participatory assessment, representing a dual focus on ecological and socio-institutional restoration practices. The aim is to develop an evidence-based understanding of how localized interventions that are based on traditional ecological knowledge and augmented through modern technologies can promote long-term eco-resilience within climate-vulnerable hill systems.

To identify ecological degradation hotspots and assess land-use transformations over time, the study employed remote sensing and GIS-based spatial analysis. A multi-temporal satellite image analysis (2005–2021) was performed to map in detail changes in vegetation (cover), soil erosion, land use changes, and water bodies. Normalized Difference Vegetation Index (NDVI), land cover classification with various hydrological layers, was used to determine important zones of deforestation, watershed degradation, and soil susceptibility. The data layers were further

Fig. 17.2 Methodology framework of the study

related to topographic (relief) and lithological maps to assess the geomorphological susceptibility zones in the Nilgiris. The quantitative results were confirmed through semi-structured interviews, focus group discussions, and participatory rural appraisal (PRA) exercises undertaken in selected panchayats within the districts. The participants in the interviews, FGDs, and PRAs included smallholder farmers, leaders of indigenous communities, officials from the forest department, representatives from NGOs, and local watershed committee members. These tools were used to examine community perceptions on climate variability, adaptation practices, ecosystem services, and obstacles to restoration. Gender representation and intergenerational exchange were prioritized, with awareness of the various roles involved in conservation practices (Fig. 17.2).

To complement the ground data, the study integrated IoT-enabled sensors and AI-based modeling tools in pilot sites to track microclimatic conditions, including soil moisture, precipitation variability, and temperature fluctuations. This data was utilized in early warning systems to forecast areas of high risk regarding landslides and drought, and allowed communities to develop adaptive land-use decisions. The study employs a socio-ecological systems perspective, characterizing the landscape as an evolving interaction between ecological processes and community actions. The analysis focused on four interaction phases: (1) ecological integrity, (2) adaptive governance, (3) livelihood security, and (4) technological empowerment. This structure enabled the study to evaluate not only environmental restoration outcomes but also the enabling conditions for sustainable community-led management.

The methodological approach was tailored to answer a central question: How can community-led and technology-supported restoration strategies be scaled to enhance eco-resilience in climate-vulnerable hilly regions? The research will develop a coherent model of land restoration that combines traditional conservation knowledge with precision-based information technology and forms of governance that are inclusive of stakeholders and engage a deliberative, participatory approach. The findings are intended to inform planning, policy development, and restoration programs at the regional level, in the Nilgiris, and other ecologically vulnerable landscapes in the Global South.

17.3 Climate Change and Land Degradation

Climate change and anthropogenic pressures are accelerating environmental degradation in the ecologically sensitive region of Nilgiris district, located inside the Western Ghats(Rajesh Prasanna et al., 2013; Sridhar, 2024). Its high-altitude plateaus, winding waterways, and varied geomorphological structure render the region highly susceptible to climate-induced disturbances such as landslides, soil erosion, and biodiversity loss (Ravichandran et al., 2020; Shiny Mariam Rehel, 2024; Sridhar, 2024). In the past few decades, increasing temperatures, irregularities in monsoon cycles, and deforestation have disrupted the 'hydrological and ecological equilibrium' of the region, resulting in landslides and soil erosion (Chandrasekaran et al., 2013b). Such environmental transformations have spelled sweeping repercussions for agriculture, water availability, and forest ecosystems—substantially compounding the socio-economic fragility of the region's local populations (Jaiswal et al., 2011; Science.thewire, 2016) (Fig. 17.3).

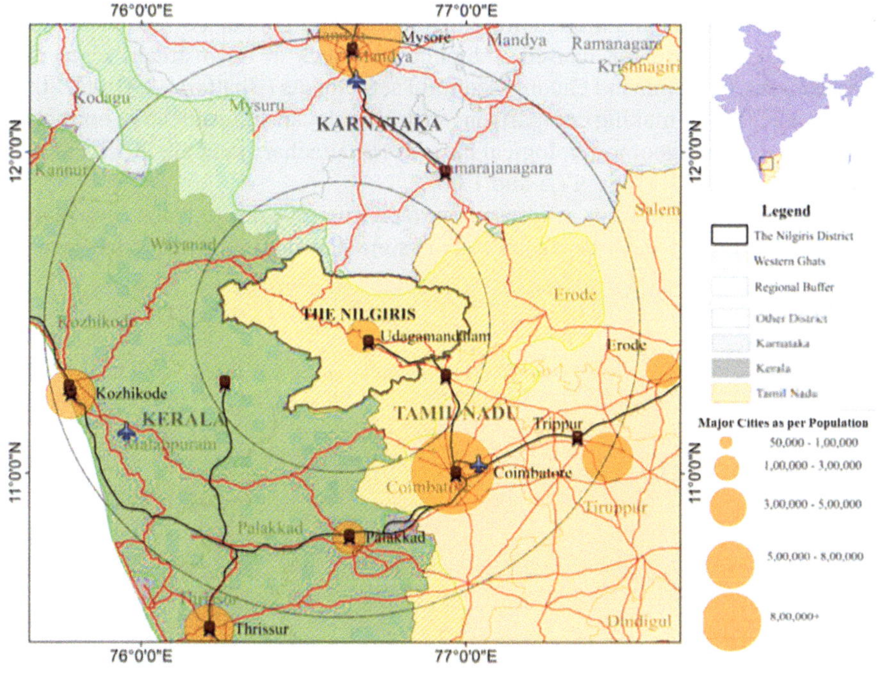

Fig. 17.3 Regional setting and connectivity of the Nilgiris region

17.3.1 Climate Variability and Its Effects

Over the past five decades, the Nilgiris has seen a steady warming trend of around 1.5 °C, resulting in drier soils, changed vegetation belts, and altered phenological rhythms of indigenous plant species (Senthilkumar & Selvaraj, 2025). Such climatic changes have not only impacted crop production but also diluted ecological interactions like pollination and seed dispersal (Lakshumanan et al., 2012; Saravanan et al., 2021). Parallelly, precipitation trends have turned highly unpredictable—characterized by abnormally long drought periods punctuated by intense monsoonal rainfalls—stabilizing local agriculture and making the productivity of major crops like millets, tea, and coffee less certain (Bonilla & Johnson, 2012; Sharma et al., 1996). These hydroclimate extremes trigger flash floods in river basins and water shortages in uplands, leaving the area vulnerable to escalating risks (Sankar et al., 2018; Sengupta et al., 2010) (Fig. 17.4).

The mounting intensity and recurrence of landslides in the Nilgiris are directly correlated with deforestation and heavy rains, which undermine soil stability in steep slopes (Anbarasu et al., 2010; Gebrehiwot & Demissew, 2020). These activities are compounded by widening monoculture plantations and unplanned land use (Sainju et al., 2006). Soil erosion caused sediment overloads in downstream reservoirs, hindered transportation, and endangered rural settlements (Bhattarai & Dutta, 2007; Fayas et al., 2019)—making comparisons with Himalayan and Andean regions characterized by similar geomorphological hazards (Bhattacharyya et al., 2015; Godwin Vasanth Bosco, 2013) (Figs. 17.5 and 17.6).

The imbalance between increased surface runoff and reduced percolation heightens the risk of flash floods, accelerates erosion, and reduces water security

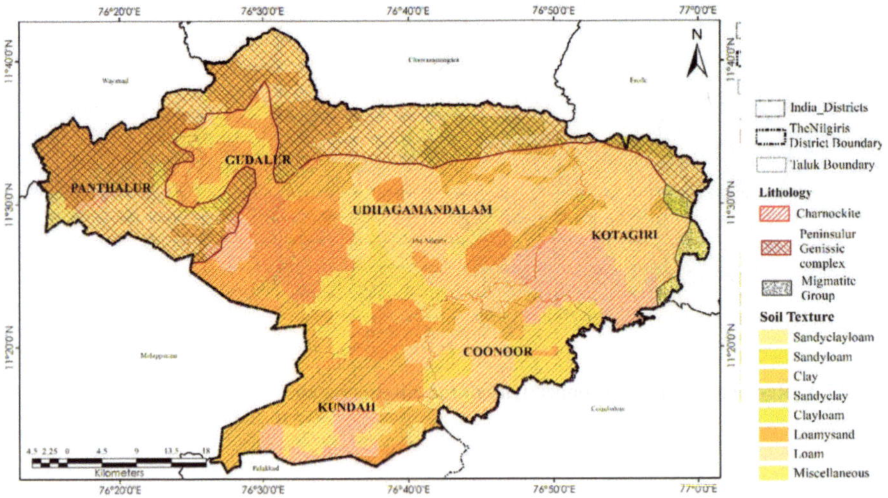

Fig. 17.4 Soil texture and lithology mapping for the Nilgiris

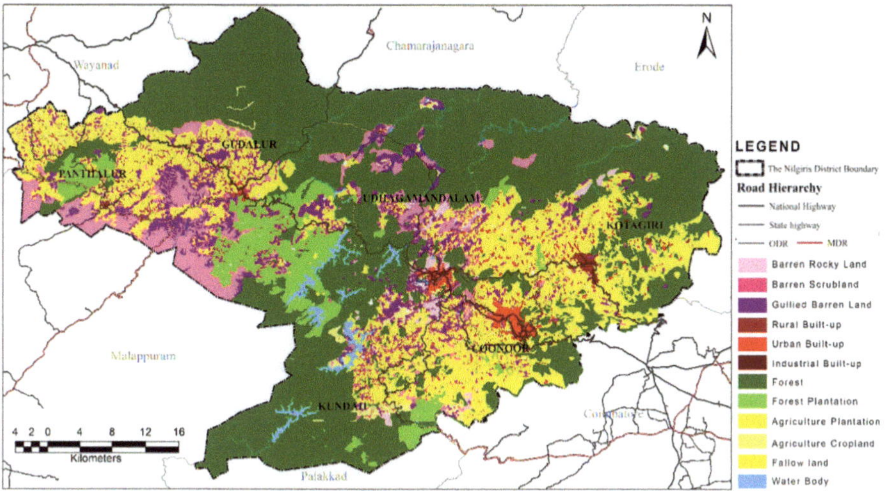

Fig. 17.5 Land use land cover, the Nilgiris (2021)

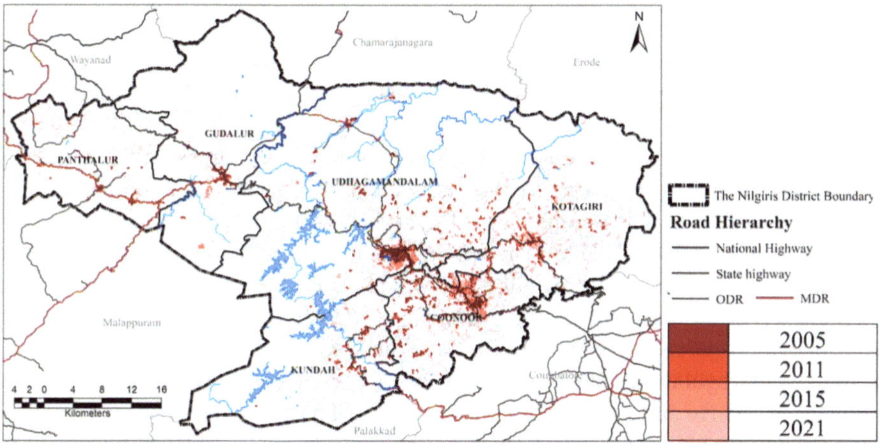

Fig. 17.6 Increase in built-up from 2005 to 2021

for both agricultural and domestic needs (Saravanan et al., 2021; Senthilkumar & Selvaraj, 2025).

Such dynamics mirror concerns raised in East African and Southeast Asian hill systems, where watershed degradation is increasingly seen as a transboundary challenge (Fayas et al., 2019; Sainju et al., 2006). This balance between excess surface runoff and minimized percolation increases flash flooding risk, speeding up erosion as well as limiting water security in agriculture and households (Borrelli et al., 2017b; Solonenko, 1997). Such processes reflect issues expressed in East African and Southeast Asian hill systems, where watershed degradation is being viewed more

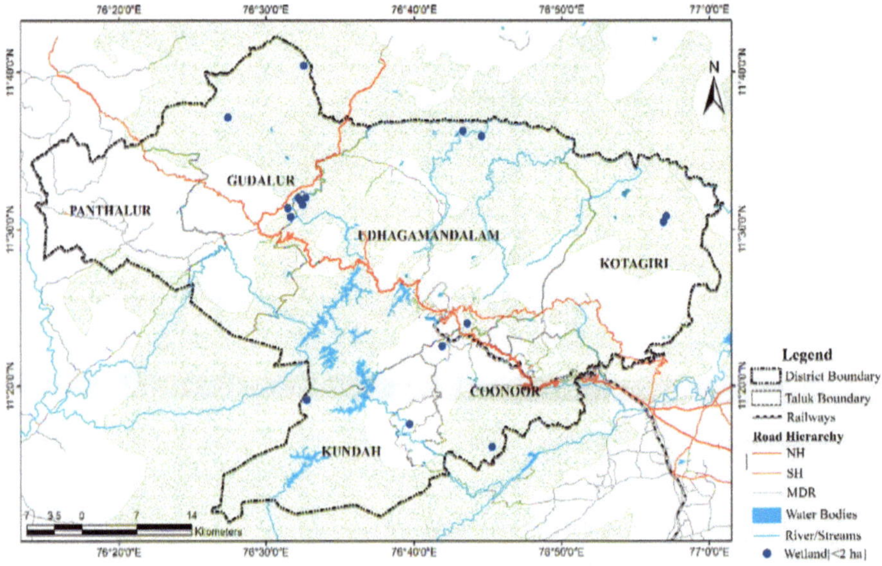

Fig. 17.7 Water bodies and wetlands in the district

and more as a transboundary issue (Bhattacharyya et al., 2015; Bhattarai & Dutta, 2007; Pradhan et al., 2012; Sainju et al., 2006) (Fig. 17.7).

The environmental stress throughout the Nilgiris' forest-grassland mosaics has increased because of increasing temperatures, land use change, and the invasion of alien species like Lantana camara and Eucalyptus (Ismail & Ravichandran, 2008; Saravanan et al., 2021). These processes are leading to habitat contraction and driving endemic species like the Nilgiri Tahr and Lion-tailed Macaque towards local extinction (Parthasarathy & Karthikeyan, 1997; Shao & Wu, 2008) (Parthasarathy & Karthikeyan, 1997; Shao & Wu, 2008). Keystone Shola Forest loss has compromised natural water regulation systems (Dregne, 2022; Pandey et al., 2007), causing microclimates to change and soil layers to dry, as reported trends in the Western Ghats and Sri Lankan highlands (Krishna Bahadur, 2009; Sainju et al., 2003; Sukhdev, 2005) (Fig. 17.8).

Climate variability and land degradation have had far-reaching socio-economic implications, particularly on indigenous and agrarian communities in the Nilgiris (Rajesh Prasanna et al., 2013). Low soil fertility and water scarcity have reduced agricultural yields, and this has made the farmers shift the crop patterns or abandon traditional farming (Arum Sari et al., 2017). Many smallholder farmers have migrated to urban centers in search of alternative livelihoods, hence contributing to rural depopulation and economic instability (Peddle et al., 1994). The tourism sector is one of the key economic activities in Nilgiris (Biswas & Pani, 2015). It has been impacted by extreme weather events, road blockages, and degradation of the ecosystem. Landslides and floods have destroyed the critical tourism infrastructure, such as heritage

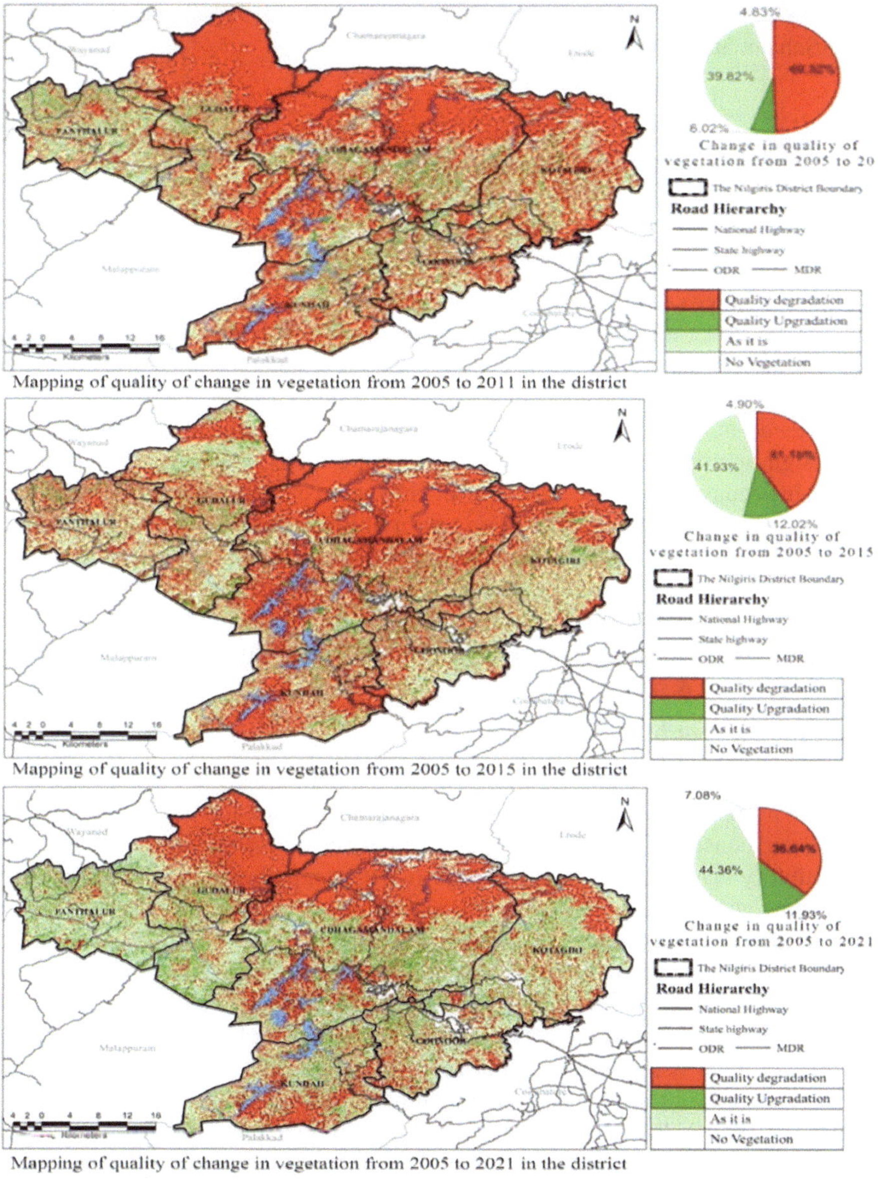

Fig. 17.8 Mapping of quality of change in vegetation from 2005 to 2021 in the district

sites, trekking trails, and eco-tourism lodges, thereby affecting the business bottom lines (Saha et al., 2002).

17.3.2 Mapping of Climate Change Effects

An integrated approach involving remote sensing, geospatial analysis, and ground-level assessments is the only way through which the spatial and temporal effects of climate change on Nilgiris can be understood to restore its sensitive ecosystems (Saha et al., 2002; Saraf, 2000). The application of GIS mapping in the identification of high-risk zones, change detection of land use, and assessing climate-related vulnerabilities is indispensable (Fu et al., 2005; Mukherjee & Singh, 2020). With spatial datasets, climate models, and field surveys, researchers and policymakers can design appropriate interventions for restoring lands, managing disasters, and conserving ecosystems. The application of GIS-based climate analysis in the Nilgiris has revealed trends related to landslide susceptibility, soil erosion, and temperature shifts, which call for immediate climate-adaptive interventions (Mamtha et al., 2016). Some of the major geospatial-related findings are landslide susceptibility mapping, soil degradation assessments, and temperature change analysis; many of them indicate rising risks from climate variability and anthropogenic pressures (Mamtha et al., 2016; Mukherjee & Singh, 2020).

Landslide Susceptibility Mapping

Among the most severe and recurrent environmental hazards in the Nilgiris, landslides are primarily exacerbated by intense monsoon precipitation, deforestation, and unregulated construction on steep slopes (Pachauri et al., 1998). GIS-based landslide susceptibility mapping has identified high-risk zones, particularly in those unstable slope areas with more than 30 degrees, where the erosion due to rainfall weakens the soil cohesion that leads to an increased likelihood of mass movements (Parihar et al., 2018). Landslides have been observed to be exacerbated in areas of deforested regions and degraded forest covers because they often lack deep-rooted vegetation that might stabilize the soil. Human encroachment into ecologically fragile terrains has increased the risk, since unauthorized settlements and infrastructural development add to surface runoff and structural vulnerabilities (Crosson, 1997; Saha et al., 2005). The spatial distribution of these landslide-prone zones, as shown by terrain roughness models and slope stability indices, indicates an urgent need for immediate intervention through vegetative reinforcement, slope stabilization measures, and the implementation of community-driven disaster preparedness programs (Arora et al., 2004). Addressing these risks will require a multi-pronged approach combining afforestation, geotechnical engineering solutions, and real-time monitoring systems to enhance slope stability and mitigate future landslide occurrences in the region (Kirschbaum et al., 2010; Senthilkumar & Selvaraj, 2025).

Implications for Land Restoration

Restoration of degraded landscapes in the Nilgiris districts, especially landslides in vulnerable areas, would need comprehensive and integrated intervention through ecological restoration, geotechnical engineering, and community-driven disaster preparedness (Owen et al., 2008; Senthilkumar & Selvaraj, 2025). Afforestation of vulnerable slopes with native, deep-rooting species can create substantial soil retention, improve the stability of the slope, and significantly reduce surface run-off, decreasing the risk from erosion and landslides (Brockerhoff et al., 2017; Chandrasekaran et al., 2013a). It develops diversified plant cover with native species, which stabilizes the terrain besides helping in conserving biodiversity and carbon sequestration. These steps would help fortify regional ecosystem resilience (Anbalagan, 1992; Pachauri & Pant, 1992). Geotechnical engineering solutions, retaining walls, bioengineering, and slope terracing also offer structural re-enforcement to resist mass-wasting events (Congalton, 1991). Engineered interventions, if implemented with natural soil conservation approaches, can significantly augment slope stability and reduce sediment displacement (Foody & Arora, 1997). Strengthening community-based disaster risk reduction is also important, ensuring early warning systems, real-time landslide monitoring, and efficient emergency response mechanisms in such risk-prone areas (Arora et al., 2004; Saha et al., 2005). The research institutions, local governance bodies, and community stakeholders can collaborate in developing a proactive risk mitigation framework for the Nilgiris to enhance climate resilience, reduce disaster vulnerability, and promote sustainable land restoration in the long term (Kirschbaum et al., 2010; Owen et al., 2008; Senthilkumar & Selvaraj, 2025) (Fig. 17.9).

Soil Erosion and Degradation Analysis

The soil erosion models in the Nilgiris bring out a gradual and alarming loss of topsoil, especially in the areas affected by intensive monoculture plantations, deforested landscapes, and unregulated infrastructure development (Sainju et al., 2006). Tea and coffee monocultures have reduced organic content and exposed topsoil to rain-induced degradation, while road cutting and slope excavation have increased sediment displacement, destabilizing entire watersheds (Bonilla & Johnson, 2012; Dregne, 2022). Lack of deep-rooted vegetation in degraded pastures enhances runoff and reduces soil moisture retention, further weakening landscape resilience. Hotspot mapping of erosion indicates intense associations between land-use intensity and altered drainage systems, emphasizing areas of severe vulnerability (Bhattacharyya et al., 2015; Ismail & Ravichandran, 2008). To challenge this, watershed restoration, vegetative barriers, contour bunding, and regenerative agriculture must be integrated as part of land management strategies in order to slow erosion rates, increase water holding capacity, and restore ecological health (Fayas et al., 2019; Sainju et al., 2003). Agroforestry systems integrating perennial tree crops into traditional cropping not only lower erosion and drought vulnerability but also enhance carbon sequestration

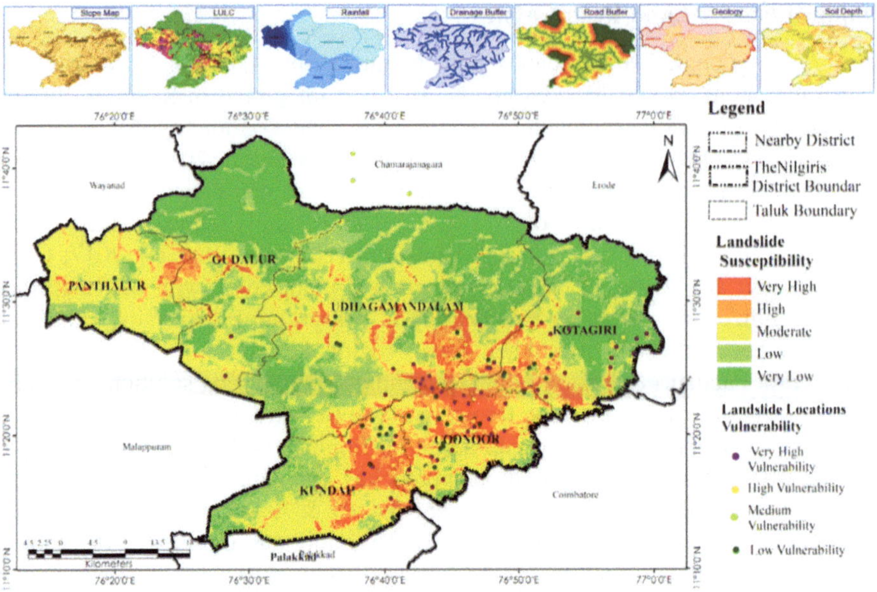

Fig. 17.9 Landslide susceptibility mapping for the Nilgiris district

and soil structure (Bhattacharyya et al., 2015; Dregne, 2022; Ismail & Ravichandran, 2008). Zero-tillage, mulching, and biofertilizer use also enhance soil microbiome regeneration and nutrient cycling—essential to restoring degraded soil systems (Bonilla & Johnson, 2012; Sainju et al., 2006). Together, they are sources of long-term food security and ecological resilience, depending on effective community participation, supportive governance, and adaptive, research-guided approaches (Bhattarai & Dutta, 2007; Ismail & Ravichandran, 2008; Sainju et al., 2003) (Fig. 17.10).

Temperature Change Analysis

Spatial temperature mapping across the Nilgiris reveals a consistent upward trend, with hotspots becoming more intense in areas with deforestation, urbanization, and changed rainfall patterns (Sridhar, 2024). Deforestation has functionally compromised the microclimatic stability of the region, compromising the natural cooling effects of indigenous forest cover and hastening soil moisture loss (Author et al., 1993). Urban heat islands—tied to infrastructure-intensive development and diminishing green spaces—have enhanced surface temperatures and increased ecological pressure on both the built and natural environments (Jaiswal et al., 2011; Sharma et al., 1996). Erratic precipitation patterns have raised evapotranspiration rates, extended dry periods, and reducing agricultural and forest systems already weakened by moisture stress (Bonilla & Johnson, 2012). The climate warming trend, coupled with receding cloud cover and mist precipitation, has also disturbed the Shola-grassland mosaic—altering species distribution, reproductive patterns, and inter-species relationships (Das et al., 2019; Sridhar, 2024). Endemic flora exhibit changed breeding

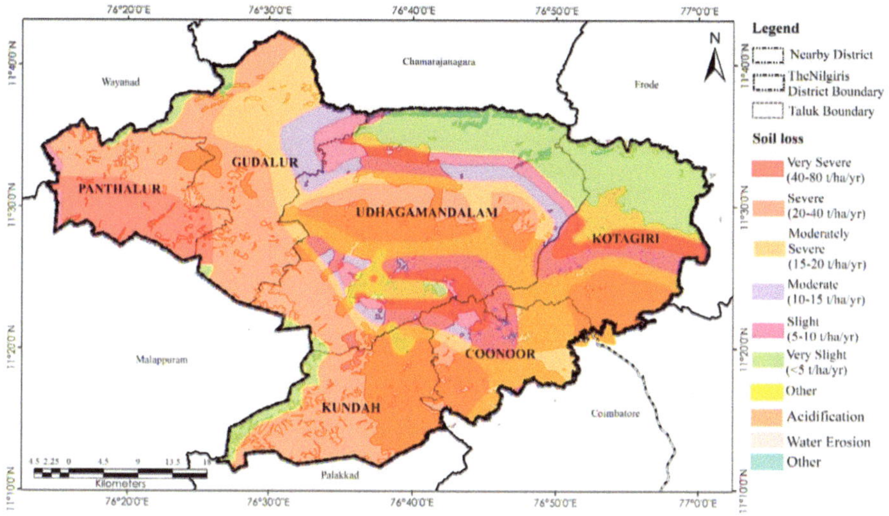

Fig. 17.10 Soil erosion and land degradation mapping for the Nilgiris

and migration behaviors, resulting in compromised pollination systems and fragmented food chains—a trend replicated in other similar high-altitude biomes in the Western Ghats and Eastern Himalayas (Das et al., 2019; Sridhar, 2024). Farmers have increasingly planted heat- and drought-resistant crops in the face of declining yields, reflecting a slow transition to adaptive agricultural practices (Ravichandran et al., 2020; Sengupta et al., 2010). In the end, community-led, technology-facilitated land restoration presents a scalable route to long-term resilience in the Nilgiris—offering a replicable model for other climate-exposed, mountainous landscapes globally (Anbarasu et al., 2010; Gebrehiwot & Demissew, 2020; Sankar et al., 2018) (Fig. 17.11).

The Urgent Need for Climate-Adaptive Land Restoration

Climate vulnerability assessments in land management policies are therefore vital to long-term environmental stability and socio-economic resilience in the Nilgiris region, which faces an ever-growing threat from landslides, soil erosion, and increasing temperatures (Bhattacharyya et al., 2015; Borrelli et al., 2017a). Findings show the imperative for adaptive restoration approaches incorporating slope stabilization and efforts toward reducing landslide risks since such efforts enhance cohesion of the soil, reduce runoff, and enhance watershed functionality (Borrelli et al., 2017b). Watershed management and sustainable agricultural practices are also vital in controlling soil erosion, preserving topsoil fertility, and maintaining hydrological balance to avoid flooding and land degradation (Dregne, 2022; Pradhan et al., 2012). Reforestation and ecosystem-based adaptation efforts must therefore be prioritized to regulate microclimates, increase carbon sequestration, and restore biodiversity as

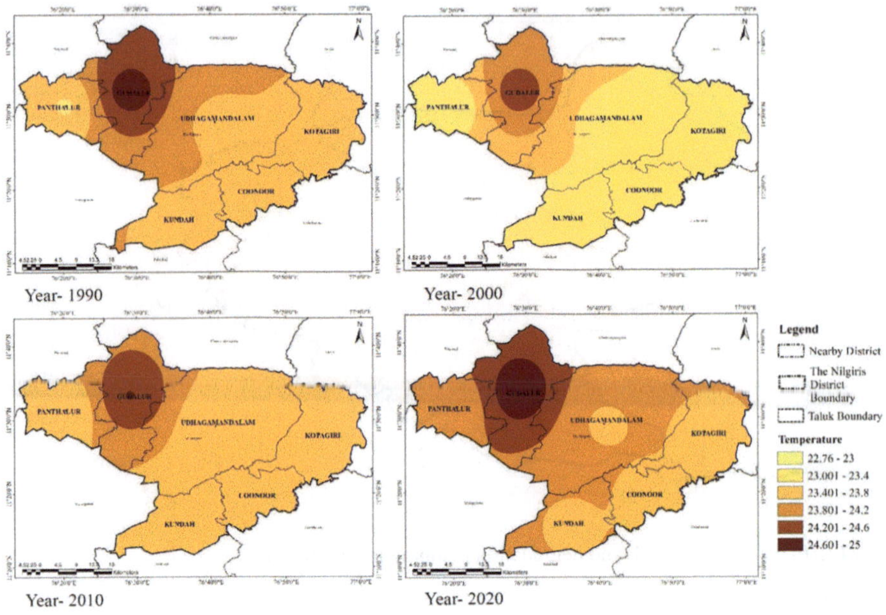

Fig. 17.11 Temperature change analysis for the district (1990–2020)

climate-induced warming exacerbates ecosystem vulnerabilities (Fayas et al., 2019; Sainju et al., 2006).

By using data analytics and predictive modeling, decision-makers can develop interventions that are site-specific, evidence-based, and strengthen climate resilience, biodiversity conservation, and sustainable livelihoods (Arora & Mathur, 2001; Krishna Bahadur, 2009). GIS-based spatial analysis, AI-driven climate simulations, and IoT-enabled real-time monitoring systems will allow for constant assessment of trends in land degradation, ensuring the effective deployment of early warning mechanisms and adaptive management strategies (Ismail & Ravichandran, 2008; Krishna Bahadur, 2009; Saravanan et al., 2021).

17.4 Community-Based Land Restoration Strategies

Climate-induced land degradation in hilly areas presents a pressing challenge that calls for localized, adaptive, and community-driven restoration initiatives (Hendersson, 2019; Pandey et al., 2007). In Nilgiris, an ecologically fragile and biodiverse landscape, severe soil erosion, hydrological disruptions, and ecosystem degradation have been observed along with increases in deforestation due to inappropriate agricultural practices and erratic climate patterns (Science.thewire, 2016; Shiny Mariam Rehel, 2024). Though large-scale governmental initiatives have provided the means of balancing these, community-led restoration efforts present critical components in the achievement of medium to long-term stability for environmental conditions and socio-economic resilience (Rajesh Prasanna et al., 2013; Sridhar, 2024). These are indigenous and agrarian community roles that generally formed the backbone of forest conservation practices, water management, and agriculture productivity persistence (Ravichandran et al., 2020; Shiny Mariam Rehel, 2024). With their traditional ecological knowledge, coupled with scientific land restoration techniques, they can achieve sustainable and climate-resilient landscape management (Chandrasekaran et al., 2013a; Jaiswal et al., 2011).

17.4.1 Watershed Management and Soil Conservation

Watershed management and soil conservation are the two cornerstones of ecological resilience in the Nilgiris, as they prevent soil erosion, maintain hydrological balance, and restore degraded landscapes (Sukhdev, 2005). The district is rugged, steep, and dependent on monsoonal precipitation, which makes it more prone to surface runoff, sediment displacement, and groundwater depletion (Lakshumanan et al., 2012). Therefore, watershed restoration must be done using sustainable techniques (Saravanan et al., 2021). It has greatly reduced the runoff velocity, increased the recharging of groundwater, and stabilized the erosive zones (Senthilkumar & Selvaraj, 2025). This has provided a sustained water supply for agricultural purposes as well as for ecological functions (Mamtha et al., 2016). The setting up of vegetative barriers using deep-rooted native plant species has strengthened the soil stability, trapped sediments, and thus reduced land degradation (Jaiswal et al., 2011). Strengthening riverine ecosystems through rehabilitating degraded riparian zones reduces flood risks and improves water quality, thereby rejuvenating critical habitats for aquatic biodiversity (Chandrasekaran et al., 2013a). Collectively, these watershed restoration initiatives from the community level have curtailed landslide susceptibility, improved soil fertility, and increased agricultural productivity, fostering a more climate-resilient and ecologically balanced system that sustains human livelihoods and conserves biodiversity in the Nilgiris (Hendersson, 2019; Rajesh Prasanna et al., 2013; Sridhar, 2024).

Check Dam Construction and Micro-Watershed Restoration

Check dams and micro-watershed restoration are critical interventions in the Nilgiris, designed to regulate the flow of water, control runoff velocity, and improve groundwater recharge, especially in areas where land degradation will continue to undermine agricultural sustainability along with ecological stability (Krishna Bahadur, 2009; Pandey et al., 2007; Saravanan et al., 2021). Driven by the sense of urgency created by these issues, community bodies and farmers' associations have engaged proactively in the construction of small reservoir check dams that have acted as successful mediums for slowing surface runoff, promoting infiltration and replenishing aquifers (Saravanan et al., 2021). These retain sediment transport, preventing siltation in rivers and reservoirs, and ensure longer-term soil moisture retention; hence delivering direct benefit to local farmers through improved crop productivity and water security (Krishna Bahadur, 2009; Sridhar, 2024).

Restoration of micro-watersheds seeks to identify degraded sub-watersheds and rehabilitate them by afforestation and soil conservation measures (Arum Sari et al., 2017). This is primarily an important activity in upper catchment areas where the resultant effects of deforestation and unsustainable agricultural practices are observed, such as increased runoff and soil loss, plus water scarcity (Ismail & Ravichandran, 2008). Major initiatives in the implementation of restoration activities have been achieved through community-led watershed committees that mobilize the local governing body and panchayats in resource mobilization and labor inputs to construct small earthen or masonry check dams in areas of high runoff (Bhattarai & Dutta, 2007). The restoration of native vegetation along the watershed boundaries has stabilized the soil, reduced erosion, and generally restored the hydrological balance (Fayas et al., 2019). The construction of rainwater harvesting ponds and percolation tanks has further supported water conservation efforts, ensuring there is enough water for agriculture and livestock even in the longest dry seasons (Bhattarai & Dutta, 2007; Pandey et al., 2007). Multi-faceted restoration strategies have shown remarkable success in improving water retention in semi-arid and degraded landscapes, reducing dependency on erratic monsoonal rains, and fostering ecosystem stability (Fu et al., 2005; Krishna Bahadur, 2009). By strengthening localized, community-based water management systems, Nilgiris has made tremendous progress toward climate-resilient land restoration, providing long-term resistance to hydrological disruptions and environmental degradation (Arum Sari et al., 2017; Ismail & Ravichandran, 2008; Vijayalakshmi Sridhar, 2024).

Vegetative Barriers for Soil Erosion Control

Soil erosion, exacerbated by heavy monsoon rains and deforestation, continues to be a major problem in the Nilgiris (Das et al., 2019; Science.thewire, 2016). In this regard, community-based soil conservation programs have successfully implemented vegetative barriers as an ecologically sustainable, cost-effective solution to stabilize the soil, prevent landslides, and restore degraded farmlands (Gebrehiwot & Demissew, 2020; Sainju et al., 2006). These vegetative barriers are carefully positioned rows of trees, shrubs, and grasses that function as natural erosion control systems (Jima et al., 2021). They reduce surface runoff, trap sediments, and retain soil moisture- particularly on steep slopes, components considered important in preventing soil degradation (Parthasarathy & Karthikeyan, 1997). Key practices in vegetative barriers include the use of deep-rooted native plants such as vetiver grass, Napier grass, and bamboo, planted along slopes and field boundaries to strengthen soil structure, reduce erosion, and increase water infiltration (Bhattacharyya et al., 2015; Hobbs et al., 2008). These include contour hedgerows and strip cropping methods where trees and perennial crops are planted beside contour lines to avoid soil erosion (Das et al., 2019). Communities also participate in afforestation, mobilizing seed dispersal and reforestation of cleared lands that work to restore the ecological balance and stabilize the soil (Lakshumanan et al., 2012; Sridhar, 2024). With respect to protected soil erosion, great vegetative barriers can reduce erosion rates by up to 50% (Ravichandran et al., 2020). Ultimately, such methods greatly enhance the long-term health of the soil (Anbarasu et al., 2010; Ravichandran et al., 2020; Sankar et al., 2018). The farmers using these conservation techniques indicate higher yields of crops and soil fertility (Gebrehiwot & Demissew, 2020; Sainju et al., 2006). As compared to engineered solutions, vegetative barriers are a less costly and more sustainable approach since they require little investment and are manageable by local communities, making them a scalable solution to soil erosion (Bhattarai & Dutta, 2007; Fayas et al., 2019; Godwin Vasanth Bosco, 2013).

Riparian buffer restoration—through native vegetation, bamboo bunds, and coir-based bioengineering (Bhattacharyya et al., 2015)—has prevented riverbank collapse, reduced agricultural runoff, and improved aquatic habitats (Chandrasekaran et al., 2013b; Rehel, 2024). Integrated farming in these buffer zones with wetland-compatible crops like bamboo and fodder grasses balances conservation with economic returns, demonstrating how watershed health directly supports sustainable livelihoods (Hendersson, 2019; Krishna Bahadur, 2009; Pandey et al., 2007). These combined strategies—rooted in community participation and traditional ecological knowledge—have not only improved hydrological stability and soil fertility but also built long-term climate resilience and ecosystem functionality across the Nilgiris landscape (Fayas et al., 2019; Ismail & Ravichandran, 2008; Ray et al., 2007; Saravanan et al., 2021; Shao & Wu, 2008).

17.4.2 Shola Forest Restoration and Biodiversity Protection

The Shola Forest of the Nilgiris, ecologically diverse montane systems of the Western Ghats, are of extreme importance to climate regulation, recharge of groundwaters, and maintenance of biodiversity (Bosco, 2013; Bhattacharyya et al., 2015; Hobbs et al., 2008). They are increasingly exposed to land use changes, climatic variability, and invasive flora, which have intervened in cycles of natural regeneration and deteriorated native habitats (Dregne, 2022; Science.thewire, 2016).

Restoration programs, initiated by local communities with assistance from NGOs and research organizations, have focused on native species reforestation, elimination of invasives such as *Lantana camara* and *Acacia mearnsii*, and application of assisted natural regeneration to enhance ecological rehabilitation (Pradhan et al., 2012; Rehel, 2024). Afforestation of the area using indigenous tree species *Michelia nilagirica* and *Syzygium arnottianum* has enhanced microclimatic stability, water recharge, and habitat integrity and helped in enhancing seedling survival and socio-ecological involvement through community nurseries (Hendersson, 2019; Ismail & Ravichandran, 2008; Saravanan et al., 2021; Shao & Wu, 2008). Invasive species management through physical removal and bio-remediation methods has recovered natural biodiversity and minimized fire hazard (Arum Sari et al., 2017; Author et al., 1993), with some locations exhibiting natural succession after removal of the eucalyptus and wattle plantations (Mukherjee & Singh, 2020; Saravanan et al., 2021; Sankar et al., 2018). These measures have reactivated the local economies through sustainable forestry of non-timber forest products and ecotourism and brought together conservation with community prosperity (Fu et al., 2005; Krishna Bahadur, 2009; Pradhan et al., 2012). Being long-term carbon sinks, Shola forests mitigate climate, and their potential is being augmented by agroforestry systems that include food crops, nitrogen-fixing species, and native trees (Biswas & Pani, 2015; Sankar et al., 2018). This shift—facilitated by carbon credit schemes and Payments for Ecosystem Services (PES)—has made farming communities engage in carbon farming while decreasing dependence on monoculture systems (Author et al., 1993; Bhattacharyya et al., 2015; Ismail & Ravichandran, 2008). Shola Forest restoration illustrates how ecosystem-based methods, based on community engagement and economic rewards, can at once conserve biodiversity, enhance climate resilience, and advance sustainable livelihoods in vulnerable hill ecosystems (Bonilla & Johnson, 2012; Borrelli et al., 2017b; Fayas et al., 2019).

17.4.3 Agroforestry and Sustainable Land Use

Nilgiris agroforestry has become a climate-resilient land-use approach that combines tree-based systems with crops and livestock, providing ecological and socio-economic advantages in degraded hill environments (Bhattarai & Dutta, 2007; Sainju et al., 2003). It tackles key issues like soil erosion, loss of fertility, and microclimatic instability by encouraging multi-layer vegetation, improving soil moisture retention, and developing carbon sinks through perennial crops and indigenous tree cover (Bhattacharyya et al., 2015; Jaiswal et al., 2011; Saravanan et al., 2021; Sukhdev, 2005). Agroforestry systems are more beneficial than monocultures as they promote biodiversity, stabilize microclimates, and provide diversified incomes through the sustainable production of timber, fruits, medicinal plants, and non-timber forest products (Lakshumanan et al., 2012; Mamtha et al., 2016). Shade cultivation, especially of tea and coffee interspersed with crops like Grevillea and jackfruit, increases the level of soil carbon and lowers evapotranspiration—a practice also observed in other mountainous countries like the Andes and Eastern Africa (Chandrasekaran, Sayed Owaise, et al., 2013a, 2013b). Grassroots cooperatives and NGOs have large-scale community agroforestry projects that integrate afforestation, watershed restoration, and organic soil revitalization to foster participatory governance and restoration of ecosystems (Ravichandran et al., 2020). Organic farming practices such as composting and nitrogen-fixing crops also enhance soil health and limit chemical runoff, thereby ensuring long-term agroecological sustainability (Chandrasekaran et al., 2013b). Together, these locally led agroforestry initiatives demonstrate the ways in which climate-smart land-use systems can improve livelihoods, restore ecosystem function, and promote landscape-scale resilience in high-risk hill environments (Bonilla & Johnson, 2012; Dregne, 2022; Hendersson, 2019; Ravichandran et al., 2020; Rehel, 2024).

17.4.4 Challenges of Community-Driven Efforts

Community-driven initiatives have garnered considerable attention in ecosystem restoration, including vulnerable and biodiversity-rich areas (Hendersson, 2019). These initiatives involve community-led mobilization, traditional ecological knowledge, and shared decision-making (Ravichandran et al., 2020; Rehel, 2024). Because these approaches are based on local images and engagement with local needs for restoration goals, these initiatives often have greater ownership, have greater commitments to sustain initiatives, and enhance resilience for communities. Ecological resilience is developed when stewards are engaged in land management activities (Bhattacharyya et al., 2015; Sukhdev, 2005).

Yet while some community-led models have transformative potential, they have their own limitations. Without adequate structural support, for example, maintenance of community-directed environmental commitment, these approaches can become under-resourced, poorly governed, and subject to local tensions and conflicts (Jaiswal et al., 2011; Saravanan et al., 2021).

Challenges of Community-Driven Initiatives

See Fig. 17.12.

Recommendations to Overcome These Challenges

See Fig. 17.13.

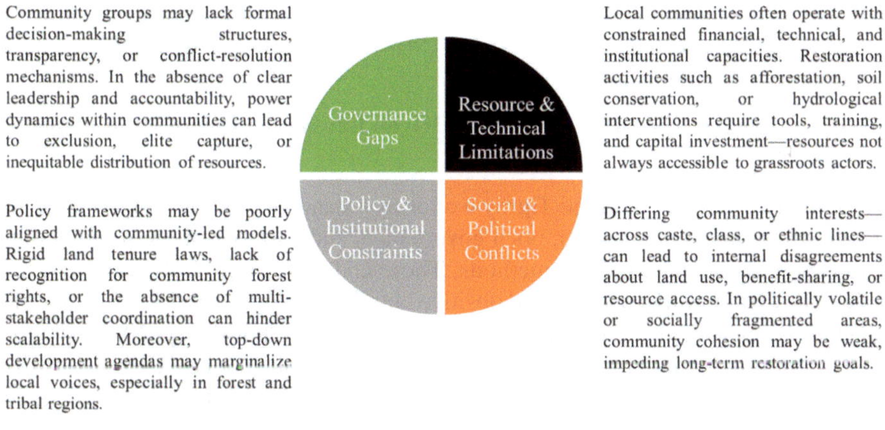

Fig. 17.12 Drawbacks of community-led initiatives

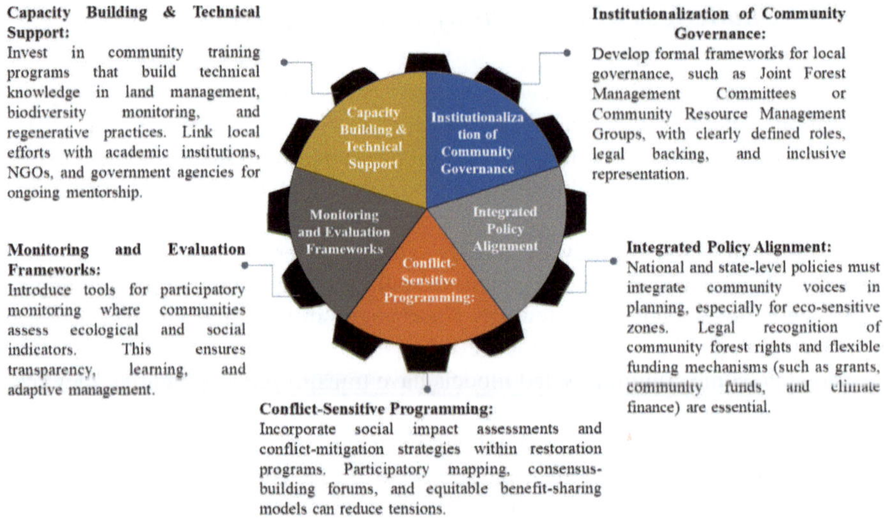

Fig. 17.13 Recommendations to overcome challenges of community-driven efforts

17.4.5 Case Studies

Community-driven ecosystem restoration initiatives have demonstrated varying degrees of success globally. By analyzing specific case studies that contribute to the effectiveness or shortcomings of these projects. The table below provides a comparative overview of selected successful and failed restoration efforts, highlighting critical community-driven indicators.

Comparative Analysis of Community-Driven Ecosystem Restoration Projects

See Table 17.1.

Both Hiware Bazar and the Tigray exclosures stand as examples of the need for active community engagement, open governance, and harmonization of indigenous knowledge with contemporary methods. These endeavors were bolstered by favorable policy frameworks and showed positive tangible ecological and socio-economic change (Birhane et al., 2018; Mekuria et al., 2015; Roy Chowdhury, 2020).

The failures of India's Joint Forest Management and Indonesia's reforestation initiatives expose the risk of top-down methods that disregard community participation. Technical planning shortcomings, insufficient local governance, and non-supportive policies were among the causes of their failures (Arfiansyah et al., 2024; Hardiono et al., 2014; Kumari & Philip, 2020; Murali et al., 2003).

Table 17.1 Comparative analysis of the case study

Indicator	Hiware bazar, India (Kumari & Philip, 2020; Roy Chowdhury, 2020)	Tigray exclosures, Ethiopia (Birhane et al., 2018; Mekuria et al., 2015)	Joint forest management, India (Kumari & Philip, 2020; Majhi, 2016; Murali et al., 2003)	Indonesia reforestation programs (Arfiansyah et al., 2024; Hardiono et al., 2014)
Community participation	High engagement through Gram Sabha meetings and voluntary labor (shramdaan)	Strong local involvement with community-established exclosures	Limited genuine participation; top-down approach	Minimal community input; externally driven initiatives
Governance structure	Transparent local governance with community-led decision-making	Community-managed with support from EthioTrees project	Centralized control by forest departments	Government-led with little local governance integration
Resource mobilization	Effective use of government schemes and local contributions	Revenue from carbon credits reinvested in community projects	Dependence on government funding; lack of local investment	Reliance on external funding; unsustainable financial models
Technical support	Integration of traditional knowledge with modern watershed techniques	Implementation of soil and water conservation methods	Inadequate technical support for communities	Poor technical planning; inappropriate species selection
Policy environment	Supportive policies under the Ideal Village program	Alignment with national restoration commitments	Conflicting policies; lack of legal recognition for community rights	Policies favoring rapid afforestation over ecological suitability
Social cohesion	Strong community unity and collective action	Shared long-term vision for land restoration	Social conflicts due to unequal benefit-sharing	Community resistance due to a lack of consultation
Ecological outcomes	Significant increase in tree cover and groundwater levels	Improved biodiversity and carbon sequestration	Limited ecological improvement; continued degradation	Low survival rates of planted trees; ecological mismatch
Sustainability	Long-term maintenance through community ownership	Ongoing benefits and reinvestment in local infrastructure	Unsustainable due to a lack of community buy-in	Projects abandoned; lack of long-term commitment

17.5 Role of AI, IoT, and GIS in Enhancing Land Restoration

Technological advancements in Artificial Intelligence, the Internet of Things, and Geographic Information Systems have dramatically reshaped the land restoration and adaptation techniques and policies in such ecologically fragile regions as the Nilgiris (Arora & Mathur, 2001; Foody & Arora, 1997). Induced by increasingly frequent climate-related disasters, including landslides, erosion, and hydrological shifts, data-driven decision tools have become the cornerstone for sustainable land management and ecosystem rehabilitation (Arora et al., 2004; Kirschbaum et al., 2010; Saha et al., 2002). Traditional land restoration approaches, however effective they are, base much of their work on manual observations and the use of historical trends (Fu et al., 2005; Krishna Bahadur, 2009; Senthilkumar & Selvaraj, 2025). This becomes inadequate and challenged under rapidly changing climate conditions. AI-powered climate models, real-time environmental monitoring through IoT sensors, and high-resolution GIS mapping, therefore, offer more precise capabilities to adapt and predict while augmenting the efficiency, accuracy, and scalability of restoration efforts (Jaiswal et al., 2011). Leverage these technologies: policymakers, environmental scientists, and local communities may be able to develop targeted interventions, optimize resource allocation, and improve ecological resilience in degraded landscapes (Das et al., 2021; Sharma et al., 1996). The diverse topography, sensitive ecosystems, and socio-economic dependence on agriculture and forestry in the Nilgiris present a critical case that can be filled by technology-enabled restoration approaches (Abdullahi et al., 2015; Gao & Jiang, 2022).

17.5.1 Technology-Enabled Restoration Approaches

AI-Powered Predictive Modelling for Climate Adaptation AI has emerged as a powerful tool for climate adaptation and land restoration, especially through an ability to analyze large databases, recognize patterns, and predict changes in the environment with high accuracy (Al-Sartawi, 2021). The use of machine learning algorithms fed with the history of climate data, satellite imagery, and soil health records can produce predictive models that anticipate climate risks, ecosystem vulnerabilities, and hotspots of land degradation (Harrington & McInnes, 2009; Stratigea et al., 2015). In the context of the Nilgiris, AI-driven climate adaptation models are used to assess landslide susceptibility by analyzing topographical, meteorological, and geospatial data to identify high-risk areas, and to forecast rainfall variability and drought risks, thus allowing for proactive water resource management and irrigation planning (Bharadiya & Bharadiya, 2023; Dhir et al., 2009; Mamtha et al., 2016). Optimize afforestation efforts by predicting which native species will have the highest survival rates under future climate scenarios (Bhattacharya et al., 2022; Rajesh Prasanna et al., 2013). By introducing AI-based distant sensing tools, researchers and conservationists can automatically monitor vegetation, calculate deforestation rates, and gauge the efficiency of carbon sequestration in restored ecosystems (Anbarasu et al., 2010; Gebrehiwot & Demissew, 2020).

17.5.2 IoT-Based Real-Time Soil and Water Monitoring

The innovation of IoT in land restoration transformed the monitoring and management of soil health and hydrological parameters (Borrelli et al., 2017b; Fayas et al., 2019). IoT-enabled sensor networks provide real-time, continuous information on soil moisture, nutrient composition, and water availability critically important factors in land degradation and ecosystem restoration (Abdullahi et al., 2015; D'Amico et al., 2020). In the Nilgiris, IoT-based monitoring systems are being installed to track soil moisture fluctuations so that the irrigation schedule of farmers can be optimized for maximum water use efficiency (D'Amico et al., 2020). To detect early signs of soil erosion and nutrient depletion so that timely interventions such as cover cropping, mulching, and organic amendments can be undertaken (Biswas & Pani, 2015). Monitoring groundwater levels and river flow rates, providing critical data for watershed restoration and flood prevention planning (Gebrehiwot & Demissew, 2020; Sridhar, 2024). By integrating IoT technology with cloud-based data analytics, real-time environmental changes can be visualized through interactive dashboards and mobile applications, thereby empowering local communities, agricultural cooperatives, and conservationists with actionable intelligence for adaptive land management (Ray et al., 2007). The long-term benefits of IoT-based land restoration approaches include enhanced resilience to climate variability, as real-time data enables early warning systems for extreme weather events (Fayas et al., 2019). Reduction in human error and reliance on manual field assessments, which will enhance the accuracy and efficiency of conservation interventions (Senthilkumar & Selvaraj, 2025). Scalability and replication, as IoT networks can be scaled across different ecological zones to support data-driven restoration strategies (Dregne, 2022; Rehel, 2024; Senthilkumar & Selvaraj, 2025; Sridhar, 2024).

17.6 Policy Recommendations for Climate-Resilient Land Restoration

In climate-sensitive regions, such as the Nilgiris, effective land restoration requires decentralizing governance and giving ecological relevance to local community self-governance, grassroots associations, and traditional governing bodies to environmental conservation and sustainable land management (Jaiswal et al., 2011; Rehel, 2024). With ecological vulnerability and high levels of socio-economic reliance on agriculture and forestry, there is much need to ensure that strengthening of local governance mechanisms can ensure land-use decisions that are environmentally sound and socially inclusive (Kirschbaum et al., 2010). Decentralized, climate-adaptive planning together with policy-driven incentives for community-led restoration efforts can significantly enhance long-term ecological resilience while fostering local ownership of the initiative (Author et al., 1993; Sridhar, 2024). One of the most effective approaches to governance in the Nilgiris has been the integration of decentralized planning institutions, such as Hill Area Conservation Authorities (HACA) and local Panchayats, into land restoration programs (Rajesh Prasanna et al., 2013; Science.thewire, 2016). They are critical institutions in formulating micro-level climate resilience strategies and enforcing land-use regulations and enabling the implementation of restoration projects at the village and district levels (Henderson, 2019; Ravichandran et al., 2020). In these regards, the delineation of ecologically sensitive zones and curbing unsustainable land-use practices have been the result of HACA (Chandrasekaran, Elayaraja, et al., 2013a). Likewise, the Panchayati Raj institutions have empowered communities to manage their own forests through afforestation, watershed management, and agroforestry programs, increasing grassroots participation in conservation efforts (Jaiswal et al., 2011; Mamtha et al., 2016; Sukhdev, 2005).

To enhance the effectiveness of decentralized climate-resilient planning, the following measures should be prioritized (Fig. 17.14).

Governance reforms, financial and policy incentives towards community-led restoration initiatives, will significantly amplify the impact of conservation and encourage wider participation in land rehabilitation programs (Lakshumanan et al., 2012; Mukherjee & Singh, 2020). Direct incentives such as payments for ecosystem services, tax benefits, and carbon credit schemes can motivate landowners and farming communities to engage in long-term ecological restoration efforts (Owen et al., 2008; Pachauri & Pant, 1992; Saravanan et al., 2021; Solonenko, 1997).

Policy-driven financial mechanisms should include (Fig. 17.15).

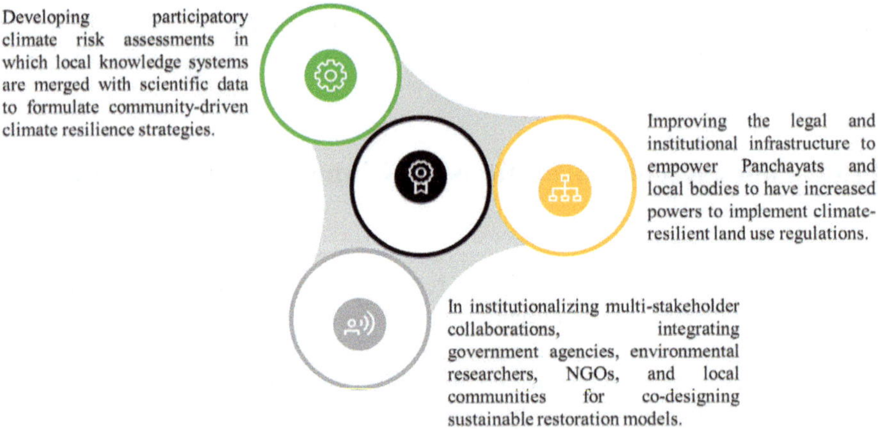

Fig. 17.14 Prioritized measures for climate-resilient planning

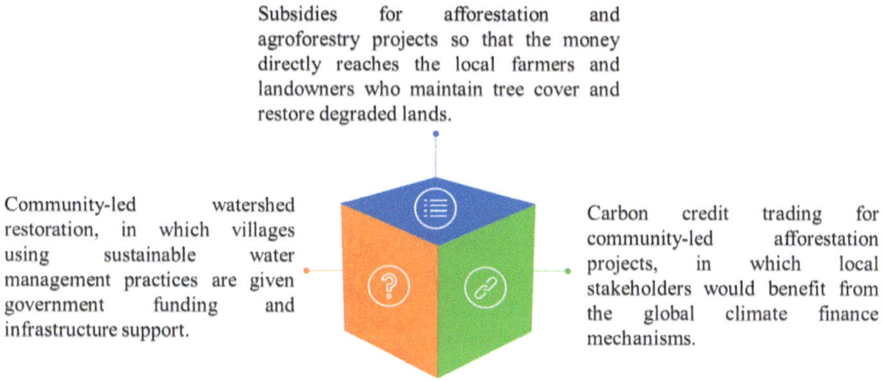

Fig. 17.15 Policy-driven financial mechanisms

17.7 Integrating Climate Finance and Carbon Sequestration Incentives

Strategically crafted climate finance instruments can improve long-term sustainability of land rehabilitation in the Nilgiris by connecting environmental protection to economic rewards for local communities (Anbarasu et al., 2010; Pachauri & Pant, 1992). Chronic loss of forests, land, and water systems imperils both environmental health and the livelihood of agrarian and tribal communities. Financial tools like Payments for Ecosystem Services (PES) and carbon credits provide key avenues to leverage restoration finance while linking economic incentives to environmental performance (Anbarasu et al., 2010; Shao & Wu, 2008; Solonenko, 1997). PES programs perform particularly well in situations such as the Nilgiris, since they pay rewards to local stakeholders for afforestation, soil protection, and watersheds restoration activities, promoting long-term ecological management (Hong et al., 2022; Jia et al., 2019). These programs also reduce rural incomes' reliance on resource-demanding agriculture and deforestation (Jia et al., 2011; Niemeijer & de Groot, 2008). Carbon markets provide another scalable option that enables community-based restoration initiatives to earn tradable credits through sequestering carbon through agroforestry and afforestation (Datri et al., 2024; Lyons et al., 2016). These credits can be marketed to industries that seek to achieve net-zero, providing a consistent source of income for conservation and helping achieve global carbon neutrality (Erwin, 2009; Mondal et al., 2017). Carbon offset projects at the local level may be able to empower rural communities by commodifying ecological work, thereby filling fiscal gaps in landscape rehabilitation (Datri et al., 2024; Lyons et al., 2016). Public–Private Partnerships (PPPs) provide an alternative financing channel, allowing cooperation between governments, NGOs, private financiers, and community groups (Er et al., n.d.; Pan & Juan, 2020). These models allow combined investment in reforestation, climate-resilient infrastructure, and land surveillance, particularly from corporations that share ESG and sustainability requirements (Hong et al.,

2022; Lal, 2008). Green philanthropy and climate bonds can also release funds for local-level ecological initiatives, enhancing community-level ownership of conservation (Parihar et al., 2018). Yet, success is dependent on sound policy design, robust accountability systems, and benefit-sharing frameworks (Chandrasekaran et al., 2013a; Lakshumanan et al., 2012; Mamtha et al., 2016).

Standardizing carbon credit validation, improving PES transparency, and targeting benefits to vulnerable groups are all essential for equity and legitimacy (Hassan et al., 2014; Vázquez-Torres & Gómez-Amador, 2022). The integration of AI and GIS tools into carbon tracking offers real-time insights into the ecological returns of restoration investments, improving transparency and adaptive planning (Sridhar, 2024). By integrating carbon incentives into inclusive financial frameworks, policymakers can make ecological restoration a sustainable economic route where communities are not just recipients but co-creators of climate resilience (Author et al., 1993; Hassan et al., 2014; Vázquez-Torres & Gómez-Amador, 2022). Strengthening these joined-up climate finance approaches will be central to upscaling land restoration in fragile hill ecosystems such as the Nilgiris—compliant with global biodiversity, climate, and sustainability agendas (Jaiswal et al., 2011; Kirschbaum et al., 2010; Kumar & Pushplata, 2013; Rehel, 2024).

17.8 Capacity Building and Knowledge Sharing

Achieving climate resilience in the Nilgiris requires a robust capacity-building ecosystem that empowers communities, farmers, and institutions with skills, technical tools, and contextual knowledge to manage land restoration sustainably (Ravichandran et al., 2020; Rehel, 2024). Education and knowledge-sharing must complement policy and finance efforts to improve adaptive land management in response to climate stress, soil degradation, and ecosystem fragmentation (Chandrasekaran et al., 2013b). Institutional multi-level collaboration and development of human capital are necessary to infuse a culture of long-term ecological stewardship (Jaiswal et al., 2011). Thematic training modules for farmers, tribal groups, and local stakeholders should focus on context-specific practices in agroforestry, watershed management, and soil conservation (Mamtha et al., 2016; Sukhdev, 2005). A combination of TEK and science-based practices by means of learning in the field can enhance restoration knowledge application on the ground (Lakshumanan et al., 2012). Peer-to-peer learning, workshops, and demonstration plots can speed up the spread of best practices, with increased productivity and landscape resilience (Saravanan et al., 2021). Gender-responsive training needs to be given priority, in cognizance of the critical role of women towards sustainable agriculture and resource management (Senthilkumar & Selvaraj, 2025). The research–implementation gap needs to be bridged through dynamic collaborations between universities, local governments, and field institutions (Saravanan et al., 2021; Senthilkumar & Selvaraj, 2025). Collaborations need to position scientific models within the context of local ecologies, such that restoration techniques are empirically valid and locally suitable (Mamtha et al.,

2016). Inculcating digital technology—such as AI-based analytics, remote sensing, and GIS—within training will enable farmers to track soil moisture, water levels, and forest regrowth in real-time (D'Amico et al., 2020; Das et al., 2021; Zhong et al., 2021). Such platforms as regional think tanks and community forums can facilitate ongoing conversation, co-strategy creation, and diffusion of innovations with local relevance (Fagotti et al., 2022). Long-term effects rely on the development of autonomous, ecologically conscious communities that can scale and lead restoration efforts on their own (Kumar & Singh, 2020; Nissen, 2016). By integrating education and participatory governance into restoration planning, local stakeholders shift from passive recipients to active designers of resilience (Garg, 2015; Sharma et al., 1996).

Augmenting these human–institutional knowledge systems will be critical in ensuring restoration efforts in the Nilgiris are inclusive, sustainable, and responsive to emerging climate realities (Abaker et al., n.d.; Ewing et al., 2013; Felice et al., n.d.; Kar et al., 2019; Zhong et al., 2021; Senthilkumar & Selvaraj, 2025).

17.9 Conclusion

The Nilgiris provides an excellent case study of how community-based land restoration can be utterly effective (as an adaptation and mitigation tool) against climate change and in creating ecological and socio-economic resilience. As one of India's most climate-sensitive and ecologically diverse regions, the Nilgiris has experienced escalating environmental degradation driven by deforestation, erratic climate patterns, and unsustainable land-use practices. Yet this study has shown that these more integrated and community-driven restoration initiatives—integrated efforts combining watershed management with agroforestry and afforestation—have been a transformative force with regard to halting soil erosion, re-energizing biodiversity, and stabilizing fragile ecosystems. Water conservation projects have been shown to mitigate high surface runoff, accelerate groundwater restoration, and stabilize landslides. Also, the role played by check dams, vegetative barriers, and soil conservation structures was instrumental in maintaining hydrological balance and agricultural sustainability so that communities could cope with climate impact, leading to water scarcity. Likewise, implementing agroforestry systems that combine native tree species and traditional crops has not only enhanced soil health and improved carbon sequestration potential but has also created sustainable economic avenues for the area's farmers. The switch from monoculture plantation systems to diversified agroforestry models has allowed landowners to increase crop resilience, boost farm yield, and promote ecosystem restoration.

Moreover, this research underscores that land restoration is not merely an ecological intervention, but a social process that must integrate community wisdom, policy direction, and technological adaptation. Advanced technological interventions such as AI, IoT, and GIS have considerably enhanced the accuracy, effectiveness, and scalability of land restoration activities. Early climate risk identification led to the early implementation of targeted adaptation strategies by the stakeholders, taking into consideration a trend in climate variability and indicators of land degradation. IoT-based real-time environmental monitoring has helped to track the soil moisture, nutrient composition, and water availability in the field to implement data-driven restoration interventions more effectively. In addition, the GIS-based mapping tools have also been very essential in the identification of high-risk zones, monitoring reforestation progress, and optimizing land-use planning to avert further ecological degradation. These digital innovations with community-based conservation efforts indicate that technology-driven environmental management is bound to achieve restoration goals in the long run.

Compared with similar restoration programs in the Andes (Peru) and Ethiopian Highlands, where community-led approaches and participatory monitoring have significantly reduced erosion and enhanced agricultural productivity, the Nilgiris case reflects a global parallel and adds further value by showcasing tech-integrated restoration practices. Despite these successes, the future of land restoration in the Nilgiris will depend on stronger policy frameworks and continued funding. Climate-resilient land management requires a governance structure that empowers local institutions, facilitates participatory decision-making, and ensures equitable access to restoration funding. The incentive of carbon credit programs, Payments for Ecosystem Services, and PPPs would provide the financial resources necessary to scale up restoration initiatives, but local communities should benefit economically from their environmental stewardship. It is important to formalize these efforts through cross-cutting policies that combine ecological, economic, and social aims and to make sure that restoration becomes and remains a major part of efforts to plan for climate adaptation and development.

The integration of climate adaptation policies into regional and national development frameworks is also a policy priority that policymakers must address. Restorative land practices must be embedded in climate resilience agendas, development planning, and ecosystem service valuation. Future work in this area should be focused on having long-term monitoring and evaluation mechanisms to better assess effectiveness and scalability over broader timeframes for restoration efforts. Elaborating on the socioeconomic benefits of agroforestry and afforestation will also be critical in demonstrating how nature-based solutions can simultaneously address climate challenges and enhance rural livelihoods. This study offers a replicable model of global relevance. The restoration frameworks developed in Nilgiris create opportunities for policy learning and knowledge sharing across other climate vulnerability hilly landscapes in Asia, Africa, and Latin America. In addition, deep research into AI and remote sensing technologies can improve the capability to predict climate conditions, allowing proactive adaptation of land use management according to changing environmental conditions. Interdisciplinary collaboration between scientists and policymakers and involving indigenous communities, can enable the development of holistic restoration models, hence integrating traditional ecological knowledge with modern conservation science. This research contributes significantly to the field of sustainable land management by offering a hybrid model of community-driven, tech-enabled, and policy-supported interventions. Nilgiris, community-led technology and policy-supported initiatives can prove for restoring lands while being a stepping stone towards developing climate resilience in ecosystems and improving ecological sustainability.

This research concludes that the critical aspect of the development of reversing degradation trends of hills with local participation, scientific innovativeness, and institutional assistance. It creates an important scientific imprint on the conversation of eco-resilience, which has immediate applicability to progress towards the SDGs, Land Degradation Neutrality (LDN) aspirations, and climate adaptation commitments under the Paris Agreement. The replicable successful restoration models through scalable integration of science-based, financially viable, and governance-related interventions would enable Nilgiris as a prototype towards sustainable land management in mountainous ecosystems, more so for areas with vulnerability towards climate.

References

Abdullahi, S., Pradhan, B., & Jebur, M. N. (2015). GIS-based sustainable city compactness assessment using integration of MCDM, Bayes theorem and RADAR technology. *Geocarto International, 30*(4), 365–387. https://doi.org/10.1080/10106049.2014.911967

Al-Sartawi, A. M. A. M. (Ed.) (n.d.). *Lecture Notes in Networks and Systems 238 Artificial Intelligence for Sustainable Finance and Sustainable Technology Proceedings of ICGER 2021.* https://link.springer.com/bookseries/15179

Anbalagan, R. (1992). Landslide hazard evaluation and zonation mapping in mountainous terrain. In *Engineering geology* (Vol. 32).

Anbarasu, K., Sengupta, A., Gupta, S., & Sharma, S. P. (2010). Mechanism of activation of the Lanta Khola landslide in Sikkim Himalayas. *Landslides, 7*(2), 135–147. https://doi.org/10.1007/s10346-009-0193-0

Arfiansyah, D., Han, H., & Zlatanova, S. (2024). Land suitability analysis for residential development in an ecologically sensitive area: A case study of Nusantara, the New Indonesian Capital. *Sustainability (Switzerland), 16*(13). https://doi.org/10.3390/su16135767

Arora, M. K., Das Gupta, A. S., & Gupta, R. P. (2004). An artificial neural network approach for landslide hazard zonation in the Bhagirathi (Ganga) valley, himalayas. *International Journal of Remote Sensing, 25*(3), 559–572. https://doi.org/10.1080/0143116031000156819

Arora, M. K., & Mathur, S. (2001). Multi-source classification using artificial neural network in a rugged terrain. *Geocarto International, 16*(3), 37–44. https://doi.org/10.1080/10106040108542202

Arum Sari, F., Rahmadi, A., & Hikmaya Aji Ningrum, dan. (2017). *Konservasi Tanah dan Air di Wilayah Cimencrang, Jawa Barat Soil and Water Conservationin Cimencrang, West Java.*

Author, B., Boardman, J., & Favis-Mortlock, D. T. (1993). Climate change and soil erosion. *The Geographical Journal, 159*(2).

Bharadiya, J., & Bharadiya, J. P. (2023). Introduction of the popular machine learning algorithm. *International Journal of Innovative Research in Science Engineering and Technology, 8*(5). https://doi.org/10.5281/zenodo.8020795

Bhattacharya, S., Jha, H., & Nanda, R. P. (2022). Application of IoT and Artificial Intelligence in Road Safety. In *2022 International conference on interdisciplinary research in technology and management, IRTM 2022—Proceedings.* https://doi.org/10.1109/IRTM54583.2022.9791529

Bhattacharyya, R., Ghosh, B. N., Mishra, P. K., Mandal, B., Rao, C. S., Sarkar, D., Das, K., Anil, K. S., Lalitha, M., Hati, K. M., & Franzluebbers, A. J. (2015). Soil degradation in India: Challenges and potential solutions. *Sustainability (Switzerland), 7*(4), 3528–3570. https://doi.org/10.3390/su7043528

Bhattarai, R., & Dutta, D. (2007). Estimation of soil erosion and sediment yield using GIS at catchment scale. *Water Resources Management, 21*(10), 1635–1647. https://doi.org/10.1007/s11269-006-9118-z

Birhane, E., Mengistu, T., Seyoum, Y., Hagazi, N., Putzel, L., Rannestad, M. M., & Kassa, H. (2018). Exclosures as forest and landscape restoration tools: Lessons from Tigray Region, Ethiopia. *International Forestry Review, 19*(4), 37–50. https://doi.org/10.1505/146554817822330498

Biswas, S. S., & Pani, P. (2015). Estimation of soil erosion using RUSLE and GIS techniques: a case study of Barakar River basin, Jharkhand, India. *Modeling Earth Systems and Environment, 1*(4). https://doi.org/10.1007/s40808-015-0040-3

Bonilla, C. A., & Johnson, O. I. (2012). Soil erodibility mapping and its correlation with soil properties in Central Chile. *Geoderma, 189–190*, 116–123. https://doi.org/10.1016/j.geoderma.2012.05.005

Borrelli, P., Robinson, D. A., Fleischer, L. R., Lugato, E., Ballabio, C., Alewell, C., Meusburger, K., Modugno, S., Schütt, B., Ferro, V., Bagarello, V., Oost, K. Van, Montanarella, L., & Panagos, P. (2017). An assessment of the global impact of 21st century land use change on soil erosion. *Nature Communications, 8*(1). https://doi.org/10.1038/s41467-017-02142-7

Bosco, G. V. (2013). *Shola-grassland restoration in the Nlgiri plateau.* Upstreamecology.

Brockerhoff, E. G., Barbaro, L., Castagneyrol, B., Forrester, D. I., Gardiner, B., González-Olabarria, J. R., Lyver, P. O. B., Meurisse, N., Oxbrough, A., Taki, H., Thompson, I. D., van der Plas, F., & Jactel, H. (2017). Forest biodiversity, ecosystem functioning and the provision of ecosystem services. *Biodiversity and Conservation, 26*(13), 3005–3035. Springer Netherlands. https://doi.org/10.1007/s10531-017-1453-2

Chandrasekaran, S. S., Elayaraja, S., & Renugadevi, S. (2013a). Damages to transport facilities by rainfall induced landslides during November 2009 in Nilgiris, India. *Landslide Science and Practice: Risk Assessment, Management and Mitigation, 6*, 171–176. https://doi.org/10.1007/978-3-642-31319-6_24

Chandrasekaran, S. S., Sayed Owaise, R., Ashwin, S., Jain, R. M., Prasanth, S., & Venugopalan, R. B. (2013b). Investigation on infrastructural damages by rainfall-induced landslides during November 2009 in Nilgiris, India. *Natural Hazards, 65*(3), 1535–1557. https://doi.org/10.1007/s11069-012-0432-x

Congalton, R. G. (1991). *A review of assessing the accuracy of classifications of remotely sensed data* (Vol. 37).

Crosson, P. (1997). Will erosion threaten agricultural productivity? *Environment, 39*(8), 4–31. https://doi.org/10.1080/00139159709604756

D'Amico, F., Calvi, A., Schiattarella, E., Prete, M. D., & Veraldi, V. (2020). BIM and GIS data integration: A novel approach of technical/environmental decision-making process in transport infrastructure design. *Transportation Research Procedia, 45*, 803–810. https://doi.org/10.1016/j.trpro.2020.02.090

Das, L., Salvi, H., & Kamboj, R. D. (2019). Phytosociological study of coastal flora of Devbhoomi Dwarka district and its islands in the Gulf of Kachchh, Gujarat. *International Journal of Scientific Research in Biological Sciences, 6*(3), 1–13. https://doi.org/10.26438/ijsrbs/v6i3.113

Das, S., Bhunia, G. S., Bera, B., & Shit, P. K. (2021). *Evaluation wetlands ecosystem health using geospatial technology: Evidence from lower Gangetic flood plain in India.* https://doi.org/10.21203/rs.3.rs-522708/v1

Datri, L. A., Lopez, M., Buchter, S., Pazcel, E. M., & Gandini, M. (2024). Functional urban wetlands in dysfunctional cities. *Current Landscape Ecology Reports, 9*(2), 21–30. https://doi.org/10.1007/s40823-024-00097-2

Dhir, B., Sharmila, P., & Saradhi, P. P. (2009). Potential of aquatic macrophytes for removing contaminants from the environment. *Critical Reviews in Environmental Science and Technology, 39*(9), 754–781. https://doi.org/10.1080/10643380801977776

Dregne, H. E. (2022). *Soil erosion: Cause and effect.*

Er, K., Zhi Qiang, S., Wright, J., Velautham, E., & Sia, A. (n.d.). *The learning forest: A habitat restoration project that ecologically connects with the Singapore botanic gardens rainforest.*

Erwin, K. L. (2009). Wetlands and global climate change: The role of wetland restoration in a changing world. *Wetlands Ecology and Management, 17*(1), 71–84. https://doi.org/10.1007/s11273-008-9119-1

Ewing, R., Pendall, R., & Chen, D. (2013). *Smart growth America.*

Fagotti, G., Proietti, S. N., Soccodato, F., Alemanno, N., Rotondi, M., Mezzi, M., Verducci, P., Bavicchi, M., Ciavaglia, C., Cincini, L., Finotto, M., Gabbarelli, E., Morosi, F., & Di Carlo, S. (2022). The integrated project for the post-earthquake rebuilding of Castelluccio of Norcia: Procedures, techniques, implementation. *Procedia Structural Integrity, 44*, 339–346. https://doi.org/10.1016/j.prostr.2023.01.045

Fayas, C. M., Abeysingha, N. S., Nirmanee, K. G. S., Samaratunga, D., & Mallawatantri, A. (2019). Soil loss estimation using Rusle model to prioritize erosion control in KELANI river basin in Sri Lanka. *International Soil and Water Conservation Research, 7*(2), 130–137. https://doi.org/10.1016/j.iswcr.2019.01.003

Felice, A., Anisa, V. E., Kirana, P., Designer, M., Muhammad, L., & Arifzqi, F. (n.d.). *From smart city to smart nation 63 March 2020.*

Foody, G. M., & Arora, M. K. (1997). An evaluation of some factors affecting the accuracy of classification by an artificial neural network. *International Journal of Remote Sensing, 18*(4), 799–810. https://doi.org/10.1080/014311697218764

Fu, B. J., Zhao, W. W., Chen, L. D., Zhang, Q. J., Lü, Y. H., Gulinck, H., & Poesen, J. (2005). Assessment of soil erosion at large watershed scale using RUSLE and GIS: A case study in the Loess Plateau of China. *Land Degradation and Development, 16*(1), 73–85. https://doi.org/10.1002/ldr.646

Gao, R., & Jiang, J. (2022). Design and implementation of environmental design based on new energy technology. *Energy Reports, 8*, 7600–7611. https://doi.org/10.1016/j.egyr.2022.05.232

Garg, J. K. (2015). Wetland assessment, monitoring and management in India using geospatial techniques. *Journal of Environmental Management, 148*, 112–123. https://doi.org/10.1016/j.jenvman.2013.12.018

Gebrehiwot, K., & Demissew, S. (2020). *Trends in plant ecology research in Ethiopia (1969–2019): Systematic analysis.* https://doi.org/10.1101/2020.09.28.316463

Hardiono, M., Widodo, S., Safik, A., & Barthel, K. (2014). *Beyond consultation: Participatory village boundary setting (VBS) and community mapping in rural Indonesia.*

Harrington, R., & McInnes, R. (2009). Integrated constructed wetlands (ICW) for livestock wastewater management. *Bioresource Technology, 100*(22), 5498–5505. https://doi.org/10.1016/j.biortech.2009.06.007

Hassan, T. Y., Majid, M. R., Davidson, S. A. A., & Medugu, N. I. (2014). Role of wetlands in mitigating the effect of climate change in Nigeria. In *Handbook of climate change adaptation* (pp. 1–13). Springer Berlin Heidelberg. https://doi.org/10.1007/978-3-642-40455-9_109-2

Hendersson, H. (2019). *The green deserts of Nilgiris.* Ecologist informed by nature.

Hobbs, P. R., Sayre, K., & Gupta, R. (2008). The role of conservation agriculture in sustainable agriculture. *Philosophical Transactions of the Royal Society B: Biological Sciences, 363*(1491), 543–555. Royal Society. https://doi.org/10.1098/rstb.2007.2169

Hong, S., Hui, E. C.-m., & Lin, Y. (2022). Relationship between urban spatial structure and carbon emissions: A literature review. In *Ecological indicators* (Vol. 144). Elsevier B.V. https://doi.org/10.1016/j.ecolind.2022.109456

Ismail, J., & Ravichandran, S. (2008). RUSLE2 model application for soil erosion assessment using remote sensing and GIS. *Water Resources Management, 22*(1), 83–102. https://doi.org/10.1007/s11269-006-9145-9

Jaiswal, P., Van Westen, C. J., & Jetten, V. (2011). Quantitative estimation of landslide risk from rapid debris slides on natural slopes in the Nilgiri hills, India. *Natural Hazards and Earth System Science, 11*(6), 1723–1743. https://doi.org/10.5194/nhess-11-1723-2011

Jia, H., Ma, H., & Wei, M. (2011). Urban wetland planning: A case study in the Beijing central region. *Ecological Complexity, 8*(2), 213–221. https://doi.org/10.1016/j.ecocom.2011.03.002

Jia, Y., Tang, L., Xu, M., & Yang, X. (2019). Landscape pattern indices for evaluating urban spatial morphology—A case study of Chinese cities. *Ecological Indicators, 99*, 27–37. https://doi.org/10.1016/j.ecolind.2018.12.007

Jima, K. B., Asfaw, Z., Demissew, S., & Dalle, G. (2021). *Medicinal plant composition, distribution, usage and conservation status in Nole Kaba district, West Wollega, West Ethiopia*. https://doi.org/10.21203/rs.3.rs-265972/v1

Kar, A. K., Ilavarasan, V., Gupta, M. P., Janssen, M., & Kothari, R. (2019). Moving beyond smart cities: Digital nations for social innovation & sustainability. *Information Systems Frontiers, 21*(3), 495–501. https://doi.org/10.1007/s10796-019-09930-0

Kirschbaum, D. B., Adler, R., Hong, Y., Hill, S., & Lerner-Lam, A. (2010). A global landslide catalog for hazard applications: Method, results, and limitations. *Natural Hazards, 52*(3), 561–575. https://doi.org/10.1007/s11069-009-9401-4

Krishna Bahadur, K. C. (2009). Mapping soil erosion susceptibility using remote sensing and GIS: A case of the Upper Nam Wa Watershed, Nan Province, Thailand. *Environmental Geology, 57*(3), 695–705. https://doi.org/10.1007/s00254-008-1348-3

Kumar, A., & Pushplata. (2013). Building regulations for environmental protection in Indian hill towns. *International Journal of Sustainable Built Environment, 2*(2), 224–231. https://doi.org/10.1016/j.ijsbe.2014.04.003

Kumar, G., & Singh, K. K. (2020). Mapping and monitoring the selected wetlands of Punjab, India, using geospatial techniques. *Journal of the Indian Society of Remote Sensing, 48*(4), 615–625. https://doi.org/10.1007/s12524-020-01104-9

Kumari, J., & Philip, E. (2020). 1 Joint forest management in India: Challenges and opportunities. *International Journal of All Research Writings, 1*(9). www.ijarw.com

Lakshumanan, C., Pradeep Kishore, V., Viveganandan, S., Krishnakumar, P., & Muthusankar, G. (2012). Landuse/Land cover dynamics study in Nilgiris district part of Western Ghats, Tamilnadu. *International Journal of Geomatics and Geosciences, 2*(3).

Lal, R. (2008). Carbon sequestration. *Philosophical Transactions of the Royal Society B: Biological Sciences, 363*(1492), 815–830. https://doi.org/10.1098/rstb.2007.2185

Lyons, J. E., Collazo, J. A., & Herring, G. (2016). Testing assumptions for conservation of migratory shorebirds and coastal managed wetlands. *Wetlands Ecology and Management, 24*(5), 507–520. https://doi.org/10.1007/s11273-015-9477-4

Majhi, B. (2016). 322 Joint forest management programme in India and community participation. *International Journal of Research in Economics and Social Sciences, 6*(3). http://www.euroasiapub.org

Mamtha, R., Jasmine Niketha, M., & Geetha, P. (2016). Analysis of deforestation and land use changes in Kotagiri Taluk of Nilgiris District. *Indian Journal of Science and Technology, 9*(44). https://doi.org/10.17485/ijst/2016/v9i44/105326

Mekuria, W., Langan, S., Johnston, R., Belay, B., Amare, D., Gashaw, T., Desta, G., Noble, A., & Wale, A. (2015). Restoring aboveground carbon and biodiversity: A case study from the Nile basin, Ethiopia. *Forest Science and Technology, 11*(2), 86–96. https://doi.org/10.1080/21580103.2014.966862

Mondal, B., Dolui, G., Pramanik, M., Maity, S., Biswas, S. S., & Pal, R. (2017). Urban expansion and wetland shrinkage estimation using a GIS-based model in the East Kolkata Wetland, India. *Ecological Indicators, 83*, 62–73. https://doi.org/10.1016/j.ecolind.2017.07.037

Mukherjee, F., & Singh, D. (2020). Assessing land use-land cover change and its impact on land surface temperature using LANDSAT data: A comparison of two urban areas in India. *Earth Systems and Environment, 4*(2), 385–407. https://doi.org/10.1007/s41748-020-00155-9

Murali, K. S., Rao, R. J., Sudha, P., Sangeetha, G., Murthy, I. K., & Ravindranath, N. H. (2003). Evaluation studies of Joint Forest Management in India: social and institutional implications. *International Journal of Environment and Sustainable Development, 2*(1).

Niemeijer, D., & de Groot, R. S. (2008). A conceptual framework for selecting environmental indicator sets. *Ecological Indicators, 8*(1), 14–25. https://doi.org/10.1016/j.ecolind.2006.11.012

Nissen, V. (2016). *Multikonferenz Wirtschaftsinformatik (MKWI) 2016 : Technische Universität Ilmenau 09.-11. März 2016*. Universitätsverlag.

Owen, L. A., Kamp, U., Khattak, G. A., Harp, E. L., Keefer, D. K., & Bauer, M. A. (2008). Landslides triggered by the 8 October 2005 Kashmir earthquake. *Geomorphology, 94*(1–2), 1–9. https://doi.org/10.1016/j.geomorph.2007.04.007

Pachauri, A. K., Gupta, P. V, & Chander, R. (1998). Landslide zoning in a part of the Garhwal Himalayas. *Cases and Solutions Environmental Geology, 36*(4). Springer-Verlag 325.

Pachauri, A. K., & Pant, M. (1992). Landslide hazard mapping based on geological attributes. In *Engineering geology* (Vol. 32).

Pan, W., & Du, J. (2020). Chapter 20—Towards sustainable urban redevelopment: urban design informed by morphological patterns and ecologies of informal settlements. *Urban Ecology*, 377–411.

Pandey, A., Chowdary, V. M., & Mal, B. C. (2007). Identification of critical erosion prone areas in the small agricultural watershed using USLE, GIS and remote sensing. *Water Resources Management, 21*(4), 729–746. https://doi.org/10.1007/s11269-006-9061-z

Parihar, C. M., Parihar, M. D., Sapkota, T. B., Nanwal, R. K., Singh, A. K., Jat, S. L., Nayak, H. S., Mahala, D. M., Singh, L. K., Kakraliya, S. K., Stirling, C. M., & Jat, M. L. (2018). Long-term impact of conservation agriculture and diversified maize rotations on carbon pools and stocks, mineral nitrogen fractions and nitrous oxide fluxes in inceptisol of India. *Science of the Total Environment, 640–641*, 1382–1392. https://doi.org/10.1016/j.scitotenv.2018.05.405

Parthasarathy, N., & Karthikeyan, R. (1997). Plant biodiversity inventory and conservation of two tropical dry evergreen forests on the Coromandel coast, south India. *Biodiversity and Conservation, 6*, 1063–1083.

Peddle, D. R., Ledrew, E. F., Foody, G. M., Zhang, A., & Franklin, S. E. (1994). Multi-source image classification II: An empirical comparison of evidential reasoning and neural network approaches. *Canadian Journal of Remote Sensing, 20*(4), 396–407. https://doi.org/10.1080/07038992.1994.10874582

Pradhan, B., Chaudhari, A., Adinarayana, J., & Buchroithner, M. F. (2012). Soil erosion assessment and its correlation with landslide events using remote sensing data and GIS: A case study at Penang Island, Malaysia. *Environmental Monitoring and Assessment, 184*(2), 715–727. https://doi.org/10.1007/s10661-011-1996-8

Rajesh Prasanna, P., Evany Nithya, S., & Allen, D. P. (2013). Remote sensing and GIS for change detection and eco degradation studies in the Nilgiris-South India. *National Conference on Industrial Pollution and Control Technology, 5*(3), 1379–1386.

Ravichandran, V., Singh, R. K., & Murugan, C. (2020). Dipcadi coimbatorensis (Asparagaceae), a new species from the Nilgiri Biosphere Reserve, India. *Kew Bulletin, 75*(1). https://doi.org/10.1007/s12225-020-9873-7

Ray, P. K. C., Dimri, S., Lakhera, R. C., & Sati, S. (2007). Fuzzy-based method for landslide hazard assessment in active seismic zone of Himalaya. *Landslides, 4*(2), 101–111. https://doi.org/10.1007/s10346-006-0068-6

Rehel, S. M. (2024). *Restoring hill wetlands of the Nilgiri Biosphere Reserve*. Happy Valley-Kotagiri.

Roy Chowdhury, A. (2020). An instance of rural development in India: A case study of a model village. *Journal of Rural Development*. https://www.researchgate.net/publication/346400962

Saha, A. K., Gupta, R. P., & Arora, M. K. (2002). GIS-based landslide hazard zonation in the Bhagirathi (Ganga) valley, Himalayas. *International Journal of Remote Sensing, 23*(2), 357–369. https://doi.org/10.1080/01431160010014260

Saha, A. K., Gupta, R. P., Sarkar, I., Arora, M. K., & Csaplovics, E. (2005). An approach for GIS-based statistical landslide susceptibility zonation-with a case study in the Himalayas. *Landslides, 2*(1), 61–69. https://doi.org/10.1007/s10346-004-0039-8

Sainju, U. M., Lenssen, A., Caesar-Thonthat, T., & Waddell, J. (2006). Carbon sequestration in dryland soils and plant residue as influenced by tillage and crop rotation. *Journal of Environmental Quality, 35*(4), 1341–1347. https://doi.org/10.2134/jeq2005.0131

Sainju, U. M., Whitehead, W. F., & Singh, B. P. (2003). Agricultural management practices to sustain crop yields and improve soil and environmental qualities. *The Scientific World Journal, 3*, 768–789. https://doi.org/10.1100/tsw.2003.62

Sankar, M., Green, S. M., Mishra, P. K., Snöälv, J. T. C., Sharma, N. K., Karthikeyan, K., Somasundaram, J., Kadam, D. M., Dinesh, D., Kumar, S., Thilagam, V. K., & Quine, T. A. (2018). Nationwide soil erosion assessment in India using radioisotope tracers Cs and 210 Pb: The need for fallout mapping. *Commentary Current Science, 115*(3).

Saraf, A. K. (2000). Cover: IRS-1C-PAN depicts Chamoli earthquake induced landslides in Garhwal Himalayas, India. *International Journal of Remote Sensing, 21*(12), 2345–2352. https://doi.org/10.1080/014311600500304933

Saravanan, S., Jacinth Jennifer, J., Singh, L., Thiyagarajan, S., & Sankaralingam, S. (2021). Impact of land-use change on soil erosion in the Coonoor Watershed, Nilgiris Mountain Range, Tamil Nadu, India. *Arabian Journal of Geosciences, 14*(407). https://doi.org/10.1007/s12517-021-06817-w/Published

Science.thewire. (2016). *Reforesting the Nilgiris, restoring ecosystems.*

Sengupta, A., Gupta, S., & Anbarasu, K. (2010). Rainfall thresholds for the initiation of landslide at Lanta Khola in north Sikkim, India. *Natural Hazards, 52*(1), 31–42. https://doi.org/10.1007/s11069-009-9352-9

Senthilkumar, H., & Selvaraj, E. N. (2025). GIS-based frequency ratio and Shannon entropy modeling for landslide susceptibility mapping: A case study in Kundah Taluk, Nilgiris District, India. *Open Geosciences, 17*(1). https://doi.org/10.1515/geo-2022-0757

Shao, G., & Wu, J. (2008). On the accuracy of landscape pattern analysis using remote sensing data. *Landscape Ecology, 23*(5), 505–511. https://doi.org/10.1007/s10980-008-9215-x

Sharma, P. K., Chopra, R., Verma, V. K., & Thomas, A. (1996). Flood management using remote sensing technology: The Punjab (India) experience. *International Journal of Remote Sensing, 17*(17), 3511–3521. https://doi.org/10.1080/01431169608949166

Solonenko, V. S. (1997). Landslides and collapses in seismic zones and their prediction. *Association Lnternationale de Geologie de Lvngenieur.*

Sridhar, V. (2024, March 5). *Nilgiris: Water shortage and climate change pose threats to biodiversity and livelihoods.* The Wire.

Stratigea, A., Papadopoulou, C. A., & Panagiotopoulou, M. (2015). Tools and technologies for planning the development of smart cities. *Journal of Urban Technology, 22*(2), 43–62. https://doi.org/10.1080/10630732.2015.1018725

Sukhdev, P. (2005). *Reversing the "Vicious Cycle" of poverty and environmental degradation two Nilgiris examples.*

Vázquez-Torres, C. E., & Gómez-Amador, A. (2022). Impact of indoor air volume on thermal performance in social housing with mixed mode ventilation in three different climates. *Energy and Built Environment, 3*(4), 433–443. https://doi.org/10.1016/j.enbenv.2021.05.002

Zhong, Y., Sun, L., & Ge, C. (2021). Key technologies and development status of smart city. *Journal of Physics: Conference Series, 1754*(1). https://doi.org/10.1088/1742-6596/1754/1/012102

Chapter 18
Climate Change Mitigation Through Ecosystem Restoration in the Bamenda III Municipality, Cameroon

Leinyuy Serge Nangsinyuy and Suiven John Paul Tume

Abstract Land cover change significantly contributes to climate change, directly impacting ecosystem services. This study aims to: (1) evaluate the impact of land cover change and urbanisation on climate change, and (2) explore ecosystem restoration as a strategy for mitigating climate change. The objectives were achieved through land cover change analysis using Landsat images, assessment of local meteorological data to determine rainfall and temperature trends, and documentation of local ecosystem restoration initiatives by civil society organisations. The findings indicated that urbanisation in the Bamenda III Municipality has led to a significant loss of vegetation and the conversion of wetlands to human settlements and agricultural land. Local meteorological records also showed a rise in temperatures alongside a decrease in rainfall. To address these issues, civil society organisations, in partnership with the local government, have implemented sustainable development and ecosystem conservation strategies in the municipality. These efforts include the incorporation of green infrastructure, urban green planning strategies, and community engagement to enhance the resilience of urban ecosystems, ultimately serving as Nature-based Solutions to mitigate climate change.

Keywords Adaptation · Degradation · Ecosystem services · Land cover change · Nature-based solutions

18.1 Introduction

Climate change and ecosystem degradation resulting from land cover change represent some of the most global challenges threatening ecosystem services, biodiversity, food security, and livelihoods, particularly in developing countries (Arneth et al., 2019; Ericksen et al., 2010; FAO, 2022; IPCC, 2023; Reed & Stringer, 2016; Tume

L. S. Nangsinyuy (✉) · S. J. P. Tume
Department of Geography and Planning, The University of Bamenda, Bamenda, Cameroon
e-mail: sergenangsinyuyl@gmail.com

U. Chatterjee et al. (eds.), *Eco-Resilience: Climate Change, Land Degradation and Sustainable Solutions*, Sustainability Solutions,
https://doi.org/10.1007/978-3-032-00708-7_18

et al., 2020; Webb et al., 2017). Rapid urbanisation and unsustainable land use have intensified the rate of forest loss, wetland encroachment, and soil degradation, contributing significantly to greenhouse gas emissions and reduced ecosystem resilience (Olsson et al., 2019; World Bank, 2011).

The Intergovernmental Panel on Climate Change (IPCC, 2019) provided comprehensive evidence linking land cover change to broader climate concerns. This special report on climate change and land emphasises that human activities directly affect more than 70% of the global ice-free land surface (Mbow et al., 2019). Sub-Saharan Africa provides evidence of how extreme climate events can rapidly transform landscapes (Boko et al., 2007; Jia et al., 2019). Nicoletti et al. (2021), Ngala and Tume (2024) examined the emergence of scale-free fires, documenting how changing climate conditions have contributed to unprecedented bushfire patterns and accelerated natural ecosystem degradation. Historical evidence shows the retreat of forests across the globe over longer time frames, documenting how colonial and post-colonial development patterns transformed landscapes (Bernard & De Konink, 2006). Tagne et al. (2022) analysed land cover changes on Manoka Island in Cameroon's Littoral Region between 1986 and 2018, finding significant transitions from natural vegetation to human-modified landscapes. Their assessment of carbon footprint implications revealed substantial carbon losses associated with these changes, highlighting the climate implications of local land cover transformations.

The groundbreaking work by Hansen et al. (2013) using high-resolution global maps documented the loss of 2.3 million km^2 of forest globally between 2000 and 2012, with substantial portions occurring in North America. Canada's vast boreal forests represent approximately 9% of the world's forest cover, making their management critically important for global climate regulation. A study by Wulder et al. (2016) analysed the Global Forest Change dataset with a Canadian perspective, revealing that Canada's forests have experienced complex patterns of disturbance and recovery. Their work documented that approximately 11% of Canada's forests underwent some form of disturbance between 2000 and 2014, with wildfires representing the dominant driver of change in northern regions and human activities more prominent in southern areas. Qin et al. (2022) provided groundbreaking evidence that carbon losses from forest degradation in the Brazilian Amazon now exceed losses from deforestation, challenging conventional wisdom about tropical forest threats.

In Sub-Saharan Africa, and Cameroon in particular, urban growth often occurs in the absence of strategic land use planning. This leads to the depletion of vegetation cover, increased surface runoff, and loss of ecosystem services such as water regulation, microclimate stabilisation, and biodiversity protection (Kimengsi et al., 2017). Cameroon's diverse ecosystems span from tropical forests to savannas, making the country a microcosm of broader African environmental challenges. Tagne et al. (2022) documented land cover changes on Manoka Island in Cameroon's Littoral Region between 1986 and 2018, finding significant transitions from natural vegetation to human-modified landscapes. Their assessment of carbon footprint implications revealed substantial carbon losses associated with these changes, highlighting the climate implications of local land cover transformations.

In recent years, the Bamenda III has undergone significant land cover transformations, with profound implications for its natural ecosystems and water resources. Ndikebeng et al. (2023), Acha and Fombe (2023), Fombe and Acha (2020), Tume and Kongnso (2022) conducted in-depth studies that revealed extensive conversion of montane forests and wetlands into built-up areas and agricultural land, largely driven by population growth and unregulated urbanisation. Their research highlights that over 70% of the area's natural vegetation and water catchments have been degraded or lost between 1980 and 2020.

In the majority of communities today in Bamenda III, there's an increase in land use due to poor agricultural systems, unplanned urban development, demand for fuel wood, demand for furniture, and a low level of sensitization, which has led to deforestation and is comparatively evident from satellite images of the year 2000 and the present. Forest naturally harbour ecological life and, as such, sustain plenty of ecology which balances the ecosystem through their interaction in the food chain. Thus, trees are the backbone of ecosystems and climate change restoration. In this regard, forest depletion due to increased land use has been and is a great threat to ecology and climate in Bamenda III and beyond, as the earth is a closed system with an interwoven ecological relationship.

Furthermore, poor agricultural systems due to the need for food have led to deforestation, causing loss of habitat and species extinction. Also, water catchment areas are a hotspot for flora and fauna in Bamenda III, as such, poor agricultural activities around water catchments have resulted in ecosystem services depletion, such as water, soils, food, culture and climate regulation. Moreover, Bamenda III has not been left out of the wave of urbanization, which is to the detriment of the ecosystem and climate due to unplanned urban development, which has led to deforestation, manifested by frequent droughts, floods, and pollution. In addition, indiscriminate wood harvesting has led to the destruction of indigenous tree species and animal extinction, causing loss of cultural values, destruction of ecosystems, loss of biodiversity, and an increase in temperatures. Poor sensitisation has also resulted in deforestation in Bamenda III, which is a result of a top-down approach that excludes local people, degrading the ecosystem because sensitisation is often carried out on radios, televisions, seminars and workshops where grassroots people are not opportune to access, thus inefficient sensitisation.

While some studies have examined climate trends or urban growth in Cameroon (Ndikebeng et al., 2023; Tume, 2021), few have analysed the role of ecosystem restoration as a localised, nature-based solution for climate mitigation. This study seeks to fill this gap by providing an empirical assessment of land cover change, climate variability, and community-led restoration efforts in Bamenda III. It contributes to the growing body of literature that explores integrative approaches to climate adaptation and mitigation in urbanising regions.

18.2 Rationale

The restoration of degraded ecosystems is increasingly recognised as a strategic intervention for climate change mitigation and sustainable urban development (Bogert, 2025; IUCN, 2021). In fast-growing areas like Bamenda III, unplanned urban expansion and unsustainable agricultural practices are driving deforestation, wetland encroachment, and biodiversity loss. These changes degrade ecosystem services and intensify climate risks such as droughts, floods, and water scarcity (Samba et al., 2019). Ecosystems, which are communities of living organisms that interact with their physical environment through nutrient cycling and energy flow, are sensitive to the changes brought about by urbanisation (Murray, 2018). Ecosystems provide numerous benefits (Fig. 18.1) through it service categorized into: provisioning services, which include essential resources like food and water; regulating services, which help manage natural processes such as flood control, drought regulation, and disease management; and supporting services, which encompass activities like soil formation and nutrient cycling; and cultural services, which offer recreational, scientific, spiritual, and other non-material benefits (Department of Environment, Water, Heritage and Arts, 2009).

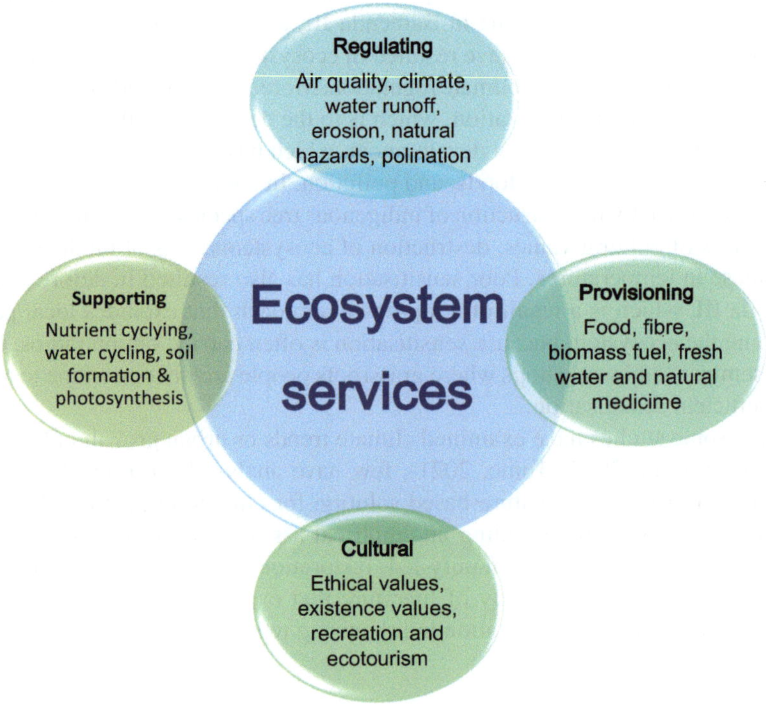

Fig. 18.1 Ecosystem services. *Source* MEA (2005)

The restoration of degraded ecosystems is increasingly recognized as a strategic intervention for climate change mitigation and sustainable urban development (IUCN, 2021). In fast-growing municipalities like Bamenda III, unchecked urban expansion and unsustainable agricultural practices are driving deforestation, wetland encroachment, and biodiversity loss. These changes degrade ecosystem services and intensify climate risks such as droughts, floods, and water scarcity (Bogert, 2025; Samba et al., 2019). Although restoration initiatives are underway in Cameroon, there is limited empirical research evaluating their effectiveness in mitigating climate impacts. This study is motivated by the need to fill this gap by assessing the impact of land cover changes on climate variability and evaluating local restoration strategies in Bamenda III. This rationale aims at assessing and providing relevant evidence for upscaling nature-based solutions in similar contexts.

Over the past two decades, the depletion of ecosystem services in the Bamenda III municipality has been exacerbated by rapid population growth, unsustainable agricultural practices, unplanned urban development, indiscriminate harvesting of wood, and the reclamation of wetlands. These activities have led to significant losses in biodiversity, water scarcity, soil erosion, livelihoods, and cultural values, and, most importantly, have contributed to climate change, as evidenced by recurrent agro-hydro-meteorological droughts and rising temperatures. This situation has caused alarm within the Bamenda III municipality, prompting urgent action for restoration initiatives led by local organisations, councils, traditional authorities, and the wider community to ensure effective management and sustainability. Therefore, nature-based solutions for mitigating climate change through ecosystem restoration include planting ecologically friendly tree species in different parts of the municipality, as well as wetland restoration.

The chapter adopts an integrative, mixed-methods approach that combines remote sensing, meteorological data analysis, and community-based solutions to assess how local ecosystem restoration efforts align with climate mitigation goals. By examining the interplay between land use dynamics, climate variability, and restoration interventions, the study contributes to both scholarly and policy debates on sustainable urban planning and environmental governance. Specifically, the study seeks to analyze land cover change in Bamenda III Municipality between 1986 and 2023, focusing on the conversion of forests, grasslands, and wetlands, assess local climate variability particularly rainfall and temperature trends, over the past seven decades (1951–2023) and finally identify the opportunities and challenges of using ecosystem restoration as a nature-based solution for climate resilience Bamenda III through community-led ecosystem restoration efforts. Although restoration initiatives are underway in Cameroon, there is limited empirical research evaluating their effectiveness in mitigating climate impacts. This study is motivated by the need to fill this gap by assessing the impact of land cover changes, documenting climate variability, and evaluating local restoration strategies in Bamenda III, based on assessments that provide relevant evidence for upscaling nature-based solutions in similar contexts across the region.

18.3 Materials and Methods

The study is conducted in the Bamenda III Municipality, located in the North West Region of Cameroon. Geographically, it lies between latitudes 5°58′30″ and 6°4′ North of the Equator and longitudes 10°10′ and 10°18′ East of the Prime Meridian (see Fig. 18.2). Bamenda III is one of the municipal councils that constitute Bamenda City, situated within the Mezam Division. The sub-division encompasses the territories of Nkwen and Dzah villages. Covering a total area of 69.6 km^2, Bamenda III has a population density of approximately 1584 inhabitants/km^2. Based on projections from the Minimum Urban Local Development Scheme (SMAUL) of February 2018, the municipality has an estimated population of 193,110 residents (Samba et al., 2019).

This study adopted a mixed-methods approach, combining both qualitative and quantitative data collection techniques to assess the impact of ecosystem restoration on climate change mitigation, which included surveys, questionnaires, interviews, focus group discussions, and Geospatial analyses with the use of GIS tools. Analyses approaches employed were thematic analyses, descriptive, and inferential statistics to measure the impact of restoration on climate mitigation indicators.

The study adopted a mixed-methods approach, integrating both quantitative and qualitative data sources to provide a comprehensive assessment of ecosystem restoration and its relationship to climate change mitigation. Land cover change was analysed using multi-temporal satellite imagery from Landsat 5 (1986), Landsat 7 (2003),

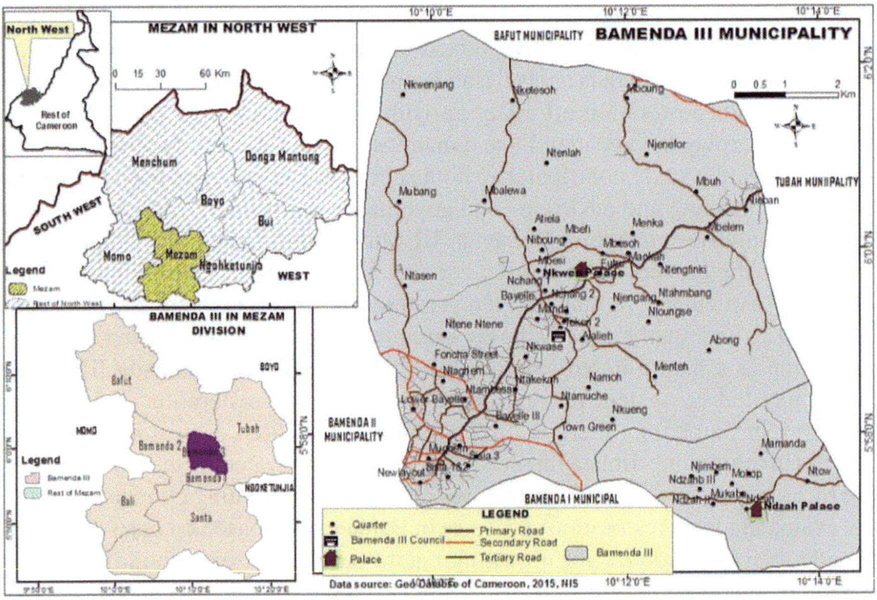

Fig. 18.2 Geographical location of Bamenda III sub-division in Mezam division

and Landsat 8 (2023). Images were preprocessed using atmospheric correction and layer stacking in ArcGIS 10.8 and QGIS 3.28. A supervised classification method was used to identify four main land cover types: grassland, settlement, degraded forest, and farmland. Classification accuracy was assessed using ground-truthing data collected through field visits and existing topographic maps. Change detection analysis was then performed to quantify transitions between land cover classes over time.

Long-term climatic data (1951–2023) were obtained from the Regional Meteorological Service for the Northwest, Bamenda. The data included monthly and annual records of rainfall and temperature. Trends were analysed using linear regression models and the Coefficient of Determination. The Standardized Precipitation Index (SPI) was calculated to evaluate inter-annual drought conditions. SPI values were computed using the WMO-recommended method, which normalises precipitation data over different time scales. SPI classification thresholds were used to identify mild, moderate, and extreme dryness events.

Qualitative data were collected through: Household surveys, selected through a stratified random sampling across the municipality. Interviews with civil society leaders, traditional authorities, and two focus group discussions with community members engaged in restoration activities. Findings from spatial, climatic, and qualitative data were triangulated to evaluate the impact of restoration efforts and their role in climate change mitigation. Descriptive and inferential statistics were applied to explore relationships among variables, while visualisations (graphs, maps, pictures) were used to support interpretation.

18.4 Results and Discussions

18.4.1 Land Cover Change

One of the indicators of land cover change, in addition to oscillating rainfall, is urbanisation. In Bamenda III, agricultural land is increasingly under pressure due to rapid urbanisation. Though commerce and settlement land uses emerged as dominant, transport routes and recreational land uses also evolved during this period. Agricultural land use also intensified because of the increasing population and the need for food to support and sustain the growing population. This has led to agricultural intensification, expansion, and encroachment of wetlands and a reduction in stream volume and deforestation. The main land cover changes assessed in this section are grassland, settlement, degraded forest, and farmland (Table 18.1).

There has been a noticeable shift in land use trends in Bamenda III Municipality over time. Specifically, there has been a significant decline in grasslands and farmland areas, with grasslands decreasing by 226.46 hectares and farmland by 1,053.69 hectares between 1986 and 2022. These changes are primarily driven by urban growth, characterised by the expansion, densification, and transformation of

Table 18.1 Land cover change in Bamenda III (1986–2023)

Land cover/use	Area (Ha)		
	1986	2003	2023
Grassland	651	506.51	424.54
Settlement	892.71	1238.1	1351.2
Degraded forest	2014.4	1964.9	2836
Farmland	3729.7	3578.3	2676
Total	7287.8	7287.8	7287.8

Source Satellite images (1986, 2003, 2023)

settlements, which have resulted in extensive forest degradation. For instance, settlement areas increased by 458.51 hectares during this period, while degraded forest areas expanded by 821.64 hectares. These changes mirror broader patterns observed in peri-urban zones of Sub-Saharan Africa, where rising land demand has led to the fragmentation of natural ecosystems and the encroachment on ecologically sensitive areas such as wetlands and riverbanks (FAO, 2022; Kimengsi et al., 2017).

The dynamics of land use are heavily influenced by population growth and the rising demand for space and infrastructure as urban development progresses. In 1986, land cover and use patterns in Bamenda III showed more temporary alterations for farming or grazing rather than the permanent modifications seen in later years. At that time, grasslands covered approximately 651 hectares, while degraded forest areas occupied 2,014.4 hectares (Fig. 18.3).

In 1986, agricultural land covered approximately 3,729.7 hectares, while settlement areas accounted for 892.71 hectares. During this time, water resources experienced minimal stress, primarily due to the predominance of farmland in the landscape. However, within less than two decades, significant changes in land cover occurred. Human activities, such as deforestation, unsustainable farming practices—especially near watercourses—the expansion of settlements, and the planting of environmentally harmful tree species like eucalyptus near waterways, along with construction along riverbanks, heavily contributed to the degradation of water resources. Additionally, the impacts of climate change, including altered discharge patterns, extreme dry-season temperatures, and the seasonal reversal of monsoon winds, further worsened the situation. These factors collectively led to a noticeable decline in both the quantity and quality of water resources, severely affecting flow characteristics and local livelihoods. Between 1986 and 2003, water discharge decreased by over 40%. The subsequent period posed even greater challenges to water resources than the one before (Fig. 18.4).

In 2023, Bamenda III Sub-Division underwent significant changes and intensification in land use over the past four decades. This transformation has been primarily driven by rapid economic growth, a rising population, and ongoing urbanisation. The evolving lifestyles of residents are closely linked to complex land use demands, creating lasting impacts due to the persistent pressure on land for essential needs such as food, housing, and infrastructure. This continuous strain has

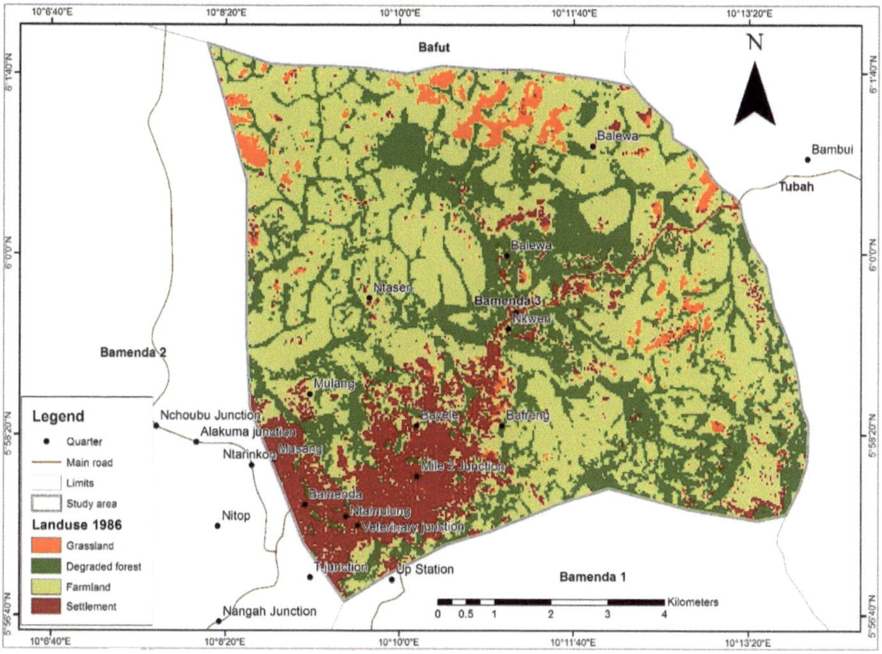

Fig. 18.3 Land cover/use in Bamenda III sub-division in 1986. *Source* Landsat 5 (NASA, 1986)

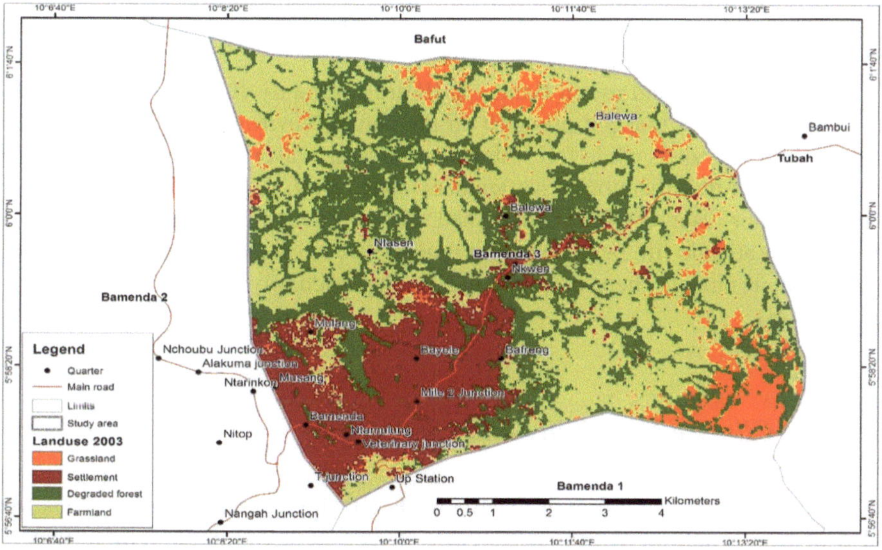

Fig. 18.4 Land cover/use in Bamenda III sub-division as of 2003. *Source* Landsat 7 (NASA, 2003)

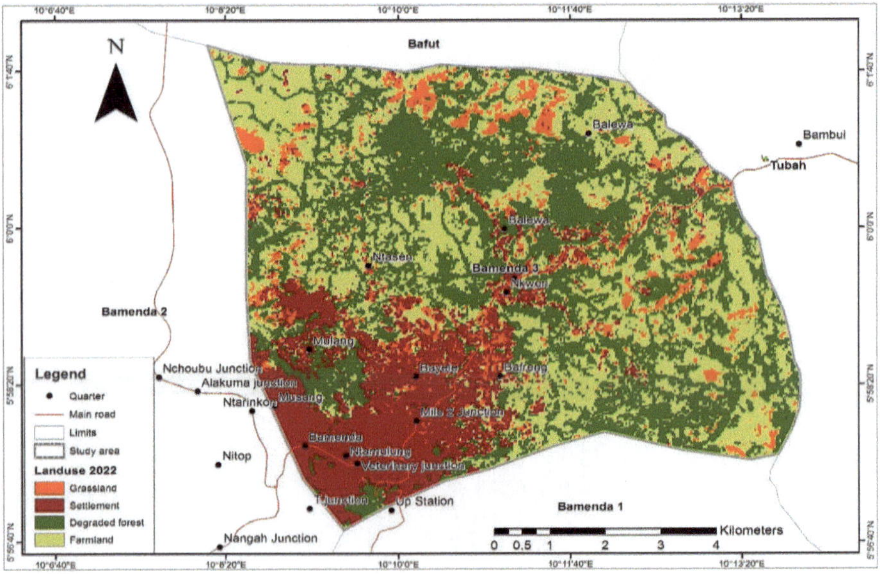

Fig. 18.5 Land cover/use in Bamenda III sub-division in 2023. *Source* Landsat 8 (NASA, 2023)

negatively affected water resources, particularly in sensitive areas like catchments, streams, and brooks. Consequently, both general and severe consequences have emerged, impacting livelihoods, agricultural productivity, and biodiversity within the municipality (Fig. 18.5).

Land cover and land use in Bamenda III are constantly changing, primarily due to human activities that put pressure on the natural environment. Grassland areas are declining steadily, while forest degradation is increasing, largely because of the expansion and densification of settlements. As a result, the availability and importance of water resources have diminished. These water resources not only serve as habitats for aquatic life but also play a crucial role in protecting and sustaining the human environment. Therefore, their decline is particularly concerning (Ndikebeng et al., 2023).

18.4.2 Climate Variability and Change

Rainfall has been decreasing in the Bamenda III Municipality against a backdrop of rising temperatures. Worth noting is the decreasing trend in the rainfall pattern. The trend has continuously been going down since 1987, with more remarkable lows since 2013. The decrease in the rainfall Coefficient of Determination (R^2) stands at 51.12% (Fig. 18.6).

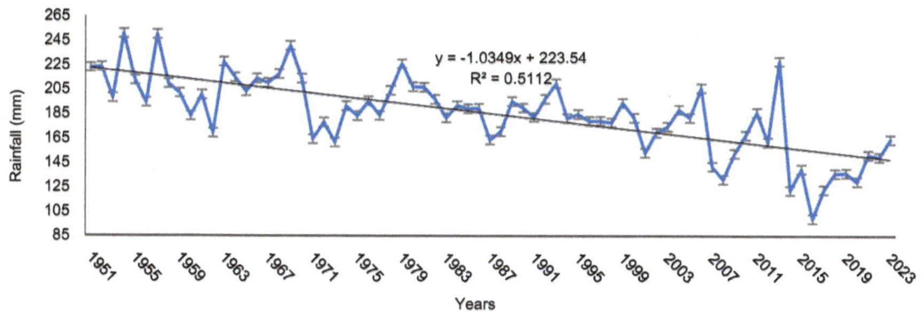

Fig. 18.6 Inter-annual rainfall trend for Bamenda (1951–2023). *Data source* Regional Meteorological Service for the Northwest, Bamenda

It can be seen that annual rainfall has been constantly fluctuating since 1951, with upward and downward movements. However, the highest rainfall that was recorded was between 1953 and 1959, amounting to about 250 mm. On the other hand, the lowest rainfall that was recorded was in 2015, when rainfall witnessed a significant drop to less than 100 mm. A summary of decadal characteristics indicates that rainfall has been reliable in Bamenda from 1951 to 2012. Since the 2013–2023 period, rainfall has been unreliable with a Coefficient of Variation of 22.88% (Table 18.2).

Despite the reliability, there were also decreasing trends during 1951–1960, 1981–2023. The decreasing trend has implications for a major ecosystem service like water supply. In all, rainfall has been decreasing in Bamenda from 1951 to 2023, with the

Table 18.2 Summary of rainfall characteristics in Bamenda

Periods	MRF	SD	Mean SPI	SPI class	CV	Reliability	RT
1951–1960	215.22	22.54	−0.05	Mild dryness	10.47	Reliable	Decrease
1961–1970	211.66	18.69	0.86	Mildly Wet	8.83	Reliable	Increase
1971–1980	190.02	19.69	0.16	Mildly Wet	10.36	Reliable	Increase
1981–1990	187.94	12.60	0.09	Mildly Wet	6.70	Reliable	Decrease
1991–2001	184.11	14.04	−0.04	Mild dryness	7.62	Reliable	Decrease
2002–2012	169.65	12.80	−0.51	Mild dryness	12.80	Reliable	Decrease
2013–2023	143.92	32.93	−1.35	Moderate dryness	22.88	Unreliable	Decrease
Average	186.07	19.04	−0.12	Mild dryness	11.38	Reliable	Decrease

MRF Mean rainfall (mm), *SD* standard deviation, *SPI* standardized precipitation index, *CV* coefficient of variation (%), *RT* rainfall trend

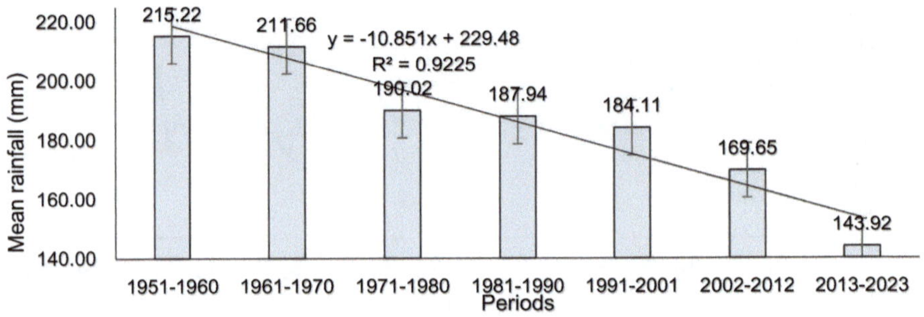

Fig. 18.7 Decreasing mean decadal rainfall in the Bamenda III municipality

least mean rainfall recorded in 2016 (95 mm), and the least decadal mean rainfall recorded during the 2013–2023 period (143.92 mm) (Fig. 18.7).

Just like rainfall, temperature is very important in the determination of climate variability, and it is one of the most visible factors for everyone. This, therefore, makes it very important to visualise the inter-annual temperature trend in the Bamenda I municipality since 1971 (Fig. 18.8).

It can be seen that temperatures have been slightly changing from 1971 to 2015. From 2015 to 2023, there has been a sharp increase in temperatures. It is also important to indicate that the lowest temperatures recorded were in 1987 when temperatures went down to about 18 °C. On the other hand, the highest temperatures were recorded in 2023, with temperatures going above 26 °C. This therefore gives an overall increasing trend since 1971.

The Standardized Precipitation Index (SPI) was also used to assess rainfall events in Bamenda. SPI was designed to map out droughts using consistent mean monthly and mean annual rainfall. A drought sets in when the mean SPI falls below zero (0). From the analysis in this study, from 1951 to 2023, Bamenda recorded 28 events of mild dryness (38.36%), where the SPI value ranges from −0.1 to −0.99, 4 episodes of moderate dryness (5.48%), where SPI ranges from −1 to −1.49 and 3 episodes

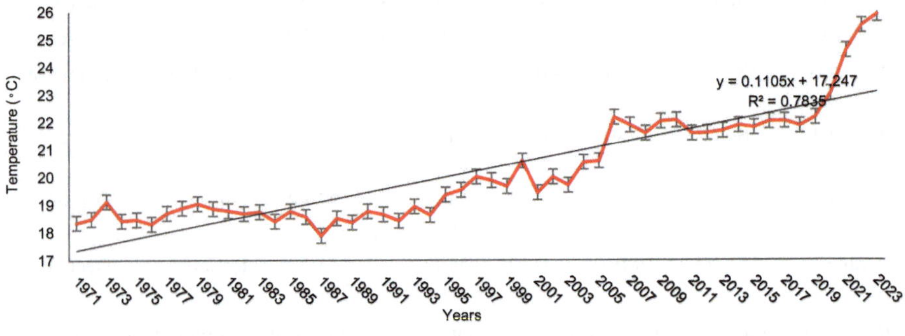

Fig. 18.8 Inter-annual temperature trend in the Bamenda III municipality (1971–2023)

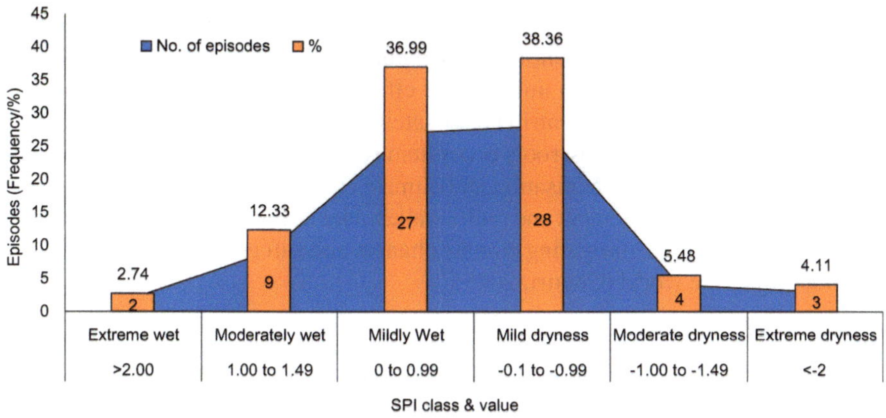

Fig. 18.9 Agro-hydro-meteorological droughts assessment by standardized precipitation index in the Bamenda III Municipality

of extreme dryness (4.11%), where SPI value ranges from <-2 (Fig. 18.9). This indicates that 47.95% of rainfall events in Bamenda have been marked by rainfall deficits since 1951, clear evidence of climate change. This calls for proactive ecosystem restoration to mitigate the situation.

These findings corroborate those of Chettri (2025); Tume (2022), who revealed that rising temperatures and oscillating rainfall are direct indicators of climate change in Bamenda. The same author also indicated that Bamenda is experiencing more dry episodes, especially after 2010. In other perspectives, Kimengsi et al. (2017), Tume (2021), Ndikebeng et al. (2023) revealed that peri-urban land dynamics have implications in the Bamenda III municipality, like water scarcity, degradation of watersheds, and farmlands. This confirms that climate variability in Bamenda III is not only evident but also accelerating, as the observed patterns reinforce the urgent need for adaptive strategies, including ecosystem-based interventions that restore natural buffers and regulate local microclimates (Tume & Nyuyfoni, 2025; Tume, 2022).

18.4.3 Ecosystem Restoration Strategies to Mitigate Climate Change

In response to the escalating environmental challenges in Bamenda III, including deforestation, declining water resources, and rising climate change, multiple stakeholders have initiated ecosystem restoration efforts. These strategies align with global calls for Nature-based Solutions (NbS) to climate change and demonstrate a growing local commitment to sustainable land management (Awazi & Tume, 2025; Golmohammadi & Kordshouli, 2025). However, while progress is notable, these initiatives face significant challenges that require further attention. There have been concerted efforts by the Bamenda Council, traditional authorities, and civil society organisations to restore ecosystems, to mitigate the effects of climate change, and enhance ecosystem services. These contribute to satellite-based landcover analyses, local meteorological data, and grassroots ecosystem restoration practices to access interconnected challenges of urbanisation and climate change by highlighting the pivotal role of community engagement and multi-stakeholder collaboration in implementing nature-based solutions, mitigating climate change, and safeguarding the ecology and contributing to the world (Chettri, 2025).

Bamenda III Council

The Bamenda III Council is an active government body committed to addressing environmental degradation and promoting sustainable development within the municipality. The Council's efforts in ecosystem restoration and climate regulation are centred on afforestation and reforestation initiatives aimed at rehabilitating degraded lands and improving biodiversity in Menteh (Mile 4, Nkwen). Enhancing community engagement and resilience to climate change through sensitization and awareness campaigns, workshops and training programs, participation in tree planting events, training of holidaymakers on environmental protection, with over 40 trained this year, supporting community-led afforestation projects, environmental policy advocacy by working in close collaboration with local organizations and traditional authorities, partnerships, and networking. With these efforts, therefore, the council helps to restore ecosystems in Bamenda III and, by so doing, regulates the climate. Yet, these efforts are constrained by limited funding, inadequate technical capacity, and insufficient monitoring. There is often no long-term plan to ensure tree survival or maintain restored areas. Integrating restoration into the municipal budget and linking it to national policies could improve sustainability and scaling.

Traditional Authorities

In Bamenda III, the role of traditional authorities, led by the Fon, in climate action has become increasingly significant. By harnessing their influence and cultural leadership, these authorities are spearheading grassroots initiatives focused on ecosystem restoration and climate resilience through tree planting. The localised nature of these efforts, which engage diverse quarters and communities, underscores the deep commitment of the people to rejuvenate their environment and mitigate the effects of climate change. This has been achieved through mobilisation and community engagement, raising awareness, establishing protected areas (secret forests), restricting illegal logging, rehabilitating water catchment areas, and promoting agroforestry practices. Consequently, the traditional authorities of Bamenda III are making substantial contributions to ecosystem restoration and climate change mitigation through their grassroots actions. Their initiatives not only revitalise the natural environment but also exemplify community-driven solutions to global environmental challenges while fostering cultural identity. In 2024, they collaborated with local organisations to plant over 21,000 tree species across various neighbourhoods: Atielah-Mbelewa, Atiesu-Mbessi, Bujong-Mambu, Futru I, Futru II,

Mbefi-Mambu, Mbelem, Mbeso-Mambu, Nibung, Manka, Mbelewa, Mubang, Njengang, Ntahbenke, Ntambang, Ntasen, Ntefinki, Ntela, Ntenetene, Ntensu-Mbelewa. However, enforcement remains weak without legal or institutional coordination with formal governance structures.

Civil Society Organisations

Re-Forestation Task Force (RETAFO)

The Reforestation Task Force (RETAFO) is a prominent environmental organisation dedicated to afforestation and ecosystem restoration in the Bamenda III Municipality. It was established in response to the escalating challenges of deforestation, land degradation, and climate change. RETAFO plays a crucial role in enhancing ecosystem recovery and regulating climate through various initiatives, including afforestation and reforestation, watershed and water catchment rehabilitation, community engagement and capacity building, biodiversity conservation, and climate change mitigation and adaptation. RETAFO's primary objective is to restore deforested and degraded landscapes by planting a diverse range of indigenous and climate-resilient tree species. These include timber, ornamental, fruit, water-friendly, bee-friendly, and medicinal trees. The organisation conducts its activities in areas such as water catchments, degraded community forests, home green spaces, and erosion-prone slopes, ensuring that tree planting serves both ecological and social purposes. Since its inception, RETAFO has planted over 200,000 trees, with more than 27,000 of those planted in 2024 alone. By increasing forest cover, RETAFO helps regulate microclimates, improve water retention, conserve biodiversity, and contribute to carbon sequestration, thereby mitigating the impacts of climate change in Bamenda III.

LoRDA Research Centre

The LoRDA Research Centre is a key local organisation focused on advancing ecosystem restoration and climate change mitigation in Bamenda III. Established to empower students and local communities through research and environmental education, the centre plays a crucial role in shaping restoration efforts and promoting sustainable development practices. LoRDA facilitates and supports research that addresses climate challenges and raises awareness about restoration and climate change through workshops and outreach programs in schools. Additionally, by collaborating with universities, environmental organisations, and local authorities, the centre ensures that its findings are directly applicable to real-world scenarios.

Community Empowerment Agency for Sustainable Development (CEMASUD)

CEMASUD is an organisation dedicated to afforestation and ecosystem restoration efforts in the Bamenda III Municipality. Its primary mission is to empower local communities to engage in environmental conservation and sustainable development through grassroots initiatives that enhance biodiversity, promote sustainable land use, and combat climate change. CEMASUD carries out various activities, including community-based tree planting initiatives. In 2024, the organisation successfully planted over 21,000 trees in collaboration with the communities of Nkwen. Other initiatives include environmental education and awareness programs, capacity building through training, promotion of agroforestry and sustainable agriculture, and establishing collaborative partnerships and advocacy efforts. Overall, CEMASUD plays a crucial role in reversing environmental degradation and promoting climate resilience in the Bamenda III Municipality.

Cameroon Gender and Environmental Watch (CAMGEW)

CAMGEW is a local organisation dedicated to promoting gender equality and empowering women in Bamenda III. While its primary focus is on gender advocacy, CAMGEW has broadened its mission to include ecosystem restoration and climate change mitigation. This approach aims to create a sustainable and equitable future. By involving women and marginalised groups in environmental initiatives, CAMGEW plays a crucial role in building resilient communities and restoring degraded ecosystems. The organisation achieves this through capacity-building programs for women, tree planting, and nursery management. It empowers women with leadership skills in community-led restoration efforts and conducts outreach programs. CAMGEW's unique strategy of linking gender equality with environmental sustainability has already empowered over 2300 girls and women. As a result, it serves as a model for inclusive restoration efforts that tackle both ecological and social challenges, ultimately contributing to climate change mitigation and enhancing the well-being of the entire Bamenda III community.

Collectively, these organisations illustrate the power of grassroots restorations, but they often operate independently. Greater synergy among local organisations, traditional authorities, and the council would improve coordination and policy-making (International Institute for Environment and Development-IIED, 2009; King, 2014). These issues mirror global findings that restoration efforts often underperform without institutional support, long-term investment, and adaptive governance mechanisms (IPBES, 2019, 2022, 2018). Integrating restoration with broader urban planning frameworks is key to ensuring effectiveness (Mbongsi et al., 2023; Ouyang et al., 2025; Tantoh et al., 2025).

18.5 Conclusion

Restoration initiatives in the Bamenda III Municipality, led by the local council, stakeholders, and community members, play a significant role in addressing climate change and enhancing ecosystem services. Despite these commendable efforts, challenges remain, particularly in scaling and sustaining restoration activities. To improve ecosystem restoration, it is recommended that the local council prioritise it as a central component of their annual budget and actively seek support from the national government and international partners. Local organisations should focus on planting a diverse range of ecologically viable and climate-resilient tree species, including fruit and ornamental trees. These trees not only promote environmental sustainability but also provide socio-economic and aesthetic benefits, thus encouraging long-term community engagement and maintenance. This study contributes meaningfully to scientific research by offering a localised, integrative approach that combines remote sensing, meteorological data analysis, and community-based ecosystem restoration to address climate change.

The findings highlight the profound impacts of land cover change and climate variability on the ecological stability of the Bamenda III Municipality. Over the past four decades, rapid urban expansion and unsustainable land use have led to the loss of grasslands, farmland, and wetlands, contributing to rising temperatures, decreasing rainfall, and increased ecosystem vulnerability. These trends reflect broader regional and global environmental challenges and underscore the need for locally adapted, nature-based solutions. The study reveals that ecosystem restoration, particularly when led by communities, traditional authorities, and civil society organisations, offers a viable pathway for climate change mitigation and ecological resilience. Initiatives such as afforestation, catchment rehabilitation, and environmental education are already making measurable contributions to landscape recovery and climate regulation.

This practical and replicable methodology can be applied in other rapidly urbanising regions around the world. By demonstrating the effectiveness of grassroots, nature-based solutions in enhancing urban resilience, this research aligns with global climate goals, such as the UN Sustainable Development Goals (SDGs), particularly SDG 13 (Climate Action), SDG 15 (Life on Land), and SDG 11 (Sustainable Cities and Communities). The findings provide critical insights for local policymakers and environmental practitioners while also contributing to the global body of knowledge on urban ecosystem restoration as a viable strategy for climate change mitigation and sustainable development. However, restoration efforts in Bamenda III are constrained by funding limitations, weak policy enforcement, and fragmented governance. Ecosystem restoration should not be seen as an isolated environmental activity but as a strategic component of climate-resilient urban planning. Thus, investing in restoration as a public good, municipalities like Bamenda III can not only address environmental degradation but also strengthen their adaptive capacity of the ecosystem and accelerate climate change.

References

Acha, M. E. & Fombe, L. F. (2023). Gentrification Patterns, Trends and Development Implications of the Bamenda Metropolis from 1972–2022. *International Journal of Social Sciences and Management Review, 6*, 139–153. https://doi.org/10.37602/IJSSMR2023.6312

Arneth, A., Denton, F., Agus, F., Elbehri, A., Erb, K., Osman Elasha, B., Rahimi, M., Rounsevell, M., Spence, A., & Valentini, R. (2019). Framing and context. In P. R. Shukla, J. Skea, E. Calvo Buendia, V. Masson-Delmotte, H.-O. Pörtner, D. C. Roberts, P. Zhai, R. Slade, S. Connors, R. van Diemen, M. Ferrat, E. Haughey, S. Luz, S. Neogi, M. Pathak, J. Petzold, J. Portugal Pereira, P. Vyas, E. Huntley, K. Kissick, M. Belkacemi, & J. Malley (Eds.), *Climate change and land: An IPCC special report on climate change, desertification, land degradation, sustainable land management, food security, and greenhouse gas fluxes in terrestrial ecosystems* (pp. 77–130). https://doi.org/10.1017/9781009157988.003

Awazi, N. P. & Tume, S. J. P. (2025). Impacts, vulnerability, resilience, and nature-based adaptation in rural communities in the bamenda highlands of Cameroon. In W. Leal Filho, G. J. Nagy, & D. Y. Ayal, (Eds.), *Handbook of nature-based solutions to mitigation and adaptation to climate change* (pp. 1–21). Springer, Cham. https://doi.org/10.1007/978-3-030-98067-2_148-1

Bernard, S., & De Konink, R. (2006). The retreat of the forest in Southeast Asia: A cartographic assessment. *Singapore Journal of Tropical Geography, 27*(1), 1–17. https://doi.org/10.1111/j.1467-9493.2006.00236.x

Bogert, J. M. (2025). Seeing is believing: The systemic filters that shape public climate change beliefs. PhD Thesis—Research and graduation internal, Vrije Universiteit Amsterdam. https://doi.org/10.5463/thesis.997

Boko, M., Niang, I., Nyong, A., Vogel, C., Githeko, A., Medany, M., Osman-Elasha, B., Tabo, R. & Yanda, P. (2007). Africa. In M. L. Parry, O. F. Canziani, J. P. Palutikof, P. J. van der Linden & C. E. Hanson (Eds.), *Climate change 2007: Impacts, adaptation and vulnerability. Contribution of Working Group II to the fourth assessment report of the intergovernmental panel on climate change* (pp. 433–467). Cambridge University Press.

Chettri, N. (2025). The ecosystem restoration in the anthropocene: Contribution towards biodiversity conservation and human wellbeing in the Hindu Kush Himalaya. In A. Nautiyal, S. Nautiyal, R. K. Maikhuri, R. C. Sundriyal, J. C. Kuniyal, & N. Rawat, (Eds.), *Indian Himalayan Region: Towards a sustainable future* (pp. 127–144). Messers Bishen Singh Mahendra Pal Singh, Dehra Dun, India.

Department of the Environment, Water, Heritage and Arts. (2009). Ecosystem services: Key concepts and applications, Occasional Paper No. 1, Department of the Environment, Water, Heritage and the Arts, Canberra. Commonwealth of Australia.

Ericksen, P., Stewart, B., Dixon, J., Barling, D., Loring, P., Anderson, M., & Ingram, J. (2010). The value of food system approach. In J. Ingram, P. Ericksen, & D. Liverman (Eds.), *2010)* (pp. 27–30). Food Security and Global Environmental Change, Earthscan.

FAO. (2022). The State of the World's Forests 2022: Forest pathways for green recovery and building inclusive, resilient and sustainable economies. *Food and Agriculture Organization.* https://doi.org/10.4060/cb9360en

Fombe, L. F., & Acha, M. E. (2020). Land use dynamics and variations in sprawl across sub-divisions in Bamenda Urban-scape 1996–2018. *Journal of Sustainable Development, 13*(4), 224. https://doi.org/10.5539/jsd.v13n4p224-234

Golmohammadi, M., & Kordshouli, F. Z. (2025). Trends and applications of nature-based solutions in ecological restoration. *11th National Congress on Civil Engineering, Architecture, Art and Urban Development*, 1–11.

Hansen, M. C., Potapov, P. V., Moore, R., Hancher, M., Turubanova, S. A., Tyukavina, A., Thau, D., Stehman, S. V., Goetz, S. J., Loveland, T. R., Kommareddy, A., Egorov, A., Chini, L., Justice, C. O., & Townshend, J. R. G. (2013). High-resolution global maps of 21st-century forest cover change. *Science, 342*(6160), 850–853. https://doi.org/10.1126/science.1244693

IBPES. (2018). Summary for policy makers of the assessment report on land degradation and restoration of the intergovernmental science-policy platform on biodiversity and ecosystem services. In R. Scholees, L. Montainarella, A. Brainich, N. Barger B. ten Brink, M. Cantele, B. Erasmus. J. Fisher, T. Gardener, T.G. Holland, F. Kohler, J. S. Kotiano, G. Von Maltitz, G. Nangendo, R. Pandit, LJ. Parrotta, M. Dpotts, S. Prince, M. Sankaran, & L. Willemen (Eds.) *IPBES Secretariat* (p. 44), Bonn, Germany.

International Institute for Environment and Development (IIED). (2009). *Participatory learning and action 60* (p. 217). IIED.
IPBES. (2019). Global assessment report on biodiversity and ecosystem services of the Intergovernmental Science-Policy Platform on Biodiversity and Ecosystem Services. In E. S. Brondizio, J. Settele, S. Díaz, & H.T. Ngo, (Eds,). *IPBES secretariat* (p. 1148), Bonn, Germany. https://doi.org/10.5281/zenodo.3831673
IPBES. (2022). Assessment report on the diverse values and valuation of nature. *Intergovernmental Science-Policy Platform on Biodiversity and Ecosystem Services.* https://doi.org/10.5281/zenodo.6522392
IPCC. (2019). Climate change and land: An IPCC special report on climate change, desertification, land degradation, sustainable land management, food security, and greenhouse gas fluxes in terrestrial ecosystems. https://www.ipcc.ch/srccl/
IPCC.(2023). Climate change 2023: Synthesis report. *Intergovernmental Panel on Climate Change.* https://doi.org/10.59327/IPCC-AR6-2023-SYR
IUCN. (2021). Nature-based solutions: A global standard for designing, implementing and verifying. *International Union for Conservation of Nature.* https://doi.org/10.2305/IUCN.CH.2020.08.en
Jia, G., Shevliakova, E., Artaxo, P., De Noblet-Ducoudré, N., Houghton, R., House, J., Kitajima, K., Lennard, C., Popp, A., Sirin, A., Sukumar, R., & Verchot, L. (2019). Land–climate interactions. In P. R. Shukla, J. Skea, E. Calvo Buendia, V. Masson-Delmotte, H.-O. Pörtner, D. C. Roberts, P. Zhai, R. Slade, S. Connors, R. van Diemen, M. Ferrat, E. Haughey, S. Luz, S. Neogi, M. Pathak, J. Petzold, J. Portugal Pereira, P. Vyas, E. Huntley, K. Kissick, M. Belkacemi, & J. Malley (Eds.), *Climate change and land: An IPCC special report on climate change, desertification, land degradation, sustainable land management, food security, and greenhouse gas fluxes in terrestrial ecosystems* (pp. 131–246). https://doi.org/10.1017/9781009157988.004
Kimengsi, N. J., Balgah, S. N., & Nafoin, A. S. (2017). Peri-Urban land use dynamics and development implications in the Bamenda III Municipality of Cameroon. *Sustainability in Environment, 2*(3), 237–288. https://doi.org/10.22158/se.v2n3p273
King, S. (2014). *Community-based adaptation in practice: A global overview of CARE international's practice of community-based adaptation (CBA) to climate change* (p. 61). CARE International and UK Aid.
Mbongsi, N. D., Fogwe, Z. N. & Tume, S. J. P. (2023). Forest potentials and threat Nexus on the Nkambe Plateau Watershed. In Z. N. Fogwe, J. N. Kimengsi, T. E. Sunjo, S. S. Kometa, & C. A. Nkwemoh, (Eds.), *Celebrating ten years of restless rest in landmark research and human resource development: Festschrift for Professor Emeritus Cornelius Mbifung Lambi: Development and environmental dynamics* (pp. 364–377). The University of Bamenda Printing Press.
Mbow, C., Rosenzweig, C., Barioni, L.G.. Benton, T.G., Herrero, M., Krishnapillai, M., Liwenga, E., Pradhan, P., Rivera-Ferre, M.G., Sapkota, T., Tubiello, F.N., & Xu, Y. (2019). Food security. In P. R. Shukla, J. Skea, E. Calvo Buendia, V. Masson-Delmotte, H.-O. Pörtner, D. C. Roberts, P. Zhai, R. Slade, S. Connors, R. van Diemen, M. Ferrat, E. Haughey, S. Luz, S. Neogi, M. Pathak, J. Petzold, J. Portugal Pereira, P. Vyas, E. Huntley, K. Kissick, M. Belkacemi, & J. Malley (Eds.), *Climate change and land: An IPCC special report on climate change, desertification, land degradation, sustainable land management, food security, and greenhouse gas fluxes in terrestrial ecosystems* (pp. 437–550). https://doi.org/10.1017/9781009157988.007
MEA. (2005). *Millennium ecosystem assessment: Ecosystems and human well-being: Synthesis* (pp. 23–103). Island Press.
Murray, M. (2018). What is an ecosystem? Australian Museum, 1–3.
Ndikebeng, K. R., Forba, C. F., Tume, S. J. P., Yenlajai, B. J. F., & Kimengsi, J. N. (2023). Land cover dynamics and implications on water resources in Bamenda III Sub-Division, North West Region, Cameroon. *International Journal of Global Sustainability, 7*(1), 37–52. https://doi.org/10.5296/ijgs.v7i1.21031

Ngala, N. E., & Tume, S. J. P. (2024). Human-induced bushfires threat to biodiversity in the Noun Valley, Cameroon. *Journal of Ecology and Natural Resources, 8*(3), 1–9. https://doi.org/10.23880/jenr-16000398

Nicoletti, G., Saravia, L., Momo, F., Maritan, A., & Suweis, S. (2021). The emergence of scale-free fires in Australia. https://arxiv.org/abs/2110.10014

Olsson, L., Barbosa, H., Bhadwal, S., Cowie, A., Delusca, K., Flores-Renteria, D., Hermans, K., Jobbagy, E., Kurz, W., Li, D., Sonwa, D.J., & Stringer, L. (2019). Land degradation. In P. R. Shukla, J. Skea, E. Calvo Buendia, V. Masson-Delmotte, H.-O. Pörtner, D. C. Roberts, P. Zhai, R. Slade, S. Connors, R. van Diemen, M. Ferrat, E. Haughey, S. Luz, S. Neogi, M. Pathak, J. Petzold, J. Portugal Pereira, P. Vyas, E. Huntley, K. Kissick, M. Belkacemi, J. Malley, (eds.), *Climate change and land: An IPCC special report on climate change, desertification, land degradation, sustainable land management, food security, and greenhouse gas fluxes in terrestrial ecosystems.* https://doi.org/10.1017/9781009157988.006

Ouyang, X., Pilz, J., Wang, X.-C., & Veintimilla, S. G.-A. (2025). Territorial spatial evolution process and its ecological resilience, volume II. Lausanne: *Frontiers Media SA.* https://doi.org/10.3389/978-2-8325-6251-2

Qin, Y., Xiao, X., Wigneron, J.-P., Ciais, P., Brandt, M., Fan, L., Li, X., Crowell, S., Wu, X., Doughty, R. and Zhang, Y., & Moore III, B. (2022). Carbon loss from forest degradation exceeds that from deforestation in the Brazilian Amazon. *arXiv.* https://arxiv.org/abs/2206.07363.

Reed, M. S., & Stringer, L. C. (2016). *Land degradation, desertification and climate change: Anticipating, assessing and adapting to future change* (p. 198). Routledge.

Samba, G., Ndoki, D., Kuma, C. J., & Nkang, M. J. (2019). Effects of urbanization on ecosystem services in Bamenda III, North West Region—Cameroon. *Global Scientific Journal, 7*(11), 1539–1551.

Tagne, T. C., Sonwa, D. J., Awono, A., Mama, M. N., Fongnzossie, E., Ngala Mbiybe, R., Essamba à Rim, L. F., & Ntja, R. D. (2022). Land cover and land use changes between 1986 and 2018, and preliminary carbon footprint implications for Manoka Island (Littoral Region of Cameroon). *Sustainability, 14*(10), 6301. https://doi.org/10.3390/su14106301.

Tantoh, H. B., Tume, S. J. P., Awazi, N. P., McKay, T. J. M. (2025). Gender and climate change education in Sub-Saharan Africa as the missing component in climate change adaptation for an effective management of natural resources. In M. F. Mbah, P. Molthan-Hill, & E. L. Molua, (Eds.), *Practices, perceptions and prospects for climate change education in Africa* (pp. 257–174). Springer Cham. https://doi.org/10.1007/978-3-031-84081-4_8

Tume, S. J. P. (2021). Impact of climate change on domestic water accessibility in Bamenda III Sub-Division, North West Region, Cameroon. *Journal of the Cameroon Academy of Sciences, 17*(2), 131–145. https://doi.org/10.4314/jcas.v17i4.3

Tume, S. J. P. (2022). Standardized precipitation index valuation of climate change in Bamenda. *Geology, Earth and Marine Sciences, 4*(1), 1–6. https://doi.org/10.31038/GEMS.2022414

Tume, S. J. P., & Kongnso, M. E. (2022). Standardized precipitation index valuation of the impact of climate variability and change on domestic water accessibility in Bamenda III Municipality, North West Region, Cameroon. *Journal of Science Frontier Research (H), Environment and Earth Science, 22*(8), 39–55.

Tume, S. J. P., & Nyuyfoni, R. S. (2025). Rainfall anomaly index valuation of agricultural production in Jakiri Sub-Division, Northwest Region, Cameroon. In A. Alam, & Ruckhsana, (Eds.), *Disaster risk, resilient agriculture and livelihood: Methods and applications* (pp. 35–55). Routledge. https://doi.org/10.4324/9781003275916-4

Tume, S. J. P., Zetem, C. C., Nulah, S. M., Ateh, E. N., Mbuh, B. K., Nyuyfoni, S. R., Ahfembombi, L. L., Kwei, J. (2020). Climate change and food security in the Bamenda Highlands of Cameroon. In V. R. Squires, & M. K. Gaur, (Eds.). *Food security and land use change under conditions of climate variability: A multidimensional perspective* (pp. 107–124). Springer. https://doi.org/10.1007/978-3-030-36762-6_6

Webb, N. P., Marshall, N. A., Stringer, L. C., Reed, M. S., Chappell, A., & Herrick, J. E. (2017). Land degradation and climate change: Building climate resilience in agriculture. *Frontiers in Ecology and the Environment, 15*(8), 450–459. https://doi.org/10.1002/fee.1530
World Bank. (2011). *Guide to climate change adaptation in cities* (p. 100). World Bank.
Wulder, M. A., White, J. C., Hermosilla, T., Coops, N. C., & Hobart, G. W. (2016). The global forest change dataset: A Canadian perspective. *Canadian Journal of Remote Sensing, 42*(5), 415–429. https://doi.org/10.1080/07038992.2016.1249493

Chapter 19
Farmers' Adaptation Strategies to Combat Degradation of Farmlands in Bafou, Menoua Division, Cameroon

Loveline Kongla Nsahlai, Suiven John Paul Tume, Nfor Delphine Mbongsi, and Ariane Nivelle Dongmo Mbougo

Abstract Soil degradation resulting from poor soil fertility management practices significantly undermines agricultural productivity in Sub-Saharan Africa. This presents a serious environmental challenge that jeopardizes the sustainability of crop yields. In Bafou, local farmers are increasingly grappling with the urgent issue of soil degradation, which poses a substantial risk to agricultural productivity and food security. This chapter explores the various adaptation strategies farmers in the region use to tackle the challenges associated with declining soil health. Utilizing a mixed research approach (qualitative and quantitative methods), interviews, surveys, and field observations, 125 questionnaires were administered to farming households, focusing on their primary livelihood activities. The findings indicate that the degradation of farmlands in the study area is caused by unsuitable agricultural practices, soil erosion, and climate change, which aggravate the situation. Farmers are adopting indigenous strategies to enhance the resilience and agricultural sustainability. The adaptive approaches put in place were crop rotation (45%), intercropping (99%), fallowing (23%), organic manuring (89%), mulching (97%), and minimized tillage/no-tillage (54%). In addition, farmers face demographic (66%), socio-economic (10%), and traditional/cultural (24%) challenges when combating the degradation of farmlands. This chapter aims to deepen the understanding of the adaptive capacity of farmers in Bafou, offering valuable insights for development practitioners and policymakers on effective strategies to mitigate the impacts of soil degradation. Eventually, the study seeks to promote sustainable agriculture and facilitate the sustainable development of Bafou and similar rural areas.

L. K. Nsahlai (✉)
Department of Geography, HTTC, The University of Bertoua, Bertoua, East Region, Cameroon
e-mail: nlove483@yahoo.com

S. J. P. Tume · N. D. Mbongsi
Department of Geography and Planning, University of Bamenda, Bamenda, Northwest Region, Cameroon

A. N. D. Mbougo
Department of Geography, University of Dschang, Dschang, West Region, Cameroon

U. Chatterjee et al. (eds.), *Eco-Resilience: Climate Change, Land Degradation and Sustainable Solutions*, Sustainability Solutions,
https://doi.org/10.1007/978-3-032-00708-7_19

Keywords Eco-resilience · Farmers' adaptation strategies · Livelihoods · Soil degradation · Soil fertility management · Sustainable agriculture

19.1 Introduction

About two-thirds of the world's population lives in the developing world, where agricultural production is not keeping pace with population increase (FAO, n.d). Agriculture is the main source of support for the majority of rural households. The degradation of farmlands causes a serious menace to food security and households' livelihoods in rural areas. Hence, sustainable agricultural practices are a call for concern to continue matching agricultural production with population increase. However, in some parts of the world, agriculture is practiced without the implementation of proper soil conservation techniques UNCCD (United Nations Convention to Combat Desertification, 2017), and the agricultural sector encounters many challenges because of soil degradation (Katilé et al., 2024). Land degradation is a universal phenomenon (IPBES, 2018) and has been a problem since humans settled on land, and started cultivating the soil and grazing domesticated animals (Sanders, n.d). This is a major challenge in agricultural production in many parts of the world, especially in developing nations (Atnafe et al., 2015). Land degradation is prevalent throughout sub-Saharan Africa and reduces plants' growth on the soil (IFAD, 1992). This is the deterioration of the physical, chemical, and biological properties of soils (Aichatou et al., 2022), leading to low crop yields and can reduce the response of soils to fertilizers or other farm inputs, and loss of sustainability (Young, 1990). It is a common problem faced by farmers; thus, good soil structure, soil moisture, and fertility are crucial for a good agricultural production (Bhuchar et al., n.d). This phenomenon has been a threat to the world's population by causing hunger and poverty (Mugume, 2014). Constant soil degradation is a serious problem and threat to food production, food security, and quality of life for poor farmers (Abay, 2011; REMA (Rwanda Environment Management Authority), 2010).

Food security is threatened by soil erosion because it results in the loss of organic matter, nutrients, and fertile soils that are essential for plant growth (Talukdar et al., 2023). Land degradation is often caused by erosion, widely recognized as an important agricultural problem in developing countries (Castaño et al., 2002); erosion washes away soil nutrients, rendering the soil infertile (California Department of Parks & Recreation, 2024; Halus et al., 2016). Soil erosion has an immediate effect on the chemical fertility of the soil (Agboola & Balogun, 2020; Jawerth & Miklos, 2018; Valentin & Christian, 2018). Thus, the UN 2030 Agenda for Sustainable Development is to end hunger and ensure sufficient food for all. However, Bafou is an agricultural hub in the west region of Cameroon that supplies Menoua Division, the west Region, and Cameroon in general with food crops. Agriculture has been a longstanding source of income for the inhabitants of this locality due to its fertile soils.

Poor agricultural practices, soil erosion, soil nutrient depletion, and climate variability/change have been menacing agricultural production in recent years. Nevertheless, to ensure food security and soil, the top layer of the earth's surface should be capable of sustaining life (Ul Zaman et al., 2018) and be in good health. According to Tegha (n.d), soil health is the ability of the soil to sustain productivity, diversity, and environmental services of terrestrial ecosystems. Soil is a limited and nonrenewable resource, vital for supporting life on Earth (Lal, 2017). It serves as the keystone of agricultural production by supplying vital nutrients, storing water, and supporting plant growth (Talukdar et al., 2023).

Soil must be protected from activities that can lead to its destruction and permanent damage by adapting practices that avoid soil contamination and degradation (Agboola & Balogun, 2020; Liniger et al., 2011) and preserving its productivity (Gupta, 2016). Soil conservation minimizes the loss of useful soil nutrients, protects soil against physical loss by erosion and/or against chemical deterioration (Maesoud, n.d), and replaces soil nutrients (Gujarati & Gandhi, 2012; Negasa, 2022). Soil conservation is a set of management strategies that prevent soils from eroding (TRFK (Tea Research Foundation of Kenya), 2012), assure their maintenance and/or improve their fertility (Gil, n.d; Sowjanya et al., 2021; Young, 1990). In sub-Saharan African countries, soil conservation should be a priority because it is the foundation for agricultural prosperity (IFAD, 1992). Conserving the agricultural system avoids further destruction of soil structure, improves the organic content in the soil, retains more water to maximize the crop yield, and prevents soil erosion (Tristi, 2021), leading to sustainable agricultural practices.

In Bafou, most farmers are smallholders (80%), cultivating small plots (0.1–1 hectare); generally, about 98% of farmers do not have farm plots of more than 5 hectares (Tchékoté et al., 2018). The farmers collaborate with their family and/or hire a few laborers to work on the farmlands using simple farming techniques. They are involved in the cultivation of food crops and market gardening. Their farm produce is used to satisfy their families' needs, while the surplus is sold, and the market gardening crops are for commercial purposes. Farmers depend on farm production for their households' livelihoods and income. To increase productivity, farmers involuntarily carried out unsustainable agricultural practices (ploughing, tillage, clearing, mono-cropping, monoculture, irrigation, fertilization, use of pesticides, and herbicides (Nsahlai et al., 2023). The goal of farmers was to increase yields; unfortunately, alongside these practices, climate change/variability aggravated the degradation of farmlands and a drop in agricultural output. Farmers, in a bid to increase yields, started adopting innovative and specific strategies to combat the degradation of farmlands. The chapter seeks to: (1) ascertain the causes of degradation of farmlands in Bafou; (2) identify the adaptation strategies put in place by farmers to combat the degradation of farmlands, and (3) pinpoint the challenges farmers face when implementing their adaptation strategies.

19.2 Study Area

Bafou is a small municipality located in the Nkong-Ni subdivision, Menoua division in the West Region of Cameroon. It is the biggest chiefdom in the division and one of the largest in Cameroon, with a population of about 91,800 inhabitants and 255 inhabitants/km^2 (Bucrep, 2005). Bafou is the agricultural production basin of Menoua, located between latitudes 5°25′30″ and 5°37′00″ north of the equator and between longitudes 10°1′31″and 10°12′30″ east of the Greenwich meridian (Fig. 19.1). The study area is part of the western high plateau of Cameroon (Bamileke plateau) with an average altitude of 1400 m. Agriculture is the main activity carried out by the population. They cultivate food crops and market gardening crops (Nsahlai et al., 2023).

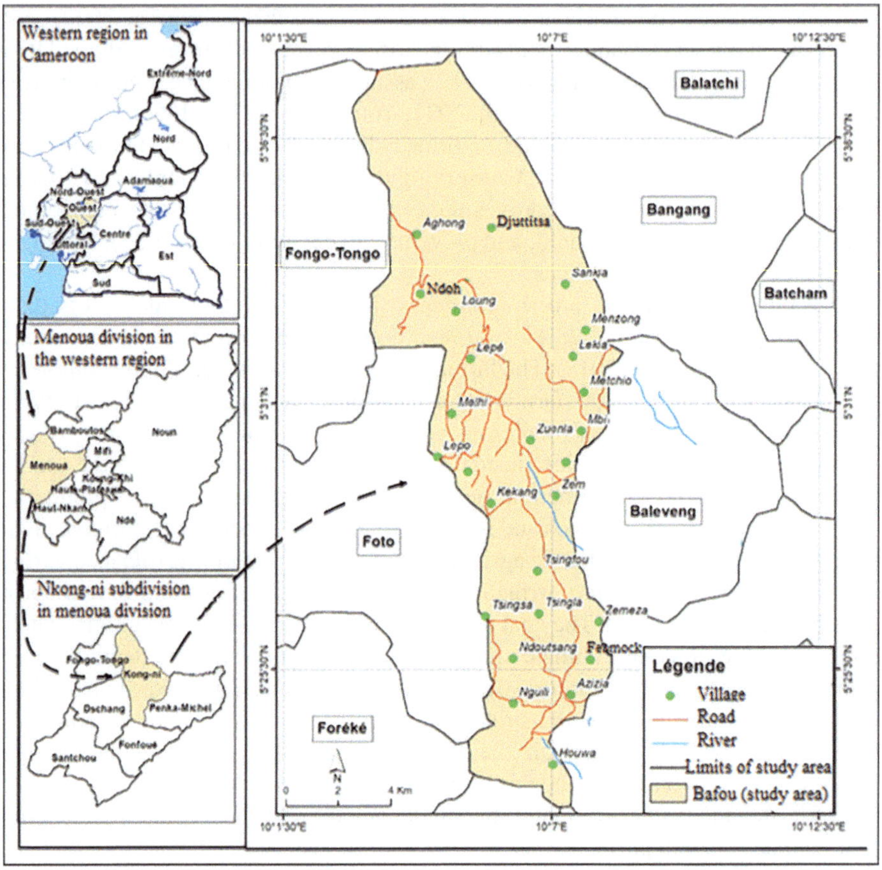

Fig. 19.1 Study area. *Source* SOGEFI GIS Database, March 2018 and fieldwork 2022

19.3 Materials and Methods

The study was conducted within five (5) localities in Bafou municipality (Djuttitsa, Feumock, Ndoh, Menzong, and Houwa). These localities were chosen because they specialize in the production of food crops and market gardening crops for household consumption and commercialization. This choice was to better apprehend the nature of soil degradation on farms subjected to market gardening practices and food crops farming practices, thus permitting the researchers to have a broader understanding of the different strategies used to combat the degradation of farmlands in the study area. A sample of 125 households out of about 206 households producing food and market gardening crops (information from the Agricultural Post in Djuttitsa) were randomly chosen from the five (5) localities (25 household heads from each) (Table 19.1), giving a confidence interval of ±5.4, which is acceptable in social sciences.

A mixed research approach was used for this study. Data were obtained via questionnaires (administered to 125 farmers), three (3) Focus Group Discussions (FGDs), interviews, and field observations (direct and participant observations). The FGDs were held in Djuttitsa (6 men and 3 women), Menzong (6 men and 3 women), and Feumock (5 men and 4 women). Focus group members were not members of households where questionnaires had been administered. These three localities were chosen because their leaders were more open and willing to collaborate with the researchers. During the FGDs, participants listed the causes of soil degradation, the consequences, adaptation strategies, and the challenges they face in combating soil degradation in a pyramid shape on A1 papers in a pyramidal manner, starting with the most prominent causes at the bottom during each FGD. At the end of each focus group, participants were asked to propose solutions for the different challenges. Semi-structured question guides were administered to the sub-divisional delegate of agriculture, 2 agricultural extension workers, and 2 quarter heads to gain reliable and comparable qualitative data regarding agricultural practices that cause soil degradation, strategies of adaptation, and factors influencing their adaptation strategies. Before collecting the primary data, we gave our identity, research objectives, and the importance of our study to the respondents. This was done to create a cordial research environment. The questionnaires were administered in a friendly manner to respondents. During the observation phases, pictures were snapped to illustrate

Table 19.1 Surveyed localities and households

No	Surveyed locality	Number of respondents
1	Djuttitsa	25
2	Feumock	25
3	Ndoh	25
4	Menzong	25
5	Houwa	25
Total		125

Source Field survey, 2022

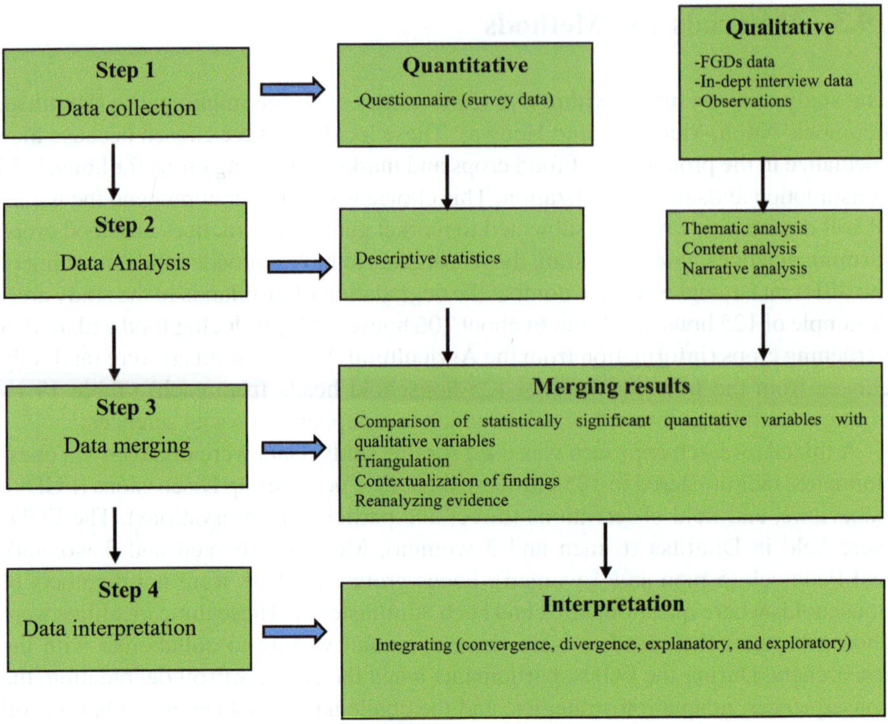

Fig. 19.2 Methodological flowchart. *Source* Adapted from Etana et al. (2020)

phenomena; notes and audio recordings (with the consent of respondents) about farmers' adaptation strategies and challenges were not left out. Secondary data was obtained from desk research (through peer-reviewed book chapters, scientific articles, and reports from Scopus and Google Scholar). The tool used to analyze data is descriptive statistics (describing the adaptation strategies and the challenges using percentages where applicable). Nevertheless, data has been collected and analysed via the following steps (Fig. 19.2).

19.4 Results and Discussion

19.4.1 Causes of Soil Degradation

The degradation of farmlands is related to factors such as unsustainable agricultural practices (99%), climate change, and soil erosion that wipes away the nutritious top layer of the soil and/or washes away the useful biological components of the soil, which favours the growth of plants. In Bafou, the agricultural basin in the West region of Cameroon, these factors diminish soil fertility and reduce agricultural output, making it a must for farmers to adopt strategies of combating soil degradation. It was observed that inappropriate farming practices such as mechanical (ploughing, tillage, clearing), cropping system (mono cropping, monoculture), practices that add ions to the soil (fertilization, manuring, pesticides, herbicides), and practices that change the hydrological soil profile (irrigation) continuously degraded farmlands (Plate 1).

Photo 1: monoculture

Photo 2: monoculture

Photo 3: Tilled land

Photos 1, 2, and 3 demonstrate inappropriate agricultural practices leading a loss of soil and soil nutrients. These practices of monoculture and tillage accelerate both physical and chemical erosion of soils (Nsahlai et al., 2023)

The different agricultural practices aimed at increasing agricultural productivity caused physical soil degradation and a loss in soil nutrients, thereby reducing agricultural output. In Rwanda, agricultural practices such as deforestation, malpractices of intensification and extensification have a major impact on reducing soil fertility and productivity (REMA, 2010). In Europe, inappropriate agricultural practices also cause the degradation of farmlands (Sustainable Agriculture & Soil Conservation fact sheets, 2009). According to the United Nations Development Programme (UNDP, 2019), the expansion of crop and grazing lands into 'natural' vegetation cover causes soil erosion. In this same line, FAO (n.d) iterates that the inappropriate application of agricultural has intensified the degradation of farmlands. These practices accelerate soil erosion, thereby increasing soil degradation (Valentin & Rajot, 2018).

19.4.2 *Farmers' Adaptation Strategies*

Farmers reduce soil erosion, improve soil fertility, decrease nutrient runoff, and increase overall soil health by implementing adaptation practices that conserve soil. The adaptation practices are clearing techniques, soil preparation techniques, cropping systems, and soil manuring techniques. These practices preserve the loss of soil nutrients, improve soil fertility, and reduce reliance on chemical farm inputs, thereby creating an eco-resilient and agriculturally sustainable. Farmers are particularly concerned about soil conservation because it is essential to use land economically, generate high yields (Catherine, 2022), and seek and employ measures that enhance the productive capacity of land (Atnafe et al., 2015). One of the most important adaptation strategies by farmers in Bafou is the farm clearing technique, which is eco-friendly and sustainable.

19.4.2.1 Clearing Techniques

Clearing farmlands in Bafou is done using cutlasses and/or fire. From experience, farmers in the study area have realized that clearing by way of fire leads to a drop in food crop production year after year. Nevertheless, this experience and information gained from extension workers has also made farmers realize that clearing farmlands by the use of fire is detrimental to soil organisms that improve soil texture. Most farmers who used to clear their farmlands by way of fire abandoned the technique in preference for clearing using cutlasses (Fig. 19.3).

Figure 19.3 shows that 88% of farmers use cutlasses to clear their farmlands, and only 12% have continued using fire for clearing. Out of the 88% who use cutlasses, 64% acknowledged that in the past, they used fire for clearing until they discovered that fire kills useful soil organisms that make the soil fertile. Farmers have abandoned

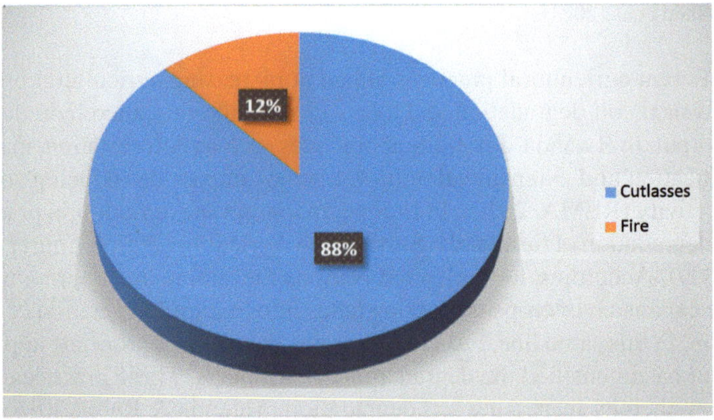

Fig. 19.3 Proportion of clearing techniques used by farmers. *Source* Fieldwork, 2022

the practice of clearing by burning and embraced the use of cutlasses to preserve soil organisms, maintain soil fertility, and increase agricultural output. This practice alone will not be sustainable if appropriate soil preparation techniques are not used.

19.4.2.2 Soil Preparation Techniques

Inappropriate soil preparation techniques cause the degradation of farmlands. Farmers in Bafou had contributed to soil degradation through inappropriate soil preparation practices that accelerated the erosion of farmlands. Having realized a drop in agricultural output, they have adapted soil preparation techniques that involve less tillage. Instead of tilling farmlands according to their ancient and traditional belief that tilling loosens the soil and helps plants to germinate easily, they have adapted strategies such as minimized tillage and no-tillage (Table 19.2) to reduce soil erosion, and maintain soil moisture, which is critical during this époque where climate change causes a lot of fluctuation in rainfall patterns.

It is revealed through field findings that all farmers in the past practiced the tilling of soils during preparation for planting despite the training received from agricultural extension workers during workshops and on-site training. Farmers did not believe in the information received from the extension workers until they began experiencing a fall in their yields, as professed by 2 farmers. The fall in agricultural productivity pushed them to seek strategies of conserving soil water and nutrients to maintain or increase yields. As seen in Table 19.2, 35% of farmers have minimized tillage, and 19% do not till their farmlands. This strategy has helped to limit the erosion of farmlands. In accordance with our findings, Unique-Kulima (2017), declares that minimized or no-tillage soil tillage and associated practices are part of land management techniques that enable carbon sequestration by the soil. In addition, 'no-tillage' practice soil lessens and/or prevents soil degradation (Jawerth & Miklos, 2018). Young, (1990), in this same line, resonates that minimum tillage is an efficient means of soil-fertility maintenance and erosion control. The cropping system on the prepared

Table 19.2 Soil preparation techniques used by farmers

No	Surveyed locality	Number of respondents	Soil preparation techniques		
			Tillage	Minimized tillage	No tillage
1	Djuttitsa	25	09	11	05
2	Feumock	25	09	10	06
3	Ndoh	25	13	08	04
4	Menzong	25	15	06	04
5	Houwa	25	11	09	05
Total		125	57	44	24
Percentage			46%	35%	19%

Source Field survey, 2022

soil also has a great role to play in conserving useful soil nutrients and texture; thus, rendering farming eco-resilient and fostering long-term sustainability.

19.4.2.3 Cropping Systems

Cropping systems that have been used for a long period by farmers in the study area are mono-cropping, monoculture, intercropping or mixed cropping, fallowing, agroforestry, mulching, and contour bunds. Intercropping is practiced by 99% of farmers in the locality on at least one of their farmlands. Contour bunds cropping system is practiced by farmers who cultivate on slopes. They till the soil to form ridges and furrows following contour lines in a bid to avoid soil erosion via runoff. Monoculture and mono-cropping are good practices that conserve the soil, improve its texture, and lead to an increase in output when appropriately carried out. Before now, these practices were not appropriately implemented, and this gradually led to a fall in agricultural productivity. In addition, the fallow period which was for about 1–2 years, has been increased to 2–4 years, and 24% of farmers practice fallowing. Fallowed lands are limited due to land scarcity in the locality.

Farmers practiced monoculture for gardening crops and monocropping for tea cultivation. Monoculture upsets the natural balance of the soil and attracts pests. As such, herbicides, fungicides, and pesticides have to be used in appropriate proportions to remedy pest and diseases. However, it was revealed that before now, about 91% of farmers did not follow the prescribed quantities of chemical farm inputs to be applied on farms, thereby causing land degradation. In bafou, intercropping is done by mixing a variety of crops on the same farmland; that is, tall, medium height, and short creeping plants (that cover the soil surface). Intercropped farms are mostly made of maize, cocoyams, sweet potatoes, groundnuts, and/or beans known as cover crops because they prevent raindrops during the rainy season from eroding the soils, and soils' fertility is conserved, output maintained and/or maximized. In line with this finding, Halus et al. (2016), in their study of soil conservation by vegetative systems, opine that cover crops increase soil organic matter. Intercropping is one of the main ways of successfully dealing with soil degradation, especially in organic farming. Overall, intercropping mitigates water and wind erosion, organic matter decline, diffuses soil contamination, and off-site damages (Sustainable Agriculture & Soil Conservation fact sheets, 2009). It was noticed in the field that 69% of farmers who practiced mixed cropping have associated fruit trees such as pears, mangoes, other edible trees, and trees to serve as fuelwood on their farms. These trees were not planted primarily for agronomic purposes; they were planted to improve household income from the selling of fruits.

However, having realized the role of agroforestry, farmers protect young trees that are growing on their farmlands. In the same accord, Tristi (2021) discourses that farmers in Rejosari, Ethpopia practice agroforestry, and intercropping to combat soil degradation and make agriculture more sustainable. However, the adoption of agronomic measures and vegetative barriers is essential for promoting sustainable agriculture and preserving soil and water resources for future generations (Gairola &

Kadagonda, 2023). According to the study carried out in Ethiopia by Negasa (2022), the trees are grown in narrow strips, and the trees enhance crop yield. The new conservation method combines growing cereal crops using no-till land management with growing fruit trees and shrub strips. The roots and leftover parts of trees, such as stems and leaves from the selected plants, improve soil structure and overall soil health, which helps hold the soil in place (Jawerth & Miklos, 2018; Young, 1990). According to studies carried out in Indonesia by Yusriadi et al. (2024), agroforestry maintains a stable microclimate and soil health. In India, it is a vegetative barrier against wind erosion in farmers' farmlands (Prabuddh et al., 2019). These methods enhance biodiversity and improve soil health, prompting agriculture to be more sustainable.

In Bafou, mulching is mostly practiced during the nursing phase by 97% of farmers, where seeds are spread on soil mixed with compost and covered with non-heavy residues from previous crops, weeds, and shrubs to favor seeds' germination. Seeds that are sown are those of vegetables and market gardening crops. The grass and/or non-heavy residue prevent soil from being swept away by wind and/or rain. Junge et al. (2008) elucidate that when residue improves soil texture and renders the soil more sustainable. Nevertheless, mulch is an effective forest cover against erosion; this makes agriculture more sustainable and ecologically resilient to soil degradation. Byamukama et al. (2019) contend that in Uganda, thanks to technical advice received from extension workers, farmers use eco-resilient farming practices such as cover cropping, mulching, and agroforestry, thus making agriculture more sustainable. According to Lal (2016), conservation tillage, mulching, windbreaks, contour bunds, diversion channels, and check dams are the most significant soil conservation adaptation strategies used by farmers to combat soil degradation in plain areas. This technique ensures water access for crops without causing soil loss. In addition to the mulching practice, farmers have reduced the use of chemical inputs that usually cause chemical soil erosion, and now use more manure that is organic.

19.4.2.4 Soil Manuring Techniques

Organic manure used by farmers is animal manure (fowl droppings, animal waste from piggeries), and decomposable kitchen waste. Animal waste from piggeries is a very important bio-fertilizer as it contains micro-organisms that increase soil nutrients and promote plants' growth. About 89% of farmers in the locality have piggeries and thus, easy access to animal waste manure for farm use. Organic pesticides are wood ash and goat dung diluted and sprayed on affected crops. All farmers apply organic manure and pesticides on their farmlands. According to Tristi (2021), organic fertilizer (decomposed manure or fresh manure that covers the land mixed with dead leaves) plays a vital role in farming. It loosens the soil structure and makes the crops grow healthier because the roots can grow deeper. Abdoulaye et al. (2022) opine that the use of organic and mineral manure is the most important soil conservation technique. A combined application of cow dung and wheat straw mulch gives the lowest soil loss and the highest yields of grain sorghum compared with a range of other practices (Morgan, 2005). In the Sudano-Sahelian zone of Cameroon, farmers adapt sustainable soil management practices, including crop rotation, intercropping, fallowing, the use of organic fertilizers, and the adaptation of conservation agriculture (Bienvenu et al, 2023). Contrariwise, Bensalem (1976) in his study in North Africa argues that soil conservation practices such as Strip cropping, Mulching, Crop rotation, and Contour ploughing are eco-friendly and sustainable.

19.4.3 Factors Affecting Farmers' Adaptation Strategies in Bafou

Farmers, in their strive to combat the degradation of farmlands through soil conservation practices, generally beneficial for soil health and productivity, face several challenges associated with the adaptation strategies. The challenges are demographic, socioeconomic, and traditional/cultural (Fig. 19.4). Also, climate change/variability continues to pose risks that prevent farmers' ability to adapt effectively.

Figure 19.4 shows that 24% of farmers are confronted with challenges that are cultural/traditional. This is because these farmers have a strong tie with their traditional method of farming that has been passed to them by their ancestors, and they find it difficult to detach from it. A majority of farmers (66%) face demographic challenges, and 10% face socio-economic challenges.

19.4.3.1 Cultural/Traditional Factors

Culture/ tradition refers to practices or skills passed down through generations within the community. These practices or skills encompass both tangible and intangible aspects that hold value in terms of cultural identity and diversity. In the study area, despite the negative effect of clearing by fire, 12% of farmers continue to practice it

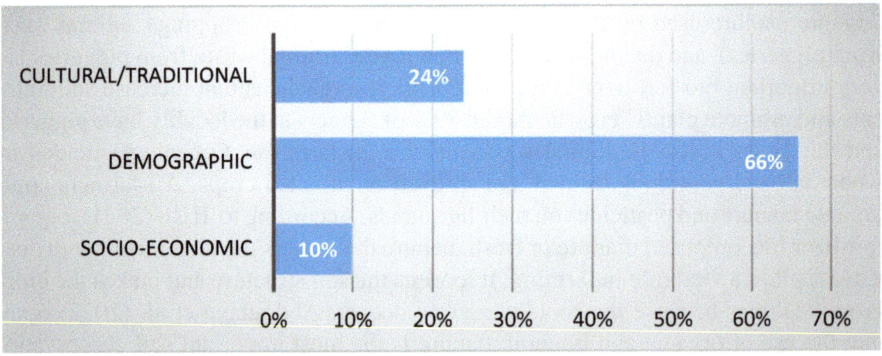

Fig. 19.4 Challenges to adaptation. *Source* Fieldwork, 2022

because, in their minds, this method is the best yielding method that has been used from generation to generation. As for them, a decrease in their yield is related to witchcraft, uprising of their ancestors, and jealousy from other farmers, and not the practice itself. According to Tristi (2021), older farmers are capable of predicting planting seasons by observing the appearance and/or disappearance of some bird species and changes within the environment. When not well predicted, output is negatively affected. Following this same angle, Lal (2017) discourses that traditional farming practices, local customs, or lack of awareness and education about the benefits of soil conservation may hinder the widespread adoption of new techniques. ATPS (African Technology Policy Studies) (2013) also opines that farmers are skeptical about shifting away from conventional agriculture due to its deep-rooted entrenchment in their cultural practice. Nevertheless, traditional knowledge and practices provide a foundation for resilience.

19.4.3.2 Demographic Factors

Demographic factors are characteristics related to household size, age, and level of education. These factors, as noticed in the field, hinder farmers' adaptation strategies. Household size entails the labor force available to transport organic manure from the house to the farms. Organic manure is heavy and not easy to transport, and therefore entails the need for physically strong persons to transport it. It was noticed that larger households (6–10 persons) having children from 10 to 18 years easily adjusted to the use of degradable household waste, fowl dropping, animal dung and waste. This is because there is an available labor force. Households of smaller sizes or bigger sizes whose children are below 10 years did not apply a huge quantity of organic manure on their farms because of a limited labor force. Those in this category who applied much organic manure on their farmlands hired laborers. This study is in accordance with the study of Byamukama et al. (2019), who highlight that households with small family sizes are less likely to adopt conservation techniques compared to households with large family sizes due to a shortage of labor force. On the other, Iliyasu et al. (2020) in their study in Nasarawa State, Nigeria, established that the use of inorganic fertilizer by farmers is eco-resilient and a sustainable agricultural practice.

Age

The parameters used to classify 'age' are young and old. Young farmers are those between 24 and 50 years, and the old are those who are above 50 years. In the study area, 67% of farmers are young and 33% are old. It was observed that their practices are still tied to clearing by the use of fire, and maximized soil tillage. The young are more flexible in adapting new strategies to improve upon yields. The old do not care about adaptation strategies because their judgement of a fall or a rise in productivity is not related to a decline in soil fertility. Tristi (2021) affirms that young farmers are more eager compared to older farmers in applying adaptation strategies to combat farmland degradation because they have no ancestral experience. Contrary to our

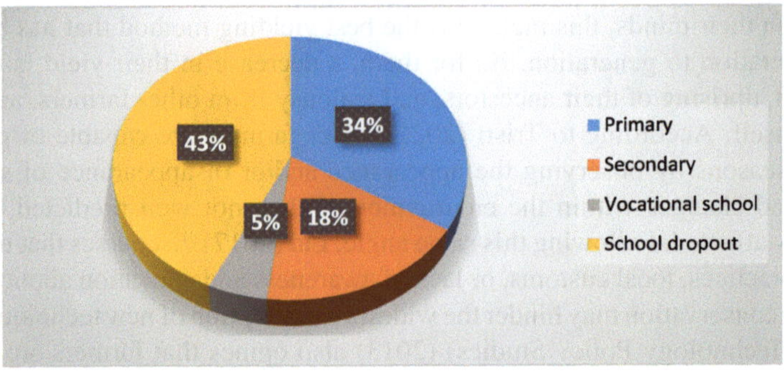

Fig. 19.5 Level of education of farmers. *Source* Fieldwork, 2022

findings, Agboola and Balogun (2020) pinpoint that the age of farmers is negatively related to adaptation strategies.

Level of Education

It is obvious that the level of education of farmers plays a role in hindering their adaptation practices (Fig. 19.5).

As seen in Fig. 19.5, 34% of farmers have primary school certificates, 5% have vocational training certificates, 18% have completed secondary school, and 43% are school dropouts (from the primary, secondary, and high schools). This explains their level of implementation of soil conservation practices. Farmers who did not receive formal education from schools cannot implement proper adaptation strategies, such as no and/or minimum tillage, and contour ploughing, because they do not easily process information provided by extension workers. They prefer to imitate what experienced farmers (above 50 years) are doing. In line with our findings, Agboola and Balogun (2020) state that education enhances the adoption of soil conservation strategies. Apart from respondents' age and the level of education, which limit their strategies of combating the degradation of farmlands, there exist socio-economic factors, which is less fundamental as ascertained during FGDs.

19.4.3.3 Socio-Economic Factors

Implementing soil adaptation strategies may necessitate investments in buying and /or transporting organic manure. Five (5) farmers in the field ascertained that intercropping or mixed cropping causes diseases and pests to spread from one crop species to another, thereby demanding more time, labor, and resources to manage the pests. However, in the absence of financial resources to combat the pests and diseases, all crops in the farm are affected, and productivity is reduced.

19.5 Recommendations

For adaptation strategies to be successful in the long term, the government and Non-governmental Organizations (in charge of soil conservation and management) should continue assisting farmers. Assistance should be in the form of access to resources, access to information, education/training about the benefits and challenges of land degradation's adaptation strategies. Farmers should be provided with technical and financial assistance to foster their empowerment. Policies promoting sustainable land management practices should be reviewed to create an enabling environment for farmers. The government should integrate newly introduced technologies with farmer-friendly indigenous measures; include farmers in any decision-making process that concerns them (the approach should be bottom-top).

19.6 Conclusion

The degradation of farmlands in Bafou is caused by unsustainable agricultural practices such as mono-cropping, monoculture, clearing by burning, and short fallow periods; soil erosion; and climate change/variability. The adaptation strategies used by farmers in Bafou to combat the degradation of farmlands show their resilience in the face of environmental challenges. The strategies used are clearing techniques, soil preparation techniques, cropping systems and soil manuring techniques. These demonstrate a proactive approach to maintaining soil health/soil fertility and enhancing agricultural productivity and sustainability. Farmers in Bafou have recognized the importance of combating land degradation, enhancing yields, and sustaining livelihoods. In the course of combating soil degradation, farmers confront challenges related to their demographic status, socio-economic factors, and cultural/traditional norms. The commitment of these farmers to adapt to soil conditions reflects a broader need for sustainable agricultural practices. Stakeholders (government bodies, NGOs, and research institutions) should enhance farmers' adaptive strategies by creating an enabling environment that promotes sustainable agriculture, and empower smallholder farmers (increased access to credit, training, and information) to invest in long-term land improvements. This will ensure a long-term sustainability in agriculture and contribute to food security and a sustainable environment.

References

Abay, A. (2011). Construction of soil conservation structures for improvement of crops and soil productivity in the Southern Ethiopia. *Journal of Environment and Earth Science, 1*(1), 21–29, ISSN 2224-3216 (Paper) ISSN 2225-0948 (Online). http://www.iiste.org/Journals/

Abdoulaye, M. A., Abdou, G. F., Soumana, B., & Barrage, M. (2022). Farmers' perceptions of land degradation and adaptation strategies adopted by farmers in the geographical area of Bagaroua in Niger. *Environment and Natural Resources Research, 12*(1), 37–47, ISSN 1927-0488 E-ISSN 1927-0496 Published by Canadian Center of Science and Education. https://doi.org/10.5539/enrr.v12n1p37

Agboola L. W., & Balogun, O. L. (2020). Soil conservation techniques and productivity of arable crop farmers in Kogi State, Nigeria. *Journal of Agricultural Science and Environment, 20*(1 &2), 78–91.

Aichatou, A. M., Fanna, A. G., Boubacar, S., & Moussa, B. (2022). Farmers' perceptions of land degradation and adaptation strategies adopted by farmers in the geographical area of Bagaroua in Niger. *Environment and Natural Resources Research, 12*(1). ISSN 1927-0488 E-ISSN 1927-0496 Published by Canadian Center of Science and Education. 37-47. doi:10.5539/enrr.v12n1p37 URL: https://doi.org/10.5539/enrr.v12n1p37

Atnafe, A. D., Ahmed, H. M., & Adane, D. M. (2015). Determinants of adopting techniques of soil and water conservation in Goromti Watershed, Western Ethiopia. *Journal of Soil Science and Environmental Management, 6*(6), 168–177. http://www.academicjournals.org/JSSEM

ATPS. (2013). Farmers' response and their adaptation strategies to climate change in Mafeteng. In T. S. T. Sekaleli, K. Sebusi (Eds.), *District, Lesotho* (p. 24). ATPS Research Paper No. 26.

Bensalem, B. (1976). *Examples of soil and water conservation practices in North African Countries: Algeria, Morocco and Tunisia; conservation practices in agriculture paper No 10 In, FAO soils Bulletin No 33* (pp. 149–158). Soil conservation and Management in developing Countries. FAO of the UN.

Bhuchar, S., Billingsley, A., Dhakal, M., Providoli, I., Sherpa, S. L., & Siddiqi, A. S. (n.d). Soil moisture retention and soil fertility options for agricultural and degraded lands. Mountain development resource book for Afghanistan. *Managing Natural Resources*, 69–81.

Bienvenu, S., Claudine, V., Arka, B., & Labara, K. B. (2023). Agriculture and soil management in the context of sustainable development in the Sudano-Sahelian zone of Cameroon. *Journal of Geoscience and Environment Protection, 11*, 171–191. https://doi.org/10.4236/gep.2023.1111013

BUCREP. (2005). Répertoire actualisé des villages du Cameroun, Troisième recensement général de la population et de l'habitat du Cameroun. Vol IV –Tome 7, p. 400.

Byamukama, W., Ssemakula, E., & Kalibwani, R. (2019). Factors influencing the uptake and sustainable use of soil and water conservation measures in Bubaare Micro-Catchment, Kabale District, South Western Uganda. *Journal of Environmental Health Science, 5*(1), 26–32.https://doi.org/10.15436/2378-6841.19.2432

California Department of Parks and Recreation. (2024). 2020 Soil conservation standard and guidelines. Off-Highway Motor Vehicle Recreation Division (Rev. 3/2024), p. 71.

Castaño, J., Meulenberg, M. T. G., & Tilburg, A. V. (2002). A new method of measuring the adoption of soil conservation practices; theory and applications, *NJAS: Wageningen Journal of Life Sciences, 50*(1), 95–114, https://doi.org/10.1016/S1573-5214(02)80006-9

Catherine, P. (2022). Soil conservation techniques and conservation tillage. *Global Journal of Plant and Soil Science, 6*(3), 001, ISSN 2756-3626. https://www.internationalscholarsjournals.com

Etana, D., Snelder, D.J. R. M., Van Wesenbeeck, C. F. A., & de Cock Buning T. (2020). Dynamics of smallholder farmers' livelihood adaptation decision-making in central Ethopia. *Sustainability 12*, 4526. https://doi.org/10.3390/su12114526

FAO. (n.d). Soil conservation and management in developing countries. *Soils Bulletin No 33* FAO of the UN, Rome, p. 208.

Gairola, A., & Kadagonda, N. (2023). Soil conservation methods: Agronomic measures and vegetative barriers. *The Agriculture Magazine, 2*(9), 235–239, E-ISSN: 2583-1755. https://www.researchgate.net/publication/376853009

Gil. N. (n.d) Role of soil conservation in watershed management on agricultural lands; conservation practices in agriculture Paper No 5. In, *FAO soils Bulletin No 33*. Soil conservation and Management in developing Countries. FAO of the UN, Rome, pp. 89–95.

Gujarati, M., & Gandhi, H. M. (2012). Soil erosion & advance techniques of soil conservation. *Indian Journal of Research, 1*(4), 75–77.

Gupta, A. K. (2016). The role of soil conservation in plain and hilly areas. *International Journal of Recent Research and Review, IX*(4), 48–52, ISSN 2277-8322.

Halus, S., Zahrul, F., & Nanda, M. (2016). Soil conservation by vegetative systems. *Oil Palm Cultivation Polish Journal of Soil Science, Xlix*(2), 223–235. https://doi.org/10.17951/Pjss/2016.49.2.223

IFAD. (1992). *Soil and water conservation in sub-saharan Africa: Towards sustainable production by the rural poor* (p. 110). Free University, Amsterdam.

Iliyasu M. A., Chunwate, B. T., & Ahmed, Y. U. (2020). Soil conservation practices among farmers in Doma Local Government Area, Nasarawa State, Nigeria. *International Journal of Science and Advanced Innovative Research, 5*(3), 12–33, ISSN: 2536-7315 (Print) 2536–7323 (Online). http://www.casirmediapublishing.com

IPBES. (2018). *Summary for policymakers of the assessment report on land degradation and restoration of the intergovernmental science-policy platform on biodiversity and ecosystem.* In R. Scholes, L. Montanarella, A. Brainich, N. Barger, B. ten Brink, M. Cantele, B. Erasmus, J. Fisher, T. Gardner, T. G. Holland, F. Kohler, J. S. Kotiaho, G. Von Maltitz, G. Nangendo, R. Pandit, J. Parrotta, M. D. Potts, S. Prince, M. Sankaran, & L. Willemen (Eds.), *IPBES secretariat* (p. 44). Bonn, Germany.

Jawerth, N., & Miklos, G. (2018). How to win the fight against soil erosion: saving fertile land and preserving water quality with the help of nuclear techniques. Atoms for peace and development, *IAEA Bulletin*, 14–17.

Junge, B., Abaidoo, R., Chikoye, D., & Stahr, K. (2008). *Soil conservation in Nigeria: Past and present on-station and on-farm initiatives* (p. 28). Soil and Water Conservation Society. https://www.swcs.org

Katilé, M. L., Bengaly, A., Traoré, S. S., & Rodrigo-Comino, J. (2024). Farmers' perception on climate change, soil erosion and adaptation strategies in small rural communities of Mali: case study of the rural municipality of Méguétan. *Investigaciones Geográficas*, (82), 43–61. https://doi.org/10.14198/INGEO.26589

Lal, B. (2016). The role of soil conservation in plain areas. *World Journal of Pharmaceutical and Medical Research, 2016, 2*(4), 227–229. https://www.wjpmr.com

Lal, B. (2017). Importance and methods of soil conservation in hilly areas. *International Journal of Recent Research and Review, X*(4), 65–69, ISSN 2277-8322

Liniger, H. P., Mekdaschi Studer, R., Hauert, C., & Gurtner, M. (2011). Sustainable land management in practice—Guidelines and best practices for Sub-Saharan Africa. TerrAfrica, World Overview of Conservation Approaches and Technologies (WOCAT) and Food and Agriculture Organization of the United Nations (FAO), p. 246. https://www.fao.org

Maesoud, F. I. (n.d). Soil conservation as a protective measure against Salinisation. Conservation practices in agriculture. Paper No 4 In, FAO *soils Bulletin No 33.* Soil conservation and Management in developing Countries. FAO of the UN, Rome, pp. 78–88.

Morgan, R. P. C. (2005). *Soil erosion and conservation* (3rd edn, p. 304). Blackwell publishing. www.blackwellpublishing.com

Mugume, E. (2014) *Methods for assessing soil degradation in East Africa. United Nations University Land Restoration Training Programme* [final project] (p. 32). http://www.unulrt.is/static/fellows/document/Mugume2014

Negasa, G. (2022). The successful soil conservation methods in Ethiopia: A review. *World Journal of Agricultural Sciences, 18*(3), 122–134, ISSN 1817-3047. https://doi.org/10.5829/idosi.wjas.2022.122.134

Nsahlai, L. K., Wakponou, D. N., & Mbougo, A. N. D. (2023). Agricultural practices as a driving factor of soil degradation in Bafou, Menoua Division. In Z. N. Fogwe, et al., (Eds.), *Celebrating Ten Years of Restless Rest in Landmark Research and Human Resource Development: Festschrift for Professor Emeritus Cornelius Mbifung Lambi. Volume I, Development and Environmental dynamics: Conference Proceedings* (pp. 119–130). The University of Bamenda printing press.

Prabuddh, K. M., Aman, R., & Suresh, C. R. (2019). Agronomic measures in traditional soil and water conservation practices in the Sikkim Himalaya, India. *American Research Journal of*

Agriculture, 5(1), 1–16, ISSN (Online): 2378–9018. https://doi.org/10.21694/2378-9018.19003. www.arjonline.org

REMA. (2010). *Tool and guideline # 5. Practical tools on soil and water conservation measures*, p. 32.

Sanders, D. (n.d). Soil conservation. Land use, land cover and soil sciences vol. IV—Soil conservation—David Sanders. *©Encyclopedia of Life Support Systems (EOLSS)*, p. 8. https://www.eolss.net/ebooklibs/sc_cart.aspx?File=E1-05-06

Sowjanya, A., Sekhar, D., Sagal, B. V., Sravanthi, S., Sandeep, B. N. N., & Rao, G. R. (2021). Soil and water conservation problems and measures in hilly area. *Just Agriculture, Multidisciplinary E-News Letter, 1*(11), 11, (e-ISSN: 2582-8223). https://www.justagriculture.in

Sustainable Agriculture and Soil Conservation fact sheets. (2009). *Linking soil degradation processes, soil-friendly farming practices and soil-relevant policy measures,* p. 40. http://soco.jrc.ec.europa.eu.© European Communities 2009. Reproduction is authorised provided the source is acknowledged. May 2009.

Talukdar, N., Harsha, H., Sarma, A. P., & Mriganko, K. (2023). *Chapter 10: Agronomic practices for soil conservation,* 159–172. https://www.researchgate.net/publication/374060109

Tchékoté, H., Melachio, N. M., & Siyapdje, E. C. (2018). Appropriation foncière, pratiques agricoles et enjeux environnementaux à Bafou-Nord dans les monts Bamboutos (Ouest Cameroun), Belgeo, 2. URL: http://journals.openedition.org/belgeo/21085. https://doi.org/10.4000/belgeo.21085

Tegha, K. C. (n.d). *Attaining enhanced soil health outcomes while improving productivity in Western Cameroon* (p. 13). African Climate Action Partnership. africanclimateactionpartnership.org.

TRFK. (2012). *Soil conservation manual* (1st Edn, p. 20).

Tristi, A. D. (2021). *Adaptation strategies of smallholder farmers using climate-smart agriculture: The case of Rejosari Village, Indonesia* (p. 64). MSc Thesis MEEM—University of Twente.

Ul Zaman, M., Bhat, S., Sharma, S., & Bhat, O. (2018). Methods to control soil erosion—A review. *International Journal of Pure and Applied Bioscience, 6*(2), 1114–1121. https://doi.org/10.18782/2320-7051.6462

UNCCD. (2017). *Threats to soils: Trends and perspectives.* A contribution from the intergovernmental technical panel on soils, global soil partnership food and agriculture organization of the United Nations GLOBAL Landoutlook Working paper, p. 28.

UNDP. (2019). *Combatting land degradation securing a sustainable future. UNDP'S work on sustainable land management and restoration.* The Global Environment Facility (GEF), the Government of Germany, the Green Climate Fund, Adaptation Fund, Lead Developed Country Fund (LDCF) and the Special Climate Change Fund (SCCF); New York, NY, 10017 USA, p. 32. http://www.undp.org

Unique-Kulima. (2017). *Agricultural adaptation: Six categories of good practices and technologies in Africa. Project Report; Adaptation of agricultural practices and technologies to climate change in Sub-Saharan Africa* (p. 80). GIZ, https://www.giz.de

Valentin, C., & Rajot, J. L. (2018). Erosion and principles of soil conservation. In C. Valentin (Ed.), *Soils as a key component of the critical zone 5: Degradation and rehabilitation* (pp. 39–82). ISTE, Wiley. (*Geosciences Series. Soils Set;* 5). ISBN 978-1-78630-219-9.

Young, A. (1990). *agroforestry for soil conservation. CAB International council for research in agroforestry* (p.318). ICRAF review of the potential of agroforestry for soil conservation.

Yusriadi, Y., Andi, C., & Firman, H. (2024). Farmer adaptation strategies through local farming systems in Enrekang, Indonesia. *Scientific Reports.* https://doi.org/10.1038/s41598-024-72953-41

Chapter 20
Thrissur's Coastal Future: A Global Model for Climate-Informed Land Suitability Assessment

S. Gayathri and Paulose N. Kuriakose

Abstract This study develops a climate-sensitive land suitability framework for Thrissur, a coastal district in Kerala, India, where urban expansion intersects with growing ecological and climate risks. Using a GIS-based Multi-Criteria Decision Analysis (MCDA), the study evaluates eight spatial parameters—including flood vulnerability, wetland degradation, slope instability, and proximity to services—to classify land for development, conservation, or restoration. Results indicate that 42.3% of land is unsuitable for development, while only 27.7% is moderately or highly suitable. These findings highlight critical trade-offs between development and environmental sustainability. The study proposes nature-based solutions and risk-informed zoning to safeguard fragile ecosystems while enabling guided urban growth. The methodology supports scalable, evidence-based decision-making and can be adapted in other coastal contexts facing similar challenges. By integrating spatial data, ecological thresholds, and climate risk, the framework offers practical tools for sustainable land management and long-term resilience.

Keywords Climate resilience · Land degradation · Eco-sensitive planning · GIS-MCDA · Land suitability

20.1 Introduction

The coastal district of Thrissur, situated on the Malabar Coast of Kerala, India, is a region of immense ecological and cultural significance (Zainulabdeen & Nagaraj, 2022a, 2022b). Its strategic location along the Arabian Sea and its lush landscapes, comprising forests, wetlands, and agricultural lands, make it a unique geographical

S. Gayathri (✉) · P. N. Kuriakose
Department of Urban and Regional Planning, School of Planning and Architecture, Bhopal, India
e-mail: s.gayathri.nanthini@gmail.com

P. N. Kuriakose
e-mail: paulosenk@spabhopal.ac.in

U. Chatterjee et al. (eds.), *Eco-Resilience: Climate Change, Land Degradation and Sustainable Solutions*, Sustainability Solutions,
https://doi.org/10.1007/978-3-032-00708-7_20

entity. However, the district's distinct natural features also render it highly vulnerable to climate change-induced challenges such as flooding, coastal erosion, and erratic weather patterns (National Centre for Earth Science Studies, 2023). Thrissur's trajectory mirrors the precarious future of coastal regions worldwide, where rising sea levels and intensified climatic events threaten human settlements, economic activities, and fragile ecosystems (Upadhyay, 2020). As a critical component of Kerala's socio-economic framework, Thrissur is characterized by its diverse land use patterns (Neelakantan, 2013). Forests occupy a significant portion of its highlands, while its midland and coastal zones host agricultural activities and urban settlements. However, the rapid urbanization and land conversion in recent decades have significantly altered the district's land-use dynamics (Muraleedharan et al., 2007). Agricultural land has declined, while urban and barren lands have expanded, often at the expense of ecologically sensitive areas such as the Kole wetlands and coastal buffers (National Centre for Earth Science Studies, 2019). These trends, coupled with a high population density and economic dependency on natural resources, underscore the urgency for sustainable land-use planning tailored to climate realities.

Globally, the need for integrating climate adaptation into regional and urban planning has gained momentum, yet practical models for climate-informed land suitability remain limited (Malczewski, 2004). Thrissur provides a compelling case study for exploring such integration. As the third-largest coastal district in Kerala, it epitomizes the challenges and opportunities inherent in balancing economic growth, environmental conservation, and climate resilience. The district has already experienced severe impacts of climate change, including catastrophic floods in 2018 that displaced thousands and caused extensive damage to infrastructure and livelihoods (Lal et al., 2020). These events highlight the district's susceptibility to natural disasters and the critical role that informed land-use strategies can play in mitigating risks. This study introduces a novel, climate-informed land suitability assessment model that integrates ecological sensitivity, socio-economic imperatives, and climate vulnerability into a single geospatial framework. While foundational works such as McHarg's Design with Nature (1969) and Malczewski's GIS-based multi-criteria analysis (2004) have shaped land-use planning, few studies have systematically aligned such models with current climate resilience frameworks, especially in vulnerable coastal regions of the Global South. This research addresses that gap by operationalizing a globally relevant, locally grounded approach using Thrissur, Kerala, as a test case. The study applies eight key spatial parameters, analysed through the Analytical Hierarchy Process (AHP) and GIS, to generate a replicable land-use suitability map responsive to evolving climate risks. The objectives are threefold: (1) To identify and spatially analyse critical land-use determinants in Thrissur district—including flood risk, wetland encroachment, coastal vulnerability, slope instability, and proximity to infrastructure—under current and projected climate conditions, (2) To develop a weighted, GIS-based land suitability model using Multi-Criteria Decision Analysis (MCDA) and the Analytical Hierarchy Process (AHP), integrating ecological, geophysical, and socio-economic indicators, and (3) To delineate zones for conservation, controlled development, and future expansion, and to propose a replicable framework for climate-resilient land-use planning applicable to other vulnerable coastal regions globally. With rising sea levels, increasing flood events, and intensified urban pressures mirrored across deltas and coasts from Southeast Asia to West Africa and Latin America, this research contributes meaningfully to international dialogues on resilient coastal urbanization (Nicholls et al., 2021; Temmerman et al., 2023; UNDRR, 2015). By connecting global planning theory with regional realities, it offers both practical tools and conceptual insight for readers, scholars, and decision-makers worldwide.

The chapter is divided into six sections. The first section introduces the coastal district of Thrissur, highlighting its vulnerability to climate change and the need for a climate-resilient land-use suitability framework. The second section provides a detailed literature review, exploring global and regional research on climate resilience, land suitability, and coastal planning, while identifying gaps addressed by this study. The third section explains the methodology adopted, including the selection of eight key parameters, the use of geospatial analysis, and the weighting process through the Analytical Hierarchy Process (AHP). The fourth section presents the results, including the land-use suitability map for Thrissur, and discusses its implications for balancing economic development with ecological preservation. The fifth section offers specific recommendations for integrating the findings into planning frameworks. Finally, the last section concludes by summarizing the chapter's contributions and proposing its applicability to other vulnerable coastal regions globally.

This study is unique in three major ways. First, it presents a climate-informed land suitability framework that integrates ecological vulnerability (forests, wetlands, flood zones), socio-economic factors (proximity to roads, urban centres), and geophysical risks (slope, coastal erosion) into a single, weighted geospatial model—a comprehensive synthesis rarely applied in the Indian context, especially for secondary coastal districts like Thrissur. Second, unlike traditional land-use assessments that are either ecologically focused or infrastructure-driven, this study combines climate vulnerability analysis with development prioritization, allowing planners to simultaneously identify areas for conservation and resilient urban expansion. Third, the model is designed to be globally transferable: its parameter set, methodology (AHP + GIS), and zoning logic are adaptable to other coastal geographies facing similar climate stress—from the Mekong Delta to the Niger River Basin. By positioning Thrissur as a global microcosm of coastal climate vulnerability, the study bridges local planning with global sustainability solutions and contributes to the emerging field of climate-resilient land use in the Global South.

20.2 Balancing Development and Conservation: A Land Use Planning Framework for Thrissur's Coast

Coastal regions across the globe are increasingly recognized as hotspots of climate vulnerability due to their geographical exposure and socio-economic significance. Approximately 40% of the global population lives within 100 kms of the coast, and these areas account for a substantial share of global GDP (Neumann et al., 2015). However, with rising sea levels, intensifying storm surges, and escalating erosion rates, these zones are experiencing unprecedented challenges. According to the Intergovernmental Panel on Climate Change (Church et al., 2013; Oppenheimer et al., 2019), global sea levels are projected to rise by up to 1.1 m by 2100 under high-emission scenarios, with severe consequences for densely populated regions like Thrissur, Kerala. Thrissur's vulnerability mirrors global trends. Its 81.2-km-long coastline, interspersed with ecologically significant zones such as the Kole wetlands and backwater systems, is emblematic of regions grappling with both rapid urbanization and climatic stress (Vidya et al., 2024). This dual challenge necessitates a structured, evidence-based approach to land-use planning, emphasizing resilience and sustainability.

Climate change significantly alters coastal systems by reshaping ecosystems, increasing disaster risks, and challenging socio-economic structures. Globally, the economic cost of coastal disasters has skyrocketed, with annual losses from coastal flooding projected to reach $1 trillion by 2050 without adaptive measures (Hallegatte et al., 2013). In India, coastal districts contribute nearly 60% to the national GDP but face annual losses exceeding $15 billion due to climate-induced disasters (Rao et al., 2014). Thrissur has witnessed first-hand the devastating impacts of climate change, most notably during the catastrophic 2018 floods. These floods displaced over 80,000 residents, damaged 30% of the district's road infrastructure, and caused an economic loss of approximately INR 595 crore (Government of Kerala, World Bank, & Asian Development Bank, 2018). Such events underscore the urgent need to integrate climate resilience into regional planning frameworks. Ecologically sensitive areas—wetlands, forests, and coastal zones—serve as critical natural buffers against climatic risks. Thrissur's Kole wetlands, a designated Ramsar site, exemplify the ecological and economic importance of such systems. Covering 32,000 hectares, these wetlands support over 40% of Kerala's rice production and act as a vital flood mitigation system (Sebastian et al., 2014). Despite this, 27% of the Kole wetlands have been converted for non-agricultural purposes over the past two decades, leading to increased flood risks and biodiversity loss (Zainulabdeen & Nagaraj, 2022a, 2022b). Urbanization has further compounded these issues. Between 2001 and 2011, Thrissur's urban population increased from 61 to 67%, contributing to the expansion of built-up land by 23% (Census of India, 2011). This growth often comes at the expense of agricultural and ecologically sensitive areas, necessitating a land-use framework that balances development with conservation. The concept of land suitability analysis (LSA) stems from Ian McHarg's seminal work Design with Nature (1969), which advocated for overlaying ecological and socio-economic factors to guide planning decisions. Recent advancements in Geographic Information Systems (GIS) have enhanced the precision and applicability of LSA, enabling multi-criteria decision-making (MCDM) models that integrate diverse parameters (Malczewski, 2004). Weighted overlay analysis of these parameters offers a robust framework for identifying zones optimized for development, conservation, or mixed-use purposes. Thrissur's diverse topography and land-use patterns provide an ideal case study for LSA (see Figs. 20.1 and 20.2).

Urbanization in Thrissur has led to significant land-use changes, with built-up areas expanding at the cost of agricultural and natural lands. As shown in Fig. 20.3, the district's built-up area increased from 19% in 2005 to 24% in 2023, while agricultural land decreased from 43 to 38% over the same period (USGS Earth Explorer, 2023). Such trends are unsustainable, particularly in a region already grappling with high population densities of 1031 persons per square kilometre (Census of India, 2011). Approximately 85% of Thrissur's coastline is classified as either eroding or highly unstable, with an average shoreline retreat of 0.5 m per year (National Centre for Earth Science Studies, 2023). Sea-level rise exacerbates this issue, with projections indicating that up to 150 square kilometres of Thrissur's coastal land could be submerged by 2130 (KSDMA, 2024). The degradation of the Kole wetlands has far-reaching implications for biodiversity and agriculture. A study by Kingsford et al. (2016) estimates that the wetlands' biodiversity has declined by 25% over the past two decades due to habitat loss and pollution. This trend undermines the wetlands' ability to support rice cultivation, fish breeding, and flood mitigation.

Thrissur's land-use dynamics are emblematic of broader global trends. The conversion of agricultural land to urban and industrial uses mirrors patterns observed in rapidly urbanizing coastal regions worldwide. For instance, China's coastal provinces have seen a 20% reduction in arable land over the past two decades, driven by urban expansion and industrialization (Liu & Zhou, 2021). Similarly, in Southeast Asia, mangrove forests—essential for coastal resilience—have declined by 30%, leading to increased exposure to storm surges (Alongi, 2002). However, Thrissur also offers unique opportunities to integrate traditional knowledge systems with modern planning tools. For example, the revival of traditional water management practices, such as the restoration of ponds and canals, can complement planning strategies to enhance water resilience. These practices align with the principles of integrated water resource management (IWRM), as outlined by the Global Water Partnership (2014).

International case studies provide valuable insights for Thrissur's land-use planning. The Netherlands' Room for the River program demonstrates the efficacy of creating floodplains to mitigate flood risks while enhancing ecological value (van Alphen, 2016). Similarly, Singapore's hybrid approach—combining hard infrastructure like seawalls with nature-based solutions such as mangrove restoration—offers a replicable model for coastal resilience (Ellison et al., 2020). Closer to home, Tamil Nadu's Coastal Zone Management Plan (CZMP) emphasizes the integration of socio-economic and ecological considerations, offering lessons on stakeholder engagement and regulatory frameworks (National Centre for Sustainable Coastal

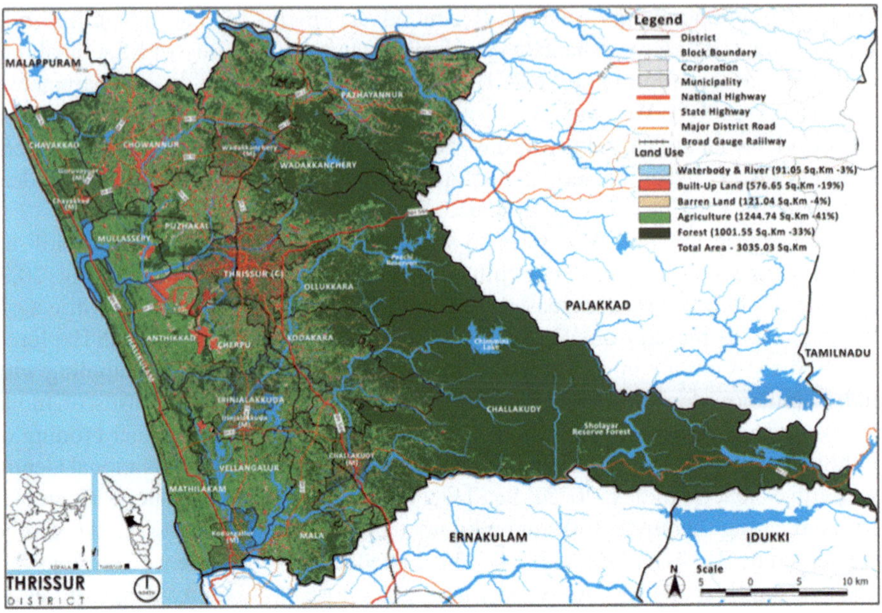

Fig. 20.1 Land use land cover map, Thrissur district-2023. *Source* USGS earth explorer (Retrieved in December 2023)

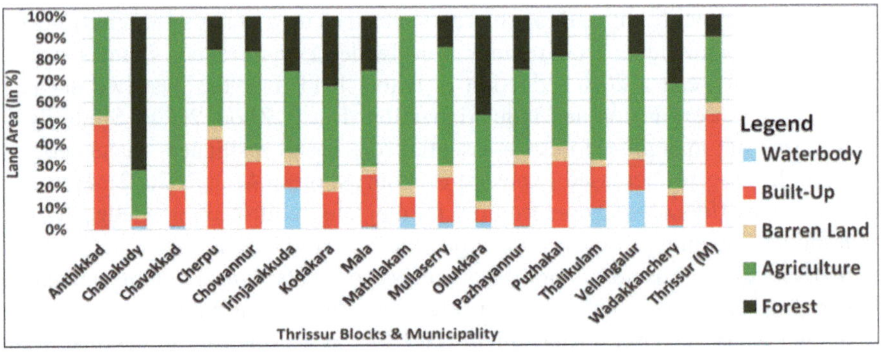

Fig. 20.2 Land cover classification, 2023. *Source* USGS earth explorer (Retrieved in December 2023)

Management, 2019). Such models highlight the importance of adopting a multi-stakeholder approach, integrating local communities, policymakers, and researchers into the planning process.

Given the district's vulnerability, there is an urgent need for a planning framework that balances development with conservation. By categorizing zones based on their ecological and economic value, planners can prioritize conservation in sensitive areas

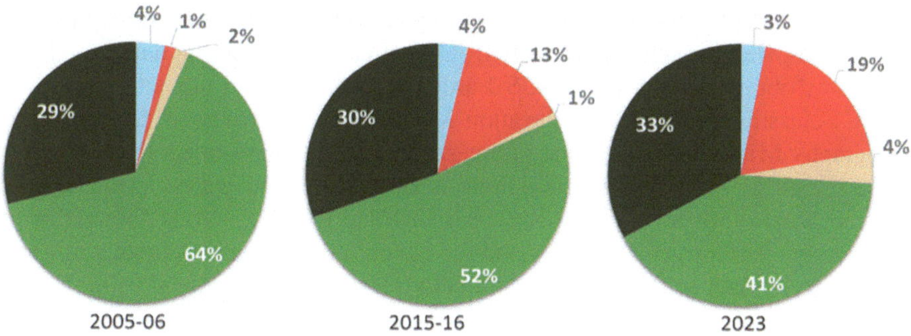

Fig. 20.3 LULC comparison, years 2005–06, 2015–16 and 2023. *Source* Bhuvan and USG earth explorer (Retrieved in December 2023)

while directing growth to less vulnerable zones. Integrating climate data into land-use assessments ensures that planning decisions are resilient to future uncertainties. Thrissur's planning process must also involve local communities, leveraging their traditional knowledge to enhance the relevance and acceptance of proposed interventions. The proposed model, combining LSA with stakeholder input, offers a scalable solution for other vulnerable coastal regions. It aligns with global frameworks such as the Sendai Framework for Disaster Risk Reduction (2015–2030) (United Nations Office for Disaster Risk Reduction, 2015) and the UN Sustainable Development Goals (UN SDGs), particularly Goals 11 and 13 (United Nations, 2015).

20.3 Methodology

This study employs a GIS-based Multi-Criteria Decision Analysis (MCDA) framework, combining spatial data analysis with the Analytical Hierarchy Process (AHP) to assess land suitability in the context of climate vulnerability, ecological sensitivity, and developmental constraints. The methodological approach was designed to integrate environmental science, spatial planning, and decision-support systems to produce a climate-informed land suitability model for Thrissur district, Kerala. The overarching aim is to identify and classify land into categories that support climate-resilient planning and policy decisions at both regional and transferable global scales.

The approach adopted in this study is structured into five methodological phases. The first phase involved the careful selection of eight spatial parameters, which serve as the core determinants of land suitability under a climate-sensitive lens. These parameters include proximity to urban centres, roads, forests, water bodies, flood-prone areas, coastal vulnerability zones, Kole wetlands, and slope conditions. The choice of parameters was guided by their established relevance in land-use planning and climate adaptation literature, and their capacity to capture both ecological risks and socio-economic development potential.

In the second phase, spatial and non-spatial data were collected from multiple reliable sources. Remote sensing imagery from Sentinel-2 and Landsat datasets (2015–2023) was obtained from the USGS Earth Explorer for land use and land cover (LULC) classification. Digital Elevation Models (DEMs) from the Shuttle Radar Topography Mission (SRTM) were processed to derive slope categories. Thematic datasets such as forest boundaries, flood hazard maps, and wetland extents were acquired from the Kerala Forest Research Institute (KFRI), Kerala State Disaster Management Authority (KSDMA), and the Kole Wetlands Development Authority, respectively. Urban settlement and road network data were digitized using Census 2011 administrative data and OpenStreetMap sources. All spatial data layers were standardized, converted to raster format with a 30-m resolution, and projected using WGS 84 UTM Zone 43N for geospatial consistency.

The third methodological phase involved the assignment of weights to each parameter using the Analytical Hierarchy Process (AHP), a structured technique for dealing with complex decision-making, originally developed by Thomas Saaty. Expert opinions from professionals in the fields of urban planning, environmental science, and local governance were solicited to carry out pairwise comparisons among the parameters. The relative importance of each factor was determined by calculating the weight matrix, followed by a consistency check to ensure the reliability of the judgments. The consistency ratio (CR) was maintained below the acceptable threshold of 0.1, which validates the logical coherence of the comparative rankings.

Following the weight assignment, the fourth phase focused on the integration of spatial layers through weighted overlay analysis in ArcGIS 10.8. Each thematic layer was reclassified into five categories of suitability: very high, high, moderate, low, and very low. These values were normalized and overlaid using the AHP-derived weights to produce a composite land suitability index. Spatial buffers were also applied to capture development proximity and ecological protection thresholds—for example, around urban centres, water bodies, and forested areas—based on relevant planning norms and environmental regulations. This process generated the final land suitability map for the entire district.

In the fifth and final phase, the preliminary results were validated through a triangulation process that included field observations, expert consultations, and comparison with known hazard-prone and ecologically sensitive areas. Ground truthing exercises were conducted in select locations within the district to verify the classification logic. Furthermore, historical flood event data, particularly from the 2018 Kerala floods, were used to cross-reference the accuracy of zones classified as unsuitable due to flood risk.

The methodological intention of this study is not limited to generating a land-use classification map. Rather, it aims to offer a scientifically grounded, climate-informed planning tool that can guide sustainable development while conserving ecological integrity. The model was designed with dual scalability: to address the micro-level planning needs of Thrissur district and to be adapted in other climate-vulnerable coastal contexts globally. By integrating climate stressors, ecological variables, and socio-economic considerations into a single spatial decision-making framework, the study provides a strategic approach to balancing development and conservation in a time of escalating climate risks.

The study's methodology reflects a fusion of environmental planning theory and practical GIS-AHP integration. It aligns closely with emerging global best practices in climate-resilient land-use planning and responds directly to the Sustainable Development Goals (particularly SDG 11 on sustainable cities and communities, SDG 13 on climate action, and SDG 15 on life on land). The resulting land suitability model is intended to serve not just as a diagnostic map but as a planning instrument capable of informing zoning regulations, infrastructure development, and ecosystem conservation strategies (Fig. 20.4).

20.4 Results and Discussion

This study presents a climate-informed land suitability assessment of Thrissur district, integrating eight parameters relevant to ecological sensitivity, socio-economic accessibility, and climate vulnerability. The results are spatially analysed to determine areas suitable for conservation, development, and restoration, offering an evidence-based planning framework for eco-resilient land use. The following discussion outlines the spatial patterns, critical findings, and implications for policy, linking the local outcomes of Thrissur with broader regional and global concerns around climate change and land degradation.

20.4.1 Land Suitability Patterns: Spatial Distribution and Development Potential

The final land suitability classification categorized Thrissur into five distinct zones: No Development Zone (42.30%), Less Suitable (0.20%), Moderately Suitable (29.30%), Suitable (27.70%), and Highly Suitable (0.40%). These results suggest a sobering reality: less than one-third of the district's area is suitable for any form of development, and only a fraction (0.40%) is highly suitable. The dominance of No Development Zones points to the extensive presence of flood-prone lowlands, ecologically sensitive wetlands, forested slopes, and unstable coastlines. Critically, regions such as Wadakkanchery (263.30 km^2), Kodakara (167.80 km^2), and Mala (171.66 km^2) possess the highest share of moderately and suitably classified lands. These areas are adjacent to urban settlements and road networks, suggesting their potential as low-risk urban growth corridors if development is strategically planned. In contrast, blocks such as Chalakudy, Anthikad, and Mathilakam are heavily constrained due to steep slopes, wetland presence, and coastal vulnerability, limiting their suitability for built-up expansion. These patterns directly fulfil

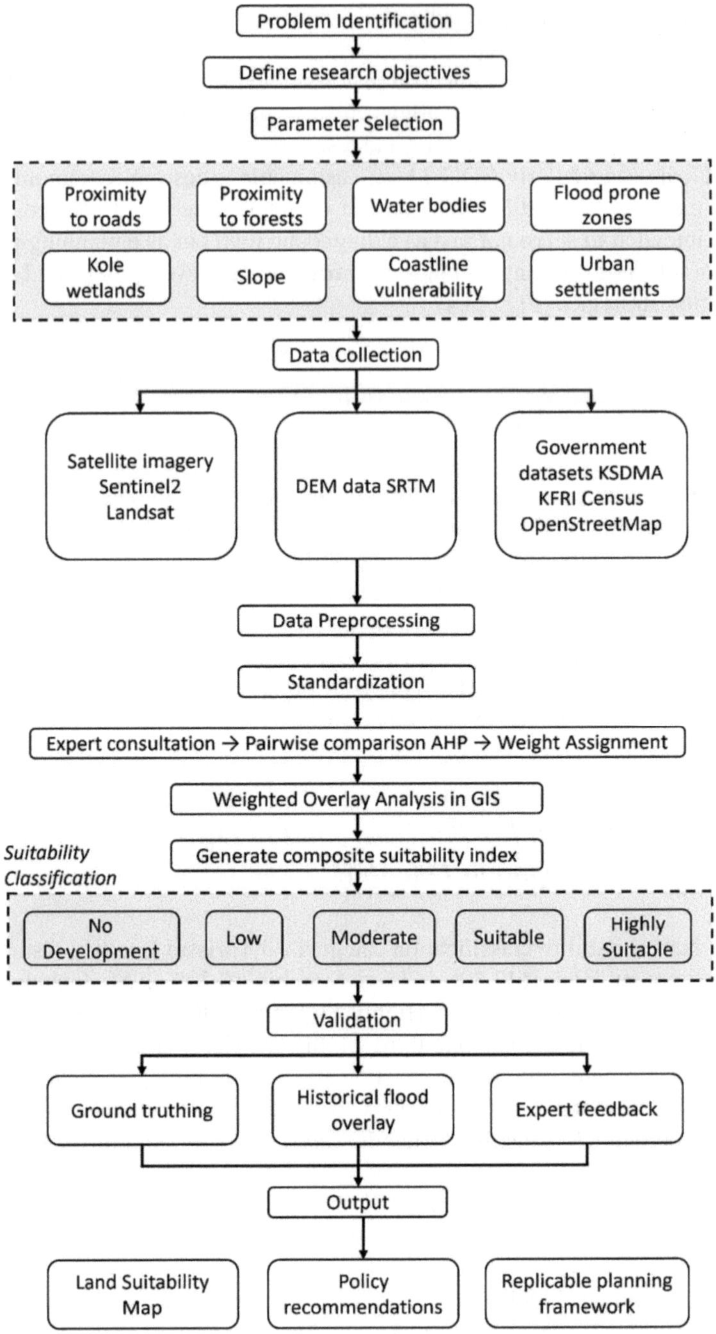

Fig. 20.4 Overview of the research methodology. *Adopted from* Malczewski (2004)

Table 20.1 Land suitability and developable area distribution

Name	No development zone (km^2	Less suitable (km^2	Moderately suitable (km^2)	Suitable (km^2)	Highly suitable (km^2)
Anthikad	82.4		39.7	61.8	0.37
Challakudy	652.83	0.75	241.96	139.33	2.25
Chavakkad	41.95		0.89	2.87	
Cherpu	53.93		9.36	20.97	
Chowannur	126.97		58.05	122.1	
Irinjalakkuda	84.65		31.46	103.37	1.12
Kodakara	65.55	0.37	167.8	91.01	3
Mala	20.97		94.01	171.66	2.25
Mathilakam	92.14		41.57	129.97	2.62
Mullassery	67.79		20.6	26.22	
Ollukarra	175.29	7.87	185.4	73.79	5.24
Puzhayannur	68.17		258.06	61.43	
Puzhakkal	117.23	0.37	44.2	74.91	2.25
Thalikulam	29.96		33.71	80.53	1.87
Thrissur M. Corp	95.51		82.03	120.23	5.62
Vellangallur	60.68		40.45	112.25	0.75
Wadakkanchery	125.1	0.37	263.3	105.25	
Total	1286.56	8.61	888.79	840.1	12.36

Source Analysis using GIS Mapping of data from Bhuvan, Irrigation Department, Kerala Forest Research Institute, Land Use Land Cover 2023, Earth Data NASA, District Spatial Plan 2011, Thrissur district disaster management plan 2015

objective 1, providing a spatialized understanding of how land-use pressures and climate-sensitive parameters interact across the district (see Table 20.1 and Fig. 20.5).

20.4.2 Flood Vulnerability and Urban Expansion Conflicts

The spatial overlay of historical flood maps, digital elevation models (DEMs), and land use change patterns reveals a high degree of overlap between flood-prone areas and expanding urban peripheries. For instance, 2018 flood data show that over 30% of Thrissur's road infrastructure and 80,000 residents were affected, with most damage concentrated in low-lying zones like Puzhakkal, Mathilakam, and Irinjalakkuda (see Fig. 20.6). Yet these same areas are witnessing high rates of land conversion from agriculture to urban use. The encroachment of flood-prone areas by residential developments is a direct outcome of informal urbanization processes, under-priced land markets in hazard zones, and a lack of dynamic risk-sensitive land use regulations.

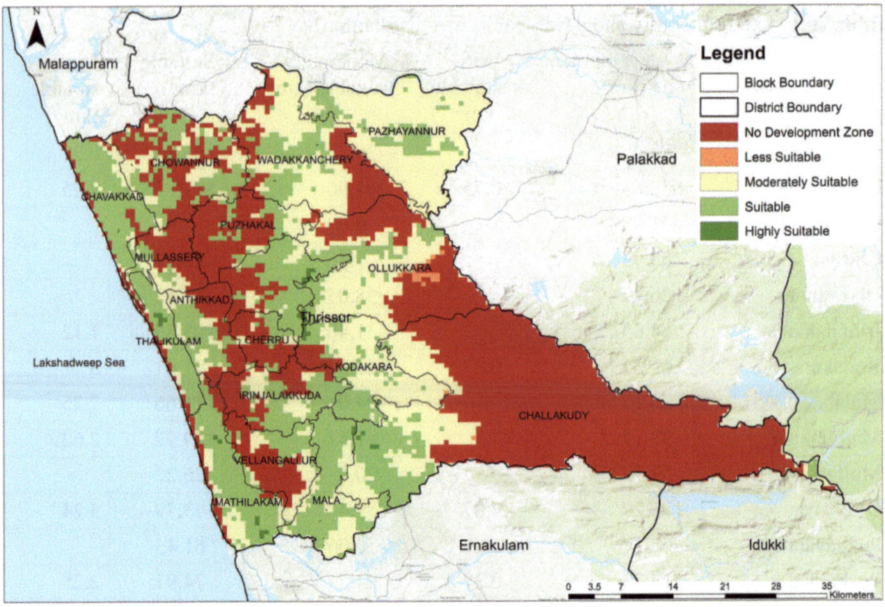

Fig. 20.5 Land use suitability map. *Source* Bhuvan, Irrigation Department, Kerala Forest Research Institute, Land Use Land Cover (2023), Earth Data NASA, District Spatial Plan 2011, Thrissur district disaster management plan (2015)

While formal planning documents exist, their implementation is undermined by political pressures and limited capacity at the local government level—a challenge mirrored in parts of Jakarta, Dhaka, and Lagos.

In Thrissur, the land suitability framework functions as a corrective mechanism, guiding planners away from climate-sensitive zones and toward more resilient alternatives. This insight advances both objective 1 and objective 2 by highlighting vulnerable urbanizing regions and guiding planners to shift development toward lower-risk areas, as identified in the suitability map.

20.4.3 Forests, Slope, and Carbon Resilience

The suitability map designates large forested regions, particularly in Chalakudy and Wadakkanchery, as zones unsuitable for development. These areas act as carbon sinks, absorbing approximately 1.2 million tons of CO_2 annually, and provide critical habitats for biodiversity. However, unsustainable land use practices—primarily mining and encroachment—have led to the degradation of over 173.56 km^2 of forest land in the past decade (see Fig. 20.7).

Excluding steep slopes from suitable development zones addresses concerns over forest loss and landslide risks, particularly in Vellikulangara and Kodassery, where

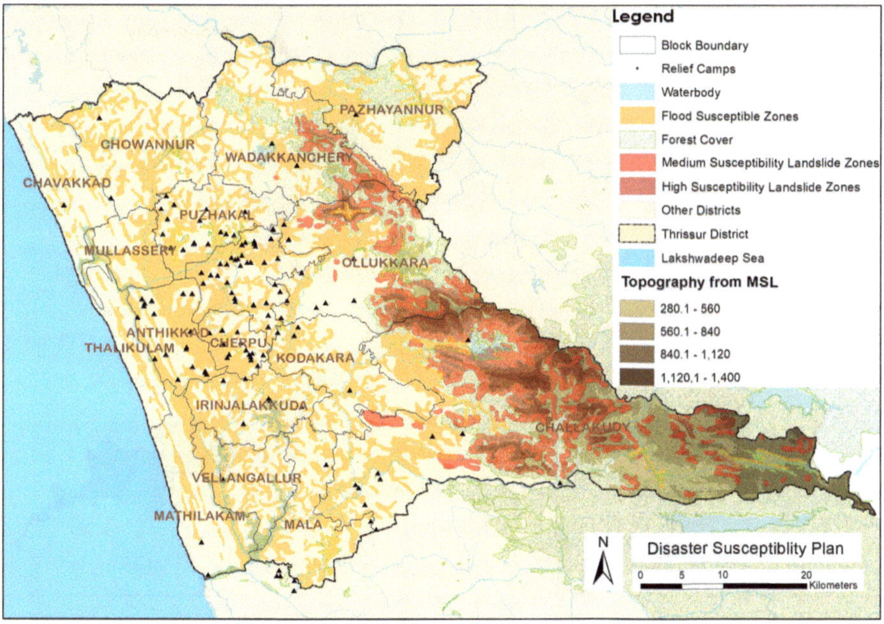

Fig. 20.6 Disaster susceptibility plan. *Source* Thrissur District Disaster Management Plan (2015)

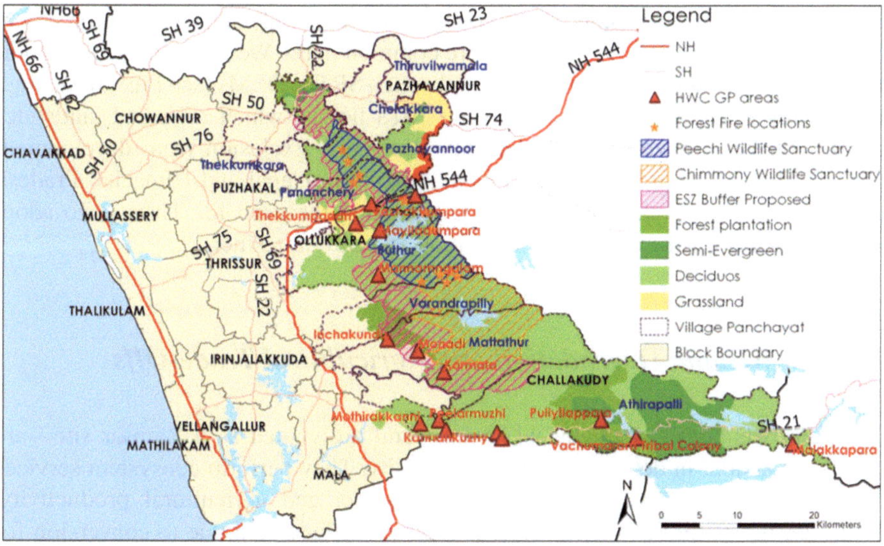

Fig. 20.7 Forest cover and typology map. *Source* Kerala Forest Research Institute (2023)

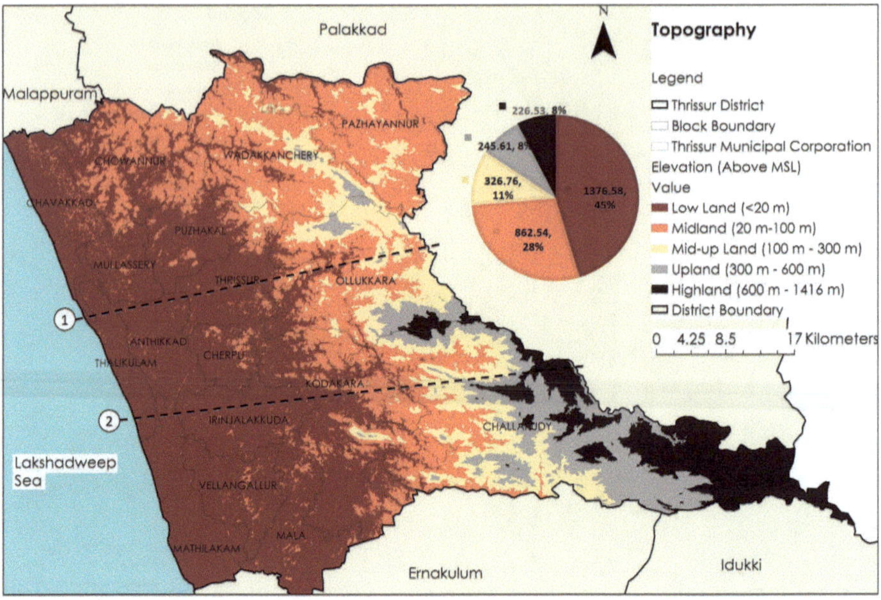

Fig. 20.8 Topography map. *Source* Earthdata, NASA (2023)

unsanctioned quarrying, monoculture plantations, and weak land tenure enforcement drive forest degradation. These issues are common in the Western Ghats and Southeast Asia. Steep slopes (>15°), identified through SRTM DEMs (see Fig. 20.8), align with erosion-prone areas, as seen in the 2018 floods, which triggered landslides across 200 hectares, displacing over 500 residents. Exclusion from development follows global best practices, like bioengineering in Nepal and Himachal Pradesh and slope zoning in the Philippines, emphasizing the need for Thrissur to adopt nature-based infrastructure for conservation and carbon resilience.

20.4.4 Wetland Degradation and Agricultural Trade-Offs

The Kole wetlands—spanning 447.5 km^2 and designated as a Ramsar site—are highly influential in determining land suitability. They perform ecosystem services such as flood mitigation, groundwater recharge, and agricultural productivity. However, the study finds a 26.8% reduction in wetland area due to conversion for built-up purposes. The observed 26.8% decline in Kole wetland area has a twofold implication: it not only exacerbates downstream flooding but also threatens local food security by reducing rice production in one of Kerala's key paddy-growing regions (see Table 20.2 and Fig. 20.9). This highlights how ecological degradation can directly translate into socio-economic vulnerability—a trend noted in Mekong

Table 20.2 List of executed protection works

Block/corporation	Kole land area (km^2)	Flood affected Kole land area (km^2)	Built-up area on Kole Land (km^2)	Kole land remaining area (km^2)
Chowannur	79.41	37.53	18.87	60.54
Puzhakkal	74.22	37.17	23.87	50.35
Vellangallur	49.29	26.27	8.69	40.6
Irijalakkuda	43.85	28.31	8.02	35.83
Mullassery	42.16	26.68	5.4	36.76
Anthikkad	37.7	29.29	15.03	22.67
Cherpu	33.35	23.04	9.11	24.24
Thrissur C	29.42	19.29	18.29	11.13
Wadakkanchery	27.34	9.81	5.31	22.03
Kodakara	14.87	9.3	3.63	11.24
Mathilakam	9.92	8.48	2.28	7.64
Chavakkad	5.98	1.7	1.69	4.29
Total	447.51	256.87	120.19	327.32

Source Kerala Disaster Management Authority 2021, National Centre for Earth Science Studies (2019)

Delta studies where wetland shrinkage correlates with reduced yield and flood exposure (Syvitski et al., 2009). In Thrissur, unchecked wetland loss could undermine both resilience and livelihoods.

The decline of Kole wetlands results from fragmented land ownership, weak enforcement, and real estate pressures, not just urban sprawl. Similar trends in peri-urban Bangladesh and China show how poor governance and speculation drive wetland loss (Liu & Zhou, 2021). Conversion of 120.19 km^2 into built-up areas highlights impacts on agriculture. In Chavakkad and Ollukara, paddy encroachment worsens food insecurity and runoff, echoing patterns in Australia and China (Kingsford et al., 2016; Liu & Zhou, 2021). Designating these wetlands as conservation zones per Ramsar norms is vital. This analysis addresses objective 2 by identifying trade-offs and informs objective 3 through an adaptable model linking food security, flood control, and land-use resilience under climate stress.

20.4.5 Coastal Vulnerability and Climate Adaptation Zones

Thrissur's 81.2 km coastline is highly vulnerable to climate change, with over 85% of the shoreline eroding at 0.5 m per year (see Fig. 20.10). Coastal Regulation Zone (CRZ) mapping suggests that 150 km^2 of land may be submerged by 2130 under high-emission scenarios. Anthropogenic stressors like unregulated construction, sand

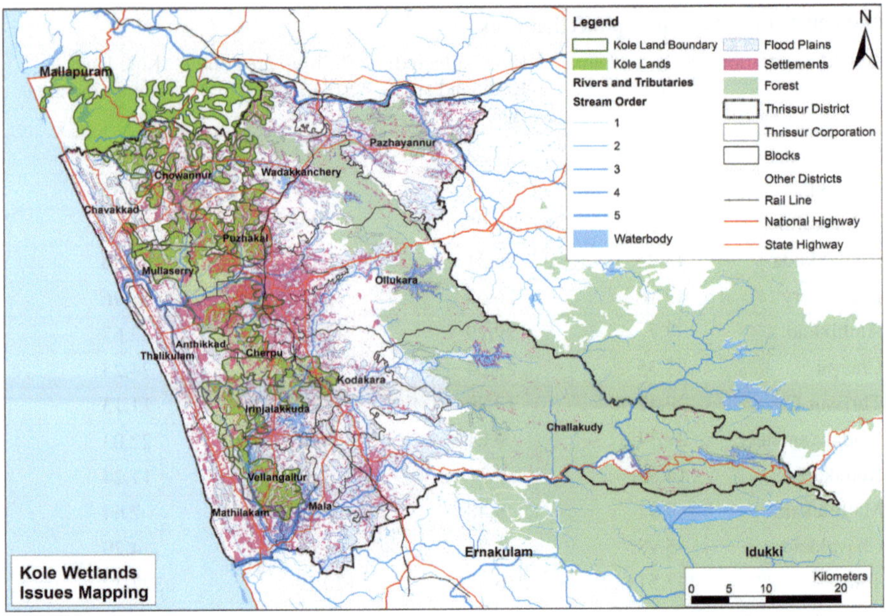

Fig. 20.9 Flood and built-up area analysis of the Kole Lands. *Source* Land use land cover map 2023 & Kerala Forest Research Institute

mining, and inadequate seawalls exacerbate this erosion, disrupting natural sediment flows and compounding the impacts of rising sea levels and stronger storms, as seen in regions like the Niger Delta and coastal Tamil Nadu (Nicholls et al., 2021).

Despite these risks, coastal villages such as Vadanappally and Engandiyur are witnessing increased infrastructure development. The land suitability framework restricts such zones from high-density development and instead recommends nature-based solutions such as mangrove restoration, living shorelines, and elevated infrastructure. These strategies draw from successful models like the Netherlands' Room for the River and Singapore's hybrid coastal adaptation framework, demonstrating Thrissur's relevance as a pilot for globally transferable resilience planning. These findings further objective 2, offering data-backed inputs for spatial planning in erosion-prone CRZ zones, and strengthen Objective 3 by mirroring global climate adaptation approaches.

20.4.6 Urban Suitability and Low-Carbon Mobility Challenges

The urban centres of Thrissur Municipal Corporation and Chavakkad are classified as highly or suitably developable due to their access to roads, services, and relatively

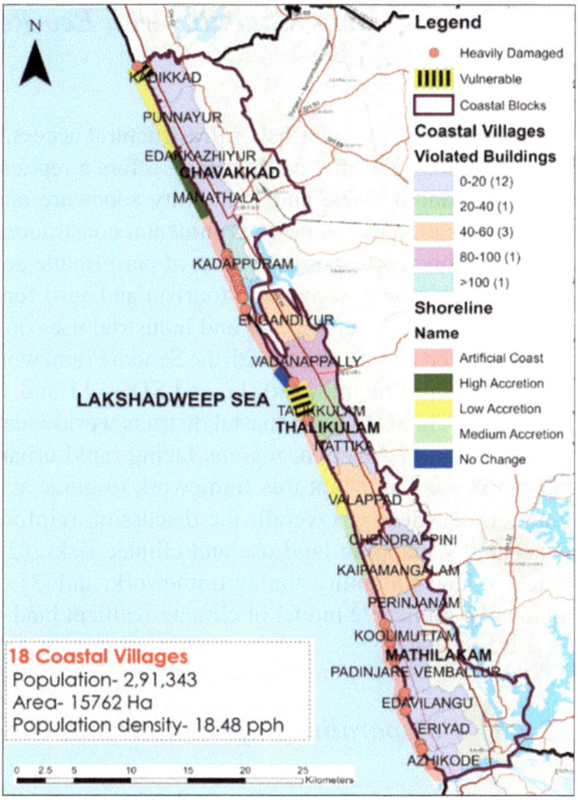

Fig. 20.10 Coastal hazard map. *Source* Coastal District Committee, 2019

stable topography. However, low public transport penetration—0.33 buses per 1000 people, compared to the national benchmark of 1 bus per 1000—limits sustainable densification. Moreover, the modal share of public transport dropped from 3.6% in 2016 to 0.05% in 2022, indicating a shift toward private vehicle dependency.

This finding implies that land suitability alone cannot determine sustainable growth unless supported by complementary investments in low-carbon mobility. The study recommends transit-oriented development (TOD) in moderately and suitably classified zones to integrate land use with mobility and minimize emissions. This mirrors approaches taken in Curitiba, Brazil, and Bogota, Colombia, where TOD frameworks have successfully aligned urban growth with environmental goals. The integration of transit-accessible development zones into the suitability framework aligns with objective 2 and reflects objective 3's intent to promote resilient, low-carbon urban growth patterns relevant to other coastal cities worldwide.

20.4.7 Integrating Suitability with Eco-Resilience and Global Relevance

By integrating ecological risk, infrastructural access, and climate vulnerability into a unified geospatial model, this study offers a replicable approach for eco-resilient land use planning. The land suitability zones are not merely outputs—they represent a strategy to reconcile environmental constraints with human aspirations. Each classified zone corresponds to a set of permissible activities: conservation-only use in No Development Zones, eco-tourism and agro-forestry in Low Suitability areas, and higher-density settlements and industrial uses only in Highly Suitable zones.

Such tiered zoning aligns with the Sendai Framework (UNDRR, 2015), the IPCC's risk-reduction principles (2021), and SDGs 11 and 13, making the model relevant not only to Thrissur but to coastal districts worldwide. Whether in South Asia, West Africa, or Latin America, regions facing rapid urbanization, wetland degradation, and flood risk can adapt this framework to guide strategic interventions and build long-term resilience. Overall, the discussion reinforces the study's objectives by: (1) spatially mapping land-use and climate risks, (2) translating those risks into a practical land suitability zoning framework, and (3) situating Thrissur's case within a globally applicable model of climate-resilient land-use planning.

20.4.8 Comparative Implications and Global Resonance

The land suitability outcomes for Thrissur reveal a landscape shaped by compounding climate and anthropogenic pressures, a trend echoed in comparable coastal settings globally. For example, similar suitability models applied in Bangladesh's Khulna district and Ghana's Volta Delta (Aerts et al., 2018; Nunoo & Agyekumhene, 2022) have revealed that over 40% of coastal land is unsuitable for high-density development due to flood risk and wetland degradation. Thrissur's 42.3% "No Development Zone" classification closely aligns with these findings, underscoring the need for nature-based interventions. Furthermore, the integration of Kole wetlands into the suitability framework mirrors strategies from Vietnam's Mekong Delta, where rice–wetland synergies are critical to balancing food security and climate resilience (Syvitski et al., 2009). This alignment supports the global shift towards adaptive zoning, where land classification not only informs infrastructure placement but also ecosystem restoration. The Thrissur model's use of spatial data to inform tiered land use resonates with the Dutch "Room for the River" initiative and Singapore's hybrid infrastructure approach (Ellison et al., 2020; van Alphen, 2020), suggesting its broader applicability in policy and planning contexts beyond India.

20.5 Recommendations

To ensure the long-term sustainability of Thrissur's coastal areas, climate-resilient land use planning must be integrated into policy frameworks. Climate vulnerability assessments should be conducted at a micro-scale level, ensuring that land suitability analysis accounts for sea-level rise, storm surges, and changing precipitation patterns (Çal & Ciravoğlu, 2024). Land-use zoning should categorize regions into high-risk, moderate-risk, and low-risk areas, with development restrictions in highly vulnerable zones (see Fig. 20.11). For instance, stringent building codes in flood-prone areas should mandate elevated structures with stilts, flood-proofing measures, and permeable pavement technologies (Kreibich et al., 2005). Coastal erosion is a significant challenge for Thrissur's coastal areas, necessitating ecosystem-based adaptation (EBA) approaches. The restoration of mangroves and coastal vegetation serves as a natural buffer against storm surges and erosion (Temmerman et al., 2023). Studies indicate that mangrove forests can reduce wave heights by up to 66%, significantly mitigating coastal flooding risks (McIvor et al., 2015). Therefore, reforestation of degraded mangrove areas, coupled with the creation of artificial reefs, should be prioritized. Additionally, green infrastructure solutions such as living shorelines, which integrate vegetation and natural materials, should replace traditional hard infrastructure like seawalls (Bilkovic et al., 2016).

Resilient infrastructure planning should incorporate climate-sensitive design principles. For transportation networks, flood-resilient connectivity routes must be developed, modelled after successful case studies from the Netherlands and Japan (Watson & Ahn, 2022). Elevated roadways and flood-resistant bridges should be constructed in flood-prone zones, while permeable pavements should be integrated into urban areas to enhance stormwater infiltration (Miguez & Veról, 2017). Additionally, stormwater retention basins should be strategically positioned in urban settlements to mitigate flash flooding events (see Fig. 20.12).

Thrissur's Kole wetlands and river systems play a crucial role in regulating water resources. However, pollution, encroachments, and unregulated sand mining have severely impacted water quality and hydrological balance. To address these challenges, stringent regulatory mechanisms should be enforced to prevent illegal sand mining and industrial effluent discharge (Nunoo & Agyekumhene, 2022). A decentralized wastewater treatment strategy, with nature-based solutions such as constructed wetlands, should be implemented to improve water quality (Castellar et al., 2022). Kole wetlands should be designated as protected ecological zones under the Ramsar Convention guidelines, ensuring their long-term conservation. Policy integration is key to effective climate adaptation. The Thrissur Regional Plan should incorporate climate resilience principles aligned with global frameworks such as the Sendai Framework for Disaster Risk Reduction (UNDRR, 2015). A multi-level governance structure involving state, district, and local bodies should be established to facilitate the implementation of resilience strategies (Meadowcroft, 2002). Community-based adaptation (CBA) strategies should be emphasized, engaging

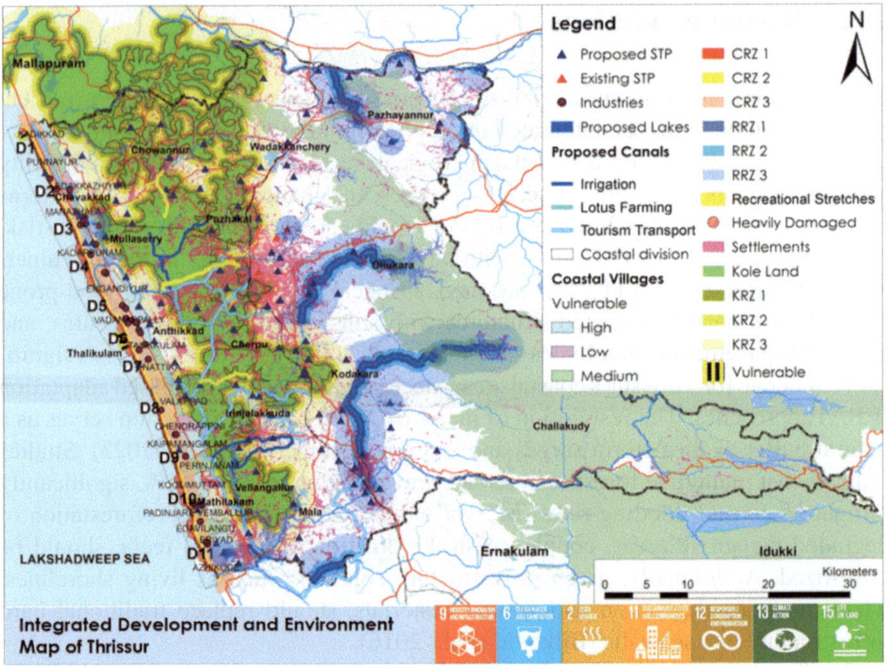

Fig. 20.11 Proposed integrated development and environment map of Thrissur. *Adopted from* Aerts et al., (2018), Bilkovic et al., (2016), Çal and Ciravoğlu (2024), Kreibich et al. (2005), McIvor et al., (2015), Nunoo and Agyekumhene (2022), Temmerman et al., (2023)

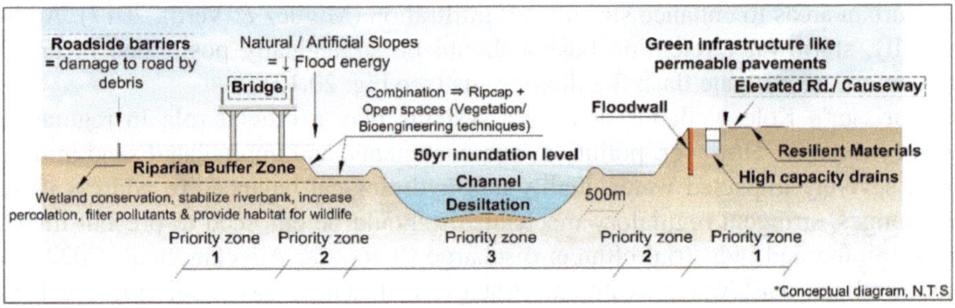

Fig. 20.12 Proposed modifications in identified connectivity routes. *Adopted from* Miguez and Veról (2017), Rosenzweig et al., (2018), UNDRR (2015), Watson and Ahn (2022)

local stakeholders in decision-making processes and enhancing disaster preparedness through early warning systems (Saleem Khan et al. 2020).

Land suitability assessments should guide economic development in climate-sensitive areas. Sustainable tourism models should be promoted in conservation zones, ensuring that eco-tourism initiatives align with biodiversity protection goals (Bushell & Bricker, 2017). Coastal fisheries, a key economic driver in Thrissur, should adopt climate-smart aquaculture practices, incorporating climate-resilient species and sustainable harvesting techniques (Tanner et al., 2014). Incentives for renewable energy adoption in commercial and industrial sectors should also be introduced to reduce greenhouse gas emissions.

Given Thrissur's vulnerability to climate-induced disasters, an integrated disaster risk reduction (DRR) framework should be developed. As shown in Fig. 20.13, the establishment of flood-resilient connectivity routes, modelled after successful international case studies such as New York's High Line flood mitigation infrastructure, can minimize disruptions during extreme weather events (Rosenzweig et al., 2018). Early warning systems should be enhanced through real-time climate monitoring, leveraging satellite-based and GIS technologies for disaster prediction (Albrecht et al., 2020). Further, Urban settlements in flood-prone zones should adopt nature-based solutions such as sponge cities, which integrate green roofs, permeable pavements, and urban wetlands to enhance flood resilience (Qi et al., 2021). Regulations should enforce setbacks for coastal developments, ensuring a buffer zone of at least 500 m in high-risk zones to reduce exposure to sea-level rise. The creation of climate-resilient affordable housing solutions should be prioritized, ensuring that vulnerable communities have access to safe, adaptive living spaces (Ziervogel et al., 2017).

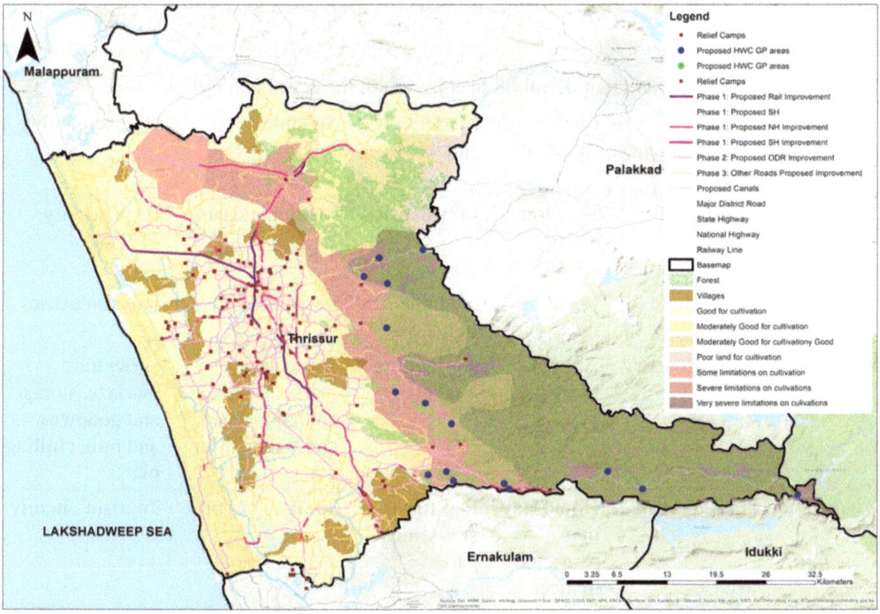

Fig. 20.13 Proposed district connectivity resilience plan. *Adopted from* Bushell and Bricker (2017), Castellar et al., (2022), Meadowcroft (2002), Miguez and Veról (2017), Nunoo and Agyekumhene (2022), Qi et al., (2021), Rosenzweig et al., (2018), Saleem Khan et al. (2020), Tanner et al., (2014), UNDRR (2015), Watson and Ahn (2022)

A robust monitoring and evaluation (M&E) framework should be integrated into Thrissur's Regional Plan to assess the effectiveness of climate resilience measures. Key performance indicators (KPIs) should be established, focusing on flood risk reduction, ecosystem conservation, and sustainable land use patterns (Kumar et al., 2020). GIS-based climate vulnerability mapping should be conducted annually to reassess risk exposure and inform adaptive policy decisions.

The LSA framework establishes a tiered approach to land use regulation, reflecting the varying ecological and socio-economic sensitivities across the designated areas. In No Development Zones, only activities that directly support the local, native community and essential ecosystem functions, such as agro-forestry, are permitted, emphasizing preservation. Progressing through zones of increasing suitability (Low, Moderately Suitable, Suitable, and High), a wider range of activities is gradually introduced, spanning from low-impact, small-scale industries and restricted tourism in Low Suitability zones to the allowance of high-density settlements, large-scale industries, and comprehensive tourism circuits in High Suitability zones (see Table 20.3). This gradient approach ensures that development corresponds to the land's carrying capacity and inherent suitability, minimizing environmental impact while optimizing resource utilization and promoting sustainable growth. The strategic distribution of activities based on suitability underscores the research's focus on evidence-based planning for long-term ecological and economic resilience.

Table 20.3 Permissible and non-permissible activities in the developable zone

Activity	No development zone	Low suitability	Moderately suitable	Suitable	High suitability
Settlements	Only for the native/local community	Very low density	Low density	Medium density	High density
Industry	Household industry	MSME	medium Industries	Large industries	large industries
Agriculture	Agro-forestry	Animal husbandry	Agriculture markets	Agriculture markets, storage and godowns, and milk chilling plants	agriculture markets, storage and godowns, and milk chilling plants
Tourism	Eco-tourism	Home stay in a restricted manner	Eco-tourism, amusement park	Tourism circuits	Tourism circuits

Adopted from Aerts et al., (2018), Bilkovic et al., (2016), Bushell and Bricker (2017), Çal and Ciravoğlu (2024), Castellar et al., (2022), Kreibich et al., (2005), McIvor et al., (2015), Meadowcroft (2002), Miguez and Veról (2017), Nunoo and Agyekumhene (2022), Qi et al., (2021), Rosenzweig et al., (2018), Saleem Khan et al. (2020), Tanner et al., (2014), Temmerman et al., (2023), UNDRR (2015), Watson and Ahn (2022)

20.6 Conclusion

This study positions Thrissur as a dynamic case for climate-resilient land use planning, offering a transferable model grounded in both spatial analysis and ecological understanding. By using a multi-criteria decision-making framework with GIS and AHP, the research identifies 27.70% of the district as suitable for future development, while over 42% is designated as a No Development Zone due to ecological or climatic vulnerability. This classification underscores the pressing need to manage growth carefully while conserving critical ecosystems. Thrissur's land suitability model bridges the gap between global environmental goals and local planning realities. Its ability to integrate flood risks, slope instability, coastal erosion, and urban expansion into a cohesive framework provides planners and policymakers with a practical tool to guide future growth. This approach is not only relevant to India but to other climate-sensitive coastal zones around the world, such as in Southeast Asia, West Africa, and Latin America. The methodology aligns with the objectives of the UN Sustainable Development Goals (SDGs), particularly SDGs 11 (Sustainable Cities and Communities) and 13 (Climate Action), and the Sendai Framework for Disaster Risk Reduction. Importantly, this research directly contributes to the focal theme of Eco-Resilience: Climate Change, Land Degradation, and Sustainable Solutions. Thrissur's case demonstrates how integrating climate risk assessments with land suitability analysis can proactively address the twin challenges of climate change and land degradation. The study proposes sustainable solutions such as the protection of wetlands, slope stabilization, flood-resilient infrastructure, and nature-based coastal defenses. These strategies foster eco-resilience by enhancing the region's capacity to absorb shocks, maintain ecological balance, and ensure sustainable livelihoods. By prioritizing nature-based planning over exploitative urban expansion, this research exemplifies a pathway to long-term sustainability and resilience in vulnerable coastal landscapes.

One of the most significant contributions of this research is its ability to translate complex geospatial data into accessible planning zones, which can then inform policies related to urban development, disaster resilience, and ecosystem management. It demonstrates that land-use planning can be both climate-conscious and development-oriented, without compromising environmental integrity. This insight has broader implications for urbanizing regions where unregulated growth has increased vulnerability to climate-induced hazards. The implications of the study go beyond regional relevance. By applying a replicable framework that prioritizes conservation in ecologically sensitive areas while enabling development in stable zones, this work sets a foundation for climate-resilient spatial planning globally. Lessons from Thrissur can be adopted in similar deltaic and coastal landscapes that face overlapping challenges of population pressure, land degradation, and disaster risk. Integrating traditional knowledge, stakeholder participation, and up-to-date remote sensing data also adds to the model's credibility and adaptability. In practical terms, the research advocates for development controls in fragile zones, stronger flood and coastal defence mechanisms, nature-based solutions like mangrove restoration, and sustainable infrastructure planning. It also highlights the importance of inclusive governance and localized data to ensure that land use decisions are responsive and forward-looking. Ultimately, this study demonstrates that land suitability analysis is not merely a technical exercise but a strategic tool to balance competing demands—economic development, social equity, and environmental protection. Thrissur's experience, therefore, serves as a microcosm of the global climate challenge. Its approach offers a scientifically grounded, policy-relevant, and globally applicable model for addressing the complexities of planning in the face of climate change. As cities and regions worldwide grapple with rising sea levels, increasing flood risks, and environmental degradation, the need for climate-informed land use planning becomes urgent. This research offers a tested, adaptable approach that can guide such efforts, making a lasting contribution to the fields of eco-resilience, land management, and sustainable spatial planning.

Acknowledgements The authors extend their sincere gratitude to the Batch of 2022–2024, M.Plan (Urban and Regional Planning), School of Planning and Architecture, Bhopal, for their active involvement in this academic exercise. We are especially thankful to Dr. Ashfaque Alam, Dr. Kshama Puntambekar, and Dr. Ajith Kalliat for their invaluable guidance and constant support throughout the study. We gratefully acknowledge the institutional support received from KILA, ESAF Bank, and the Thrissur Management Association, which was instrumental in facilitating this research. Our appreciation is extended to the domain experts whose insights significantly enriched the Analytical Hierarchy Process (AHP) evaluations. Finally, we wish to thank Aditi Jain, Sandeep Singh, Anjali Netam, Reeti Choudhary, Shervin Azmina, Tarishi Kumawat, Dhruvil Godhani and Sunidhi Tiwari for their assistance during the course of this work.

AI Disclosure: AI tools were used to assist with language editing and grammar check only. All analyses, interpretations, and graphics are solely the author's own.

Authors' Contribution: S. Gayathri: Methodology, formal analysis, writing – original draft, writing – review & editing, visualization; Paulose N. Kuriakose: Conceptualization, methodology, supervision, funding acquisition, resources, project administration.

Funding: The research was supported by funding from Thrissur Management Association and ESAF Bank.

Potential conflicts of interest: On behalf of all authors, the corresponding author states that there is no conflict of interest.

Data Availability Statement: The datasets during and/or analysed during the current study are available from the corresponding author on reasonable request.

References

Aerts, J. C., Botzen, W. J., Clarke, K. C., Cutter, S. L., Hall, J. W., Merz, B., Michel-Kerjan, E., Mysiak, J., Surminski, S., & Kunreuther, H. (2018). Integrating human behaviour dynamics into flood disaster risk assessment. *Nature Climate Change, 8*(3), 193–199.

Albrecht, C. M., Elmegreen, B., Gunawan, O., Hamann, H. F., Klein, L. J., Lu, S., Mariano, F., Siebenschuh, C., & Schmude, J., (2020). Next-generation geospatial-temporal information technologies for disaster management. *IBM Journal of Research and Development, 64*(1/2), 1–5.

Alongi, D. M. (2002). Present state and future of the world's mangrove forests. *Environmental Conservation, 29*(3), 331–349.

Bapuji Rao, B., Santhibhushan Chowdary, P., Sandeep, V. M., Rao, V. U. M., & Venkateswarlu, B. (2014). Rising minimum temperature trends over India in recent decades: Implications for agricultural production. *Global and Planetary Change*. https://doi.org/10.1016/j.gloplacha.2014.03.001

Bhuvan. (2023). *Land use land cover (LULC) map of Thrissur, Kerala*. Earth Data, NASA. Retrieved January 16, 2024.

Bilkovic, D. M., Mitchell, M., Mason, P., & Duhring, K. (2016). The role of living shorelines as estuarine habitat conservation strategies. *Coastal Management, 44*(3), 161–174.

Bushell, R., & Bricker, K. (2017). Tourism in protected areas: Developing meaningful standards. *Tourism and Hospitality Research, 17*(1), 106–120.

Çal, İ, & Ciravoğlu, A. (2024). Determining vulnerability indicators of buildings for sea-level rise and floods in urban coastal areas. *Sustainability, 17*(1), 27.

Castellar, J. A., Torrens, A., Buttiglieri, G., Monclús, H., Arias, C. A., Carvalho, P. N., Galvao, A., & Comas, J. (2022). Nature-based solutions coupled with advanced technologies: An opportunity for decentralized water reuse in cities. *Journal of Cleaner Production, 340*, Article 130660.

Census of India. (2011). *Population of Thrissur District, Kerala*.

Church, J. A., Clark, P. U., Cazenave, A., Gregory, J. M., Jevrejeva, S., Levermann, A., Merrifield, M. A., Milne, G. A., Nerem, R. S., Nunn, P. D., Payne, A. J., Pfeffer, W. T., Stammer, D., & Unnikrishnan, A. S. (2013). Sea level change. In T. F. Stocker, D. Qin, G.-K. Plattner, M. Tignor, S. K. Allen, J. Boschung, A. Nauels, Y. Xia, V. Bex, & P. M. Midgley (Eds.), *Climate change 2013: The physical science basis. Contribution of Working Group I to the Fifth Assessment Report of the Intergovernmental Panel on Climate Change*. Cambridge University Press.

Ellison, A. M., Felson, A. J., & Friess, D. A. (2020). Mangrove rehabilitation and restoration as experimental adaptive management. *Frontiers in Marine Science, 7*, 327.

Global Water Partnership (GWP). (2014). *What is integrated water resources management (IWRM)?* Global Water Partnership. https://www.gwp.org/en/GWP-CEE/about/why/what-is-iwrm/

Government of Kerala, World Bank, & Asian Development Bank. (2018, September). *Kerala floods and landslides 2018: Joint rapid damage and needs assessment report*.

Government of Kerala. (2015). *Thrissur district disaster management plan 2015*. Kerala State Disaster Management Authority. Retrieved January 16, 2024.

Hallegatte, S., Green, C., Nicholls, R. J., & Corfee-Morlot, J. (2013). Future flood losses in major coastal cities. *Nature Climate Change, 3*(9), 802–806.

Kerala Forest Research Institute (KFRI). (2023). *Land Use Land Cover 2023, forest cover map*. Kerala Forest Research Institute. Retrieved January 16, 2024.

Kerala State Disaster Management Authority (KSDMA). (2024). *Hazard maps*. Kerala State Disaster Management Authority. Retrieved January 16, 2024, from https://sdma.kerala.gov.in/hazard-maps/

Kingsford, R. T., Basset, A., & Jackson, L. (2016). Wetlands: Conservation's poor cousins. *Aquatic Conservation: Marine and Freshwater Ecosystems, 26*(5), 892–916.

Kreibich, H., Thieken, A. H., Petrow, T., Müller, M., & Merz, B. (2005). Flood loss reduction of private households due to building precautionary measures–lessons learned from the Elbe flood in August 2002. *Natural Hazards and Earth System Sciences, 5*(1), 117–126.

Kumar, P., Debele, S. E., Sahani, J., Aragão, L., Barisani, F., Basu, B., Zieher, T., et al. (2020). Towards an operationalisation of nature-based solutions for natural hazards. *Science of the Total Environment, 731*, Article 138855.

Lal, P., Prakash, A., Kumar, A., Srivastava, P. K., Saikia, P., Pandey, A. C., Srivastava, P., & Khan, M. L. (2020). Evaluating the 2018 extreme flood hazard events in Kerala, India. *Remote Sensing Letters, 11*(5), 436–445.

Liu, Y., & Zhou, Y. (2021). Reflections on China's food security and land use policy under rapid urbanization. *Land Use Policy, 109*, Article 105699.
Malczewski, J. (2004). GIS-based land-use suitability analysis: A critical overview. *Progress in Planning, 62*(1), 3–65.
McHarg, I. L. (1969). *Design with nature* (pp. 7–17). American Museum of Natural History.
McIvor, A., Spencer, T., Spalding, M., Lacambra, C., & Möller, I. (2015). Mangroves, tropical cyclones, and coastal hazard risk reduction. In *Coastal and marine hazards, risks, and disasters* (pp. 403–429). Elsevier.
Meadowcroft, J. (2002). Politics and scale: Some implications for environmental governance. *Landscape and Urban Planning, 61*(2–4), 169–179.
Miguez, M. G., & Veról, A. P. (2017). A catchment scale Integrated Flood Resilience Index to support decision making in urban flood control design. *Environment and Planning b: Urban Analytics and City Science, 44*(5), 925–946.
Muraleedharan, P. K., Kallarackal, J., Menon, A. R. R., Balagopalan, M., Sasidharan, N., & Rugmini, P. (2007). Land use change and its impact on selected biophysical and socioeconomic aspects of Karuvannur river basin in Thrissur district of Kerala. *Kerala Forest Research Institute Research Report* (p. 298).
National Centre for Earth Science Studies. (2019). *Coastal zone management plan: Kerala (Draft map, prepared as per Coastal Regulation Zone Notification 2019).* Department of Environment, Government of Kerala.
National Centre for Earth Science Studies. (2023, March). *Draft report on coastal zone management plan (CZMP) for Thrissur district—Kerala (Prepared as per CRZ Notification 2019).* Department of Environment, Government of Kerala.
Neelakantan, R. (2013). Detection of land use pattern changes and management priorities for Thrissur district, Kerala, India. *Forest, 873*, 28–79.
Neumann, B., Vafeidis, A. T., Zimmermann, J., & Nicholls, R. J. (2015). Future coastal population growth and exposure to sea-level rise and coastal flooding-a global assessment. *PLoS ONE, 10*(3), Article e0118571.
Nicholls, R. J., Lincke, D., Hinkel, J., Brown, S., Vafeidis, A. T., Meyssignac, B., Hanson, S. E., Merkens, J. L., & Fang, J. (2021). A global analysis of subsidence, relative sea-level change and coastal flood exposure. *Nature Climate Change, 11*(4), 338–342.
Nunoo, F. K., & Agyekumhene, A. (2022). Mangrove degradation and management practices along the coast of Ghana. *Agricultural Sciences, 13*(10), 1057–1079.
Oppenheimer, M., Glavovic, B. C., Hinkel, J., van de Wal, R., Magnan, A.K., Abd-Elgawad, A., Cai, R., Cifuentes-Jara, M., DeConto, R.M., Ghosh, T., Hay, J., Isla, F., Marzeion, B., Meyssignac, B., & Sebesvari, Z. (2019). Sea level rise and implications for low-lying islands, coasts and communities. In H.-O. Pörtner, D. C. Roberts, V. Masson-Delmotte, P. Zhai, M. Tignor, E. Poloczanska, K. Mintenbeck, A. Alegría, M. Nicolai, A. Okem, J. Petzold, B. Rama, & N. M. Weyer (eds.) *IPCC special report on the ocean and cryosphere in a changing climate* (pp. 321–445). Cambridge University Press. https://doi.org/10.1017/9781009157964.006.
Qi, W., Ma, C., Xu, H., Chen, Z., Zhao, K., & Han, H. (2021). A review on applications of urban flood models in flood mitigation strategies. *Natural Hazards, 108*, 31–62.
Rosenzweig, B. R., McPhillips, L., Chang, H., Cheng, C., Welty, C., Matsler, M., Iwaniec, D., & Davidson, C. I. (2018). Pluvial flood risk and opportunities for resilience. *Wiley Interdisciplinary Reviews: Water, 5*(6), Article e1302.
Saleem Khan, A., Sabuj Kumar, M., Sudhir Chella, R., & Devdyuti, B. (2020). Chennai city and coastal hazards: addressing community-based adaptation through the lens of climate change and sea-level rise (CBACCS). *Climate change, hazards and adaptation options: Handling the impacts of a changing climate*, 777–798.
Sebastian, M. K., Azeez, P. A., Shree, C., & Arun, P. R. (2014, December). *Ecological and ethnocultural examination of the rise and fall of rice culture in Southern India with special reference to the Western Ghats: Final report.* Submitted to Indira Gandhi National Centre for Arts, New Delhi. Sálim Ali Centre for Ornithology and Natural History.

Syvitski, J. P., Kettner, A. J., Overeem, I., Hutton, E. W., Hannon, M. T., Brakenridge, G. R., Day, J., Vörösmarty, C., Saito, Y., Giosan, L., & Nicholls, R. J. (2009). Sinking deltas due to human activities. *Nature Geoscience*, *2*(10), 681–686.

Tanner, T., Mensah, A., Lawson, E. T., Gordon, C., Godfrey-Wood, R., & Cannon, T. (2014). Political economy of climate compatible development: artisanal fisheries and climate change in Ghana. *IDS Working Papers*, *2014*(446), 1–30.

Temmerman, S., Horstman, E. M., Krauss, K. W., Mullarney, J. C., Pelckmans, I., & Schoutens, K. (2023). Marshes and mangroves as nature-based coastal storm buffers. *Annual Review of Marine Science, 15*(1), 95–118.

U.S. Geological Survey. (2023). *Land use/ land cover (LULC) map of Thrissur, Kerala, India.* Earth explorer. Retrieved December 2023, from https://earthexplorer.usgs.gov/

United Nations Office for Disaster Risk Reduction (UNDRR). (2015). *Sendai framework for disaster risk reduction 2015–2030.* https://www.unisdr.org/we/inform/publications/43291

United Nations. (2015). *Sustainable development goals (SDGs): Goals 11 and 13.* United Nations. https://sdgs.un.org/goals

Upadhyay, R. K. (2020). Markers for global climate change and its impact on social, biological and ecological systems: A review. *American Journal of Climate Change, 9*(03), 159.

Van Alphen, S. (2020). Room for the river: innovation, or tradition? The case of the Noordwaard. *Adaptive Strategies for Water Heritage, 309.*

Van Alphen, J. (2016). The Delta programme and updated flood risk management policies in the Netherlands. *Journal of Flood Risk Management* 9(4) 310–319. https://doi.org/10.1111/jfr3.2016.9.issue-4, https://doi.org/10.1111/jfr3.12183

Vidya, P. V., Rajathy, S., & Ratheesh Kumar, C. S. (2024). Hydrogeochemical characteristics of Thrissur Kole Wetland, Southwest India. *International Journal of River Basin Management*, 1–14.

Watson, G., & Ahn, J. E. (2022). A systematic review: To increase transportation infrastructure resilience to flooding events. *Applied Sciences, 12*(23), 12331.

Zainulabdeen, Y. P., & Nagaraj, H. (2022a). Anthropogenic Impacts on Wetlands of Kerala, India: A Review of Literature. *Journal of Geography, Environment and Earth Science International, 26*(6), 28–38.

Zainulabdeen, Y. P., & Nagaraj, H. (2022). *Spatio-temporal analysis of wetlands of Thrissur District, Kerala, India: A study based on remote sensing and geographic information system.* International Journal of Novel Research and Development.

Ziervogel, G., Pelling, M., Cartwright, A., Chu, E., Deshpande, T., Harris, L., & Zweig, P. (2017). Inserting rights and justice into urban resilience: A focus on everyday risk. *Environment and Urbanization, 29*(1), 123–138.

Chapter 21
Wetland Dynamics-Changing Climate Nexus: Between Conservation and Livelihoods Sustainability in the Ndop Plain, Cameroon

Moye Eric Kongnso, Suiven John Paul Tume, Tiomo Dongfack Emmanuel, Tsafack Ngoufo Louis Serge, and Mbinkar Godlove Suika

Abstract Wetlands are vital ecosystems that regulate climate and provide essential resources and services to riverine communities. This chapter aims to explore the nexus between climate change and wetland resource dynamics and how conservation efforts can improve the livelihoods of local populations. Data was collected from both secondary and primary sources using a mixed-methods approach. The rainfall anomaly index indicates a declining trend in rainfall over the past three decades, accompanied by an increase in extreme weather events. Additionally, the normalized difference water index (NDWI), groundwater potential index (GWPI), and a change detection model show a reduction in floodplains, shrinking underground aquifers, and changes in land use that negatively affect wetlands in the Ndop Plain. Between 2001 and 2020, surface and groundwater potentials decreased by more than 30%, with a significant reduction in wetland vegetation. Understanding the dynamics and the climate-regulating functions of wetlands is crucial for conservation efforts and sustainable livelihoods.

Keywords Climate change · Resource · Degradation · Livelihoods · Sustainability · Wetlands

M. E. Kongnso (✉) · T. D. Emmanuel · M. G. Suika
Department of Geography, Environment and Planning, University of Dschang, Dschang, Cameroon
e-mail: moyeeric@yahoo.com

S. J. P. Tume
Department of Geography and Planning, The University of Bamenda, Bamenda, Cameroon

T. N. Louis Serge
Department of Geography, University of Yaounde I, Yaounde, Cameroon

U. Chatterjee et al. (eds.), *Eco-Resilience: Climate Change, Land Degradation and Sustainable Solutions*, Sustainability Solutions,
https://doi.org/10.1007/978-3-032-00708-7_21

21.1 Introduction

Wetlands make up 6% of the world's ecosystems and are highly valuable as they regulate the climatic system, provide cultural services and sustain the livelihoods of the population (Barrosn & Albernaz, 2013). According to the Ramsar Convention Secretariat (2006), wetlands are areas of marsh, fen, peatland or water, whether natural or artificial, permanent or temporal, with water that is static or flowing, fresh, brackish or salt, including areas of marine water the depth of which at low tide does not exceed six meters. They equally include riparian and coastal zones adjacent to the wetlands, and islands or bodies of marine water deeper than six meters at low tide lying within the wetlands (Ramsar Convention Secretariat, 2006). Wetlands have a rich biodiversity and play a vital role as they retain water in the landscape, maintain local climate and hydrological sites, and reduce extremes of temperature. Wetlands store water from precipitation and release it slowly in the surrounding environment, thus making it possible to recharge groundwater and maintain the atmospheric hydrological cycle. Evaporation and transpiration of water by plants have a local cooling effect (Fennessy & Lei, 2018). Their richness in valuable resources has attracted human exploitation, causing degradation. The rapidly growing world population and increasing exploitation of wetlands for agriculture, fishing, construction and craftworks have exacerbated their exposure to the whims of climate and there is a need for conservation.

The global climate change, characterized by rising temperatures, falling rainfall trends and a heightened frequency of extreme weather events, especially dry spells, has had negative impacts on wetland resources (Wang et al., 2023). Climate change has led to changes in atmospheric temperatures and water courses, an increase in evapotranspiration and the intensification of extreme events leading to sudden variations in flow and water levels (Records et al., 2014). Water is an important resource that conditions the sustainability of wetlands, but climate-related disaster risks are increasing, and it is estimated that 90% of disasters are linked to water (UNISDR, 2015). Wetlands have suffered a marked decline over the last century due to anthropogenic pressures to which climate change has been added more recently (Malekmohammadi et al., 2015; Zhang, 2016). Extant scholarly literature on wetlands has focused on the importance of wetlands as carbon sinks and climate regulators (Fennessy & Lei, 2018; Taillardat et al., 2020) and their dynamics that result from anthropogenic pressure (Nguh & Kimengsi, 2016; Yemele et al., 2023), but little has assessed the vulnerabilities of wetland resources to the whims of climate using the nexus approach, which is crucial for sustainable management. This chapter seeks to bridge this scientific gap by contributing to knowledge on eco-resilience and sustainable management of wetlands.

This decline in wetlands has led to resource degradation with negative impacts on the population. Wetlands are at the interface between terrestrial and aquatic environments and among the most sensitive habitats to climate change (IPCC, 2007). They impact livelihoods differently depending on the landscape in which they are located, their water resources and related hydrological regimes. Wetland resources such as flood plains, swamps and raffia bushes are exploited for livelihood sustenance. There is a source of food and income to the riverine communities and also provides recreational and cultural services (Chen et al., 2018; Moye et al., 2024). During the dry season and periods of droughts, wetlands provide fresh water for domestic use and for agriculture as well as fresh pasture for cattle (Moye et al., 2024). Nevertheless, Chen et al., (2018) have shown that climate change affects all productive sectors of rural development, including the wetlands sector. This sector is strongly influenced by climatic factors, particularly precipitation, evapotranspiration, extreme temperatures and extreme phenomena such as drought (Xu et al., 2023). As such, there is a need for wetland restoration to ensure effective climate change adaptation and mitigation.

Our capacity to adapt to climate change depends on our ability to put in place a wide range of responses, including rational exploitation of wetlands and the restoration of those that are degraded (Moomaw et al., 2018). In harnessing the natural capacities of wetlands to protect communities from the negative effects of climate change, we can amplify climate resilience and ensure sustainable livelihoods. Protection and restoration of wetlands for climate change mitigation and adaptation are key principles of the Strategic Plan of the Ramsar Convention that equally advance the achievement of the Sustainable Development Goals and that of the Paris Climate Agreement (Fennessy & Lei, 2018). Also, efforts to protect and restore wetlands as well as promote their rational use will help countries achieve nationally determined contributions under the Paris Agreement (Ramsar Convention Secretariat, 2022). Flood plains, rivers, lakes and swamps function as sponges, absorbing and storing the excess precipitation and thus reducing floods. During the dry season, they release stored water, delaying the occurrence of drought and water shortages (Fennessy & Lei, 2018; Wang et al., 2023). Wetlands of the Ndop plain that occupy 11.5 km^2 (which is approximately 22.4% of the total surface area) are of national importance as they provide a livelihood base for more than 70% of the population (Kometa et al., 2017; Nzembayie, 2016). It has an economy dominated by swamp rice production and irrigated rice from rain-fed streams. However, climate change is posing a threat to this activity (Moye et al, 2020; Nkiene et al., 2016).

The objective of this chapter is to explore how climate change impacts the degradation of wetland resources in the Ndop plain. Using water and pasture as the main wetland resources that ensure livelihoods in the Ndop plain, this work reveals a dynamic situation as these resources fluctuate depending on climatic trends. Adopting a nexus approach, this chapter demonstrates the cause-and-effect relationship between climate change and wetland evolution and proposes recommendations that ensure the sustainable exploitation of wetland resources in the Ndop plain.

Theoretical Framing: The Nexus Approach in Environmental Resource Management

> *"The Nexus Approach to environmental resources management examines the interrelatedness and interdependencies of environmental resources and their transitions and fluxes across*

spatial scales and between compartments. Instead of just looking at individual components, the functioning, productivity, and management of a complex system are taken into consideration"

(United Nations University Institute for Integrated Management of Material Fluxes and of Resources, UNU-FLORES, 2017).

The term nexus was introduced in the 1980s by the UNU in a program called the food-energy nexus program, but the approach gained prominence in 2011 during the nexus conference in Germany. It was presented as a strategy that promotes linkages and synergies and reduces trade-offs in the management of natural resources, especially in ensuring food, water and energy security (UNI-FLORES, 2018). It is a conceptual approach that thinks across sectors, but the ecosystem occupies a central stage. The interconnectedness is complex and demands that stakeholders involved in the management of such resources must come together to determine the most efficient and sustainable management strategies (Avellan et al., 2017). Wetlands are important ecosystems with multi-dimensional functions, and this work argues that the nexus approach is valuable in understanding the sustainable exploitation of their resources as well as trade-offs, especially in the context of climate change adaptation and mitigation. In wetland ecosystems, there are interlinkages as indicated in (Fig. 21.1).

Figure 21.1 shows three major linkages that are observable in wetland ecosystems. They include; water-to-food, water-to-energy and energy-to-food linkages. UNU-FLORES (2018) stated that a healthy ecosystem is an essential requirement for the sustainability of these networks in the context of climate change and globalization. In wetlands, water plays a central role in the nexus, as corroborated by

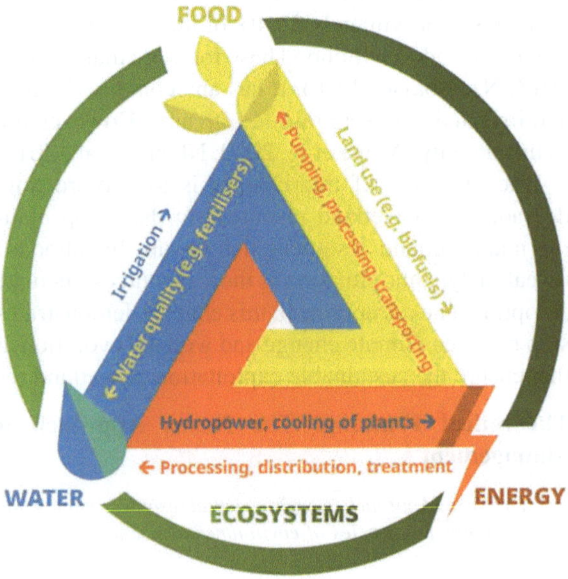

Fig. 21.1 The water-food-energy nexus approach in an ecosystem. *Source* Global water partnership-mediterranean (GWP-Med), 12, Kyrristou str.10556 Athens, Greece, https://www.gwp.org/en/GWP-Mediterranean/WE-ACT/Programmes-per-theme/Water-Food-Energy-Nexus/the-nexus-approach-an-introduction/

Balaican et al. (2024). Swamp rice cultivation, which is the main livelihood activity of the population, depends solely on water from flooded plains and is irrigated from rain-fed streams (Nkiene et al., 2016), as it provides food and revenue to the community. Nevertheless, climate change, characterized by rising temperatures, with higher rates of evapotranspiration, irregular rainfall and extreme events such as frequent droughts and floods and has reduced the ability of wetlands to supply water for irrigation and fresh water for domestic use (Lal, 2015). This affects the water-food linkage. In an attempt to increase agricultural productivity, wetlands are destroyed to increase farm sizes with intense mechanization and energy consumption. Machines are used to till, pump water and transport products within the rice fields by the UNVDA. In the Ndop plain, the government constructed a reservoir along the river Nun in 1975 to augment the water level at Edea for the generation of hydroelectric power in Cameroon (Mbih et al., 2014). The regulation of water levels in the reservoir affects the size of the flood plain and human activities such as fishing. This food-energy linkage is corroborated by Hoff et al. (2013). These interactions have trade-offs, which may be the negative consequences on the ecosystem and sustainable livelihoods. The nexus approach in natural resource management enhances the achievement of sustainable development goals (Caucci & Hettiarachchi, 2018; Hülsmann & Jampani, 2021). Besides food, water and energy, and its vulnerabilities to climatic aberrations, wetland ecosystems are great carbon sinks that regulate micro temperatures and mitigate climate change (Moomaw et al., 2018; Xu et al., 2023). However, in order to reduce trade-offs, maximize gains and ensure sustainability, there is a need for synergies between stakeholders to ensure a better understanding of these linkages that have been complexified by climate change.

21.2 Materials and Methods

21.2.1 The Study Area

This study was carried out in the Ndop Central Subdivision, which is one of the subdivisions of the Ngokentujia division in the North West region of Cameroon. It is located between latitudes $5° \, 15^1$ and $6^0 \, 10^1$ North of the equator and longitude $10^0 \, 15^1$ and $10^0 40^1$ East of the Greenwich Meridian (Fig. 21.2). Situated within the western highlands agro-ecological zone, the Ndop central has an estimated surface area of 448.5 km^2 out of which 81.3 km^2 consists of wetlands (Kometa et al., 2017) and an average altitude of 1100 m above the sea level with vast marshy swamps (Moye et al., 2020). Ndop has vast fresh water wetlands generally found along river valleys and lowlands, and a Bamendjin reservoir in the South Eastern part, whose construction in 1975 greatly influenced the size of the flood plain and its river regimes. It submerged more than 442 km^2 upon completion of the dam, thus affecting human activities on the entire flood plain (Ndzeidze et al., 2016).

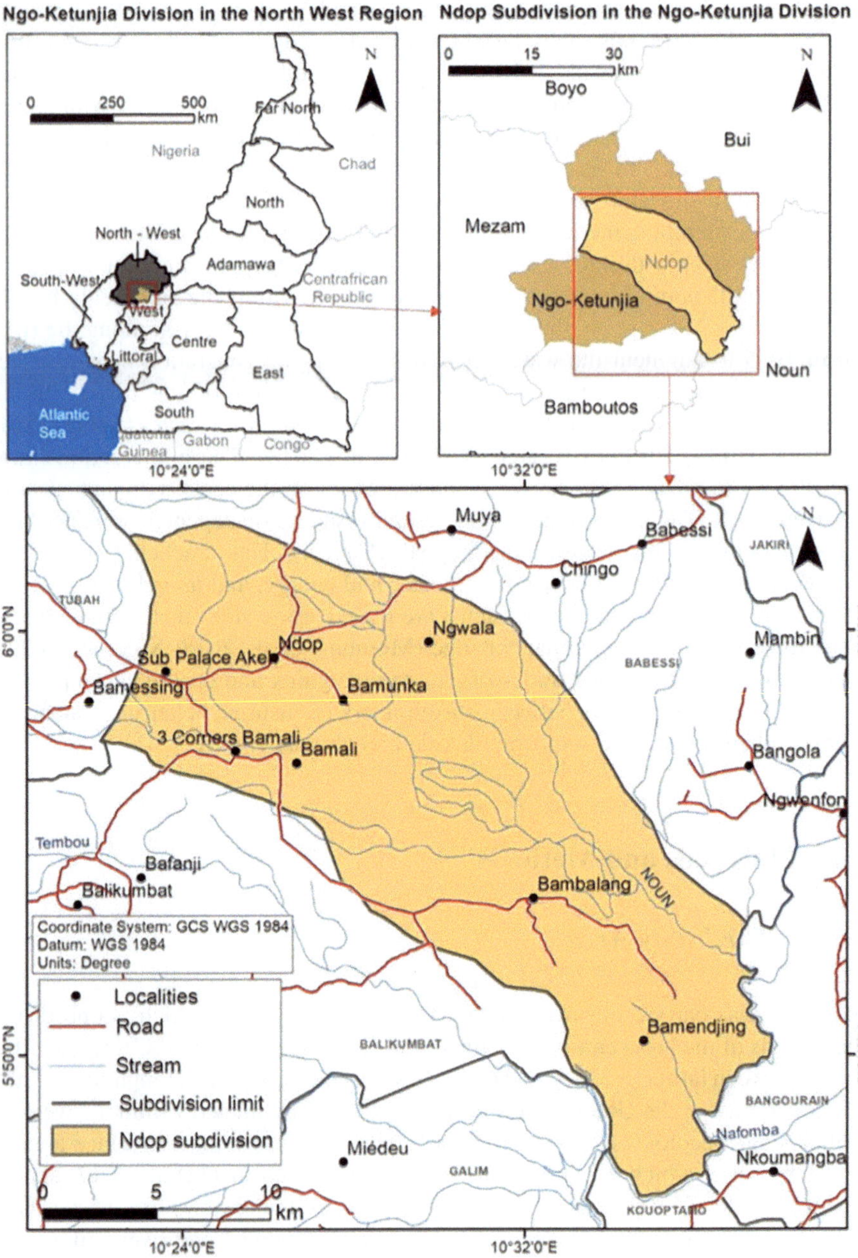

Fig. 21.2 Location of the Ndop Central subdivision. *Source* Streams, localities, subdivision limit (Interactive forest atlas, 2020)

Ndop plain has an agro-ecological zone suitable for crop cultivation and cattle rearing. It has a humid tropical climate with a short dry season and a long rainy season that runs from mid-march to late October. It had an estimated population of 107,580 inhabitants in 2017, of which 70% depend on the exploitation of wetland resources for their livelihoods (Kometa et al., 2017). The creation of the Upper Noun Valley Development Association (UNVDA) in 1985, with the main objective of developing and distributing paddy rice fields to farmers, has significantly changed the ecological character of wetlands. Besides swamp rice cultivation, the wetlands of the Ndop central subdivision attract herders from the surrounding highlands during periods of transhumance. They provide fresh pastures and water during the dry season and period of dry spells, which correspond to periods of acute scarcity in the mountains and up hills. The combined effects of pressure from human activities and climate change on wetlands have led to dynamics that call for an investigation.

21.2.2 Data Collection and Treatment

The mixed methods approach was used in the collection of secondary and primary data. Archives and reports were consulted at the Divisional delegations of Environment and Nature protection, Agricultural and Rural Development, livestock and animal husbandry, and the UNVDA, while online sources were also consulted. Climate data was obtained from the North West regional center for Meteorology from 1987 to 2023. Landsat images were collected via the https://earthexplorer.usgs.gov/ platform for three different years: 2001, 2013 and 2024. SRTM images with a spatial resolution of 30 m were also downloaded from the same platform. The Landsat images retained were taken between the dry season months of December and February, a period which corresponds to a low level of cloud Cover. Table 21.1 shows the characteristics of the different Landsats that were used to study wetland dynamics.

Primary data collection was via in-depth interviews and a household questionnaire survey. Key informant interviews were conducted with knowledgeable persons at the divisional delegations that intervene in wetlands management (3), municipal authorities (3), UNVDA (2), traditional authorities (2) and rice farmers' associations (3). During these discussions, voice notes were recorded for exploitation. To

Table 21.1 Image characteristics

	Acquisition date	Spatial resolution (m)	Path & row
Landsat 7 ETM+	05/02/2001	30	186–56
Landsat 8 OLI	15/12/2013	30	
Landsat 8 OLI	13/02/2024	30 T	

Source https://earthexplorer.usgs.gov/ Software used: ArcGIS for spatial analysis and layout, Envi 5 for the processing of satellite images assess the impacts of wetland exploitation on livelihoods, a household questionnaire was randomly administered to 150 household heads in the main wetland areas (Table 21.2). The simple random sampling technique was used to select participants because more than 70% of the population is peasants with direct and indirect links with wetlands.

Table 21.2 shows that the number of questionnaires administered per village or locality is proportionate to the surface area of its wetland. Bambalang, with 44.6% of wetlands, had 37.5% of participants; Bamunka had 28.7%, Bamali had 18.7% while Bamessing, with only 6.8% of wetlands, had just 16.7% of the total participants. Questionnaire administration was done on the spot, which justifies the 100% return rate.

Data collected was treated quantitatively and qualitatively. Questionnaires were edited to correct omissions and inaccuracies in the responses, while responses to open questions were coded and treated using the Statistical Package for Social Science (SPSS) to get the required information. On the other hand, voice notes from interviews were transcribed, coded and analyzed qualitatively according to themes and content. Rainfall and temperature data were analyzed using the anomaly index and trends established over the past few decades. The variation in wetland resources has been compared with climatic trends and land use patterns to establish the nexus. The treatment of cartographic data began with a preprocessing phase that includes layer stacking, resizing and composition coloring. Proper treatment involved the calculation of indices to detect changes in surface water, underground water and changes in land use patterns over time.

Normalized Difference Water Index (NDWI)

Proposed by McFeeters in 1996, NDWI is a tool used to detect changes in surface water from satellite images using colour palettes. It makes use of the near-infrared and short-wave infrared (NIR-SWIR) combination to detect the presence of water in leaves and the visible green and near-infrared (Green-NIR) combination to detect changes in the content of surface water bodies. It is calculated using the formulae;

$$\text{NDWI} = (\text{Green} - \text{NIR})/(\text{Green} + \text{NIR})$$

For Landsat 7 data: NDWI = (Band 2 – Band 4)/(Band 2 + Band 4)

Table 21.2 Distribution of questionnaires in localities with wetlands

Wetland area	Surface area (km^2)	% of wetland	% of respondents
Bambalang	36.3	44.6	37.5
Bamunka	29.3	36.0	28.7
Bamali	10.2	12.5	18.7
Bamessing	5.5	6.8	16.7
Total	81.3	100	100

For Landsat 8 data: NDWI = (Band 3 – Band 5)/(Band 3 + Band 5).

The value of NDWI varies between −1 (non-acquired surface) to 1 (water extension). Values of water bodies are larger than 0.5 while vegetation has much smaller values, which permits to distinguish between vegetation from water bodies.

Groundwater Potential Zone Mapping Using GIS and Remote Sensing with Multi-Criteria Approach (MCA) and Analytical Hierarchical Process (AHP) Approach

The groundwater potential in this study was assessed using six (06) criteria, including soil, land cover, slope, drainage density, rainfall, and geology. For each criteria, thematic maps were created and utilized to assess the groundwater potential of the study area. The AHP method was applied to determine the weight for each data layer and the MCA was used to determine the weights of each criterion. The weighted overlay analysis was used to calculate the value of potential indices of underground water using the equation proposed by Malczewski in 1999, according to which

$$\mathrm{GWPI} = \sum_{w=1}^{m}\sum_{j=1}^{n}\left(W_j^{*}X_i\right)$$

where

W_j represents the normalized weight of the j parameter, X_i refers to the weight of the i parameters, m denotes the number of the parameters, and n denotes the number of classes within a specific parameter. This index permits to know if an area is excellent, very good, moderate, bad or very bad in terms of underground water availability.

The Normalized Difference Vegetation Index (NDVI)

Wetland vegetation is an important resource in the Ndop Plain that has been affected by human activities and climate change. The NDVI is the most popular vegetation index in analyzing greenery globally. According to Rouse et al., (1973), this index shows the ratio of the variance between the near-infrared band (NIR − RED) and the red band (NIR + RED) as indicated in the formula.

$$\mathrm{NDVI} = (\mathrm{NIR} - \mathrm{RED})/(\mathrm{NIR} + \mathrm{RED})$$

For Landsat 5, NDVI = (Band 4 − Band 3) / (Band 4 + Band 3)
For Landsat 8, NDVI = (Band 5 − Band 4) / (Band 5 + Band 4).

The NDVI has values that vary from −1 to +1. Greater value of NDVI signifies a dense and greener vegetation cover, while lesser values indicate other land use patterns. As such, this index is used to evaluate the change detection resulting from land use and land cover change.

Climate variability and change is analyzed using the anomaly index and the seasonality index to demonstrate seasonal fluctuations that have implications on rainfall distribution and water availability. The seasonality index (SI) is used to determine the intra-annual monthly distribution of precipitation and assess the precipitation regime in the region. The index is the sum of the absolute deviations of the monthly rainfall from the mean monthly rainfall, divided by the total annual precipitation of the given year.

$$\mathrm{SI} = 1/\mathrm{Ri} \sum_{n=12}^{n=1} |X_{in} - R_i/12|$$

The variable R_i is the total annual precipitation in a particular year, and x_{in} is the actual monthly rainfall in month n. Figure 21.3 shows the methodology flowchart of this work.

21.3 Results

21.3.1 Observed Climatic Trends and Anomalies in the Ndop Plain

Climate variability in the Ndop plain is established using rainfall and temperatures due to direct impacts on wetland resources. Rainfall is characterized by a mean of 2362.71 mm, standard deviation of 259.69 mm and coefficient of variation of 9.09%. Fluctuations in rainfall amounts portray positive and negative anomalies over the past three decades (Fig. 21.4a). Positive anomalies are years or periods that recorded rainfall above the normal, while those with negative anomalies had rainfall below the normal. For instance, the periods 1964–1968, and 1978 to 1981 recorded positive anomalies with a likelihood of experiencing floods. The exceeding rainfall amount went up to about +400 mm, as was the case in 1999. On the other hand, the period 1970–72, 1982–1988, 1990–2000, and 2011–2018 recorded rainfall below the normal. Years like 2019 had as low as −600 mm below the mean for the period. This is an indicator of dry spells and potential water shortages.

Figure 21.4 shows that as the rainfall trend is falling, temperatures are rising steadily over the same period. A rise in temperatures is an indicator of climate variability that reflects the global rise in temperatures (Fig. 21.4b). Increased temperatures affect evapotranspiration and water availability in wetlands. A decadal analysis revealed fluctuations in time and space, considering the period from 2001 to 2024. In the year 2001, the highest rainfall amount was 2167.6 mm and the lowest was 2016.4 mm, with significant spatial disparity from the South East to the North Eastern part of the Plain. High rainfall quantities dropped to 1697.32 mm in 2013 and rose to 2669.1 mm in 2020. Within the same period, there has been a falling

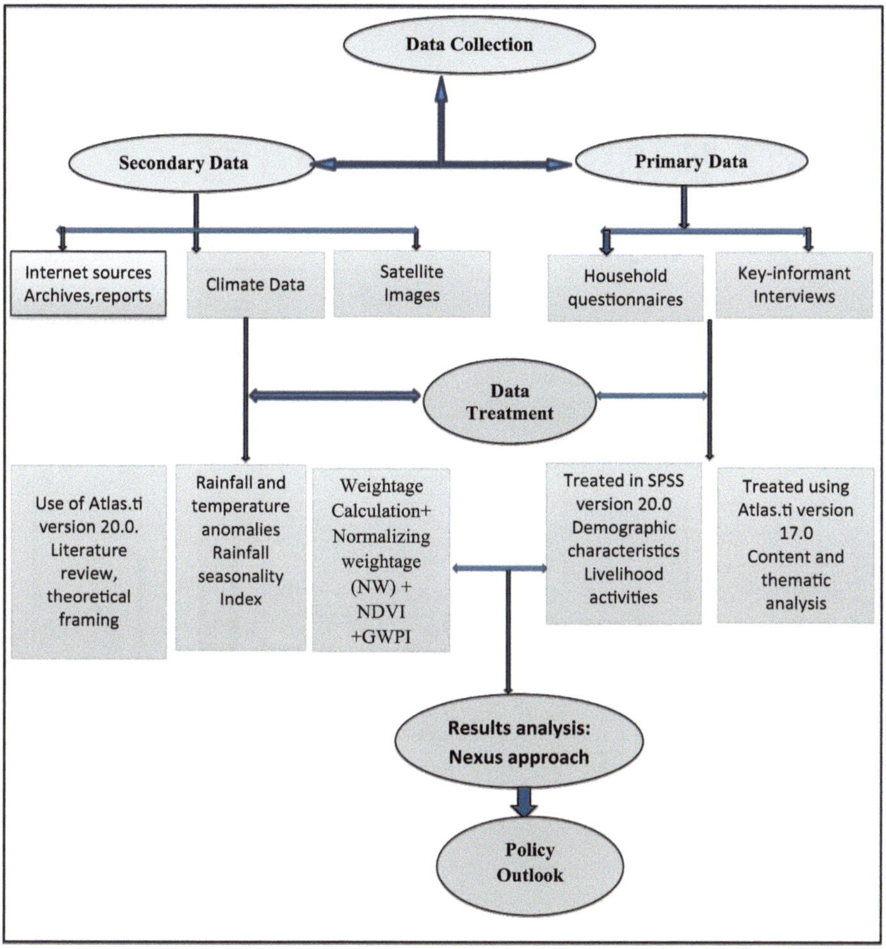

Fig. 21.3 Methodological flowchart

trend in annual rainfall amounts with an increasing number of dry spells. Another indicator is a shift in the seasonal pattern of rainfall caused by fluctuations in the dates of onset and cessation of rains (Fig. 21.5).

Rainfall distribution in the Ndop plain is seasonal with two distinct seasons: a long wet season and a short dry season (Fig. 21.5a). The rainy season begins in mid-march and ends in mid-October while the dry season runs from November to February, as in all humid tropical climates. Nevertheless, monthly anomalies reveal a disruption in the distribution pattern (Fig. 21.5b). The months of March and April are wet months, but recorded rainfall amounts are below the normal. However, the seasonality index of Walsh and Lawler depicts a shift in the seasonal pattern of rainfall in the Ndop plain. The computed value of 0.97 falls within the limit 0.80–0.99 of the seasonality index class limit (Table 21.3).

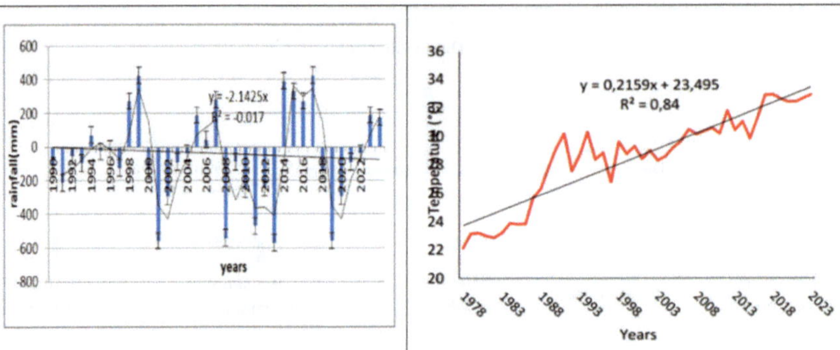

a. Inter-annual rainfall anomalies and trends

b. Inter-annual temperature anomalies and trends

Fig. 21.4 Rainfall and temperature anomalies and trends in the Ndop Plain

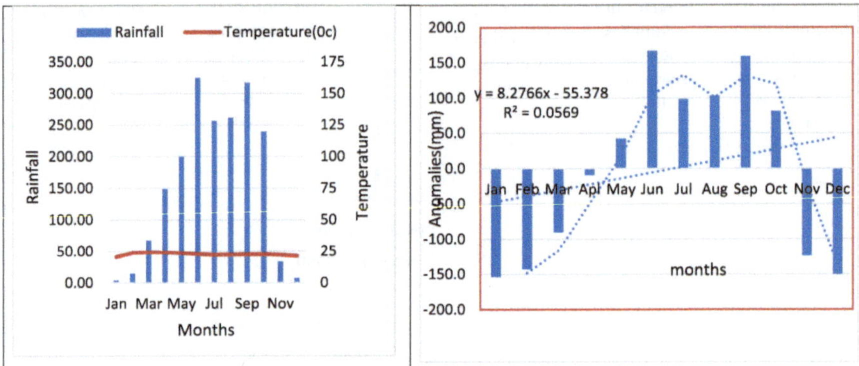

a.Ombrothermic diagram

b. Monthly rainfall anomaly and trend

Fig. 21.5 Monthly rainfall anomaly and seasonality in the Ndop plain

Table 21.3 The Rainfall seasonality scale

Seasonality index class limit	Rainfall regime
≤0.19	Very equable
0.20–0.39	Equable but with a definite wetter season
0.40– 0.59	Rather seasonal with a short drier season
0.60–0.79	Seasonal
0.80–0.99	**Markedly seasonal with long drier season**
1.00–1.19	Most rain in three months or less
≥1.20	Extreme, almost all rain in one to two months

Source Walsh and Lawler (1981)

Table 21.3 shows that the rainfall distribution pattern has been disrupted. The rainy season normally starts by mid-march and ends in October, and therefore, we would have expected the rainfall pattern for this area to be seasonal as in most humid tropical climatic regimes. However, the shift from "seasonal" to "markedly seasonal with a longer drier season" is an indicator of rainfall variability. This shift is justified by the reduction in rainfall totals with the increasing occurrence of dry spells. This entails that natural systems, including wetlands and pasture lands, will be vulnerable to water shortages, which is the most important wetland resource in the Ndop plain.

21.3.2 State of Wetland Resources and Change Over Time

21.3.2.1 Groundwater Evolution

The Ndop plain has significant groundwater resources with several aquifers located beneath the surface. Groundwater aquifers are primarily recharged through rainfall, with the region's high precipitation levels and favorable hydro-ecological conditions that enhance groundwater replenishment. However, the groundwater potential index reveals fluctuation in groundwater over different potential zones (Fig. 21.6 and Table 21.4).

Figure 21.6 shows fluctuations in groundwater potential zones from 2001 to 2020. In 2001 and 2013, there was much water beneath the ground, while in 2020, we observed a marked reduction. The evolution is not unidirectional, but the synthesis from 2001 to 2020 shows an overall loss in groundwater in the Ndop plain. In the high potential zone, for instance, there has been a decline from 5.52% in 2001 to 3.98% in 2013 and to 4.51% in 2020 (Table 21.4). In the low potential zone, we observe a rise from 1.22% in 2001 to 1.40% in 2013 and 2.63% in 2020. The high and moderate potential zones show fluctuating groundwater levels but with an overall falling trend. The high zone fell from 51.28% in 2001 to 43.59% in 2020, although there was a slight increase in 2013 (55.07%). These fluctuations can be attributed to human activities such as agriculture and climate variability.

21.3.2.2 Surface Water Evolution

Surface water is an important wetland resource that supports livelihoods in the Ndop plain. Figure 21.7 and Table 21.5 show a fluctuation in surface water between 2001 and 2020.

Wetlands had an estimated surface water area of 5173.741 ha (41.30%) in 2001 and in 2013, the surface water was reduced to 3723.559 ha (29.73%) and to 3628.637 ha (28.97%) in 2024 (Table 21.5). The synthesis shows that losses are more than the gains. This change in surface water can be a result of human activities and climate change. Human activities such as irrigation farms and swamp rice cultivation depend

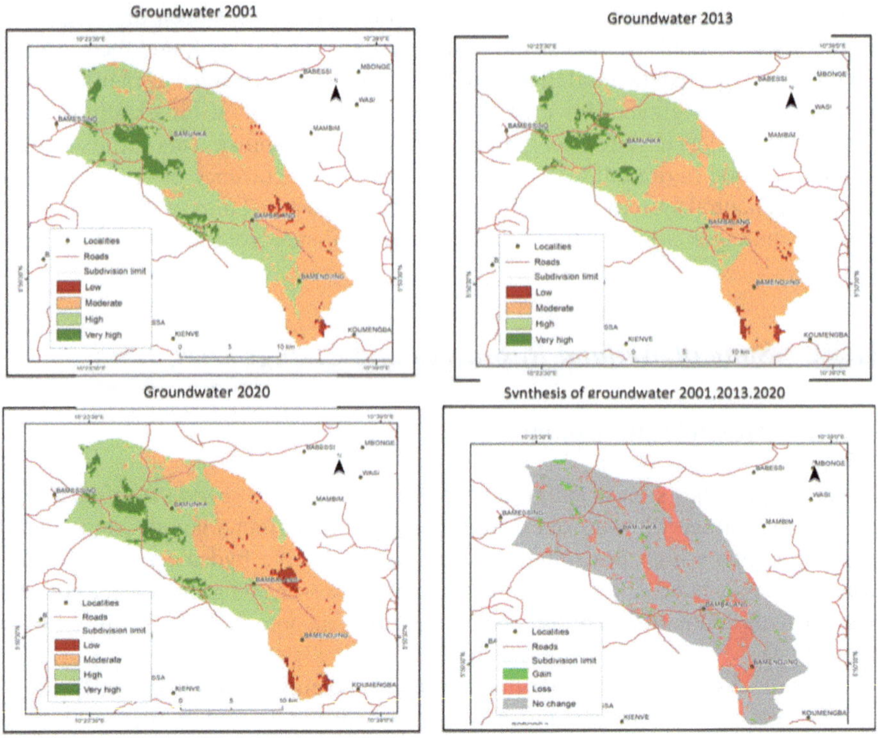

Fig. 21.6 Groundwater potential NCSD. *Source* streams, localities, subdivision limit, (Interactive forest atlas, 2020)

Table 21.4 Groundwater potential zones in NCSD

		2001		2013		2020	
Value	Potential zone	Area in ha	%	Area in ha	%	Area in ha	%
2	Low	418.7344	1.22	484.0416	1.40	906.6176	2.63
3	Moderate	14,463.624	41.99	13,606.9472	39.50	16,968.3472	49.25
4	High	17,663.6768	51.28	18,973.6624	55.07	15,016.8144	43.59
5	Very high	1901.592	5.52	1371.4512	3.98	1555.848	4.51
Total	34,447.6272	34,447.6272	100	34,447.6272	100	34,447.6272	100

on surface water, and its fluctuation downstream is greatly influenced by activities taking place upstream.

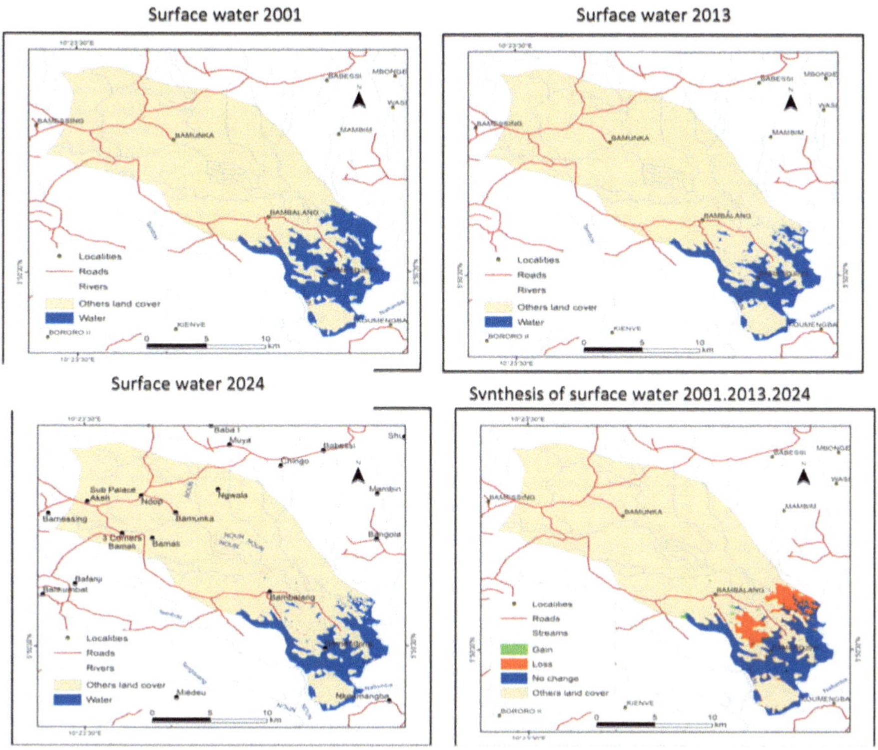

Fig. 21.7 Surface water changes in NCSD. *Source* streams, localities, subdivision limit, (Interactive forest atlas, 2020)

Table 21.5 Surface water evolution

Year	Surface water ha	%
2001	5173.741	41.30
2013	3723.559	29.73
2024	3628.637	28.97
Total	12,525.94	100.00

21.3.2.3 Livelihood Activities, Land-Use Dynamics and Conservation Challenges

The spatio-temporal analyses of the land-cover and land-use dynamics and the change detection model from three Landsat images (2001, 2013 and 2024) show considerable changes in wetland vegetation and other land use patterns over the past three decades (Fig. 21.8).

Figure 21.8 shows fluctuations in wetland vegetation over the years, with the overall trend portraying a remarkable loss in wetland vegetation from 2001 to 2024.

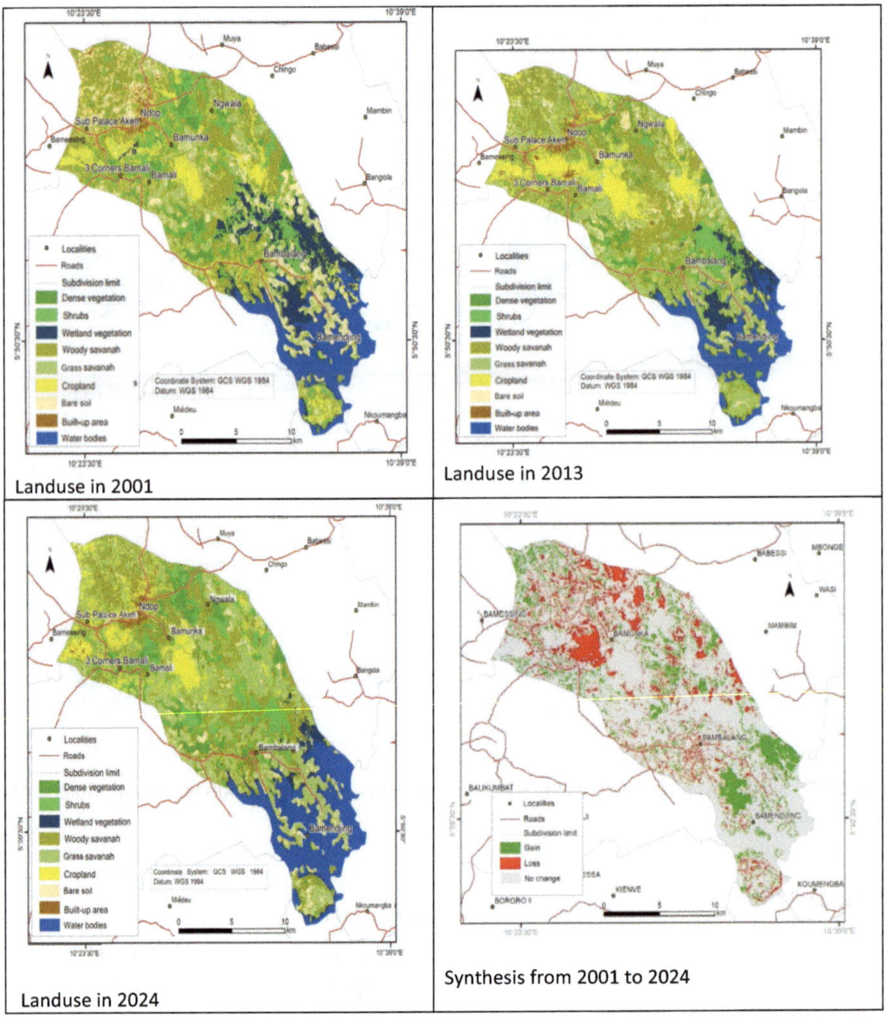

Fig. 21.8 Change detection in land-use and land-cover from 2001 to 2024. *Source* Streams, localities, subdivision limit, (Interactive forest atlas, 2020)

In the year 2001, wetland vegetation occupied 3731,659912 ha, which is 10,89% of the total land surface, and it dropped to 2108,189941 ha (5.99%) in 2013 and to 1154.709961 ha (3.28%) in 2024. The reduction in wetland vegetation is caused by the extension of human activities such as built-up areas and farmlands (Table 21.6).

The analyses in Table 21.6 show a steady increase in built-up areas and croplands from 2001 to 2024. In 2001, the built-up area was 957.6489868 ha (2.75%) and rose to 968.7470093 ha (2.93%) in 2013 and 1126.439941 ha (3.20%) in 2024, while farmlands equally increased from 2700.860107 ha (7.68%) in 2001 to 3979.850098 ha

Table 21.6 Change detection in land-cover and land-use patterns in the Ndop plain from 2001 to 2024

	2001		2013		2024	
CLASS_ NAME	Area ha	%	Area ha	%	Area ha	%
Built-up area	957.6489868	2.75	968.7470093	2.93	1126.439941	3.20
Water bodies	4767.490234	13.56	3477.419922	9.89	3457.629883	9.83
Cropland	2700.860107	7.68	3314.48999	9.43	3979.850098	11.32
Woody savannah	8966.269531	25.50	10229.59961	29.09	10054.59961	28.59
Grass savannah	10038.09961	28.54	3919.330078	11.14	8612.650391	24.49
Bare soils	1788	5.08	4164.75	11.84	2143.290039	6.09
Dense vegetation	1813.280029	5.16	1541.969971	4.38	1044.630005	2.97
Wetland vegetation	3731.659912	10,61	2108.189941	5.99	1154.709961	3.28
Shrubs	2970.540039	8.45	3828.97998	10.89	2639.629883	7.51
Total	35166.8985	100	35166.8985	100	35166.8985	100

(11.32%) in 2024. The encroachment of wetlands by human activities was identified by key informants as the main driver of wetland resource degradation, as close to 65% of the population depends on wetlands for their livelihoods. Equally, the adverse effects of climate variability and change on the surrounding uplands have pushed the population into wetland exploitation. The Ndop plain is mainly rural, and the main livelihood activities are cattle rearing and crop cultivation. These activities depend on wetland resources, especially water. The questionnaire survey revealed that about 45% of the population are involved in swamp rice cultivation, 17% in freshwater fishing, 15% in cattle rearing, 13% in gardening, weaving 5% and palm wine tapping 5% (Fig. 21.9). These activities have been exerting pressure on wetlands and causing rapid degradation.

Interviews with the authorities of the Upper Nun Valley Development Authorities (UNVDA), a government Institution in charge of rice cultivation, revealed that Ndop has lost more than 45% of its wetlands over the past few decades. This loss is caused by both human and natural stressors and if sustainable solutions are not put in place, the crop farming population may lose their sources of livelihoods. There is therefore a need for conservation and sustainable exploitation.

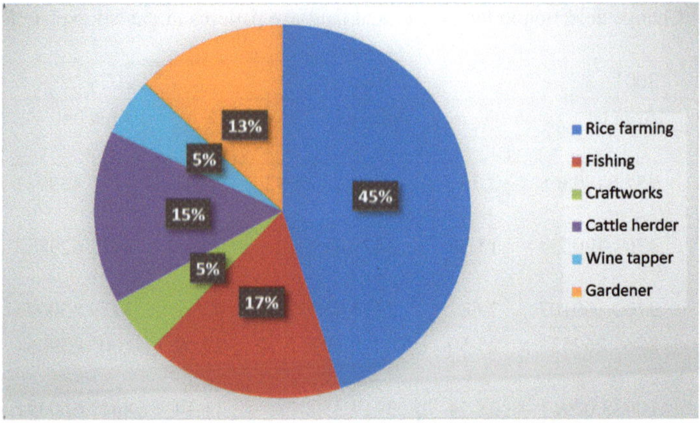

Fig. 21.9 Livelihood activities exerting pressure on wetlands in the Ndop plain. *Source* Fieldwork (2024)

21.3.3 Role of Climate Variability and Change in Wetland Dynamics

Given that wetland dynamics is influenced by a number of factors, the AHP method was applied to determine the weight for each factor using the Multi-Criteria Decision Making approach. To establish the relationship between factors, the hierarchical structure of all the factors was created as the first stage in the AHP approach. Findings revealed that a plethora of factors contribute to wetland resource evolution, but climate plays a predominant role (Table 21.7).

Table 21.7 shows that climate elements, especially rainfall, have the highest contribution in wetland dynamics, as it weighs (38.2%), followed by drainage density, which accounts for 29.9%, Land use/Land cover change (10.0%), slope 8.4%, geology 7.4% and soil 6.2%. The high rainfall in this area, which can reach over 2000 mm per year, is a significant factor in groundwater recharge. As such, fluctuations in rainfall amounts have implications for groundwater potentials. The relationship between rainfall variability and groundwater potential and acknowledged during key informant interviews, as captured in the following excerpt:

Table 21.7 Weights of factors contributing to wetland resource evolution

Values	Factors	Weights (%)
1	Drainage density	29.9
2	Climate	38.2
3	Slope	8.4
4	Land use/land cover	10.0
5	Soil	6.2
6	Geology	7.4

> ……rainfall variability has led to the rising frequency and intensity of floods occurrence in the Ndop plain. Intense rainfall reduces infiltration and causes runoff and floods. This reduces ground water supply but increases surface water potentials….. (Interview, 2024)

The rate of rainfall determines the amount of water on the surface. Most streams in and around the Ndop plain are rain-fed and seasonal. As such, periods of low rainfall characterized by negative anomalies and dry spells are bound to witness a fall in surface water potentials. On the other hand, periods with more intense rainfall will experience more run-offs, less infiltration and floods. If the amount of rainfall is less than the soil can absorb, it will all infiltrate; there will be no run-off. The change in surface water is reliant on independent variables of change in inland surface rainfall intensity. In most locations, rainfall replenishes surface water and transitory water bodies and contributes to surface water potentials. However, the rainfall pattern and intensity have altered dramatically as a result of climate change. Changes in surface temperature also have effects on surface water levels. High temperatures cause higher rates of evapotranspiration and dry spells. This has equally reduced flow in streams from the surrounding highlands of Sabga and Oku, resulting in a substantial effect on water levels in the plain. Climate variability and change equally have an impact on wetland vegetation. From the questionnaire survey, 65% of the population ascertained that climate variability has led to a reduction in pasture lands on the surrounding highlands of Wainamah and Sabga, leading to migratory movements down the plain. Interviews revealed that the degrading rangeland conditions have been exacerbated by anthropogenic activities and frequent droughts that make moisture availability a problem. The river courses of the uplands surrounding the Ndop plain have been shrinking considerably, while some dry off completely during the dry season. This situation has led to transhumant pastoralism in the Ndop Plain with consequence for wetland vegetation as captured in the excerpt.

> ……the natural conditions and environmental characteristics put in place by climatic fluctuations have brought great discrepancy in pasture endowment between the Ndop flood plain and the surrounding low highlands. Wetlands usually contain fresh water and pastures all year round but the concentration of human activities on wetlands in recent times is an adaptive strategy that has negative outcomes on wetlands… (Interview, 2024)

Cattle rearing on wetlands and rice cultivation have greatly reduced wetlands. Equally, another prominent factor that plays a vital role in wetland dynamics on this plain is the Bamindjin Dam constructed in 1975 across the river Nun. According to Ngwa (2003), before the construction of the Bamindjim dam and the introduction of rice cultivation, the Ndop plain had a total grazing land of 30,000 ha, but it has drastically reduced to about 10,000 ha due to the ever-expanding Bamindjim. Rainfall variability has a direct effect on the fluctuation of water levels in the Bamindjim reservoir. Intense rainfall will increase flow into rivers and finally into the reservoir, thereby expanding its surface area. This situation which is common during the rainy seasons, extends the flood plain, but it reduces during periods of dry spells. Hence, wetland dynamics have a direct and direct relationship with climate variability and change.

21.4 Discussion: Understanding the Climate-Wetlands Nexus

Climate variability and change a major problems affecting both natural and human systems nowadays. Wetlands are one of the world's most important ecosystems not only because of their resources but equally due to their ability to act as a carbon sink, thus mitigating climate change (Taillardat et al., 2020; Wang et al., 2023). However, these wetlands are rapidly degrading and due to their global importance, the Ramsar Convention Secretariat (2006) has stressed the need to conserve these important ecosystems. This chapter on climate change and wetland degradation is interlinked with the focal theme 'Eco-Resilience: Climate Change, Land Degradation and Sustainable Solutions' and the nexus approach is crucial in the integration of climate change, land degradation and sustainable livelihood practices to foster long-term eco-resilience and sustainability. Using the specific case of the Ndop plain, results show that climate variability and change manifest in the Ndop through anomalous events. As a major contributor to ground water and surface water recharge, rainfall has been falling over the past decades with a rise in the number and intensity of dry spells. The high rates of evapotranspiration caused by rising temperatures have had negative outcomes for wetland resources and activities that depend on them (Nkiene et al., 2016).

Wetlands are very rich ecosystems essential for both the natural environment and livelihoods sustenance. Globally, these resources have been shrinking and this work has established the link between climatic aberrations and the shrinking resource base. Multiple factors and interactions can influence wetlands, but climate change, especially changes in rainfall patterns has the highest contribution in wetland dynamics (38.2%) despite the role of other factors. This result is corroborated by Zhang (2016). Using specific resources of water (underground and surface) and wetland vegetation, this work shows evolutions that reflect changes in climate patterns. Wetlands had an estimated surface water area of 5173.741 ha (41.30%) in 2001, and in 2013, it dropped to 3723.559 ha (29.73%) and to 3628.637 ha (28.97%) in 2024. This change in surface water can be a result of human activities and climate change, but many works have ascertained the contributions of climate change on wetland surface water reduction, especially in dry areas (Wang et al., 2023). On the other hand, it is revealed that there has been a fluctuation in the high underground water potential zone from 5.52% in 2001 to 3.98% in 2013 and to 4.51% in 2020, but with a slight increase in the low potential zone, from 1.22% in 2001 to 1.40% in 2013 and 2.63% in 2020. The overall scenario depicts a fall in groundwater potential over the past three decades. Groundwater is an important wetland resource but very sensitive to environmental changes (Wang et al., 2023). Underground water in the Ndop plain is recharged by percolating rainfall water, and as such, the quantity at any given time depends on the input from rainfall. Rainfall irregularity and falling trends reduce surface flow and cause the shrinking of water courses. Many river courses in and around the Ndop plain have become seasonal. They dry up completely in the dry season, while in the rainy season, floods become very common. These have implications for the fluctuation in the size of flood plains (Ndzeidze et al., 2016).

Another remarkable element is the fluctuations in wetland vegetation over the years, with the overall trend portraying a remarkable loss in wetland vegetation from 3731,659,912 ha (10.89%) in 2001 to 1154.709961 ha(3.28%) in 2024. The reduction in wetland vegetation is caused by the extension of human activities such as built-up areas and farmlands. The Ndop plain had an estimated population of 107,580 inhabitants in 2017, with 70% depending on wetland exploitation for their livelihoods (Kometa et al., 2017). The change detection model shows a steady increase in built-up areas from 957.6489868 ha (2.75%) in 2201 to 1126, 439941 ha (3.20%) in 2024, while farmlands rose from 2700.860107 ha (7.68%) in 2001 to 3979, 850098 ha (11.32%) in 2024. The expansion of the Ndop town, coupled with the creation of the Upper Noun Valley Development Association (UNVDA) in 1985, with the main objective of developing and distributing paddy rice fields to farmers, has significantly changed the ecological character of wetlands. Equally, transhumant pastoralism has been revealed to be a significant contributor to wetland degradation in the Ndop plain. Many studies have stipulated that the degradation of Ndop wetlands is largely caused by human encroachments, thereby undermining the role of climate variability and change (Nguh & Kimengsi, 2016; Mbanga & Dingha, 2022; Yemele et al., 2023). This work bridges the empirical gap by establishing the direct and indirect role played by climatic fluctuations in the observed dynamics using the nexus approach to establish the linkages and trade-offs. The interactions between wetland resources and climatic aberrations have trade-offs, which may be the negative consequences on the ecosystem and sustainable livelihoods. According to Caucci and Hettiarachchi (2018) and Hülsmann and Jampani (2021), the nexus approach in natural resource management enhances the understanding of linkages and achievement of sustainable development goals.

21.5 Conclusion

This chapter had as objective to examine the link between climate change and wetland resource evolution in the Ndop plain. Climate change and wetland resource degradation is interlinked with the focal theme 'Eco-Resilience: Climate change, land degradation and sustainable solutions and understanding the interactions between these concepts requires a nexus approach. Given that wetland exploitation is the mainstay of the Ndop economy, it is imperative to assess the rate of degradation and inform policies that ensure sustainable exploitation practices. As such, using a mixed methods approach in data collection and analysis, the nexus approach was employed to conceptualize the complex interconnectedness that shapes natural resource management in the context of growing human pressure and climate change. It is argued that the nexus approach is key in understanding the sustainable use of the environment and the achievement of the Sustainable Development Goals. Findings revealed that the Ndop is rich in wetlands distributed all over the plain with 44.6% in Bambalang,

28.7% in Bamunka, 18.7% Bamali and 6.8% in Bamessing. These wetlands have not been static over time, and their evolution is caused by many factors, including climate variability and change. Among the plethora of factors, climate variability and change account for a 38.2% change in wetlands. Surface water has evolved negatively from 41.30% in 2001 to 28.97% in 2024, while underground water potential dropped from 5.52% in 2001 to 4.51% in 2024. The linkages between the dynamics and climate variability are observable and reported by the local population. The shrinking of water courses and frequent floods either increase or decrease the size of flood plains that constitute the major type of wetlands in Ndop. The exploitation of wetlands for rice cultivation, cattle rearing and construction of urban infrastructure has negative consequences on wetlands, but this chapter opines that climate variability and change have exacerbated their impacts and created new dynamics. It is therefore primordial to understand these complex dynamics so as to reduce trade-offs, maximize gains and ensure sustainability by synergizing actors and practices. The sustainability of wetlands in this area requires an integrated management approach that will involve the population and local authorities. A coordinated exploitation will sustain existing livelihood outcomes and ensure eco-resilience in the context of climate variability and change.

Declarations

Funding: The research was not supported by any funding.

Potential conflicts of interest: On behalf of all authors, the corresponding author states that there is no conflict of interest.

Authors' Contribution: All authors have contributed equally to this work.

Data Availability Statement: The datasets used for this chapter are available from the corresponding author on reasonable request.

References

Avellan, C., Ardakanian, R., & Gremillion, P. (2017). The role of constructed wetlands for biomass production within the water-soil-waste nexus. *Water Science and Technology*. https://doi.org/10.2166/wst.2017.440

Balaican, D., Teubner, K., Tudor, I.-M., Nichersu, I., Burada, A., Trifanov, C., Tudor, M., Iticescu, C., Moraru, L., & Georgescu, P.-L. (2024). Water energy food nexus in urbanizing wetlands: Challenges and solutions explored through choremes and focus maps. *Water, 16*, 922. https://doi.org/10.3390/w16070922

Barros, D., & Albernaz, A. (2013). Possible impacts of climate change on wetlands and its biota in the Brazilian Amazon. *Brazilian Journal of Biology, 74*. https://doi.org/10.1590/1519-6984.04013

Caucci, S., & Hettiarachchi, H. (2018). *The nexus approach and safe use of wastewater in agriculture: A workshop on policy and implementation for Tunisia*. United Nations University Institute for Integrated Management of Material Fluxes and of Resources (UNUFLORES).

Chen, H., Wanchang, Z., Huiran, G., & Ning, N. (2018). Climate change and anthropogenic impacts on wetland and agriculture in the Songnen and Sanjiang Plain, Northeast China. *Remote Sensing*. https://doi.org/10.3390/rs10030356

Fennessy, S. M., & Lei, G. (2018). La restauration des zones humides contribue à la résilience climatique. Note d'information Ramsar no 10. Gland, Suisse: Secrétariat de la Convention de Ramsar.

Hoff, H., Fielding, M., & Davis, M. (2013). A 'nexus' approach to soil and land management. *Rural, 21*(47), 10–12. http://www.rural21.com/english/current-issue/detail/article/a-nexus-approach-tosoil-and-land-management-0000833/

Hülsmann, S., & Jampani, M. (2021). The nexus approach as a tool for resources management in resilient cities and multifunctional land-use systems. https://doi.org/10.1007/978-3-030-57530-4_1

Kometa, S., Kimengsi, J., & Petiangma, D. (2017). Urban development and its implications on wetland ecosystem services in Ndop, Cameroon. *Environmental Management and Sustainable Development*. https://doi.org/10.5296/emsd.v7i1.12141

Lal, R. (2015). The nexus approach to managing water, soil and waste under changing climate and growing demands on natural resources. In M. Kurian, R. Ardakanian (Eds.), *Governing the Nexus*. Springer. https://doi.org/10.1007/978-3-319-05747-7_3

Malekmohammadi, B., Blouchi, L., Ghehi, N., & Shakib, F. (2015). *Investigating the effects of climate change on wetlands using risk assessment and remote sensing (case study: Choghakhor Wetland, Iran)*. U.S.-Iran Symposium on Climate Change: Impacts and Mitigation, Irvine, California.

Mbih, R., Ndzeidze, S., Steven, L., & Bamboye, G. (2014). The Bamendjin Dam and its implications in the upper noun valley, Northwest Cameroon. *Journal of Sustainable Development*. https://doi.org/10.5539/jsd.v7n6p123

Moomaw, R., Chmura, G. L., Davies, G. T., Finlayson, C. M., Middleton, B. A., Natali, S. M., Perry, J. E., Roulet, N., & Sutton-Grier, A. E. (2018). Wetlands in a changing climate: Science. *Policy and Management, Wetlands, 2018*(38), 183–205. https://doi.org/10.1007/s13157-018-1023-8

Moye, E., Nsahlai, L., & Kiming, I. (2020). Agricultural extension approaches and climate change communication within the Ndop rice sector, north west region, Cameroon. *International Journal of Social, Political and Economic Research*. https://doi.org/10.46291/IJOSPERvol7iss2pp125-141

Moye, E., Tiomo Donfack, E., & Nwinifor Kawah, A. (2024). Urban political ecology of wetland resources in Bamenda town, north west region, Cameroon. *Eco Cities*. https://doi.org/10.54517/ec.v5i2.2736

Ndzeidze, S., Mbih, R., & Bamboye, G. (2016). Using remote sensing to detect change in the Ndop floodplain wetlands of Cameroon. *International Journal of Remote Sensing Applications (IJRSA), 6*. https://doi.org/10.14355/ijrsa.2016.06.015

Nguh, B. S., & Kimengsi, J. N. (2016). Land use dynamics and wetland management in Bamenda: Urban development policy implications. *Journal of Sustainable Development, Canadian Center of Science and Education, 9*(5), 141–141.

Ngwa, C. (2003). Development authorities as agents of socio-economic change: A historical assessment of the upper nun valley development authority (unvda) in the Ndop region of Cameroon, 1970–1995. *Nordic Journal of African Studies, 12*(2), 220–237.

Nkiene V. A, Nkwemoh, C., & Tchawa P. (2016). The vulnerability of swamp rice production to the observed effects of rainfall and temperature variability in Ndop sub-division, Cameroon. *Canadian Journal of Tropical Geography*. Available on line at http://laurentian.ca/cjtg. Accessed on the 2nd of June 2017.

Nzembayie, J. (2016). The degradation of fauna diversity and its socioeconomic impact: The case of Ndop Central SubDivision, N.W. Cameroon. *Journal of Environmental Research and Management, 7*(1). https://doi.org/10.18685/EJERM(7)1_EJERM-15-021

PCC. (2007). Climate change 2007: Synthesis report. Contribution of Working Groups I, II and III to the Fourth Assessment Report of the Intergovernmental Panel on Climate Change [Core Writing Team, Pachauri, R.K and Reisinger, A. (eds.)]. IPCC, Geneva, Switzerland, 104 pp.

Ramsar Convention Secretariat. (2022). 14th session of the conference of the contracting parties (COP14) of the convention on Wetlands (Ramsar Convention) 5–13 November 2022 | Wuhan, China, and Geneva, Switzerland, http://enb.iisd.org/convention-wetlands-ramsar-cop14#:~:text=COP14%20will%20take%20place%205,Conference%20Centre%20Geneva%20(CICG).

Ramsar Convention Secretariat. (2006). Guidelines for the rapid ecological assessment of biodiversity in inland water, coastal and marine areas. Secretariat of the Convention on Biological Diversity, Montreal, Canada, CBD Technical Series no. 22 and the Secretariat of the Ramsar Convention, Gland, Switzerland, Ramsar Technical Report no. 1.

Rouse, J. W., Haas, R. H., Schell, J. A., & Deering, D. W. (1973). Monitoring vegetation systems in the Great Plains with ERTS. *NASA Spec. Publ, 351*(1), 309.

Taillardat, P., Thompson, B., Garneau, M., Trottier, K., & Friess, A. (2020). Climate change mitigation potential of wetlands and the cost-effectiveness of their restoration. *Interface Focus, 10*, 20190129. https://doi.org/10.1098/rsfs.2019.0129

United Nations International Strategy for Disaster Reduction (UNISDR). (2015). *The human cost of weather-related disasters 1995–2005*. UNISDR Office. Disponible à: https://www.unisdr.org/2015/docs/climatechange/COP21_Weather_Disasters_Report_2015_FINAL.pdf

UNU-FLORES. (2017). *Annual report: The nexus approach budding. annual report*. United Nations University Institute for Integrated Management of Material Fluxes and of Resources (UNU-FLORES). https://collections.unu.edu/eserv/UNU:6092/AnnualReport2016.pdf

UNU-FLORES. (2018). *From isolation to integration: Five years of advancing a nexus approach to environmental resource management*. United Nations University Institute for Integrated Management of Material Fluxes and of Resources (UNU-FLORES), https://collections.unu.edu/view/UNU:6404#relatedLinks

Walsh, R. P., & Lawler, D. M. (1981). Rainfall seasonality: Description, spartial appterns and change through time. *Weather*. https://doi.org/10.1002/j.1477-8696.1981.tb05400.x

Wang, R., Jian, D., Xiangyu, G., Jinjie, W., Shaofeng, Qin., Jiao, T., Lijing, H., & Zhang, Z. (2023). Impacts of climate change on the wetlands in the arid region of Northwestern China over the past 2 decades. *Ecological Indicators, 149*, 11016. https://doi.org/10.1016/j.ecolind.2023.110168

Xu, D., Gautam, B., Tan, Z., Sinha, E., Di Vittorio, A. V., Zhou, T., Ivanov, V. Y., & Leung, L. R. (2023). Climate change will reduce North American inland wetland areas and disrupt their seasonal regimes. Nature Communications. https://doi.org/10.1038/s41467-024-45286-z

Yemele, O. M., Ngaba, M. J. Y., & Ajonina, G. N. (2023). Stakeholder perceptions of wetlands management effectiveness in Cameroon. *Journal of Ecology and The Natural Environment, 15*(3), 50–60.

Zhang, X. (2016). Climate change impacts on wetlands of the yellow river headwaters. *Nature Environment and Pollution Technology*, ISSN: 0972-6268

MIX
Papier aus verantwortungsvollen Quellen
Paper from responsible sources
FSC® C105338

If you have any concerns about our products,
you can contact us on
ProductSafety@springernature.com

In case Publisher is established outside the EU,
the EU authorized representative is:
Springer Nature Customer Service Center GmbH
Europaplatz 3, 69115 Heidelberg, Germany

Printed by Libri Plureos GmbH
in Hamburg, Germany